THE RICH GET RICHER

To my son and daugher-in-law,
Max and Jeanice

CONTENTS

PREFACE

The worldwide debt crisis, together with the economic slide that the United States has experienced within the global economy, has been the subject of countless books and articles within the past decade. This book does not attempt to turn over such already plowed ground. Rather, my goal is to show how such international and domestic forces have worked to reduce real income and drive up relative income inequality on a personal level. Economic expansion by multinational corporations and bank loans from industrial countries to Third World countries have been called "development." The research here indicates the stark opposite. Where penetration by corporations from industrial countries in poor nations has occurred, actual declines in meeting basic human needs have been the result. Only small minorities have seen their real incomes go up, while the great bulk of the people in less developed countries (LDCs) have become more impoverished or seen their incomes stagnate.

While these income cuts took place in poor nations, most people in the United States did not benefit either. The record overseas investments of American corporations led to capital flight, starvation of our industries at home for want of urgently needed investment (deindustrialization), and loss of high paying manufacturing jobs. The bottom line has been a great increase of income inequality within America and a severe erosion of the U.S. middle class. The rich have become much wealthier in the past two decades, but especially during the 1980s. A drastic slash in the nation's social safety net also set in during the 1970s and reached epidemic proportions under the Reagan administration during the 1980s. The poor grew even more desperate as they saw their real income deterioriate at an even faster pace. U.S. defense spending nearly reached wartime levels, which further drained resources away from productive manufacturing. Commercial research and development dropped like a rock, making our industrial exports more dated and of inferior quality. Fiscal policies pursued by the federal government during the past decade were like salt being poured on a wound. Unheard-of trade deficits, budget gaps, and personal/corporate debt piled up as a result. Despite the false appearance of economic growth, the pillars of the American economy became rotten with neglect.

Yet not all sectors of the economy nor all groups within the United States have witnessed declining income. Nor do we need to feel helpless in the face of declining income. There is still hope of reversing the growth of inequality in America. We do not have to be victims. To accomplish this, a number of strategies and recommendations to stop the income decline are suggested in the final chapter. For the most part, it is a political task, since much of the income inequality growth was either engineered or aggravated by politicians.

To help build a greater awareness of the income slide and the dangers it entails, this book makes a major attempt to reach out to a broad audience of concerned citizens. *The Rich Get Richer* is, hopefully, not just another academic tome. I have actively sought to avoid difficult professional concepts and overly complex explanations. Simplicity has been a major goal in this effort to build understanding.

At times, attempting to write a simple, easily understood book on income inequality with enough academic substance to give it weight seemed impossible. I owe a great deal to the help of my wife, Cathy, who served as audience and gently constructive critic. Although it seems almost mandatory for an author to express thanks to a spouse, words simply cannot convey Cathy's help with this enterprise. She was a source of emotional support as well as intellectual aid, when the faceless statistics took on a life of their own. She was at my side when we were mugged in Rio, feeling the fear, anger, and compassion that firsthand contact with desperate income inequality breeds.

Important technical assistance in making maps was provided by my son Dennis. My colleague Steve Buechler provided a much appreciated critique of portions of this manuscript. Karen Purrington, in our sociology department, helped gather materials, took care of travel details, and coordinated office help. Mankato State University has been consistent in its efforts to assist this project by providing a variety of research grants and by reducing my teaching load. In particular, thanks are due to sociology department chairman Joe Davis, College of Social and Behavioral Sciences Dean Barbara Keating, and Graduate Dean Miriam Lo. For my talks with Brazilian economists, I owe thanks to J. Ray Kennedy (Center of Brazilian Studies, School of Advanced International Studies, Johns Hopkins University) and especially to Professor Edmar L. Bacha (Pontificia Universidade Catolica do Rio de Janeiro) in helping me to network with his colleagues. I owe a special debt of gratitude to José Marcio Camargo of this university for consenting to a fairly lengthy interview. Regis Bonelli and his staff at the Instituto de Planejamento Economico é Social in Rio de Janeiro provided invaluable assistance and insight into Brazilian income inequality. Material and findings from Helga Hoffman at the United Nations (Office of the Director General for Development and International Economic Cooperation) were also very useful.

My book relies on government documents, congressional studies, and federal data for its analysis of American income inequality. U.S. Representatives David Obey (D-Wisconsin) and Gerry Sikorski (D-Minnesota) were originally quite helpful in locating information for this purpose. A tremendous amount of assistance came from Representative Tim Penny (D-Minnesota) and his staff, who located and forwarded material that would have been difficult to get otherwise. Tim Penny has been an outspoken advocate for combatting world hunger in his activities in Congress. David Obey has been in the forefront of numerous efforts to deal with the increasing levels of low income and poverty in the United States. Although none of these congressmen necessarily endorse the conclusions reached in my book, I am confident that they share a common concern for preserving an equitable and just society for all Americans.

In preparing the second edition, Christopher Chase-Dunn (Sociology Department, Johns Hopkins University) helped me network with colleagues who provided additional data and information. One such was Professor Cornelis Peter Terlouw (Utrecht, the Netherlands), who shared his scores used to rank nations in a Modern World Systems perspective. Terry Boswell (Sociology Department, Emory University) and William J. Dixon (Political Science Department, University of Arizona) kindly shared their newly derived penetration scores for multinational corporate dominance. Particular gratitude is due to the three reviewers who contributed suggestions for revisions to use in the second edition: Terry Boswell (Emory University), Laurie Wermuth (California State University, Chico), and Jim Herrick (University of Washington).

WHY INCREASING INEQUALITY
IS A DANGER TO US ALL

It seems that people have always argued about how resources in society are to be shared. Disagreement usually centers around whether the share we get is an equitable and fair reward for our hard work, effort, and sacrifice. One school of thought believes that those who receive more must somehow deserve their higher incomes. Another view sees society as basically unfair. Scarce resources are passed out on the basis of inheritance by the wealthy to their offspring. This depiction asserts that persons in a ruling class or powerful elite maintain their positions of influence by use of coercion, oppression, naked aggression, and force. Such an elite preys upon the weak of society in maintaining and increasing its entrenched, high position.

This privileged class is often accused of cruel indifference toward the suffering of the masses they dominate. A famous quote attributed to Marie Antoinette just prior to the bloody French Revolution in the eighteenth century illustrates this attitude. The story is told that as queen of France, at the apex of an incredibly opulent and decadent court society, she was informed that the peasants had no bread, that they were literally starving to death. Her alleged offhand quote, "Let them eat cake," earned her a place of infamy in history.

There is evidence that she never voiced such incredible indifference.[1] Yet even if she lacked concern for the poor, this may not have stemmed from genuine cruelty. Marie Antoinette was simply ignorant of the true condition of the poor within French society. Being so out of touch with her starving subjects literally cost her her head! The parallel message for any society with an economic elite which bleeds the very

1

financial life from its citizenry ought to be crystal clear. To continue—through chicanery, fraud, and naked force—to steal from the poor to enrich those who are already abundantly wealthy eventually carries an extreme penalty. To follow a path of economic gluttony while those around us literally starve all but guarantees some form of "French Revolution" for contemporary society.

THE SHRINKING MIDDLE CLASS

A more subdued but similar debate began simmering in the popular press in the United States during the last decade. Two interrelated threads that are not easily separated continued to reappear as major themes: (1) America has lost its economic dominance and leadership among the nations of the world and (2) principally as a result of our country's precipitous decline, middle- and low-income Americans are less well off than the economic reforms of the Reagan and Bush administrations had promised. It is especially around the issue of a "shrinking middle class" that the debate has been fiercely fought. Concern was initially kicked off in 1982 with the publication of Bluestone and Harrison's book, *The Deindustrialization of America.*[2] The authors documented a decline in employment in unionized smokestack and goods-producing industries along with a parallel growth in nonunionized high-technology and service-producing industries. Critics have surmised from these developments that the middle class is disappearing. A bipolarization of the earnings structure can be the only result. The substantially higher pay of manufacturing jobs has been replaced with low-paying service sector jobs—the McDonaldization of the work force.

Early commentary by the mass media threw cold water on any notion that structural causes in the American economy could be to blame. *U.S. News and World Report,* for example, cited demographic trends as the primary cause of declining family income (the growth of new households, baby boomers glutting the job market and depressing wages, increasing divorce resulting in poor, female-headed households, etc.).[3] *Newsweek* presented some brief counterevidence to reject this notion,[4] while the business community's *Forbes* magazine descended to the level of blaming the victim, concluding that "the most important contributor to the poverty statistics . . . is the fact that so many unmarried teenagers choose to have children they cannot support."[5] The same judgmental tone is echoed in a later article in *Forbes*:

> So the statistics may be telling us more about families than about economic failure. If couples get divorced, if unmarried teenagers in a Chicago slum have babies, if a flood of inexperienced workers join the work force, that

will drive down the income statistics. But it doesn't in itself mean the economy is falling apart.[6]

Less biased and more thorough, objective articles have been published in the popular press as well.[7] An article in *Business Week* briefly concluded that the income erosion was probably real, and that to dismiss it as "merely" demographic was of little solace to a young family just getting its start.[8] The author concludes that unless economic growth produces an equitable distribution of benefits, something is deeply wrong with democratic capitalism. A *Time* magazine article sought to underline the importance of a healthy middle class:

> Any substantial decline of the middle class—even if it is partially psychological—would be ominous for the U.S. as a whole. It is the middle class whose values and ambitions set the tone for the country. Without it the U.S. could become a house divided in which Middle Americans would no longer serve as a powerful voice for political compromise. . . . Virtually everyone agrees that America needs to maintain its middle class.[9]

Press coverage in the 1990s was more sympathetic to the dangerous consequences of growing income gaps between the rich and the poor. Perhaps this was because of the nearly unanimous opinion (discussed in a later section) given by a retinue of sociologists, political scientists, and economists that income inequality has increased. The dire effects are no longer possible to ignore. The very magazines that once lampooned the idea that growing inequality could be a threat now speak of its dangers. *The Wall Street Journal* reports that the widening rich-poor gap is a "threat to the social fabric."[10] In turn, the normally conservative *U.S. News and World Report* now states that:

> The growing income disparity between the rich and just about everybody else in America helps explain the country's current economic malaise and the deep pessimism that most citizens feel today about their future prospects. . . . Our rivals in Japan and in most of Europe . . . have also managed to achieve better economic growth than America, and they have spread it in such a way that most people benefit. . . . You don't have to be a Marxist to believe that America as a whole would be a happier place if the fruits of economic growth had been spread a little more widely. . . . It is time we faced up to the fact that something enormous and unattractive is happening to what we still imagine to be a middle-class nation.[11]

A *Business Week* article concludes that heightened income inequality in the United States is undermining the ideal of equal opportunity, and that our country will continue to suffer both socially and economically if this is allowed to continue.[12] The *Washington Post* has dubbed this "the rich and poor problem," and traces the themes of inequity and income monopolization through current popular media reports—while arguing that the trends deserve even more attention.[13] The *Post*'s Robert Kuttner argues that such inequities can reach deep levels and become entrenched as unfairness is perpetuated from one generation to another in unexpected ways. A case in point is that affluent parents can subsdize their college-age children in low-paying internships with high career promise, while fledgling graduates from middle-class or poor families are shut out because they cannot afford an apprenticeship position below a living wage.[14] The growth of income disparities, for Kuttner, is class warfare at its worst. He worries that politics today is dominated by a favored economic elite that promotes ugly social divisions and scapegoating, while ignoring the real pocketbook concerns of ordinary people.[15]

A DECLINING CONSUMER BASE

Barbara Ehrenreich adopts a parallel theme echoing the grave consequences of growing income inequality in a *New York Times Magazine* article. She points out that in the area of consumer goods we have already become a two-tier society.[16] The middle is disappearing from the retail industry (e.g., Korvettes, Gimbels) while remaining middle-income retailers such as Sears and Penney's scramble to reposition for a more upscale market. The stores and chains that prosper tend to serve clientele at either extreme of the income spectrum: Saks and Bloomingdale's for the affluent; K-Mart and Woolco for the underclass. While this may initially appear frivolous to those of us who are not involved with marketing, it should also be remembered that America is probably the most consumer-oriented country in the world.[17] No one knows for sure what types of reaction would ensue if today's relatively plentiful flow of goods became tomorrow's trickle. Moreover, the newly working poor who now staff the shops in malls may have their class-consciousness expanded as they sell to the wealthy luxury goods they themselves can no longer possibly afford. The outcome may be a nation cut in half. Religious sectarianism may increase along with political extremism, perhaps culminating in the rise of a reactionary, militant right.[18]

A THREAT TO DEMOCRACY

Is this overly dramatic? Can we expect a fundamental alteration in the American way of life, a threat to our basic democratic institutions, if economic inequality gets worse? One is tempted to dismiss popular media articles as sensational and exaggerated, no matter how persuasive they seem. But throughout the past two decades it has become more apparent to average citizens that there is little hope of capturing the American dream of a stable and secure middle-class income. In a country that has defined itself as "the land of opportunity," reality now provides a jarring contrast of institutionalized low income, a theme which will be developed later in this book. Continued erosion of income seems guaranteed to leave many unhappy. When viable means no longer exist to achieve financial security, tensions will build to a boiling point.

Under such stressful conditions, it should be expected that a negative backlash will eventually develop. A long-standing theory holds that revolutions, collective violence, and social unrest will follow in the aftermath of generalized feelings of relative deprivation. Revolutions are said to be more likely after a long period of economic development followed by a sharp reversal of fortune.[19] In short, continuous progress breeds high expectations, whereas an abrupt economic reversal highlights the gap with reality, thereby increasing discontent. Using a Frustration-Aggression approach, Gurr sees revolution growing from sharp feelings of relative deprivation among large numbers of dissatisfied people.[20]

Such social-psychological approaches have come under heavy criticism in the past two decades. Among shortcomings listed are the lack of good empirical evidence, the well-known gap between attitude and behavior, vague concept/measurement specification, and questions regarding the causal order between relative deprivation and collective action.[21] No doubt the most basic criticism is the leap of faith researchers must take between macro events—such as revolutions—premised upon micro conditions—such as feelings.[22] In the end, these psychological theories disregard the way political mobilization takes place via organized elites contending for power.[23]

Yet, on a case-study basis, how income inequality begets political violence is clearly illustrated by events in Panama. In this example, income deterioration led to mobilized collective action as opposed to individualized random violence. Weeks of violent demonstrations plagued the city of Colón as the poor made an attempt to focus public attention on their declining living standards in comparison to those of the wealthy who had not suffered any setbacks.[24] Although Panama has one of the highest per-capita income levels of any developing country ($1,935), it is cursed by outrageous maldistribution of income. Over one-third of its population lives in extreme poverty as officially defined by the government, and this proportion had increased by 59 percent during the year leading up to the riots.

A more familiar example of inequality begetting violence unfolded closer to home. In the wake of the acquittal of the white police officers who beat black motorist Rodney King while arresting him, Los Angeles erupted in an explosion of violence, arson, looting, and death. After it was all over, the death toll stood at 58, recorded injuries tallied 2,383 victims (including 228 who were in critical condition), and property damage had soared above $785 million.[25] The Rodney King verdict was the proverbial spark that set off the powder keg. While police brutality is a serious problem, the underlying issue has more to do with racism coupled with severe income inequality. Los Angeles is without doubt one of the most racially and ethnically heterogeneous cities in the United States, if not in the world.[26] Projections indicate that non-Hispanic whites will become a minority of its inhabitants by the turn of the century. Thus, the underlying conditions for racial conflict were in place long before the latest explosion of violence.

But since its inception, the City of Angels has also been under the heel of upper-class domination by wealthy and powerful business elites.[27] As of 1980, income disparities among households in Los Angeles County put the city within the top 5 percent of the most unequal counties in the nation.[28] Even to a casual visitor, the contrasts offered by Rodeo Drive and Beverly Hills as opposed to East Los Angeles are quite dramatic. The picture of Porsches driving by large clusters of homeless people in the downtown area form vivid and lasting snapshots which reflect the depravity of such piercing inequality.

Preceding the riots, economic deterioration for broad segments of Angelenos had risen during the 1969 to 1987 period. Median family income went down in constant dollars. The poverty rate climbed from 11 percent to 15.6 percent—a rate rise of 42 percent in less than two decades. Although the poverty rate for white persons actually declined in the 1969–1987 period, ending at 6.9 percent, it went up sharply for all minorities. By 1987, rates were double and triple that of whites: Asians had a 14.2 percent poverty rate while one in four black or Latino residents was poor.[29] The ratio of income going to the poorest 20 percent of families went down compared to the percentage going to the richest quintile.[30] In short, while this Pacific Rim city went through an unprecedented economic boom, large segments of its inhabitants (many of whom are minorities and/or recent immigrants) actually became worse off than before the explosive growth began. The beating of Rodney King was reprehensible. But the beating was merely the catalyst for an explosive brew of income inequality and racial injustice.

The ability of a given society to meet the basic needs of its citizens would seem a major indicator of its health and staying power. The very foundations of democracy may depend upon a government's ability to maintain the economic well-being of its population. Yet case studies, in the end, provide only anecdotal proof of an income inequality/political violence relationship. Evidence from sociology of a more quantitative nature does indicate, however, that political extremism and violence will result when income inequality increases. For example, income inequality has been found to be the most important predictor of police-caused homicides. U.S. states which were most economically unequal were also most likely to have the largest rate of killings committed by policemen.[31]

In an important comparative study, Edward Muller analyzed over fifty countries where income inequality data existed. It was discovered that the death rate from political violence in countries (regime repressiveness, coups, revolutions, disappearances, etc.) actually goes up as income inequality increases.[32] As a result, countries which follow a strategy of development which ignores distributional equality are likely to experience higher levels of mass political violence. On the international scene, this is called the "Brazil Model"—named after the country so closely identified with rapid accumulation of wealth through government efforts to aid rich landowners and industrialists while virtually ignoring the welfare of the poorer masses.

Further development of this research, examining the impact of income inequality upon democracy, led to an even more important discovery. Muller's latest research of sixty-four countries clearly shows that progressive economic development within LDCs will not necessarily yield greater increases in democracy. Greater income inequality among countries at intermediate levels of economic development has actually led to substantial declines in levels of democracy, despite gains in Gross Domestic Product per capita.[33] In essence, no matter how wealthy a country becomes, it is still vulnerable to political violence and instability if its distribution of income is fundamentally unequal. Conversely, even if a country is relatively poor, democratic institutions will survive and flourish if income is distributed in a fair manner. It may have become a cliché to state that the very survival of our democratic way of life is dependent upon how a government treats its citizens, but concrete evidence now gives new meaning to this axiom.[34]

It is important to note that the income inequality/political violence thesis has also been seriously challenged, and that some scholars dispute its very existence. A moderate caution by Zwicky states that research results may vary, depending upon the time period used in various studies.[35] In essence, the income inequality/political violence relationship evident among nations during the 1960s seems to have disappeared in the 1970s. Zwicky does identify a more potent predictive variable of political violence, however, by measuring the degree of increasing inequality. In the end, for Zwicky it is not necessarily the level of income inequality but the rate of change in this variable that feeds violence. Although his panel design is an improvement over typical studies, Zwicky's research does have the drawback of being limited to developing countries.

One of the most outspoken and persuasive critics disputing the income inequality/political violence connection has been Erich Weede. In cross-national research which attempted to use improved data and measurement techniques, Weede found that income inequality did not contribute to political violence in countries.[36] In questioning Muller's research, Weede has pointed out that relevant predictive variables were left out (together with the country of Taiwan), which would have radically altered Muller's conclusions.[37] Muller's research has also been criticized on the basis of the way democracy was measured and on its use of a questionable data source for deaths due to political violence.[38] Nonetheless, Muller has capably defended his findings, proving they remain strong using a variety of different data sets and substituting diverse independent variables.[39]

Mark Irving Lichbach has offered an even broader and more thorough critique of the income inequality/political violence theme. After reviewing an exhaustive array of forty-three studies in several different disciplines, Lichbach concludes that research results frequently contradict one another, while the belief that economic inequality produces political violence remains unproven.[40] Much of the confusion and disagreement has been caused by differences in the way economic inequality and political violence are measured, which nations are included in the analysis, variation in control variables employed, dissimilarity in time frames, and the absence of explanatory theory.

While Lichbach does a good job of identifying contradictory findings, this weakness remains true of nearly all research. In the social sciences, it is rare to get complete agreement—let alone a reasonable consensus—on any causal association. Scholars may approve of his call for clearly defined theory to more firmly guide this type of research in the future. Yet anyone who has witnessed the endless nitpicking and bickering among theorists will realize such a path may not necessarily lead to enlightenment either. In the end, Lichbach does not explain the persistence of the income inequality/political conflict nexus which has tended to surface in most sociological studies over the past few decades. He is also unbiased in identifying the strengths of this approach: data are now more comprehensive, a consensus on measuring the dependent variable has evolved (deaths due to domestic political conflicts), control variables are more relevant, and statistical techniques are better.[41]

Such improvements can be seen in the most recent study of political violence. Terry Boswell and William J. Dixon utilize a sophisticated empirical model in their sixty-three-nation study which employs a plethora of competing causal variables to explain political violence. The measurement of independent variables is built with complex mixtures of variables found to be predictive in past research. Among the independent predictive variables is income inequality, which is again found to be significantly related to political violence.[42] Lastly, the very latest study (again by Boswell and Dixon) introduces a Marxist component to the equation, essentially meeting Lichbach's call for a more developed theoretical base.[43] Once again, the effect of higher income inequality on the rise of political violence proves strong and robust. Rising income inequality—as measured by the percent of income received by the richest 20 percent of a country's population—effectively explains mounting deaths due to violent rebellion, no matter what combination of competing independent variables are used. The ongoing debate over whether there really is a connection between growing income inequality and political violence will continue in the future. Up to this point, however, income inequality has endured as a compelling variable predicting political violence. Its negative impact persists in study after study. Thus, to dismiss the importance of income inequality as a contributor to political violence would be unwarranted and rash.

AN INCREASE IN CRIME

The income inequality/violence relationship has received particular attention domestically with reference to crime rates. Nearly all research distinguishes between relative income inequality (usually measured by the Gini ratio, which will be explained later) versus absolute income inequality (generally measured by the percentage of persons below the poverty line). Judith and Peter Blau have been the most lucid in framing the essential research question, which alludes to frustration-aggression and relative deprivation.[44] In particular, the Blaus point out that income inequality need not necessarily spawn political rebellion. Inequities can be so great as to deprive the lower strata of the strength to organize successful collective action such as a strike or revolution. For the Blaus, great inequality spells a potential for violence. When collective action is blocked, unrest remains diffuse and finds its outlet in criminal violence such as murder, rape, robbery, and assault.

Another strand of theory is also generally included in the inequality/crime relationship via the concept of anomie.[45] Racial and ethnic discrimination, layered together with growing income inequality, spawn prevalent disorganization and distrust. This discontent is reflected in a sense of social injustice brought on by the gap between real life and the egalitarian values of American society. In short, excessive income inequality undermines the legitimacy of society and leads to general demoralization. Under conditions of great income inequality, respect for the law goes down as crime goes up.[46]

Thus, there is good theoretical reason to suspect a strong relationship between income inequality and criminal violence—whether this stems from relative deprivation or from the disorganization/anomie tradition. On a case study basis, Brazil provides a fitting example. That country's Gross National Product (GNP) per capita—$2,770 in 1992—is relatively high for less-developed countries.[47] But Brazil also has long held the record for the world's highest recorded level of relative income inequality—with over one-half of all household income going to the wealthiest 10 percent of its population while the poorest fifth of households receives only 2.1 percent of all income.[48] Nowhere is this extreme inequality more evident than in the city of Rio de Janeiro, which could be described as two blocks of Paris that wind along breathtakingly scenic beaches—but which is surrounded by Ethiopia. If poverty is the norm in the hilly favellas, opulence rules in the nightclubs of Rio for the privileged few who are wealthy. The contrast spawns violence, which has emerged in Rio's homicide rate.[49]

The overall empirical evidence documenting the income inequality/criminal violence relationship is quite impressive as well. In a global context, a remarkably consistent finding is that income inequality is directly related to homicide levels.[50] This relationship is particularly lethal when overlayered with economic discrimination against racial, religious, or ethnic groups.[51] When U.S. data are examined, the same trends identified on a global level repeat themselves inside our boundaries. Although there are a few exceptions, most research studies corroborate the positive relationship between relative income inequality and high crime rates[52] as well as between absolute deprivation (high poverty rates) and increased crime.[53]

The effect of income inequality is evident in the increase of rape, robbery, and murder among states in the United States. The greater the relative income inequality in a state, the higher its murder rates are.[54] The rate of increase for income inequality is directly related to higher rates of robbery (where it is the most important predictor) and rape (which may reflect the anomie and disorganization produced by a rapid change in relative income). Whether income inequality is regarded as reflective of societal disorganization or relative deprivation, its ill effect upon American society is obvious when measured by violent crime rates. The increase of inequality is the highest predictor of robbery and significantly predictive of rape, while high levels of income inequality significantly predict higher murder rates. Figure 1.1 graphically illustrates the rise in murder rates as poverty levels among American families increase.

Given the affinity between income inequality and criminal violence, it will come as no surprise that the two trends have been rising in tandem over the past two decades. The Federal Bureau of Investigation (FBI) categorizes violent crimes as murder, forcible rape, robbery, and aggravated assault. The rates are measured as the number of offenses in these categories known to the police, per 100,000 population. In 1979, 549 persons

FIGURE 1.1   Poverty Increases Murder Rates in U.S. States: 1990

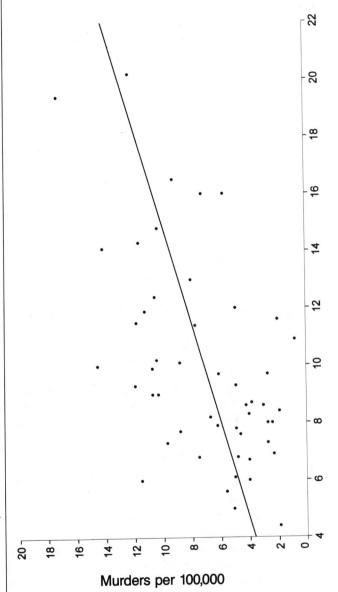

Percent of Families Below Poverty Level

Murders per 100,000

Source: Data from 1990 Census data and *U.S. Statistical Abstract*, 1990.

per 100,000 had encountered at least one of these crimes; by 1992 this composite criminal violence rate had risen to 758 per 100,000— for an increase in the violent crime rate of 38 percent.[55] While the overall murder rate is down slightly by 4 percent in this time period, there are ominous signs of future growth. A whole new generation appears to be undergoing an apprenticeship in homicide. Murder rates by teenagers have increased at an explosive rate in the last decade, doubling for white male teens and tripling for black male teens.[56] In fact, the number of homicides among juveniles involving handguns increased fivefold between 1984 and 1993. With such an explosion of youthful murders and the projected teen increase of 20 percent over the next decade, arrests of juveniles for violent crimes will double by the year 2010.[57] Most would agree that even now the United States has an unacceptable level of criminal violence, especially in comparison to other industrial countries. The U.S. murder rate of 12.4 per 100,000 in the 1975–1980 period compares dismally with Canada's (2.7), Australia's (2.5), Britain's (1.6), Germany's (1.2), and Japan's (0.9).[58]

The American propensity for violence ultimately shows up in our crowded prisons. The incarceration rate in the United States, now at 519 per 100,000 persons, is five to fourteen times higher than that of our trading partners: England (93), France (84), Germany (80), and Japan (36).[59] Indeed, America has the dubious record of having the second highest imprisonment rate in the world, barely below that of Russia (558) but well ahead of South Africa (368). By 1995, nearly 5 million people in our country were under some correctional supervision. One and a half million persons were actually behind bars, with another 2.8 million on probation and 671,000 on parole.[60]

While no one can feel good about the deluge of violent crime in our nation, many might derive comfort from our elevated imprisonment rates. Yet this may be a delusion. Little evidence exists that imprisonment actually lessens crime. Criminal justice experts point out that "get tough" efforts will not work. During the past two decades, the added billions spent for additional police and more prisons resulted in a fivefold increase in prisoners, yet our violent crime rates have continued to soar.[61] If locking up those who break the law truly contributes to a safer society, the United States should be the safest country in the world. It is clearly not.

While persons who commit violent crime should not go free, our criminal justice system is not without a bias that may enflame crime even further. Not only are have-nots resentful and shut off from any hope of achieving success, but income inequality is also associated with harsher punishment. Those who are less well-off are more likely to do time and generally receive longer sentences.[62] This could create a downward cycle for those in poverty. Since the poor experience discriminatory sentencing, their anger could feed even more violent crime in an effort at retribution.

Without doubt, our national crime policies have been reactive rather than proactive. The cost associated with containing crimes after they have been committed is exorbinant. At this point, the United States is spending $163 billion per year as a result of crime:

> Crime costs include better than $31.8 billion at the state and federal level for police, $24.9 billion for corrections, $36.9 billion in retail losses, $20 billion in insurance fraud, and $17.6 billion for individual property losses and medical expenses. Still $15 billion more is spent on private security, $9.3 billion on court costs and $7.2 billion on prosecution and public defense. . . . [In total] this is nearly two-thirds of what America spends on national defense and more than five times as much as the federal government spends on education.[63]

The average cost to house an inmate is now $25,000 per year. Even the mounting toll of homicide victims costs a large amount. In Washington, D.C., an outright killing costs $7,000 in various services. If the victim lives and recovers without complications, the price tag goes up to an average of $21,000. Keep in mind that these are public costs, since most victims are uninsured.[64]

The economic crime ledger does not even begin to address the cost in human suffering, which cannot be assigned a dollar amount. But the focus on money brings us to the heart of the matter. Income inequality breeds violent crime which costs all of us, even if we are not directly in the line of fire. The price is not set only in dollars. It also increases our fear level and stunts our personal freedom to appear in public wherever we may choose. It decreases the general quality of life even within the safe confines of suburbia, because threats are always lurking on the freeway, around a corner, at a shopping mall, and so on. The punitive approach to crime seems to focus more on the symptoms than on the disease. By spending more money on programs designed for prevention, investing in education, creating opportunities for training, providing meaningful jobs, guaranteeing health insurance for all, and fighting the poverty that breeds crime, we attack the insidious nature of crime at its root. Above all, by reducing income inequality, we reduce crime.

DETERIORATING FAMILY LIFE

We thus have a multitude of reasons for being concerned with preventing and lessening income inequality, both in our own society and in those of our neighbors in this one-world community. If collapse of basic democratic institutions in our own country or world unrest still seems remote or unlikely, there is much reason for worry closer to home. Beginning in 1973, the average real wage of American workers in constant (inflation adjusted) dollars began to drop sharply.[65] A 10.5 percent drop in hourly wages occurred in the 1977–1989 period alone, whereas average wages fell an additional 2.8 percent from 1989 through 1994.[66] A study done by Danziger and Gottschalk shows that the share of national income going to American families with children declined by nearly 20 percent within one decade.[67] American families with children have lost a large amount of real income, especially in the bottom three quintile groups (the poorest 60 percent of families), but families with children in the lowest group lost proportionately more income. In fact, the poorest fifth of families lost over 30 percent of their income (in constant dollars, after inflation is subtracted) between 1973 and 1990.[68]

Given these trends, in the past two decades women have increasingly entered the labor force in an effort to maintain living standards. Although there are many positive consequences to this trend, an unrecognized fact is that the doubling of female participation in the labor force since 1940 is equivalent to a 20 percent increase in the average family's paid workweek. People work more and have less time for leisure or family activities because of economic deterioration.

Disaster can also ensue with divorce, which has grown sharply within the past several decades. For example, disposable income falls 73 percent for ex-wives within a year of the divorce whereas it rises by 42 percent for ex-husbands.[69] While the "feminization of poverty" can be easily documented statistically,[70] such trends bode poorly for the children who are assigned to the custody of these women, as they are in 90 percent of the cases. A vicious downward spiral is also apparent with divorce. To begin with, as relative income inequality increases, so does the divorce rate.[71] Parents that are least well-off economically are the most likely candidates for divorce.[72] Two-parent families in poverty have twice the risk of breaking up as nonpoor two-parent families.[73] Once divorced, women with children have at best a three-out-of-four chance of getting a child-support award; of those supposed to receive support, only three out of four women do. In the end, barely one-half of divorced women with children in the home actually receive any financial support from their ex-husbands. When they do get support, it tends to be paltry (the mean award, which typically covers more than one child, is $277 per month).[74] At rock bottom, income inequality is both a cause of divorce and a consequence of divorce. Others have eloquently addressed the emotional pain and psychological consequences of the explosion in divorce.[75] The impact on children forced to grow up in poverty, however, is dealt with below.

The grave effects of income inequality and poverty upon children led the Carnegie Corporation of New York to form a task force on meeting the needs of young children. In their report, the task force identified a "quiet crisis" unfolding in America for the especially susceptible children under three.[76] The crisis was made up of the explosions in unmarried mothers, divorce, children in poverty, foster care, eroding health care, abuse, neglect, and so on. Although to some degree, children from all walks of life may be exposed to these threats, such dangers are magnified for poor families and their children.

Probably the best advocate for children who are forced to struggle with the very real perils of poverty has been the Children's Defense Fund, based in Washington, D.C. This organization has been tireless in its zeal to protect poor children through a variety of channels: political pressure, education, research, and prayer. Without its existence, the state of poor children in America would be even worse. Yet the evidence this group brings to bear, documenting in detail the harmful effects of poverty, paints a grotesque picture. When parents experience a job or income loss, they are more prone to stress and depression. Having a depressed and stressed mother increases the likelihood that children will have more medical, sleep, behavioral, and attention deficit problems as they grow up. Poor mothers tend to use more harsh and inconsistent discipline with their children.[77] Parents under stress value obedience more and are more likely to use physical punishment—which can frequently lead to outright child abuse. Without doubt, poverty has been the single most predictable risk factor for child abuse and neglect. The rate of abuse for children in families with annual incomes below $15,000 was 4.5 times higher than in families above this level.[78] Current research finds a consistent and disproportionately large rate of abuse and neglect among lower-class families, with the most severe cases of maltreatment among the poorest of poor families. Among families reported for physical abuse, 29 percent included an unemployed caregiver (42 percent among neglect cases).[79]

Poor children experience a higher incidence of conduct disorders, behavioral problems, depression, low levels of self-confidence, and poor social adaptation. Recent research shows that the longer a child exists under poverty, the more likely the child is to suffer from frequent mood changes and from feelings of being unloved, afraid, confused, and worthless. Such children are reported more often by poor mothers as being high strung, obsessive, unhappy, withdrawn, clinging, too demanding, and overly dependent.[80] For those children currently living under poverty, the same study found more disruptive behaviors with the following symptoms: cheating, arguing, disobeying, trouble getting along, impulsiveness, not liked, restlessness, stubbornness, strong temper, and destructiveness. Poor mothers are significantly more likely to spank and to be less emotionally responsive to their children (scolding, not answering children's questions, tone not conveying positive feelings).

Not only are you cursed if you are a child who must grow up poor, but also once you can bear children, you end up with a greater probability of bringing into this world offspring who are also likely to live in poverty. It is a fact that poor teenage women who have below-average basic skills are more than five times as likely to have children than nonpoor teenage women with average or better basic skills.[81] Fully 83 percent of teenagers who give birth are from economically disadvantaged households.[82] Nearly one-half of teenagers who give birth end up on welfare within four years; the same fate awaits approximately three out of four unmarried teens.[83] And thus the cycle of poverty is perpetuated.

ILL HEALTH AND EARLY DEATH

Poverty and rising income inequality are likely to produce both physical and social-psychological stress for individuals and families who must suffer its ill effects. This, in turn, can lead to increasing susceptibility to illness and to an early death. Poverty and absolute deprivation bear directly upon our chances of staying alive. For example, the mortality rate among American children in families eligible to receive Aid to Families with Dependent Children (AFDC) is three times higher than among nonpoor children.[84] A study in Maine found that poor children were at a greater risk of disease-related deaths (3.5 times as high). They were 2.6 times more likely to die of accidents and 5 times more likely to die of homicide than nonpoor children. Applying these findings to the United States as a whole would mean that ten thousand children die from poverty each year.[85]

This higher death rate exists partly because women in poverty are much less likely to get prenatal care, and that lack leads to complications in pregnancy and low birth weights of infants upon delivery. Pediatricians agree that an infant born with a birth weight below 2,500 grams (about 5.47 pounds) poses an exceptional health risk, where survival becomes very tenuous.[86] Because of increasing financial barriers and inadequate health care provision, more than one-third of pregnant women (1.3 million per year) receive insufficient prenatal care. One-fourth of these women have no insurance coverage at the start of their pregnancy, and 15 percent are not covered at the time of delivery.[87] This ultimately means that 7 percent of all American babies (and 13 percent of all African-American babies) have low birthweights. The situation has become even worse: a baby born in 1992 was more likely to have a low birthweight than a baby born in 1980.[88]

The infant mortality rate is determined by tabulating the number of deaths of infants below age one, dividing by the number of live births in a calendar year, and multiplying that number by 1,000. It has been called the most sensitive single index of the overall well-being of a country, since it readily reflects the poor provision of basic human needs such as sufficient food, adequate shelter, universal health care, immunization against disease, and so on. All of the above translates into a higher infant mortality rate in the United States, which plummeted from sixth lowest in the world in the mid-1950s to twenty-first lowest in 1994.[89]

It is not hard to understand why. Immunizations for children are far below acceptable standards in the United States. Only two out of three two-year-olds were completely immunized against preventable childhood diseases in 1993, which left 1 million toddlers without protection from tetanus, polio, hepatitis, measles, etc.[90] There are 37 million Americans today who are not covered under any health insurance plan, mostly because they cannot afford to enroll and/or their employers do not provide such a job benefit. In 1993 alone, the number of children without health insurance increased by 806,000. By that year, one in seven children (9.4 million kids) nationally did not have health insurance, while one-half million pregnant women were uninsured.[91] If current trends continue, fewer than half of the nation's children will be covered by employer-based health insurance plans in the year 2000. The impact of free or low-cost comprehensive, universal health care upon the health status of low-income families cannot be over-emphasized. Gregg Olsen, for example, reports that while the United States and Canada spend an approximately equal percentage of their Gross Domestic Product (GDP) on social welfare programs, average health standards of Canadians are much higher than for Americans because of universal health coverage in Canada.[92]

Although children are more vulnerable to the ravages of poverty and inequality, adults also suffer. Not surprisingly, poverty shortens life. Even within categories of illness, such as cancer or heart disease, affluent men have a lower risk of dying than poor men. Controlling for economic conditions also removes most of the gap between white and African-American mortality rates.[93] When Americans as a whole are ranked into wealth deciles, the ratio of those who died in the poorest group exceeds the ratio in the richest 10 percent by three to one. Of course, we all must die sometime, but being wealthy helps a person postpone the "day of reckoning." Both the positive effect of economic wealth and the negative effect of poverty remain strong and consistent, even when race, marital status, age, education, geographic region, and small town residence are held constant.

Despite the major effects of poverty and wealth upon mortality, and contrary to practices followed in Great Britain and other European countries, data are lacking on income and health within the United States. What knowledge we have of their high correlation comes from specialized, periodic surveys. The last such study, done by the National Center for Health Statistics, looks at self-assessed health status and bed disability days per year. Not surprisingly, low-income people have worse health. Low-income people are three to four times more likely to encounter disease than persons of moderate- to high-income levels. Low income carries a risk factor of ill health twice as large as lifestyle threats more frequently championed in the media, such as diet, exercise, alcohol, and cholesterol. Yet the impact of low income is ignored. Comparing families that in 1985 had income under $5,000 to families with income over $25,000, analysts found that nearly 30 percent of low-income persons must limit their activities due to chronic conditions (only 8.7 percent of well-off persons must do this). Low-income persons report an average of 13.2 bed-disability days per year, compared to only 4.5 for persons of moderate income.[94] In 1993, over 44 percent of low-income persons (earning less than $15,000 per year) reported "poor" or only "fair" health, while only 12 percent of those with moderate income ($25,000 and over) believed they were in these categories. Again, almost one-fourth (23 percent) of low-income persons reported that they were hospitalized in the previous year *and* sometime in the four years preceding the last year, compared to 9 percent of those with moderate income.[95]

The dynamics of how disease and death spread as a result of poverty can be seen in a research study by Mary Merva and Richard Fowles.[96] Death rate measurements were taken in the thirty largest metropolitan areas in the United States between 1976 and 1990. This research—representing the communities of 80 million Americans—showed a clear link between deteriorating economic opportunities and rising death rates. A 1 percent rise in the unemployment rate resulted in a 5.6 percent increase in deaths due to heart disease and a 3.1 percent rise in deaths due to stroke. Although this may sound insignificant, the actual 2 percentage point rise in unemployment occuring from mid-1990 to mid-1992 caused 35,000 more heart attack deaths and nearly three thousand more stroke deaths in these thirty metropolitan areas. The reasearchers also found a statistically significant relationship between rising poverty rates and increases in deaths due to suicides and accidents.

Corroborating research in Great Britain found that members of the economically active population who become unemployed or move into the lowest income category suffer a 30 to 40 percent increase in death rate.[97] Yet, it is more than a matter of absolute poverty or low income. Among the world's twenty-three wealthiest countries that belong to the Organization for Economic Cooperation and Development (OECD), relative income inequality is equally important in predicting morbidity and death. Relative income distribution is frequently measured as the percent of all income going to the wealthiest 20 percent of households, wealthiest 10 percent, and/or poorest 20 percent. What this means is that those nations where income differentials between rich and poor are smallest also have the highest average life expectancy. Almost two-thirds of the variation in national mortality rates may be accounted for by differences in income distribution alone. In the end, if the United States were to adopt an income distribution more like that of Japan, Sweden, or Norway, it might add two years to average life expectancy.[98]

INADEQUATE, DANGEROUS, AND CROWDED HOUSING

During the last decade, a large number of low- and moderate-income American families experienced downward mobility in their housing status. Frank Levy offers a depressing statistic which sharply illustrates just how bad the housing situation is for newly formed families. The typical father of today's baby-boomer faced housing costs that were equivalent to about 14 percent of his gross monthly pay. Even as long ago as 1983, a thirty-year-old man had to allocate a staggering 44 percent of his income for house payments. Indeed, Levy reports that for males in previous generations, the average increase in earnings was 30 percent between ages forty and fifty. By 1983, however, this had actually changed to a minus 14 percent for men who turned forty in 1973.[99]

Despite more women working, increasing proportions of younger families are being frozen out of the housing market. Tracing the changes in housing between 1980 and 1990, one study discovered:

1. A disappearance of modest, affordable housing for those with lower incomes—delaying the ability of young persons to establish independent households
2. An increase in overcrowding (the first since the Great Depression of the 1930s)
3. Growing problems in affordability of home ownership and especially of renting
4. A decline in the overall rate of home ownership (the first since the Great Depression)
5. A major reduction of home ownership rates concentrated among the young with soaring ownership rates for the elderly, culminating in a gap increase from 18 percentage points in 1980 to 30 points in 1990
6. Even more severe affordability and home ownership problems for minorities, low-income families, and females who headed their own households
7. Increasing polarization between affluent (mainly middle-aged to elderly) persons, who are *generously housed* in spacious, affordable homes, and the young, who are *precariously housed* in rental units they cannot afford.[100]

It is particularly for young first-time home buyers (the twenty-five-
to thirty-four-year-old group) that problems abound. During the last
decade especially, stagnating or eroding income among the young,
together with sharp increases in the costs of owning relative to renting,
exerted a downward pressure on home ownership rates. While this age
cohort's home ownership rate stood at 51 percent in 1980, it was down
to 45 percent in 1990 (a 12 percent drop in the rate over the entire
decade). The lessened home ownership potential of these younger
cohorts represents a retreat from the American Dream of having your
own house. Advantages accrue to owning your own home, which is why
so many of us want to buy one. There is more security of tenure, outdoor
and indoor space, amenities (garages, dishwashers, fireplaces, etc.), equity
buildup, and the important tax advantage of being able to deduct mortgage interest.[101] In short, although owning your own home makes the best
financial sense, an entire generation is being frozen out of this option.

Many persons are too poor to buy a home. What is seldom realized
is that many families may also be too poor to rent an apartment—which
has without doubt contributed to the explosion of homelessness Over the
past fifteen years, the gap between low-income renters and the availability of low-rent units has been rising. The number of low-income renters
(the poorest 25 percent of all households) was 5.3 million in 1970, but
there were 5.8 million low-rent units (costing 30 percent or less of the
income of low-income renters) available. By 1991, there were 8 million
low-income renters but only 2.8 million low-rent units.[102] Much of this
was due to deliberate policies instituted under the Reagan presidency in
the last decade. Programs in the U.S. Department of Housing and Urban
Development (HUD) which encouraged the construction of subsidized
housing for the poor were slashed to the bone. A large part of the deterioration in housing is also due to the shredding of the social safety net during the 1980s, a process that is continuing unabated in the 1990s. Cuts to
Aid to Families with Dependent Children (AFDC), food stamps, and
other programs to assist low-income families in providing for their basic
human needs will be discussed in a later section. It is noteworthy that
while the poor have always struggled to achieve adequate housing, it
became much more difficult because of the retrenchment in social programs that began in 1980. The major impact of these cuts was to reduce
the already paltry income of our very poorest citizens. Thus, by the end of
the 1980s, HUD estimated that the national average fair-market rent for a
modest two-bedroom apartment was more than half the monthly income
of a family of three at the poverty line. By 1987, nearly two out of three
poor families spent more than half of their income on housing—a doubling in the proportion since 1974, when only one in three poor families
were forced to do so.[103] Almost three out of four poor married couples

with children were paying 50 percent or more of their income for housing in 1993.[104]

Even when poor families are able to obtain housing at an exorbitant cost to them, their distress does not diminish. What follows is a grim "rule of three." Overall, poor children are more than three times as likely to live in inadequate housing than nonpoor children. Because of poverty, poor families move twice as often, are more likely to go without heat and electricity because of broken equipment or utility shutoffs, and are more exposed to leaky pipes and water damage. Thus, poor children are more exposed to damp, moldy housing which leads to higher rates of asthma and respiratory problems. The cockroaches which infest and flourish under such conditions further aggravate allergies. Poor children are three times as likely to live in homes with rats and mice, exposing them to contagion from bites and more allergies from rat urine. Poor children are three times more likely than nonpoor children to be exposed to peeling paint, which can cause lead poisoning if eaten, and are three times more likely to suffer from overcrowding. Finally, poor children are 50 percent more likely to live in mobile homes than nonpoor children. House fires in such homes are three times as likely to result in a child's death.[105]

BETTERING ONESELF IS DIFFICULT

Given the conditions that people of low income must face, it will be no surprise that poor children experience a learning deficit. Problems that occur disproportionately among poor children lead to a lack of behavioral and cognitive skills upon entering school, low achievement and motivation in school, lower math and reading aptitude scores, a greater probability of not finishing high school, and less likelihood of attending college.[106] Even before school, problems start with the inability of poor parents to afford stimulating toys, children's books, and good quality child care. Once in school, poor children in central cities are almost certain to attend inferior schools that have inadequate textbooks and learning materials. These schools frequently lack funds for enriching learning experiences that require field trips, laboratory or studio materials, and so on. Students are less likely to have access to computers, educational magazines, or even encyclopedias at home. For every year a child must live under poverty, his or her chances of falling behind a full grade in school increase by 2 percentage points—and the probability of the child becoming a dropout increases by 3 percentage points. In fact, students whose families rank in the bottom fifth of income are almost eleven times more likely to become high school dropouts than those coming from families in the richest fifth of income.[107]

Lest we mistakenly believe this a problem for "them," a current estimate is that one-third of American students are poor.[108] In short, a lot of youngsters fall into the ranks of the disadvantaged. Furthermore, lack of access to education strikes the middle class as well. This is especially true with regard to the basic chance a person has of going on to college. A study tracing higher education trends of high school graduates from 1970 through 1988 has found that only 27 percent of eighteen- and nineteen-year-old high school graduates from the bottom fifth of family income enroll in four year colleges—compared to 60 percent from the richest fifth of families.[109] Predictably, a young adult's probability of enrolling goes up with each income quintile—but the gap in attendance between the poorest and richest widened between the late 1970s and the late 1980s. And as any college student knows, being admitted and registering do not guarantee that one will finish. Nationally, half who enroll never receive a bachelor's degree. Recipiency of a bachelor's degree is even more stratified by income than is college enrollment. A youth in the top fifth of family income has more than three times the chance of graduating as a youth in the lowest family income quintile (39 percent versus 12 percent). In the final analysis, while there is still some opportunity for low-income youngsters for upward mobility through higher education, the dice are loaded against the poor. It is also true that chances for educational success are less generous for middle-income students than for those from the wealthiest fifth of U.S. families. Put simply, once you encounter limiting economic opportunities, such as being born to a low- or middle-income family, your climb up the ladder of success becomes more precarious. Given all the limitations reviewed to this point, it becomes obvious that pulling yourself up by your own bootstraps is largely an impossibility for those in poverty or from low-income backgrounds.

SLIDING ECONOMIC STABILITY CAUSED BY WEALTH INEQUALITY

A chilling picture is offered by Ravi Batra, an economist on the faculty at Southern Methodist University, who presents evidence tying in growth of wealth inequality with economic depressions. His thesis is that, since the American Revolution, especially deep depressions (when unemployment, for example, reaches 25 percent) have occurred periodically when the growth of wealth inequality reaches a magnitude where it destabilizes our basic economy.[110] Before explaining his conclusions in detail, it is important to note that wealth and income inequality are not the same, although they are related.[111] Basically, income inequality derives from mostly job-related earnings, although other sources of income can be very important (alimony and child support, rent, interest on savings, stock dividends, transfer payments such as Aid to Families with Dependent Children [AFDC], Social Security, Medicare, etc.). Wealth income within the United States is almost totally derived from ownership in stocks, bonds, and capital goods—-that is, it is the *capitalist* dimension of our society.

An average citizen may have some wealth equity, such as a house or IRA or pension investment, but the great majority do not own stocks, where the comparison becomes one of giants and dwarfs. For example, the top 1 percent of all U.S. families own over 60 percent of all corporate stock in the country.[112] The inequality in wealth within our society—what we receive in income plus the worth of the assets we own—is very much more lopsided than that of income alone. One of the easiest ways to remember wealth distribution in the United States is by "the rule of thirds." In essence, if we rank all Americans in terms of their wealth in 1986, the bottom 90 percent have 33.4 percent of all wealth; those in the ninetieth to ninety-ninth percentile control 35.1 percent of all wealth in the country; the top 1 percent of the population owns the remaining 31.5 percent of wealth.[113] Ranking persons or households by either income or wealth shows extreme disparities, but the differences are much higher for wealth inequality than for income inequality. Indeed, the top 20 percent of families own nearly all of the financial net wealth in the United States (94.3 percent), although they receive only slightly more than half of all money income (55.5 percent).[114] What does all of this mean?

> There is some correlation between wealth and income but . . . at the lower end of the income scale we find that increases in income bring only small increases in wealth. This is because with higher income more money remains (after purchasing necessities) to purchase things that can be held as wealth (such as a home or corporate stock). But at the higher end of the income scale we find a significant jump in wealth. Quite simply this is because substantial wealth often brings higher income. . . . In other words, we find that the causal relation between income and wealth becomes reversed as we reach higher levels of income and wealth—great wealth brings a high income. And as we find that great wealth is more likely to be inherited in the United States today, . . . we find that the base of many high incomes today is also inherited.[115]

The role of inheritance plays a crucial role in Batra's theory of how wealth inequality causes economic depression. He points out that great inequality in wealth does not develop overnight, because it derives from inheritance. Thus, it usually takes one or two generations before wealth distribution becomes critically unequal—making depressions relatively rare in the American experience (generally, they occur on a sixty-year cycle). It should be noted, however, that the rate of wealth concentration is likely to be curvilinear—increasing at astronomical rates just before the onset of depression. Batra notes that in 1922, 1 percent of U.S. families owned 31.6 percent of national wealth, but just seven years later this had risen to 36.3 percent.[116] Although this may not seem on the surface to be a startling figure, it represented the highest concentration in history for our country and a gigantic leap in the rate of concentration of wealth, from the typical glacial pace at which the rates change.

Wealth inequality causes two other factors to come into play which set the stage for panic and depression: speculative fever and shaky loans made to high-risk customers. The number of persons with no or few assets rises because of increasing income disparity. Because of this, the borrowing needs of the poor become more pronounced. The banks, in turn, become awash in deposits from the very rich. They cannot afford to pay interest on the deposits without lending them out. Therefore, as the concentration of wealth increases, the number of banks with less credit-worthy loans also rises. This increases the potential number of bank failures if panic ensues (over two thousand banks suspended operation in 1931 alone). But what could kick off a rush on the banks?

> A side effect of the growing wealth disparity is the rise in speculative investments. As a person becomes wealthy, his aversion to risk declines. As wealth inequality grows, the overall riskiness of investments made by the rich also grows. It essentially reflects the human urge to make a quick profit. It means margin or installment buying of assets and goods only for resale and not for productive purposes. It means, for instance, increasing involvement of investors in futures markets. When others see the rich profiting quickly from speculative purchases, they tend to follow suit. . . . Speculative fever tends to feed on itself, and by the time the general population rushes to join the bandwagon, the venture is usually nearing its last stage. . . . Eventually even those normally too cautious for such ventures are tempted by "easy" profits.[117]

Batra emphasizes that the speculative fever cannot begin in the absence of wealth disparity, since it is only the very rich who can afford to take such heady risks with potentially high but uncertain return. It is the concentration of wealth—great inequalities in the distribution of income—that is the centerpiece for his assertions. He ultimately concludes that high wealth disparity is responsible both for the surge in speculative mania and the fragility of the banking system. The real cause of great depressions is extreme inequality in the distribution of wealth.

One could argue against this negative scenario by stressing the real divergence between wealth and income, but it should be remembered that the two are closely related. Extreme wealth always ends up generating extremely high income. Thus, inequality in the U.S. income distribution picture is an indicator of whether this growth in wealth inequality has reached the blast-off stage in its rate of growth. This point is quite important, because accurate and recent wealth data are much harder to get than income data.[120] One source of wealth estimation is estate tax records, which are filed only after the person dies. Although inheritors wealth can be estimated from such archives, there is an obvious lag in up-to-date information. As we shall see later, income information is gathered and published yearly by the Census Bureau. It is thus possible to trace income inequities and their immediate consequences in our society more readily than to analyze disparity in wealth holding, which has a more historical, *ex post facto* character.

Batra's work shows the rate of wealth concentration steepening abruptly just before the onset of the Great Depression. If his theory is correct, it is important to monitor wealth accumulation with as much up-to-date data as possible. Fortunately, we do have a few recent special studies that tap this dimension of economic inequality. To begin with, there is great unevenness when wealth is looked at by itself. A handy way of measuring the distribution of wealth is to think of the Richter scale. This well-known tool measures the severity of earthquakes, and is on a log-normal distribution. There is a massive difference between an earthquake with a magnitude of 5.0 and one of 8.0—much greater than a three-unit difference on a linear scale (such as a ruler) would indicate. In short, quakes at the high end of the Richter scale have much more serious consequences than do those at the low end.

So it is with wealth. A person who is one rung above you on the wealth ladder can have many times as much wealth. Carroll has ranked persons in the top 60 percent of wealth (since the bottom 40 percent have no wealth to speak of).[119] If these persons are ranked in a cumulative percentile distribution, we find that an individual at the halfway point (50th percentile) has an average net worth of $79,214. Fascinating things begin to happen by the time we analyze persons in the 90th percentile. Here, average net worth is $354,060. The 95th percentile yields $608,944, the 99th averages nearly $2 million, the 99.5th percentile reports a net worth slightly over $3 million, and the 99.75th percentile rings in slightly over $5 million. The scale compresses dramatically at the upper end. The more wealthy you are, the greater your distance in net worth from persons just below you on the wealth ladder.

How rich are the truly wealthy? As with the distribution inequities, the actual amounts of money involved are staggering. The 1986 Survey of Consumer Finances found the total net worth of the richest 1 percent of Americans was $3.72 trillion. In that year, all consumers in the United States spent $2.8 trillion. Even in 1991, the wealthiest 1 percent owned well over the total amount of coin and currency in circulation ($257 billion).[120] By the end of the 1980s, the United States had become, in terms of wealth, the most unequal industrialized country in the world. While the top 1 percent of our wealth holders controlled 39 percent of all household wealth in 1989, comparable figures for France (26 percent in 1986), Canada (25 percent in 1984), Great Britain (18 percent in 1986), and Sweden (16 percent in 1986) are much smaller.[121]

There is no doubt that very wealthy Americans did profit from all the political and economic bones tossed to them during the 1980s. One group of scholars, adopting a relative approach, measure the rich as persons in families with income over nine times the poverty line (about $95,000 for a family of four in 1987). By this measure, the percent who are rich nearly doubled in a fifteen year period—going from 3.1 percent in 1973 to 6.9 percent in 1987.[122] There are some who might argue that an income of $100,000 does not go very far for a family of four in today's world. But few would dispute that a yearly income of a million dollars would probably suffice. The ranks of millionaires rose more than four-teen-fold during the last decade. Nearly sixty thousand more persons became millionaires during the 1980s going from 4,414 taxpayers at the beginning of the decade to 63,642 at its end. But looking exclusively at taxpayers, which only measures the amount of money made during a year, underestimates the true number of millionaires in our nation today. When net worth (assets minus liabilities, or wealth as we have been discussing) is looked at, there were 2.1 million households with a net worth of $1 million or more as of 1991. This represents an increase of 62 percent in just seven years.[123]

Edward Wolff has compared findings from the 1983 and 1989 Surveys of Consumer Finances to discover who benefited from the run-up of wealth in the United States during the 1980s. In this period, 55 percent of the increase in total real household wealth went to the top 0.5 percent of U.S. families, while 29 percent went to the next richest 9.5 percent of families. In essence, 84 percent of the increase in wealth during this period accrued to the wealthiest 10 percent of our population. The only other time during this century to witness such a skyrocketing rise in wealth was the period from 1922 to 1929—just before the stock market crash.[124]

It is also important to note that while there was a large growth in wealth during the 1980s, much of it due to an explosion in stock prices, most of us benefited only slightly and many of us actually lost net worth (figure 1.2). In short, there was a huge transfer of funds—$256 billion—from moderate- and low-income people to the very wealthiest. The poorest fifth of American families actually had a negative net worth of –$2,000 in 1983, and this increased to -$14,000 by 1989; our next-poorest fifth saw their net worth slide from $10,000 to $7,000. What this means is that a large minority of our population has gone deeper into debt. As income has fallen, a large segment of the American population has borrowed and has sold off the few assets they had to maintain an

FIGURE 1.2   Percent Change in Wealth by Income Class, 1984-1991

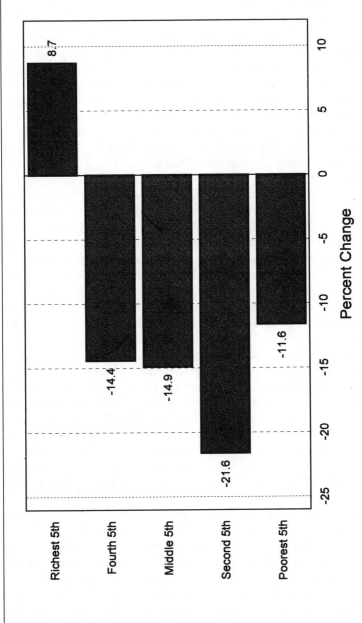

Source: Data from Mishel and Bernstein, 1995, p. 248.

adequate standard of living. Savings have entirely disappeared for many, while there are no longer any funds for the proverbial rainy day brought on by a layoff or illness. The bottom line: life for U.S. families has become very financially unstable.[125]

Huge gaps in wealth also exist between whites and minorities. For example, for every dollar of median net worth held by a white household, African-American households have a little over eight cents and Hispanic households about eleven cents.[126] This lack of wealth hinders minority families from advancing the economic prospects of their children. This is primarily because there are fewer resources to pass on to succeeding generations to give them a boost in accumulating more revenue at a later time. Wealth begets wealth. When you do not have it, you cannot play the game. High status African-American families not only have more trouble than white families in transmitting their status to their children, African-Americans are also more likely to experience a steeper "fall from grace." In essence, African-American offspring from white-collar backgrounds are more than twice as likely as white offspring from white-collar backgrounds to fall back into blue-collar work. Finally, African-Americans from lower-income white-collar families are much less likely to make the climb to upper-white-collar jobs than white offspring are. For minorities, then, lack of wealth inheritance is a two-edged sword: it serves to hinder upward mobility and offers no protection from downward mobility. The existing pattern of wealth distribution in the United States serves to intensify racial inequities and tensions while passing them on to the next generation.

Because of increasing wealth inequities, the future retirement plans of many Americans are also in great danger. The disparities in wealth among those approaching retirement are simply enormous. If those fifty-one to sixty-one years old are ranked by wealth holdings, average household wealth (at the fiftieth percentile) is $97,506 before pensions and expected Social Security benefits are factored in. But those at the thirtieth percentile have less than $40,000 for retirement. At the twentieth percentile, there is only $16,000 on average.[127] While Social Security will help those with less wealth, it was never intended as a full-scale income maintenance plan. Moreover, coupled with declining savings rates and the large drop in employers who offer pension plans to workers under thirty-five today, there is plenty to worry about in the future.

It is important to keep in mind the extreme inequality at the upper end of the wealth distribution, because persons in this rarified sanctum of privilege literally form the tail that wags the dog. According to Batra and other social scientists, this select group has so much financial clout they can either destabilize the U.S. economy by their investment behavior *or* be its saving grace. The general thrust of the twelve year Reagan/Bush administrations (1980–1992) was to pander to this select group with pro-rich policies such as cutting their taxes, encouraging a better business climate, easing regulatory restrictions, and so on. The theory was that this privileged group would then reinvest its newly gained money to start new businesses, invest in product development, build factories, expand trade activities, and create a lot of new jobs for the rest of the population. Benefits would ultimately "trickle down" to everyone.

Comparative research shows that this did not happen, although the data do not yet reflect the full impact of the 1980s. The assets and investment decisions of the wealthiest U.S. families in 1950 and in 1983 have been examined to see if any significant shifts toward a "reinvestment" in America took place.[128] Unfortunately, the opposite trend occurred. No matter what slice of the very wealthy is looked at—the top 0.5, top 1.0, or top 10 percent—the rich have increasingly failed to invest productively. The proportion of total liquid savings held by the nation's wealthiest 10 percent dropped from 11.3 percent (1953) to 4.6 percent (1983). The decrease in savings was even greater for the top 1.0 percent. In summary, there has been a slackening of the role that the very wealthy have played in the U.S. economy over time. They do not seem as concerned with investments directed toward rebuilding and growth of the economy—particularly by starting and/or owning their own businesses. In contrast, the very wealthy have increasingly tied up their money in consumption-oriented purchases or in tax-haven investments not directly concerned with business growth. The rich were passed the torch that was to light the way toward reinvigoration of our economy, ostensibly so that all of us could benefit. Unfortunately, they allowed the lamp to go out, leaving nearly all middle- and low-income Americans in the darkness of surging wealth inequality, stagnating income, and swelling poverty.

There is an increasing awareness of the danger that the growing gap between rich and poor may eventually pose for the economy as a whole. There is now an expanding group of economists who believe that increasing inequality within the United States must inevitably result in lower economic growth. Income disparities discourage or prohibit workers from being educated and trained to their fullest capacity, which translates directly into a less productive labor force. Mathematical models developed by economist Paul Romer at Berkeley, for example, clearly illustrate how income inequality hurts Gross Domestic Product by lowering efficiency.[129] Other studies show that the growth of jobs and income is slower in cities with wide wage inequities and faster in more egalitarian communities. In short, inequality can act as a brake on the economy that forces everyone to lose out, including the rich. Among metropolitan areas, for example, suburbanites forgo $690 in annual income for every $1,000 difference between their earnings and those of workers in the central city.[130]

According to a new book by Andrew Glyn and David Miliband, countries which have more income equality also have higher productivity and economic growth rates (Japan, Germany, Norway, France, Belgium, and Sweden).[131] Countries with slower economic growth rates, such as the United States, New Zealand, Australia, and Swizerland, also have much higher levels of income inequality. Finally, if more proof is needed, a sophisticated study using historical panel data for advanced industrial countries and post–World War II cross sectional data of many nations has found that inequality lessens growth.[132] In the study, the proportion of income going to the middle class (forty-first to sixtieth percentile) was revealed as a significant variable that predicts greater increases in Gross Domestic Product per capita. Deciphered, it means that a healthy and wealthy middle class is good for the country. Equality leads citizens to invest more in their country and leads to appropriate redistributive government policies that promote growth.[132] These policies as measured include income transfers, such as Social Security, pensions, and unemployment compensation, but not health and education expenditures. The caveat is that the relationship holds true only for democracies, where the emergence of more egalitarian policies can be influenced by voters.

What all this comes down to in the end is a hidden cost for every citizen. Great inequality always levies a high price for society as a whole. Where gigantic differences in wealth and income are allowed to fester, costs mount via underutilization of human capital, drops in productivity are caused by reduced access to health care and education, expenses that are associated with surging crime go up, political decisions are made to reduce helpful social programs such as Medicare, Social Security, Head Start, and the like.[133] The economy becomes unstable, and begins to lurch from one crisis to another. There is difficulty in shoring up the failing economy caused by lack of faith or funds from the middle class. Conclusively, the United States will continue its startling slide to economic loss, and will persist in failing to compete effectively in the international arena, if its great wealth and income inequities continue unabated.

HOW MUCH IS ENOUGH? SOCIAL JUSTICE AND THE VERY RICH

Lastly, an important reason for being greatly concerned with the distribution of income is simply a matter of social justice. It can well be asked, "How much is enough?" In essence, what extremes of income disparity are we willing to tolerate in our society? The answer, of course, will lie in the eye of the beholder and will no doubt be greatly influenced by that person's relative position on the income spectrum. But the question can be answered, albeit with some degree of imprecision. There is evidence that Americans do have a normative value set with which they judge a person's earnings as either fair, too high, or too low.[134] More vitally, there is a mix of both merit and need factors that come into play when people consider what is a fair or unfair salary. The topic of how we perceive inequality is of such major importance that a large portion of chapter 7 will address the issue.

Above all, however, is this basic truth: the question "How much is enough?" is rarely asked in American society. It should be. Despite the great increases in wealth going to only a tiny percentage of our population, there are many who will still argue that it is the inalienable right of every American to earn as much as he or she can in a "free" and "open" competitive market economy. Time does not permit a specific refutation of this position, although any good sociology textbook on social stratification will do the job.[135] As the evidence unfolds in the following pages, it should also become apparent that the playing field we compete on to get higher income is uneven. Simply put, the game is fixed in advance, with the wealthy and influential determining the rules of access and reward (income) within U.S. society. Most of us operate in a limited market of educational and employment opportunity (so we can hopefully avoid McDonald's in favor of McDonnell-Douglas), but we are excluded from large incomes because we failed the first litmus test—being born as sons and daughters of the truly wealthy in our society. These offspring are doubly blessed, both because of the money they will inherit and because of the advantages which accrue to their social class that will permit them to take over the helms of our major corporations and banks in the future—eventually giving them still higher incomes.

One mechanism which creates such immense inequality in our society is overpayment of business executives. They are lavished with great salaries. It is important to point out that total compensation for these corporate executives includes much more than just a salary figure—although this is usually quite generous as well. In addition to salary, a bonus plan is nearly always included in a chief executive officer's pay packet—this can include cash, deferred salary, and bonus payments, as well as director's fees and commissions. Other added incentives include payments from long-term compensation plans, restricted stock awards, thrift plan contributions, etc. One can easily see that a huge source of compensation comes directly from stock ownership, which includes holdings of the chief executive officer (CEO), his or her spouse, and their children.[136] Stock gains in this category thus include net value realized in shares or cash from the exercise of stock options or stock appreciation rights granted in previous years. Typically, CEO pay is handsomely buttressed by such stock options. For the CEOs of large American firms with annual sales of $2 billion in 1992, stock options yielded an additional $735,000 to a base pay of

$746,000. It is typical to add increases for performance, restricted stock, and bonus pay. This ultimately yields an average compensation of $2,505,000 for top executives—which translates to $1,300 per hour (forty-hour work week/one month vacation).[137] Again, these are averages. Donald Pels—as CEO of LIN Broadcasting—reaped $186 million from stock options, which were added to total a staggering compensation of $217 million in 1990 (the equivalent of $113,000 per hour).[138] During the same year, Chairman Stephen M. Wolf of UAL Corporation (parent company of United Airlines) earned $17 million in stock options while his company's earnings dropped by two-thirds. Steven J. Ross, CEO of Time-Warner, raked in average earnings of $16 million per year from 1973 through 1989.[139] Recently, his company laid off 605 workers because of "hard times" while Ross took home $78 million. This amount is two and one-half times the wages of the laid-off workers. To translate this into more understandable terms, if a person earned $50,000 per year for forty years (which could be called lifetime earnings for most middle-class people), the sum would still be only a thirty-ninth of what Ross made in that one year.[140]

How much is enough? *Forbes* magazine proudly monitors the pulse of corporate America by publishing salaries of its leading executives each year. During the last decade, the very *worst* year for corporations was 1987. Despite the October market collapse of 1987, when the Dow-Jones industrial average lost over 500 points in a single day, it was still a good year's pay for nearly all of the top chief executive officers (CEOs) in America's largest eight hundred corporations. A full 273 CEO's earned $1 million or more, while the median compensation for all eight hundred was $762,253—an 8 percent increase over 1986.[141] Lee Iacocca, whose name has become a household word because of his frequent nice-guy, "just an average Joe" advertising appearances as Chrysler's CEO, wrapped up third place in the mega-million-dollar sweepstakes with a wage of $17,656,000 in that year. In 1990, Iacocca took a 25 percent raise although earnings had fallen 17 percent at Chrysler and workers were being asked to sacrifice.[142] In actuality, CEOs of automotive firms are the worst offenders in this contest of greed, frequently insisting on huge salaries despite running their companies into the ground. Roger Smith, while CEO of General Motors during the 1980s, engineered the collapse of market share for GM. This led to an implosion at GM causing the loss of $500 million per month in its North American operation and the closing of twenty-one plants in North America, which effectively cut in half the number of GM workers between 1985 and 1995. For this stellar performance, Smith was rewarded with an increase in his pension plan when he retired in August of 1990, from $700,000 to $1.2 million a year. Every sixteen days in his retirement, Smith makes what an auto assembly-line worker makes in one year on the job.[143]

In actuality, even when executives do a poor job, they are given "golden parachutes" and luxuriant retirement packages to buy them off and ease them out the door. Examples abound. J. P. Bolduc, ousted as CEO of W. R. Grace & Co., received a $20 million severance bonus, which was $5 million more than his contract specified. Joseph E. Antonini, when forced out as K-mart's CEO, gained $3 million in severance despite the poor performance of his company. Robert J. Morgado was given $50 million to $75 million for going along with his forced retirement. Such beneficent severance packages are on the rise and are now routinely negotiated before hiring is formalized. Although many are not contractual, they are still freely given despite poor executive performance. Why? According to Graef Crystal, the leading expert on CEO compensation in the United States, it is due to guilt. Crystal contends that directors of corporations personally know their CEOs, have golfed with them, and had them in their homes. It is a lot more difficult for them to fire someone they know personally than to close a plant and throw ten thousand people out of work that they do not know.[144]

The feeding frenzy of CEO salaries took a breather at the end of the 1980s in the wake of several Wall Street scandals, the savings and loan default crisis, record bankruptcies, ensuing recession, and an angry public awakened to these lavish payouts while their own wages remained stagnant and their jobs were threatened by cutbacks. By 1995, avarice was unabashedly back in style. Two surveys of top CEO pay revealed huge gains for upper management. *The Wall Street Journal*, which commissioned a special survey of America's largest 350 corporations, found an 11.4 percent increase in CEO pay—about triple the 4.2 percent increase of their white collar employees.[145] The median compensation for this elite group, including all the bonuses, stock options, and the like, was $1,779,663 in 1994. *Forbes'* annual survey of CEO pay in the nation's eight hundred largest corporations also found that nearly two-thirds of America's top executives now receive more than $1 million per year in total compensation.[146] The highest paid CEO in 1994 was Stephen Hilbert of Conseco Inc. at $39.6 million. Being second (Lawrence Coss, Green Tree Financial Corporation, $28.9 million) or third (James Donald, DSC Communications, $25.2 million) was not exactly painful either. Even tenth place finisher Louis Gerstner Jr. at IBM was not entirely ill-served with a compensation of $12.3 million. Analyzing the total compensation for the twenty-five highest paid business executives over the past five years, *Forbes* calculates that all together these men reaped $1.5 billion.[147]

It is impossible to spend money at the rate many of these executives rake it in. In 1994 the average CEO compensation of the top eight hundred was just short of $1 million (to be exact, $993,000—up 11 percent over the previous year). For executives of America's largest companies (averaging revenues of $21 billion), pay, bonuses, and stock options climbed by a whopping 23 percent in 1995.[148] This translates to over $2,100 per hour, or slightly less than $84,000 per week. A person can eat only so many meals per day, own so many cars, live in so many houses, buy so many suits, ad nauseam with this deluge of money. What do they do with it? As shown in the previous discussion, they invest it in stocks so they can realize even greater profits. The singularly important point to note is that many of these top executives "earn" in less than a week what it takes the rest of us an entire year to make. Are we to conclude that these CEOs possess talents and skills so rare as to warrant such extravagant pay? Are the jobs they perform and the positions they command so crucial to the country's well-being that they somehow deserve these exorbitant sums? It is reasonable to point out the other side of the coin as well: by awarding such lavish sums to top corporate executives the message is also unmistakably being communicated to us that the tasks we perform in our economy are relatively unimportant in the scheme of things.

Experts believe that compensation for U.S. corporate executives has reached scandalous proportions.[149] This has happened despite evidence that there is almost no relationship between compensation for these chief executive officers and their performance. Are these corporate executives really worth the money they receive? Surprisingly, even the business magazine *Fortune* thinks not.[150] In a sophisticated multiple regression study,[151] compensation analysis revealed that *rational* factors one could expect to be used as justification for CEO pay accounted for only 39 percent of the variation in ultimate compensation. In essence, almost two-thirds of corporate executive pay for the largest 170 U.S. industrial and service corporations cannot be logically accounted for. The good news was that, among the rational factors examined which should figure as important in predicting executive pay, the corporation's financial performance was the most salient. Other, less important objective factors driving up executive pay were larger size of company, higher risk for corporation in volatile markets, and absence of government regulation of the industry (transportation and utility companies pay lower because of this).

On the other hand, factors that ought to matter in total compensation—such as the age of the executive or whether the corporation's board of directors came from within or outside of the company—were found to have no relationship to CEO pay. The most remarkable finding was that it did not matter how much corporation stock was owned by the CEO in predicting "rational pay." The study found a multitude of executives who were lavishly awarded with stock options and who were even permitted to swap old option shares for newer, lower priced shares when the corporation's market performance was dismal.

Our industrial competitors did a much better job in the last decade than American business executives in building up their companies, earning increased market share, and enhancing their profits. This happened despite the fact that they do not pay their corporate executives nearly as much. Corporations in Europe and Japan pay their CEOs at a more sensible rate. In 1990, the average CEO of an American manufacturing company with an annual revenue of $250 million was paid $633,000 in salary and other compensation.[152] This was two-thirds more than the second place average, which went to CEOs of comparable German companies and over twice as high as what typical Japanese executives earn ($308,000). What is more, U.S. Labor Secretary Robert Reich points out that an American CEO's money goes much further.[153] A typical CEO can buy three times as much in America with the same dollar as a similar Japanese executive in Japan and about twice as much as a German CEO in Germany. Within the past fifteen years top American executives increased their pay by 12 percent per year, which widened the gulf between their compensation and that of their typical workers. Put differently, the average American CEO in 1960 had an income only twelve times greater (after taxes) than the average income of his company's workers. In the United States as recently as the mid-1970s, this same ratio was "only" 40 to 1. But by the start of the 1990s typical CEO income was seventy times greater than the average line worker's. Compensation specialist Graef Crystal found that among the 292 executives of America's largest corporations who remained CEOs of their companies between 1992 and 1994, the typical CEO pay was 172 times greater than that of the average worker in the companies they headed. Crystal pulls no punches in his reaction to such outrageous overpayment: "The system is rotten. Pay is going crazy. It's a never-never land."[154]

A European corporation's chief executive officer typically receives pay six to eight times that of an entry-level professional employee.[155] Put simply, most European and Japanese managers believe that an organization suffers when the CEO receives an astronomical multiple of the average employee's pay. An eloquent summary of these findings helps us take a step toward answering the question: "How much is enough?"

> Some would even say that CEO pay at its most stupendous is just wrong. Plato, apparently the world's first compensation consultant, suggested that the highest paid person in the community should earn no more than five times the lowest. Management writer Peter Drucker relates approvingly that J. P. Morgan raised that ratio to around 20 and maintained it in his enterprises. There is nothing scientific about those numbers, but they reflect a society's instinctive sense of fairness. . . . The CEO's fortunes should rise and fall with his company's. But at many corporations the board has adopted only half the principle. The CEO gets a terrific reward when

the company does well . . . but he still gets a pretty good reward when it does badly.[156]

Finally, after reviewing the huge wealth and pay disparities discussed above, there appears to be an emerging answer to our question of "How much is enough?" It seems safe to conclude that there is never enough for the very rich in our society—whether they are the chief executives of America's largest corporations, scions of the country's wealthiest families, or major recipients of the great bulk of stock dividends parcelled out every year. Yet the crucial issue is not whether income has become more unequally distributed within the United States, but rather just how bad the situation is. There are many who would argue that such a trend is both necessary and desirable for a society to develop economically (especially with reference to Third World countries), that the poor are always with us,[159] that the pace of change is glacial and consequently irrelevant, that the pattern will reverse itself when certain demographic trends play themselves out, and so on. All of these points will be addressed in subsequent sections and, in most cases, refuted. The economic news is exceedingly and unjdeniably bad, so it makes little sense to fiddle with rationalizations while Rome burns. Nonetheless, there may still be time to turn the decline around before many of the tragic consequences described above come to pass. A framework to accomplish this will be offered in the concluding section.

It is also the intention of this book to describe income inequality between and among populations, rather than to address only the trends revealing economic deterioration. It is a fact of life that if one area of the world or a given country is doing poorly, prosperity will be found elsewhere. Different countries of the world will be compared on income inequality, as will U.S. regions, states, and counties. Looking at income inequality both geographically and descriptively will help in isolating common factors which tend to produce it or, conversely, to promote greater egalitarianism in the distribution of income. Before turning to these tasks, however, the next chapter will review some of the theories advanced to explain income inequality between and within societies. Such an exercise can provide a better perspective with which to understand the mass of studies dealing with income inequality. In our contemporary era of information overload, an adequate theoretical grounding can help us avoid overlooking the forest because of the trees.

Chapter

2

THE BATTLEGROUND OF IDEOLOGY: THEORIES AND UNDERLYING VALUES SURROUNDING THE EXPLANATION OF INCOME INEQUALITY

This chapter looks at a variety of reasons why some groups receive more income than others. Why do some nations seem blessed with great affluence while others suffer under the curse of poverty and underdevelopment? There are many ideas that seek to account for the vast income differences prevalent in the past and that are still with us today. Unfortunately, sharp disagreement among scholars is more often the rule than the exception. Many of these disputes stem from competing values and beliefs. All involve attitudes about what is right and just in a society, and what is unfair and basically wrong with the system we live in.

Most theories eagerly claim to know the ultimate truth about how wealth is passed out. Thus, there is a very real danger when studying income inequality that ideological and political overtones will bias our view. Yet social reality is complex. This will guarantee that one theory alone will never have an absolute and final answer explaining why some are more equal than others. In truth, all theories contain some basic values which are often unexpressed. Such values greatly affect how we view inequality and the system that generates and maintains it. In essence, these ideas carry with them implied assumptions about whether income inequality is "good" or "bad." All theories can be said to encapsulate underlying value assumptions and ideological premises accepted as axioms of truth.

Because of this, certain theories regarding income inequality are more easily accepted than others. This is especially likely when theories are based upon beliefs which are dear to our country. The values and beliefs surrounding capitalism provide the starting foundation, in the

United States, upon which inequality is viewed. One common American view is that poor people deserve their lowly place in our perfectly functioning, competitive market system. Those who suffer from low income are by definition somehow "not plugged into the system." This may be caused by a number of things: lack of education, outmoded job skills, minimal contacts, poor information about labor markets, personal laziness, moral depravity, and so forth.

In essence, relative failure to achieve a decent income in our society is seen as somehow due to personal failure. Few people believe that low income could be caused by environmental conditions. Research on attitudes of Americans about their economic position in life is eye-opening. There is a marked tendency to see lack of advancement and success as one's own fault.[1] It is assumed without question by most people that the United States is the land of opportunity where anyone can succeed by applying a little effort. The system or structure is seen as basically sound, so that personal shortcomings must be to blame for those classes of people who do not benefit from it. It is this type of social bias that we must guard against as we try to come to grips with income inequality.[2]

There are basically two major sides to the debate surrounding income inequality.[3] The *conservative* position holds that economic inequality is both necessary and functional. The *radical* view sees inequities as neither fair nor just. The two opposing views separate on a number of different points. Conservatives believe that the system of distribution we now have is just. The system rewards those who are most deserving (because of their individual effort and hard work), so that defending the status quo in society works to protect the interests of all.

Radicals view the existing distribution of wealth as unjust. The powerful simply expropriate more riches by means of force, coercion, inheritance, discriminatory tax laws, and other unfair methods. As part of this scheme, governments are said to work on behalf of the wealthy to maintain the interests of the rich over those of the common people. Thus, conflict in society is inevitable because of the unfair means of distribution. An exorbitant degree of inequality may in this way be damaging to the functioning of a nation and the well-being of its inhabitants.

Reality lies somewhere in between for most societies. Thus, it is probably more helpful to think of nations in positions along a continuum of income inequality. The specific causes of a country's stratification system will undoubtedly vary through time and by region. Income inequality may be due to coercion and traditional inertia. It may also stem from the operation of a competitive free market. More likely, inequality will result from a combination of both. There are elements of truth in both radical and conservative thought, as we shall discover as we go along. Analysis gets muddied too often by simply insisting that one model is correct while the other is false. Thus, Princeton University Professor Robert Gilpin simply labels theories of inequality as "ideologies," intellectual acts of faith that cannot be proved.[4] He believes such theories are based upon assumptions about people and society that cannot be put to a fair and impartial test.

MARX AND CAPITALISM

Concern about ideological bias is justified. Until recently, Americans, as an example, could get hysterically anticommunist. There has tended to be a total rejection of anything termed "Marxist" without any evaluation.[5] Much of this compulsive overreaction has stemmed from fear and ignorance. We were in a cold war with Russia and other communist nations between 1947 and 1992. Many in the United States found it threatening that over one-third of the world's population was at one time living under communism. Communism still reigns in China, with over a billion people. Yet most Americans have no clear idea what Marx had to say about inequality, economics, and the role of social classes in society.

Some would find it surprising, for instance, that Karl Marx admired a number of things about capitalism in modern nations. Marx wrote at the dawn of heavy industrialization in the mid-1800s. On the basis of Europe's experience, he concluded that capitalism was a truly dynamic, modernizing force. It would soon wipe out the older, rigid, autocratic systems of inequality.[6] Capitalism meant doom for the nobility and aristocracy of feudal regimes. He regarded the technology of capitalism as far superior to, and more efficient than, the way work had previously been organized. Ironically, he thought this would eventually lead to the downfall of capitalism.

At the base of Marxian thought is how people in a society organize their work. Early bands set up their means of production by hunting and gathering. There was fairly equal sharing in a communal arrangement. Systems of slavery were next, in which the means of production were determined by ownership of human beings. A feudal order came next. Here the means of production was centered around ownership of land by the nobility. Rent was paid by serfs, who returned a portion of their harvest to the aristocracy. Finally, the era of capitalism overthrew the old feudal arrangement. The new means of production became primarily industrial, owned in the form of factories and capital by a small class of entrepreneurs called the bourgeoisie. For the most part, the rest of society formed the proletariat or working class who sold their labor to the factory owners in order to survive.

A major component of Marxism is the idea of unequal exchange. Briefly, the exploitive nature of capitalism is obvious from the subsistence wages paid to workers, which are greatly below the value of the products they manufacture. The difference between the wage paid to labor for manufacturing the product and the much larger value of what it is actually worth is "surplus value" that is taken by the capitalist for profit. This profit is so excessive that it can only be spent by reinvestment in more industrial expansion. This is true especially in the purchase of capital goods such as factories and machines, which again add to the buildup of more profit at a later time. Thus, capitalism is by its very nature inherently expansionary. It produces a thirst for new markets and continuous investment in new industries. The wage-earning class gets larger when previously undeveloped areas are industrialized.

Central to all societies—regardless of how their economies are organized—are two classes of people: the haves and the have-nots. Whether these be free men and slaves, lords and serfs, bourgeoisie and proletariat, they are always in the nature of oppressor and oppressed and invariably in periodic conflict with each other. In essence, each historical era is marked by rather extreme inequality and thus carries within itself its own demise—in Marxian terms "the seeds of its own destruction." Feudalism gave birth to early capitalism. Basically, it arose by the forming of towns, trade, and craft production. The new business class soon successfully challenged the power and privileges of the landed nobility.

And so it is to be with the contemporary era of capitalism. It is the very efficiency and relentless, growing energy of the capitalist system that may breed its own downfall. Early capitalism was marked by fierce wars of survival between industrialists. Often this led to hostile takeovers, mergers, and business failures. The eventual outcome left a few big firms in control of most of the industry. It is under monopoly capitalism that the proletariat are gathered *en masse* to man the giant factories. As ownership becomes more concentrated in fewer hands, more and more people are forced into wage labor to survive. The remaining firms, engaged with one another in death struggles for survival, cut all possible costs of production in their drive for greater profit. Wages paid to labor are slashed to bare subsistence as greater work demands and larger quotas are instituted. Incomes start to shrink. Workers begin to realize that they are oppressed. Since they work and live in the same place, organization for resistance is comparatively easy. Strikes and violence become normal as open war against factory owners sets in. Such clashes grow more virulent and bloody until revolution takes place. At this point the proletariat forcibly takes over the means of production. A new socialist state is ushered in that will ultimately be replaced by the perfect communist society. Income will then be divided on this basis: "From each according to his ability, to each according to his need." In a word, those who have greater need for income will be given more by the state.

The end point is clearly utopian. Yet Marx has been unfairly criticized for predicting a perfectionist outcome. Actually, he had little to say about ultimate communism, preferring to dwell upon the more pronounced traits of capitalism in his day. A more valid weak point is that communist revolutions did not occur in modern states. The world instead witnessed agrarian, nonindustrial countries such as Russia, China, Cuba, and Vietnam become communist. In advanced capitalist countries, workers have been less militant and have been able to win major concessions from giant corporations.[7] This took place while a healthy middle-income sector arose. The predicted increases in poverty simply failed to materialize. In fact, many were helped when deprivation and want did arise. Welfare capitalism came about in an effort to provide a safety net for the poor. This was something Marxists clearly believed would never happen.

Another flaw of Marxism is its mono-determinism. It tends to see everything in society as ultimately caused by the mode of production. Marx believed the very ideas and ideology of capitalist society were formed and controlled by the bourgeoisie. He included such areas as intellectual theory, science, art, and even religion. His famous quote—"Religion is the opiate of the masses"—essentially stems from this line of reasoning. By this he meant that religious dogma supports values of the upper class at the expense of the lower class. Poor laborers were urged by the clergy to remain docile and obedient in order to reap true riches in heaven.

For Marx, all aspects of society are reduced to the economic. He virtually ignored the role and potential power of the state, which he saw only as a pawn serving capitalist interests. Neo-Marxists today argue that the state does have an independent power with some free rein apart from the corporate class.[8] A major defect of Marxism is its failure to see the role of political factors in international relations.[9] For example, national rivalries occur regardless of common economic systems. The hatred between Russia and China, between Vietnam and China, and between Vietnam and Kampuchea are prime examples.

A CAPITALIST VIEWPOINT: THE MARKET MECHANISM IN SOCIOLOGY

The role of values as an independent force in their own right is often said to be neglected in Marxist theory. Max Weber was a scholar contemporary with Marx. Weber is thus frequently touted as supplying a refutation to Marxian thought. Weber's *The Protestant Ethic and the Spirit of Capitalism* argued that the values in religion could change the way the economy works.[10] For example, early Calvinists would look to material success and prosperity as a sign of grace from God. This would subtly impel them toward more industriousness in a self-fulfilling prophecy. In actuality, however, Weber built very closely upon Marxian concepts. This was very true in his definition of social class in "Class, Status, and Party."[11] Weber largely agreed with Marx that class is determined by the ownership of economic property. As such, class ultimately determines a person's life chances—including even health and longevity. But Weber added a component to the definition of class that is similar to today's description of a market system. Weber pointed out that a person could better his class position by taking advantage of opportunities for income—essentially by raising the skill level of the person's occupation. The implied idea is that a person could advance in social class by training for skilled, highly paid occupations.

Within sociology, the idea that a market economy is at the bottom of much of today's work world was carried to an extreme by Davis and Moore.[12] They believed that rewards (essentially income) for a given occupation were set by its functional importance. In other words, how does the job increase the survival and well-being of society? More important positions must carry with them higher rewards to induce the most talented persons in that society to train for such jobs and to perform adequately in them. In the end, inequality in rewards is positive, functional, and inevitable in any society.

This explanation of income inequality has been sharply criticized. The attack was very hostile with respect to the underlying market mechanism which is said to reward us on the basis of merit. A major question is: who determines what jobs are more important than others in a society?[13] On the surface, the answer seems very simple—work that pays more must be more important. A moment's reflection shows this to be circular thinking. In essence, if we argue that pay is dependent upon importance, then importance cannot be dependent upon pay. The theory is very conservative and status-quo oriented. It assumes that given income inequalities must be rational and deserved because those with top paying jobs would not receive the salaries they get if they did not deserve them. Nothing is said about the role of power in perpetuating or increasing income inequality. It remains true that persons who receive larger rewards in a social system also end up with more power. They use this lopsided influence to insure that they continue to receive larger rewards, no matter what function they serve for society.[14]

Carried to its logical extreme, the theory should lead to perfect equality in the distribution of wealth.[15] In a market economy, if labor is free to move to higher paying jobs, such occupations should attract a surplus of workers—leading to a decline of income for these jobs. Positions paying low salaries would also attract fewer workers, which should ultimately lead to higher wages to entice more people to enter them. Frankly, perfect mobility does not exist in society because many are barred from even starting the race. There are many curbs to the effort to build scarce skills. Serious gaps in educational quality and opportunity have been shown to exist. At the same time, mediocre offspring of wealthy families receive the best training that money can buy. Their connections for job and business networking become plus marks in the race for more pay. A review of eleven panel studies which looked at jobs and earnings of American men found the same thing: family background explains about half the difference in men's work success.[16] The most important single predictor of a son's occupational status is his father's occupational status. Put in money terms, 57 percent of economic status is due to family background.[17] A man born into the top 5 percent of family income had a 63 percent chance of earning over $25,000 per year in 1976 (the top 17.8 percent of family income). But a man born into the bottom 10 percent of family income had only a 1 percent chance of attaining this level.

Lastly, even if one were to seriously accept the market explanation underlying functional theory, we are still left with the question: How much is enough? If, for example, we believe a medical doctor is actually functionally more important than a garbage collector, how much more should the doctor be paid? Ten times more? Twenty times more? Does a ratio of 20 to 1 mean that the doctor is twenty times more important than the garbage collector? One can see how easily this line of reasoning tends to end in subjective bias.

CAPITALISM CONTINUED: THE MARKET IN ECONOMICS

A neoclassical economist will see the health of a free market as crucial.[18] Many Western economists believe it is in the very nature of humans to be self-seeking, to barter, and to trade for goods and security. Such activity will lead to the natural evolution of a market which will set prices of goods by the laws of supply and demand. Individual consumers will respond to price changes, buying at the lowest price with all other things (e.g., quality) being equal. Firms must grow by greater innovation and more efficiency to remain viable. This drives the whole system ever upward to achieve greater and greater levels of wealth and productivity. Progress means increasing per capita wealth. The fact that some groups or persons will not benefit as much from a country's growth in affluence is seen as unfortunate but also unavoidable. In short, some persons will contribute more to the economic productivity of society than others, and more productive people will be more amply rewarded. Only a few will risk more in the hope of getting an even larger amount of money.[19] Income inequality is thus seen as both rational and positive. It serves as a reward for those who contribute the most and as a way to urge others to try harder. The end result, in this view, can only be to advance the whole system as we all work to achieve higher productivity.

How do we achieve higher productivity? A basic factor in classic economics is human capital.[20] The theory is that people will voluntarily defer income. They will invest in themselves by staying out of the labor market while going to school. The results work to enhance their skills and lifetime earning potential. Immediate sacrifices are made. Yet people will see such costs balanced by greater pay in the long run. The model assumes perfect rationality and a good knowledge of the labor market. Equality of education and similar native talent are also basic assumptions of this model. Finally, money income is viewed as the main motivator pushing us into given jobs in an open, competitive labor market where wages are flexible.

Obviously, this scenario does not exist in reality. Just a decade ago the average pay of an American college professor was less than the average wage of a coal miner or an auto worker.[21] Perhaps some of this income rift can be explained by unions. Noneconomic rewards may also work to get people to take lower paying jobs. Yet such factors as intrinsic job interest or an opportunity to work for social justice cannot be safely used to prove the theory. To begin with, people vary quite a bit in the value they attach to psychic rewards in lieu of pay. Taken to the extreme, as Osberg points out, this argument quickly becomes absurd:

> If one really believed in the theory of compensating differentials, one could, for example, argue that a Mississippi sharecropper with a money income of $8,000 and a New York office worker earning $35,000 are equally well off, since the sharecropper has nicer weather and lives in a more tightly knit community.[22]

Although noneconomic interests play some part in what line of work a person may enter, there is no proof that such psychic rewards cause income inequality. It is hard to actually test such values since they are vague and amorphous by their very nature.[23] On a broader level, market theory errs in its basic assumption. There is some doubt that economically equal actors always behave in a rational manner in a free and competitive exchange. We know that exchanges are seldom free or equal. Elements of coercion, vast power differences due to monopoly, and politics all get in the way to "rig the game" in advance.[24] A major fault in the reasoning of classical economists is the failure to recognize that the economy functions within a very real social class system. The same criticism for fixation on one major variable said to explain all things, levelled at Marxist theory, can also be applied to market economics. Many contemporary economists behave as if everyone enters the starting block as equals in a fair test of competitive prowess. Unfair disparities caused by inheritance and favoritism toward the well-off are too easily ignored. Generalized class bias promoting the interests of the wealthy seem not to exist. Capitalist economists simply fail to see the havoc caused by market forces. This is especially true of the grossly unequal distribution of wealth and income among societies. The unquestioned orientation of classical economics is geared toward maximizing profit for persons and firms, which is assumed to benefit everyone in the long run. The economic status quo becomes the conservative bedrock of market theory. Stability and equilibrium form the frame of its foundation.[25]

Rather than having such absolute free choice with an equal starting point, in reality people mostly make limited choices. Some options simply do not exist for certain segments of the population. In essence, unless we are lucky enough to come from a wealthy background, alternatives can be more restricted for us. Even for those with outstanding academic talent, an education at Harvard would be simply unavailable. Its average yearly cost of about $25,000 makes it impossible for poor and middle-class Americans to attend this university. The point is that we still have the choice to go to college, but the option of attending a higher quality Ivy League university is simply not there for most of us. It then follows that we are not completely responsible for our position on the income distribution. This is exactly the reverse of what market economics would have us believe.

As a case in point, one study estimates that over 80 percent of American workers are members of "internal labor markets."[26] This is a system where pay and assignment of workers are set by administrative rules and bureaucratic procedures. Wages are determined by customs and patterns which have evolved from the past. Such pay scales have been gradually codified and set through labor/management agreements. Wide pay differences from one department to another may exist in jobs that have the same responsibility and work difficulty. The system may not be rational, but it is seen as fair and just by workers because this is the way it has always been.

Much of the skill within the internal labor market is gained by on-the-job training. One employee informally helps to train another. The knowledge upon which a person's income may be based is thus obtained in a work-setting group whose norms may operate outside of official company policy. In short, social skills and congeniality may determine much of our income over and above any formal educational preparation we went through to get the job in the first place. Lester Thurow believes that 60 percent of U.S. employees derive all their work skills on the job.[27] The labor market which acts to set our income is really a market where supplies of trainable labor are matched with training opportunities. It is the job openings that exist rather than the formal training we undergo that determine our pay. The locus of this "choice" is firmly fixed by the decision of a given corporation to create openings or of a particular industry to expand. Once we are hired at the entry level, we must further pilot our way among the shoals of management and coworkers to ingratiate ourselves. Success will mean our coworkers' willingness to impart the informal skills that lead to higher pay. A major point is that if you are outside the internal labor market, you are simply shut out.

We have been endlessly told that the best way to increase our earnings is through more formal education. But there are many questions about the effectiveness of this strategy. Some argue that the major function of education is to screen workers for future jobs.[28] In a society based upon credentials, having a master's degree is more important than having a bachelor's degree. This is quite apart from the actual skills needed on the job. Access to better jobs will be restricted to those with higher levels of education, even if more advanced education is not needed to perform the work. Thus, the norm for landing the really good jobs changes rapidly. Whereas a decade ago this may have meant earning a bachelor's degree in business administration, it now means a master's degree (M.B.A.). Education then becomes a necessary but not a sufficient prerequisite for a better paying job. The basic minimum to qualify for any decent job in today's society would be a baccalaureate degree, but achieving this level of education will not necessarily get you a nice job. Not having a college degree, on the other hand, is a guarantee of poor employment. If more education actually did produce greater real income, the explosion in college education since World War II should have reduced income differences. This it clearly has not done.[29] Although there is some room for debate, it seems that education has little or no impact on earnings.[30] At best, the effect of education upon income is very different among business firms and industrial contexts. In the end, there is no guaranteed payoff.[31]

The Dual Economy thesis forms another prong of the attack on the rational functioning of the market economy.[32] A great deal of American wage research is tainted by the individualistic bias of market theory. That is, a person's starting place in the system and eventual lifetime earnings are claimed to be a reflection of that worker's basic value to the system. From this perspective, income inequality due to earnings is caused by our personal attributes (or lack of them) such as education, motivation to achieve, commitment to work, and so on. Thus, variables used to explain income differences tend to focus upon characteristics of individuals such as work experience, amount of formal education, age, and the like.

A better explanation would focus upon industrial characteristics as they change workers' earnings. A core industrial sector of huge monopolistic firms which arose in the era of early capitalism is commonly described by Dual Economy supporters.[33] This is always compared to a periphery made up of smaller firms operating in an open, competitive environment. Bluestone provides an excellent summary of these basic differences:

> The core economy includes those industries that comprise the muscle of American economic and political power. . . . Entrenched in durable manufacturing, the construction trades and to a lesser extent the extraction industries, the firms in the core economy are noted for high productivity, high profits, intensive utilization of capital, high incidence of monopoly elements, and a high degree of unionization. . . . The automobile, steel, rubber, aluminum, aerospace, and petroleum industries are ranking members of this part of the economy. Workers who are able to secure employment in these industries are, in most cases, assured of relatively high wages and better than average working conditions and fringe benefits. . . . Concentrated in agriculture, nondurable manufacturing, retail trade, and subprofessional services, the peripheral industries are noted for their small firm size, labor intensity, low profits, low productivity, intensive product market competition, lack of unionization, and low wages. Unlike core sector industries, the periphery lacks the assets, size, and political power to take advantage of economies of scale or to spend large sums on research and development.[34]

In the end, two American labor markets are created by a sector's dominant technology. The primary labor market is made up of high-paying jobs. These jobs carry reasonable security, good fringe benefits, and nice working conditions. Such work is typical of industries geared toward long, stable production runs. The secondary labor market is made up of subcontractors, service workers, secondary manufacturers, and suppliers. The welfare of these firms springs directly from the well-being of the primary industries. Thus, such jobs are marginal and low paying. They tend to be unstable and haunted by poor working conditions. Many minorities—especially blacks, Hispanics, the foreign born, and some women—tend to be thrown into this area. The job experience of the work force in this market is branded by low pay and poor prospects for advancement. Morale is low. Insecurity, absenteeism, and turnover are very high. Certain categories of workers find themselves trapped in this job market for a variety of reasons. Its existence is guaranteed because of fixed conditions that must be met for capitalist economies to do well. Overproduction, business slowdowns, seasonal shifts, and technological changes can lead to bouts of unemployment for those who hold jobs here.[35]

Since the sector one works in has a noticeable effect upon pay, differences in the impact of personal assets such as education will also be present. Core industries with minutely tuned wage scales will use education to assign individuals to specific jobs that they are said to be "trained for." In the peripheral sector, there is a smaller and simpler opportunity structure with fewer differences between jobs. Thus, individual traits such as education, skill, motivation, and so forth play a much smaller role in earnings. The Dual Economy theory thus offers an alternative to the individualistic bias present in the human capital/market explanation of income inequality.

Analyzing data representative of the U.S. work force in 1975–1976, Beck and his colleagues found major differences between the two sectors, especially with regard to pay. At that time, core workers earned over $3,000 per year more than periphery workers. In 1995, after adjusting for inflation, this amount would be roughly equal to a $8,350 pay difference. Core workers also showed many more gains: better education, more high-status occupations, more full-time work, and higher numbers of workers in unions. Lastly, they were more likely to be male and white—which explains some of the gap in earnings. The effect of these variables upon earnings is quite important, so they were held constant through regression analysis. The researchers were still able to conclude that workers in the core would receive over $1,000 more per year than periphery workers ($2,785 in 1995 dollars). This was apart from the effects of sex, age, race, education, and occupation. Their study ends with a strong message to those seeking to explain income inequality using market-level explanations:

> What does all this mean? It means that we should be very suspicious of any attempts to build models of occupational or earnings processes in industrial society which consist exclusively of individual level variables. Our analysis suggests that the rules which govern the distribution of socioeconomic benefits to individual workers are not uniform across all sectors of the economy. Analysts in the human capital and status attainment tradition have tended to interpret income and status differences as due to the application of a fixed set of rates-of-return to different mixes of individual background characteristics, skills, and experience. We have shown that these rates are not fixed, and that one important determinant of their variability is a distinction between core and periphery sectors derived from the theories of the structure of industrial capitalism. In our view these findings constitute a challenge to models of the social and economic order which underlie the human capital and status attainment research traditions.[36]

OF MARX AND MARKETS: THE DYNAMICS
OF THE INTERNATIONAL POLITICAL ECONOMY

Contrary to market explanations, how much income people receive in a society is not a result of the rational "choices" they make to "earn" higher pay. The social structure we are born into and our family's class position can give us a head start. The schools we attend and the values we learn from our parents, teachers, and friends affect our later pay packet. The part of the dual economy we may land in plus aspects of the internal labor market (social factors on the job) work to set wages. Our sex, age, and race also enter into the equation of what income we will eventually receive. Despite what we have previously learned, there is one crucial point to remember. The idea that the system is basically fair and that we receive income in direct proportion to the real value of our labor is simply not true. This does not mean that effort has nothing to do with the income we receive in life. As noted above, it is necessary to get a college degree in the right field to qualify for a decent job in today's labor market. There are some things we must do to remain generally competitive and in the running to qualify for a middle-class income. Riches do not beckon to us all, however, if we somehow earn the right college degree from one of the more prestigious universities. It matters less what innate abilities we are lucky enough to have, such as high intelligence, or how hard we work on the job. What really matters is the social structure of the economy and labor market that we attempt to plug into.

Americans may have some trouble seeing the impact and great importance of structural variables. Our training from early childhood includes learning values which fit a competitive labor market system. We have come to believe that our personal economic well-being is due entirely to our own work and effort.[37] For example, a recent Gallup Poll found that an overwhelming majority of Americans (71%) rejected the idea that U.S. society is divided into two groups: the "Haves" and the "Have-Nots." On the other hand, 73 percent in Great Britain believe that the United Kingdom is so divided.[38] Among those Americans not recognizing class polarization, the largest group believes that it is "lack of effort that is more often to blame if a person is poor." Only one-third see poor circumstances as a cause of poverty and want.

One reason for the popularity of such beliefs is that there is little talk in schools or in the media about how large-scale economic forces serve the interests of the wealthy. A more penetrating query would look at how multinational corporations are tied in with U.S. financial policy. Defense spending, initiatives (or their absence) in basic research and development, funds for education, and so forth cannot fail to have some effect on our personal financial situations. Many studies show a large and direct upper-class bias among top government office holders.[39] Extravagant contributions to political action committees (PACs) have funded payoffs for wealthy individuals and corporations.[40] Economic elite dominance of semi-official policy-research groups (Council on Foreign Relations, Trilateral Commission, Committee on Economic Development to name a few) works to sway top federal decisions.[41] We have known for quite some time that intensive congressional lobbying by big business can have repercussions for our economic and physical well-being.[42] The rearguard actions fought against corporations to reduce physical pollution, to increase car safety, and to recognize hazards of smoking all serve as major examples.

What should be kept in mind is that there is little we can do on a personal level to increase the pay we earn. Perhaps we cannot even keep our current income from going down. The real power which skews the distribution of income stems from more macro-oriented variables. These can include major moves of multinational corporations, the political and economic drama among nations, changes in domestic labor and business laws, the level of defense spending, and so forth. It is this point that is the most damning flaw in market explanations for income growth and distribution.

The unproven assumption is also made by free-market advocates that increasing affluence of the well-to-do in any nation will ultimately benefit all of its citizens. This is the bottom line behind the questionable idea of "trickle-down" economics. This fiscal policy, dubbed Reaganomics during the 1980s, appeared to wane with the election of Bill Clinton as president in 1992. But as of 1994, with the Republican takeover of both houses of Congress, the fiscal imperative of supply side economics is now back with a vengeance. With the exception of a two-year respite, this perspective has dominated American society for nearly fifteen years. After living with the consequences of supply side economics for this long, do most Americans still really believe that "a rising tide lifts all boats"? Is it not by now obvious that there is a need for some democratic control over how any increase in wealth is to be more fairly divided? The dubious assumption is too often made that it is possible for the economic pie to keep growing without limit. It is too easily accepted that the American model of development can be repeated in other nations willing to make the necessary sacrifices to follow our lead.

It is certainly true that there is an inherent tendency for market-driven economies to expand rapidly. Marx wrote admiringly of the many ways in which capitalism tended to organize and improve industry. But it always ended up with periodic gluts of overproduction, which then led to economic downturns. It was this very efficiency that led capitalist industries to expand in an insatiable search for new markets to sell their goods, which in turn led to greater profits for the owners. Again, in turn, this led such owners to invest in even more productive facilities. But Marx foresaw only collapse. In the long run, overproduction would lead to a satiation of markets. Slack demand for goods, less need for manufacturing, a spurt in unemployment of workers, further poverty, and eventual violence would well up within a sinking economy.

A great defect of Marx's theory was his focus on domestic economies, although to be fair this was the *modus operandi* of his day. What developed after his death was a profound globalization of world markets that had reached epic proportions by the start of World War I. By 1917 Lenin had published his first extension of Marx in *Imperialism*.[43] His book described a world economy led by giant industrial combines working closely with huge banking interests. In effect, Lenin's revision of Marx argues that industrial nations had escaped working class revolutions and were able to enjoy a higher standard of living because of overseas imperialism. By seizing colonies, capitalist economies were able to get rid of extra goods manufactured at home. They could also import cheap raw materials for their own firms while investing any added profit in their various spheres of influence. In this stage of advanced capitalism, the world becomes one great battleground for markets, profit, and global dominance. Inequality among nations continues and may even get worse. But this is not only because of colonial exploitation. The advanced countries are also unequal in their ability to compete with one another. The international contest of wills arising among them breeds instability, political change, and war. Gilpin nicely summarizes the meaning of the new revisions:

> In Marx's critique of capitalism, the causes of its downfall were economic; capitalism would fail for economic reasons as the proletariat revolted against its impoverishment . . . the actors in this drama [were] social classes. Lenin, however, substituted a political critique of capitalism in which the principal actors in effect became competing mercantilistic nation-states driven by economic necessity. Although international capitalism was economically successful, Lenin argued that it was politically unstable and constituted a war-system. In summary, Lenin argued that the inherent contradiction of capitalism is that it develops the world and plants the political seeds of its own destruction as it diffuses technology, industry, and military power. It creates foreign competitors with lower wages and standards of living who can outcompete the previously dominant economy on the battlefield of world markets. Intensification of economic and political competition between declining and rising capitalist powers leads to economic conflicts, imperial rivalries, and eventually war.[44]

Lenin can be criticized, of course, for being as deterministic as Marx. Lenin's rigid view allows no scenario other than cataclysm to result from the expansion of market forces. But no matter how the revolutionary scene plays out, humans still stand accused of being inherently self-centered. They may continue to fight over scarce resources under communism as much as under any other system. From witnessing the horrors of Stalinism to those of the Khmer Rouge in Kampuchea, we hear a ring of truth to this point. It is also a fact that Lenin ignored the existence of state autonomy as much as Marx did. It is possible that a government will pursue national goals that are not in the best interest of its top economic class. Nevertheless, many of the inherent tendencies that Marx and Lenin identified as typical of capitalist market economies are quite valid. There is an undeniable urge to expand through trade and to export capital, investment, and technology in an effort to get more profit. It thus creates a true world economy. The outcome of this process has yet to be fully grasped. Its operation has spawned two contending perspectives of the world economy. Each claims to show a greater impact than the other on the level of personal income we receive.

TOWARD A WORLD SYSTEM: DEVELOPMENT AND DEPENDENCY THEORIES

A recent analysis offers both similarities to and differences from the ideas advanced by Marx and Lenin.[45] The *modern world system theory* of Immanuel Wallerstein seeks to explain today's inequality by paying heed to the past.[46] The idea of an emerging global economy we now hear so much about is not new. Simply put, the erroneous belief that all persons start the race for income equally also applied to nations. Many countries start from a better position in a world economic hierarchy, which usually provides higher levels of income for its citizens. Those in poor countries are doomed to remain in poverty since they must compete under the burden of many initial disadvantages. One such drawback arises from their nation's low rank in a global economic pecking order.

Wallerstein begins to build his theory by posing a distinction between a single "world economy" and a "world empire." The single world economy forms a global division of labor made up of many nations with separate governments. The world empire rests upon a single, dominant state made up of many national economies which do not mesh well with one another. In the past, much of the world was, from time to time, organized under empires via military dominance (e.g., the Roman Empire). By the end of the sixteenth century, however, there had arisen in Europe a new type of commercial capitalism which was resistant to world empire. This newly emerging world economy was to have an even greater global impact than world empire.

Wallerstein describes a number of countries in northwestern Europe as the core of this world economic system. England and Holland came to dominate Latin America and Eastern Europe, which were said to occupy a periphery. Midway between these two extremes were nations of the "semiperiphery" (Mediterranean Europe, including Spain and Portugal). Such nations were somewhat free and were not completely subjugated by the economic system. Nevertheless, they lacked the power to become truly influential on an international level. The basic pivot of this world economy was trade. The global market arises out of the exchange of manufactured goods from core countries for raw materials from periphery countries. The major division of labor is now seen from a world perspective, not as internal to a particular country. The gap between factory owners and wage laborers in Marxist thought is less important. The major actors on the world stage are nation states, although the ideas of surplus value and exploitation still operate. In the modern world there is now one mode of production in which all nations engage: production of goods for sale and profit on the world market. The basic point is that a world market has emerged as an all-powerful force. The system integrates all nations *economically* through global trade, production, and a new division of labor among nations.

As always, some nations are more equal than others. Powerful economic actors in each state, while pursuing profit in this single world market, seek to influence their governments to dominate and distort the market to their benefit. Business interests in core nations seek easy access to cheap raw materials from periphery countries. They also try to build a guaranteed market for their manufactured goods within these poor nations. The strong countries of the core joust with one another in colonization, in military sallies, in diplomacy, and in drawing up trade treaties. But no single nation is influential enough in the end to completely dominate the world market.

Different areas of the world end up specializing in different economic roles for a variety of reasons. The world geographic division of labor is for the most part dependent upon resources. This is very true for the extraction of raw materials. The time when an area joined the world system also dictates its relative power. Recent arrivals tend to be weaker and more easily exploited. The geographic division of labor has two consequences. First, rings form across nations, uniting parties with the same economic interests. Economic elites often align with one another despite their national home bases. Second, core economic elites try to influence and control what happens in other countries. A nation in the core has an advantage here since it generally has an advanced and highly productive home economy. A country in the periphery does have such modern attributes as mines, ports, and plantations that specialize in raw material export. But a traditional sector of tribes and subsistence agriculture also exists along with the modern sector. A country in the periphery is export oriented because of core influence and demand. As a result, its own internal economy is poorly developed and depends upon imports from core countries. Its government is consequently weak and is often subjected to great influence from the outside by core countries. Because of this outside force, local manufacturing is not allowed to develop. Core firms fear that the market for selling their manufactured goods would be lost if local businesses were allowed to develop. In Wallerstein's view, the march toward the creation of a total world economy made up of all nations on earth is unavoidable.

Modern world system theory has been further refined by Chase-Dunn and Rubinson to include these elements:

1. Unequal exchange is a permanent feature of the system. The basic rank of the nations in the core, periphery, and semiperiphery will remain the same despite any other type of change. Shifts in trade goods, manufacturing, newly discovered resources, and advances in technology have no effect on this world hierarchy. Productivity in core nations always remains high. More capital ends up flowing into them because of unfair trade advantages. Goods bought from periphery nations never make up for this unequal inflow. A core country has highly advanced plants, machines, and plenty of money to invest to ensure greater profit. Its human capital in the form of a well-educated and skilled labor force is also very high. Although productivity and wages may go up in the periphery, they will also do so in the core. The core is able to drain more money from peripheral lands because core nations are powerful enough to control world prices for the goods peripheral countries export. Hence, there are higher rates of exploitation and lower rates of capital buildup in the periphery.

2. While there are many political nations in the system, there is only one world market. The states vary in terms of power, and they tend to both compete and make deals with one another based upon their economic self-interest.

3. There are different forms of production relationships, with free wage labor in the core and different types of coercion or forced labor in the periphery. Governments often work to prevent unions from forming in less-developed countries. Their goal, at the urging of core countries, is to keep the average wage low.

4. Although inequality in power between core and periphery tends to remain constant, its form will shift through time. From the end of WW II to about 1970, the structure of political control was more relaxed because all economies were growing. Dominance was achieved more through economic multinational corporate influence. In times of economic contraction, such as today, attempts to influence the periphery are less subtle and much more direct. Fear engendered by saber rattling and military force is a time-honored

tradition. Economic browbeating is an old standby. Clandestine paramilitary ventures and secretly organized coups have unfortunately become too common. All such indirect sanctions brought to bear against periphery nations are meant to keep them in the system and in their "proper" place of subordination. Above all, core nations are fans of free trade, since they have the competitive advantage and power to set world prices in their favor.

5. A single top power can emerge among the core nations. This country may occupy a central position in the world economy for a time. The example of the United States after WW II is a case in point. At that time, America had the only intact economy left after the war's devastation. There is also a tendency for other core nations to emerge and overtake the head power (Japan). Much of this is caused by the huge overhead the hegemonic power must support to keep world stability through military and political means. This erodes its own ability to continue investing in itself in order to stay competitive through advancing technology. Competition eventually gets worse between core countries as the lead power fades. Military conflict and trade wars usually result until another strong power emerges to keep the economic peace.

6. The world economy is prone to periods of expansion and contraction. The long-term trend, however, is one of continued growth. Rarely does a downturn go back to an earlier low point in these economic waves. During such skids, military conflict and trade wars again tend to prevail.

7. The system results in greater riches for all participants. This does not mean that all benefit equally, as will be seen in the next chapter. There are some who argue that periphery countries would be better off without being part of the world system. Core nations tend to become more wealthy. Another trend toward massing of capital in huge banks or corporations also develops, feeding the expansion of business firms.

8. The number of nations participating in the world system tends to grow. Much of this is a result of the decolonization during the 1950s and 1960s of previously subjugated countries. A more contemporary example can be seen with the breakdown of the former Soviet Union in the early 1990s. Many new nations were spawned in theis process of devolution; nearly all of them sought immediate entry into the capitalist world order of trade. When this conversion gets under way, a growth in core dominance of newly freed countries results because of investment, loans, and the location of branch plants in foreign lands.

9. As the entire globe is now spoken for in terms of spheres of influence by various core nations, competition among core countries for dominance of the periphery has increased.[47]

Although it bears similarities to some Marxian ideas, the modern world system theory is different—especially with regard to the dynamics of social class. For example, social classes cut across national boundaries in a system very different from that envisioned by Marx. In the not-too-distant past, core economic elites would often enter into quiet agreements with core labor interests. The two groups worked with economic elites in peripheral countries to exploit the labor and resources of those countries. It is this powerful alliance that has prevented the "international proletarization of the working class." Marx predicted a growth in working class consciousness that was to develop on a worldwide basis. The past collusion of core labor with economic elites acted to prevent this.

It is with the semiperiphery countries of the system that some hope remains for independent economic growth and development. It is within these nations that there is a mix of core and periphery activities. Due to such advantages as natural resources and internal markets, a chance does exist for these countries to grow freely:

> Because of the mix of core and periphery activities, dominant classes tend to have very opposing interests. . . . Thus it is often the case that the state apparatus itself becomes the dominant element in forming the power block that is able to shape the political coalitions among economic groups. . . . This may take either a leftist or a rightist political form. Those upwardly mobile countries that rely on alliances with core powers tend to develop rightist political regimes (e.g., Brazil since 1964) while those that attempt autonomous development move toward the left (e.g., China and the Soviet Union). Whether leftist or rightist, upwardly mobile semiperipheral countries tend to employ more state-directed and state-mobilized development policies than do core countries. . . . Antiimperialist movements may take state power and try to mobilize for development (e.g. Angola, Cuba, Vietnam, Mozambique, Cambodia, etc.), but the development of core-type activities requires resources that small periphery countries most often do not have. Internal market size, natural resources, and sufficient political will to isolate the country from core powers are necessary if such antiimperialist mobilization is not to become either an isolationist backwater or a CIA countercoup. Those areas that escape the system but do not economically develop are soon reconquered (e.g., Haiti, Ghana, Chile).[48]

It is possible only within the semiperiphery to avoid the collusion that occurs across nations between economic elites in core and periphery countries—and not always possible there. A national consensus strongly opposed to core domination can develop. Such a fear can breed an alliance between domestic capitalists and labor in semiperiphery countries. This partially explains why many socialist movements have come to power in Russia and China rather than in the periphery, where conditions for workers and peasants are actually worse. In the end, however, it is the very isolation of such leftist movements within the semiperiphery that works to stabilize the exploitation of the periphery by the core. According to Wallerstein, the contamination is then quarantined to an area where it can do no harm while still keeping in place the overall status quo of world economic inequality.[49] In this way, periphery countries remain rigidly locked up within a system of subordination and subservience.

A drawback of the modern world system theory has been its vagueness regarding which nation belongs to what category. Snyder and Kick examined 118 countries to identify them as core, periphery, or semiperiphery.[50] In doing so, four factors were considered: (1) the net value of trade between countries (exports minus imports); (2) military interventions into other countries; (3) diplomats posted within other countries; (4) participating in international treaties. The authors believe their method explains the power dimension within world system theory:

> Theories of imperialism typically emphasize the domination of the world economy by core powers . . . and the use of threat of superior armed forces as the means of ensuring economic domination. We also consider that diplomatic exchanges constitute a salient form of information flow in the world system. For sending nations, diplomatic missions provide regular and ostensibly reliable information concerning local economic opportunities, political conditions, etc. in the host country. Such missions may also facilitate trade agreements and sometimes serve as a base for attempts to manipulate local conditions. . . . Finally, we included the treaty data on the rationale that they would address networks of defense commitments or reflect attempts by some nations (e.g., core powers) to legitimate potential military intervention in others (peripheral countries).[51]

It is important to note that locating a nation within the three-tier ranking of the modern world system is not done only by use of economic terms. Of the four criteria used by Snyder and Kick, only the information on superior trade advantage could be called economic. The other three networks have less direct financial results. This distinction will have some bearing when developmental models of income inequality are discussed below. Snyder and Kick did go on to measure the impact of world system position on subsequent economic growth for the 118 countries. They looked at change in Gross National Product per capita (GNP) from 1955 to 1970. Strong support for modern world system theory was evident from their results. In essence, they report the "cost" of location in a peripheral or semiperipheral nation is roughly $500 per capita over this fifteen-year period. A full *twenty-nine countries had a per capita GNP of less than this amount* in 1992.[52] Core countries always experienced greater economic growth, and at a faster pace, than peripheral or semi-peripheral countries.[53]

Research by Kenneth Bollen also indicates that the position a country occupies within the world system theory is predictive of whether democratic institutions will flourish or not.[54] This finding stayed the same whether a nation was affluent or poor. But affluence, in and of itself, was important in encouraging democracy. Taken in conjunction with Snyder and Kick's findings, what this means is that peripheral and semiperipheral countries are less democratic than core nations. They are also poorer but they will not develop economically as fast as core countries because of their world system position. Since they are less well off, they will also be less democratic. Yet since they are less democratic, there will be less encouragement of economic development. In essence, peripheral countries are truly locked into a Catch-22 situation. They seem doomed to endless subservience, political tyranny, and economic quagmire.

Edward Kick continued to revise and perfect his typology of nations into modern world system classifications by expanding the criteria for ranking to eight transnational networks: trade flows, economic aid and assistance treaties, transportation/communication treaties, sociocultural treaties, administrative/diplomatic treaties, political conflict between nations, arms transfers to other countries, and military conflicts.[55] Rather than remaining with a three-category system, he partitions his data into eleven blocks of countries in a roughly descending order of strength and importance on these eight indicators of national strength. In his article, however, he also advocates a fourfold typology (core, semicore, semi-periphery, periphery) that he believes would more accurately portray the varying strengths of all nations in the system. Neither change, whether into eleven groups of nations or four enclaves, fits the extensive prior research of the modern world system. Hence, following Boswell and Dixon,[56] the three-category world system approach is maintained using Kick's most recent data from the 1970–1975 period. The position of these countries is presented in table 2.1.

While the basis for these rankings is more recent than information originating as far back as 1965, when the original typology was cast, the data are still relatively old. World system theorists would argue that this really does not matter, since change occurs very slowly—if at all— within the system. For example, one study that measured mobility in the world system over the past half century found that 95 percent of the nations remained classified in their original rank—whether core, periphery, or semiperiphery.[57] Although a few nations did manage to move up the

TABLE 2.1 Division of Countries into Core, Semiperiphery, and Periphery in Modern World System Theory

Core	Semiperiphery		Periphery		
United States (3.75)	Russia (2.95)[a]	Saudi Arabia (.44)	Kenya (–.31)	Honduras (–.63)	Haiti (–.72)
Germany (2.31)[b]	Taiwan (n.a.)	India (.42)	Trinidad (–.39)	Somalia (–.64)	Sierra Leone (–.73)
France (2.10)	China (1.09)	Kuwait (.37)	Cyprus (–.39)	Bolivia (–.64)	Guyana (–.74)
United Kingdom (2.00)	Australia (.63)	Egypt (.30)	Sudan (–.40)	Afghanistan (–.64)	Madagascar (–.75)
Japan (1.74)	Austria (.61)	Argentina (.15)	Ivory Coast (–.40)	Paraguay (–.65)	Mali (–.75)
Italy (1.42)	Norway (.54)	Iran (.14)	Jordan (–.40)	Burma (–.65)	Togo (–.75)
Belgium (1.25)	Finland (.38)	Singapore (.13)	Ecuador (–.41)	Liberia (–.66)	Mauritania (–.77)
Canada (1.22)	Poland (.37)	South Korea (.11)	Panama (–.42)	Mozambique (n.a.)	Benin (–.79)
Netherlands (1.17)	Brazil (.29)	Mexico (.08)	Costa Rica (–.44)	Bahrain (n.a.)	Niger (–.79)
Switzerland (.82)	Greece (.22)	Iraq (.07)	Ethiopia (–.44)	U. Arab Emirate (n.a.)	Malawi (–.80)
Sweden (.81)	Hungary (.14)	Turkey (.02)	Zaire (–.45)	Libya (n.a.)	Chad (–.81)
Spain (.73)	Bulgaria (.13)	Venezuela (.01)	Gabon (–.45)	Bangladesh (n.a.)	Gambia (–.87)
Denmark (.67)	Romania (.13)	Pakistan (.01)	Malta (–.45)	Qatar (n.a.)	Vietnam (n.a.)
	Portugal (.09)	Malaysia (–.02)	Tanzania (–.48)	S. Yemen (n.a.)	Sri Lanka (n.a.)
	Israel (.04)	Algeria (–.02)	Colombia (–.18)	Congo (n.a.)	Cuba (n.a.)
	New Zealand (.01)	Thailand (–.03)	Morocco (–.19)	Rwanda (n.a.)	Angola (n.a.)
	Ireland (–.10)	Indonesia (–.05)	Syria (–.22)	Albania (n.a.)	Oman (n.a.)
	Lebanon (n.a.)	Nigeria (–.12)	Zambia (–.53)	N. Korea (n.a.)	N. Yemen (n.a.)
		Philippines (–.17)	Guatemala (–.54)	Nepal (n.a.)	Burundi (n.a.)
		Peru (–.23)	Cameroon (–.55)	Senegal (–.49)	Zimbabwe (n.a.)
		Tunisia (–.25)	Dominic. Repub. (–.56)	Nicaragua (–.50)	Upper Volta (n.a.)
		Chile (–.26)	Guinea (–.58)	Ghana (–.51)	Kampuchea (n.a.)
		South Africa (–.28)	El Salvador (–.59)	Uganda (–.67)	Laos (n.a.)
		Uruguay (–.31)	Jamaica (–.60)	Cen. Afr. R. (–.71)	

a. The score for the former Soviet Union was given to Russia following its devolution, reasoning that this country was the core power within the former U.S.S.R. This should be regarded as only a temporary estimate until a new categorization is made.

b. The score for West Germany was substituted for Germany since the reunification clearly led to its economic and political dominance over what was once East Germany. This is only a rough approximation until new scores can be calculated.

hierarchy (Japan and Italy into the core, South Korea and Taiwan into the semiperiphery), nearly all countries remained firmly in place. Nonetheless, given the wholesale breakup of the former Soviet Union and the Eastern Bloc, which spawned more than a dozen new nations, there is a need for scholars to continuously update these standings. Since the early 1990s, East and West Germany have unified. The opposite process of breakup dominated, however, as countries spun off from the former Soviet Union, and then a few of these new nations broke up as well. The former country of Yugoslavia spawned four new countries, spurring the ongoing war between Bosnians, Serbs, and Muslims in that area. Czechoslovokia formed two new nations (Slovakia and the Czech Republic). From the U.S.S.R. came Lithuania, Latvia, Estonia, Tajikistan, Azerbaijan, Armenia, Kyrgyz Republic, Georgia, Uzbekistan, Turkmenistan, Moldova, Kazakhstan, Ukraine, Belarus, and Slovenia. In short, given the tremendous influx of new nations in this decade, all more or less clamoring to become part of the global capitalist economy, there is an urgent need for world system scholars to update the rankings of these countries.

A more serious drawback has emerged around the difficulty in defining the exact nature and role of the semiperiphery. There continues to be disagreement about which nations ultimately fall into this middle ground. The point has also been made by Dutch scholar Cornelis Terlouw that both Wallerstein and Chase-Dunn, original and enthusiastic proponents of modern world system theory, never intended states to be clustered into groups.[58] Rather, countries were to be regarded as existing along a continuum between core status and periphery status.[59] Professor Terlouw has kindly allowed his scores to be published here (in parentheses in table 2.1).

Professor Terlouw's methodology produces a composite Z-score indicating the degree of a nation's "coreness" based upon its standing in the following areas:

1. Gross Domestic Product (GDP) per capita as part of total world GDP
2. Part in world trade, 1978–1983
3. Stability of trade relations, 1961–1983
4. Number of embassies sent and received by a state in 1985
5. Number of diplomats sent and received by a state in 1985
6. Military power in 1985[60]

Terlouw, in short, uses roughly the same criteria as Kick and other scholars in making his categorizations. The big advantage in his approach, however, is twofold. His scores are based upon the most recent data. As important, however, is the continuous nature of his scoring. Z-scores are degrees of departure from the norm or average of all countries and will thus vary from roughly +3.0 to –3.0 at their extremes. Most scores will tend to cluster around zero, resembling a normal or bell-shaped curve. Terlouw's scores foster more precise measurement and help scholars avoid tedious arguments over whether or not certain countries really belong in the semiperiphery. Another important point to be made is that when the countries are coded in Kick's typology (1=core, 2=semi-periphery, and 3=periphery), the correlation with Terlouw's scores is –.80, which is very high. In other words, countries with high negative scores in Terlouw's ranking are closely associated with scores of 3 (periphery status) in Kick's rankings. The two sets of scores are measuring pretty much the same phenomena.

A glance at table 2.1 shows the complete dominance of European countries, together with the United States and Canada, in the lineup of core countries. Japan is the only Asian core country according to the three-part ranking offered by Kick. Such a situation has resulted in the criticism that the theory is ethnocentric, being ultimately Euro-centered. It is hard to agree that any bias arises merely from Anglo scholars narrowly focusing on only their own experiences. Ranks were drawn up by using a number of objective dimensions of power. More recently, the end product of world economic inequality is increasingly being posed as a North/South question. Such a polarity takes note of the immense power these core countries have within the system.[61]

Scholars of what has become known as the dependency school offer a view of this structure through the eyes of intellectuals within the periphery (most of whom are Latin Americans). Dependency theory actually developed before modern world system theory was created, but it does fit quite nicely within the world system framework. A social scientist named Theotonio Dos Santos is one of the better known advocates of this approach.[62] He points out that the world is in its third period of dependency today (the prior two being colonialism and imperialism). Today's world has seen money spent by multinational corporations in areas which cater to the internal markets of less developed countries (LDCs). Simply put, the possibility for economic growth in an underdeveloped country depends on having the money to buy machinery and raw materials not available within that nation. Yet the foreign dollars needed for industrial growth must come from exports of raw materials. This in turn increases the power of elites who have long been in control of these businesses. Such elites have no interest in seeing local industry increased. They may even see such development as a threat to their continued influence. It is also true that in many countries these very export businesses are themselves owned and controlled by foreign capital, so that high profits return to the home bases of multinational firms. Trade relations take place in a highly monopolized international market anyway, which tends to lower the price of exported raw materials. In such a system, advanced industrial countries can simply raise the prices of industrial products to ensure greater profits. A chronic budget deficit results which then finances development. This occurs because so much capital is drained from poor countries by multinationals. These countries are thus forced into repeated borrowing from core nations in order to continue economic expansion.

The real value of foreign investment and aid to poor countries, however, is doubtful. Much of the money is used to finance North American investments in Latin America and to subsidize foreign imports which dominate or replace local products.[63] The net effect when foreign capital penetrates an underdeveloped country is to freeze it at the existing level. More gains become impossible. All benefits now go to foreign firms who get red carpet treatment from governments who wish to persuade them to make new investments. Dos Santos calculates that for each dollar entering Latin America as investment, $2.70 leaves in the form of loan repayments, interest charges, profit to corporations, and the like.[64] Not only does growth stagnate, conditions may actually worsen. Politics and the day-to-day running of government come under the thumb of multinationals. Income inequality increases between the ever richer political, military, and industrial elites and the mass of workers. Real wages are lowered through intimidation of unions which in turn limits labor's purchasing power. This acts to retard the development of an internal consumer market which could spawn new businesses. Job creation is further reduced by bringing in high-tech production techniques with less labor demand. Most important, however, the profits which return to core firms carry away much of the economic surplus that the country could use to invest in itself. The unavoidable result is backwardness, misery, and social marginality wherever dependent capitalism is introduced. It rewards only a narrow segment of society while stopping true economic growth. The never-ending pile-up of trade deficits in turn generates greater exploitation through more loan and investment dependence.

A graphic portrayal of this relationship can be seen in figure 2.1, which plots the degree-of-coreness scores developed by Terlouw against an income-inequality measure known as the Gini ratio. Although the Gini ratio and associated inequality measurements will be discussed in detail in the next chapter, for now it is important to know that it ranges from a theoretical minimum of zero (all households receive exactly the same income) to +1.0 (only 1 percent of households have all the income, and no one else has any). In reality, this number for nations around the world will range from an actual low of .25 (most egalitarian) to about .60 for a high (most unequal of income distributions). The graph clearly shows

FIGURE 2.1   Core Dominance Increases Inequality (1980s)

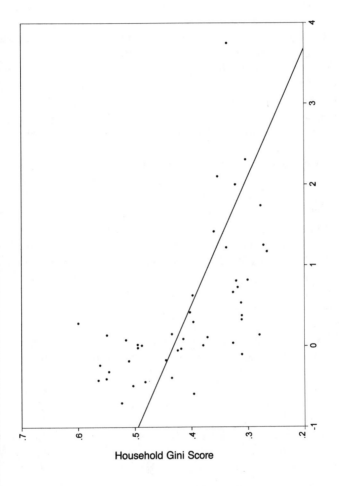

that countries with low coreness scores (i.e., those nations which tend to cluster more in the periphery) tend to have much higher degrees of income inequality than the stronger, richer countries of the core.

Despite the face validity of world system and/or dependency theories, a number of criticisms have been leveled at them. One weakness of these ideas is their lack of logically related points that researchers can test.[65] When conflicting results appear in dependency studies, supporters tend to quickly fall back on rhetoric to explain away the findings that do not fit. The study of dependency among nations is haunted by ideological bias.[66] To be fair, however, such a criticism can also be applied to development theory (discussed below). No theory is ever completely removed from ideological bias, even if it claims to be free of such values.

World system theory tends to ignore internal forces acting within a country that could lower income inequality. Nationalist movements and political conflict are of this ilk. The facts of history upon which the world system was built are open to challenge as well.[67] The theory cannot explain the success cases in the periphery. Some of these countries (Taiwan, South Korea) have gone into economic hyperdrive within the last decade or two.[68] New forces such as strong and free state governments have risen in the periphery, allowing national growth to be started and kept going on its own.[69] It is also claimed that certain parts of capitalism are progressive, which may at the least make economic growth possible in LDCs.[70] The theory is also accused of trying to explain three different outcomes (underdevelopment, marginalization, and dependent development) by using the same independent variable (the nature and shape of the world economy).[71]

There is now emerging evidence that many newly industrializing countries (NICs) that have followed an export path are not necessarily under the harsh heel of core domination. Gary Gereffi points out that the path to development, although involving enormous industrialization for all NICs, can employ greatly varying strategies. Many NICs have moved away from cheap labor-intensive manufacturing to high technology exports, whereas local private firms rather than foreign multinationals are the main exporters within Third World nations today.[72] Moreover, it is important to distinguish among periphery nations when making generalizations, since the experience and strategies of East Asian NICs differ markedly from those of Latin American countries. The entire nature and consequence of dependency and foreign development have differed tremendously between these two regions. For example, both areas started with commodity exports of raw materials, such as food (coffee, bananas, etc.), minerals, and oil. Countries in both areas then moved to primary import-substitution policies in their second stage—developing home-grown industries in an attempt to avoid heavy importation from core countries in basic consumer goods such as textiles and clothing. But their paths soon diverged. NICs in East Asia, such as Taiwan and South Korea, went on to pursue primary export-oriented industrialization. This involved the manufacture of higher value-added items that are skill intensive (e.g., electronics). Developing countries such as Brazil and Mexico instead went into secondary import substitution, producing steel, petrochemicals, cars, and so on. Such manufacturing is more labor-intensive and requires a less fully developed local industrial base.

A better understanding of the nuances among NICs can also be gained by examining the global commodity chains that are rooted in today's transnational production system. What we have today are products developed in one country, partially built in another, shipped to other nations for more work, and then distributed throughout the world. The role of nations can differ immensely in this process. While core countries such as the United States and Japan have typically developed original brand-name manufacturing and chain-store distribution, farming out manufacturing to NICS, this is no longer universally true. East Asian countries (especially South Korea, but also Taiwan and Hong Kong) have been successful at developing and manufacturing products under their own brand names in such areas as automobiles, computers, electronics, and household appliances. The products of these nations sell well in North America, Europe, and Japan as well as at home.[73] In the last analysis, therefore, our theories must acknowledge that the development experience for Third World countries is not universally similar. By failing to heed these distinctions, we create the ultimate danger of lumping apples and oranges together in a theoretical hodge-podge unrelated to reality.

CAPITALISM IN RETREAT: WORLD MARKETS AND DEVELOPMENT THEORY

The flip side of world system/dependency theories is a clear, no-nonsense form of market capitalism. Development theory claims that only good things will result for nations as a global free market economy comes to dominate our world. The market is seen as a tireless engine of economic growth for Third World countries. Enthusiastic fans of international capitalism make no apologies for its pervasive effect and influence upon poor countries. They actually see its long-range impact as leading to real-income growth for the people within these countries.

It is surprising, therefore, that there is no single comprehensive theory of the effects of development upon income distribution.[74] A major area of consensus, however, does exist. All seem to agree on the need to remove political and governmental roadblocks to the operation of free markets. In a sense, this is the bottom line of development theory, and it gets universal support. The major assumption is that an interdependent world economy based upon free trade, export specialization, and an international division of labor will produce economic growth within nations. Developed nations invest money, knowledge, and business skills in LDCs. They help LDCs establish industries suited to their particular natural resources. Trade will then order itself in a natural way so that goods, capital, and technology flow in the most efficient manner possible.

If a country fails to develop, the fault does not lie with the operation of the international market system. Frequently the blame can be laid at the feet of poor national leadership, obstacles raised by traditional culture, a poor rate of domestic savings, and so on. In a word, the ultimate task of economic development must rest with each nation itself. It is precisely those nations which have followed marketing prescriptions most aggressively that have benefited the most (Hong Kong, Singapore, Taiwan, South Korea). For the earth's economic basket cases, the real reason for continued poverty is lack of integration into the world market. Nations unfortunate enough to be consistently poor tend to rely on irrational state policies which block the free market. In this view, the poor are poor because they are inefficient, not because they are being bullied by core countries. The similarity to functional theory at this point is stunning. Both views conservatively assume that economic actors—whether individuals or nations—cause poverty. It is not due to the way the system works, which is seen as basically just and fair.

A high per capita gross national product is always seen as the mark of success in the development model. Many areas of the periphery have witnessed their per capita GNPs rise at historically high rates since World War II,[75] and even during the past century.[76] Many studies conclude, however, that the chasm in per capita income between poor and rich countries is actually growing wider in real dollars. In the twenty-five years from 1950 to 1975, for example, the absolute GNP income gap between LDCs (including the "miracle economies" of Taiwan and South Korea) and developed countries *doubled*. Assuming that these rates continue, the great majority of developing countries will never catch up to the developed countries. It is possible, however, for sixteen LDCs to close the gap within one thousand years.[77] It would take China, with one-fifth of the world's population, over twenty-nine hundred years to close the gap. In terms of relative wealth, the total income of the world's poor countries declined from 4.3 percent of what was made by industrialized countries in 1950 to a mere 2.5 percent in 1980.[78]

Development theorists argue, however, that economic growth must by its very nature be unequal since it does not start in every part of the economy at the same time. Economies grow by using more and better technology. But only a limited number of people at a given time can be transferred from the older techniques of subsistence agriculture to newer industrial modes. The remainder must wait their turn. While they do so, they are likely to fall behind.[79] Simon Kuznets has explained the reasoning behind the apparent contradiction of economic growth in a nation and increasing income inequality. He states that as economies mature in LDCs, factors such as rapid population growth, more industry, and a surge of people moving to cities come into play.[80] These forces at first act to lessen the well-being of lower-income groups and to raise general income inequality. As development later gets going at a faster pace, social and political factors emerge to improve the poorer sector of the economy. These can include unions, populist political parties, or even a newly developed savvy about urban living gained through experience. The upper-income group at the same time experiences less spectacular growth than under early industrialization. Hence, after a certain threshold is reached, income inequality begins to decline once again. This formula has been called the inverted-U hypothesis. The importance of Kuznets's hypothesis cannot be overestimated. It has been at the center of nearly all subsequent research and discussion of economic development.

Development scholars go so far as to argue that this initial surge in the share of income going to an upper industrializing elite is desirable. Since this group saves more, it will also invest more.[81] Thus, the entire population will benefit from subsequent economic growth. That this is a twin of supply-side arguments offered by political conservatives is quite obvious. In the United States, it has been claimed that allowing the rich to receive more income via lower taxes with less governmental interference will lead to increased investment and economic growth. Development theory also states that growing nations must go through a stage of greater income inequality at the outset before things improve for everyone.[82]

Development theory has been seen as ethnocentric and biased in favor of a particular economic path (market-driven capitalism). It is biased in its Western assumption that democracy is needed for growth to occur. The belief in never-ending progress, which can occur only through clear historical stages, is both dogmatic and naive.[83] The assumption that it is necessary for poor countries to pass through a long transitional period of worsening inequality is open to question. The belief that economic growth will inevitably lead to greater income and less inequality continues to be seriously debated. Virtually no one denies that there has been stupendous economic growth on a worldwide basis since the end of World War II, but the distribution of rewards is another story. The increased number of poor people in peripheral countries during the 1960s and 1970s was equivalent to well over 40 percent of all persons living in advanced capitalist nations.[84]

Development theory tends to downplay or completely ignore adverse reactions such as those described. Other negative results are the rise of cutthroat business practices, the destruction of older trades, and geographical polarization. A speed-up in population growth caused by a lower death rate can also increase the drain on resources.[85] There is more and more evidence to show that the conversion of agriculture from subsistence farming to export-driven agribusiness results in pauperization. A decline in living standards and outright starvation occur within many Third World countries following this path.[86] Indeed, the very nature and intent of development "aid" have been seriously challenged as inadequate. Aid is often disguised militarism. It has provided the means and motive for LDC dictators to remain in power to do the bidding of core elites.[87]

The politics within which economic growth takes place is all but neglected in this theory. Influential domestic and international elites form the financial status quo. Since their power is unchallenged, such interests can set the rules of the game in which victory leads to economic success.[88] A number of international agencies exist that channel the political and economic agenda of wealthy business elites from core countries into poor countries undergoing development. One such is the World Bank. Although this entity is comprised of 176 member countries, it is firmly controlled by the largest contributors: the United States (17.9%), Japan (7.4%), Germany (5.7%), France and Britain (5.5% each). The World Bank's president is always nominated by the U.S. president and is invariably a citizen of the United States.[89] The primary purpose of the World Bank has been to assist poor nations in development of their economies, always by enhancing exports. By strategic loans and providing technical expertise, supporters of the World Bank claim it reduces global poverty while enriching Third World nations.

Critics maintain the opposite. To begin with, U.S. Treasury officials estimate that for every $1.00 the United States lends, our country gets back $2.00 in bank-financed procurement contracts.[90] In other words, poor countries receiving the loans are forced to buy products from us. The result is a system of very shaky loans to LDCs for projects of dubious purpose. Unneeded dams and nuclear reactors always seem to top the Third World wish list of grandiose projects. Many experts believe these loans will never be fully paid back.[91]

In its effort to advance the interests of businesses in core nations, moreover, the World Bank actually creates poverty under the guise of development. Even interest payments due on these huge debts cannot be paid by some periphery nations, forcing them to borrow more from the Bank's sister institution—the International Monetary Fund (IMF)—to service their debts. The price for this accommodating gesture on the part of core countries is the introduction of austerity programs into recalcitrant nations. Seventy countries in the past fourteen years were subjected to 566 stabilization and structural adjustment programs as conditions for retaining the flow of loans. What these packages typically do is dismantle the role of governments in their economies. Remedies normally insisted upon are cuts in the value of their currencies and a drastic rollback in government spending. The axe tends to fall especially hard where money is spent on social needs, such as food or transportation subsidies. Higher taxes, caps or rollbacks on wages for labor, selling-off of government operations to private businesses (privatization), deregulation of industry, and removal of investment restrictions from abroad fill out this bleak portrait of want. On the whole, such policies have been utter failures. The economies subjected to adjustment are invariably sentenced to stagnation and decline. Low investment, increased unemployment, reduced social spending, reduced consumption, and low output all interact to create a tangled snare of poverty.[92]

Despite such brutal austerity, development has simply not been working for LDCs.[93] For a period during the 1980s, as an example, Latin American countries paid back four-and-one-half times as much in debt service to core countries as they were receiving in new capital inflow (aid and investment). Austerity programs were indeed successful—for wealthy core countries. Between 1984 and 1990 a net transfer of $178 billion in loan repayments was made by the Third World to commercial banks in the United States.[94] And despite the emphasis on exports by the World Bank, the share of world trade for LDCs declined from 28 percent in 1980 to 19 percent by 1986. Developed countries increased their share of world trade from 63 percent to 70 percent in the same period. Put differently, in 1980 developed countries bought 29 percent of their imports from LDCs and 66 percent from each other. By 1986 they bought only 19 percent of their imports from these poor countries and 77 percent from each other.[95]

The crucial question that is never asked is what kind of development are we about? Development for whom, and with what kind of consequences? The most desirable outcome is always seen as growth in Gross National Product (GNP) per capita, but what does this actually mean? George sees a great many drawbacks to this model:

> Industrialization is frequently its centerpiece, sometimes export agriculture relying on industrial inputs. The rich countries of the North nearly always built up their own industries on a strong agricultural base; the model conveniently forgets this and favors instant industrialization over food security. . . . The model is costly. It neglects resources that the local environment could provide and the skills that local people could supply, counting rather on imports, at escalating prices. It neglects not only peasants but anyone who does not belong to a thin layer at the top of society, identified as the "modernizing" elements. . . . The model is outward looking. It never seeks to enhance the specific, generic, original features of "undeveloped" countries and their peoples, treating them rather as if they

were a kind of undifferentiated clay to be moulded to the standard requirements of the world market and of world capital, to the uniform tastes of international bureaucrats and national ones trained in their image. Hunger is one result. People who . . . cannot become consumers in the global food system will not get enough to eat. Militarization is another. Masses of miserable people with little to lose are prone to revolt. Armed forces (including the police) in third world countries are used as often internally as against outsiders. Debt is a further outcome of the mal-development model. Elites borrowed to put it into practice and now expect their poorer compatriots to bail them out.[96]

Ultimately, it can be seen that growth in GNP per capita is too narrow and rigid a measure of economic development. While there may be growth, it is questionable whether there are any beneficial consequences for the masses in Third World countries. Reviews of studies documenting the way development works have found that any benefits to growth that may occur have been unevenly parceled out. Development has simply failed to raise the low income which the bulk of Third World inhabitants suffer from.[97] In essence, this brings us back to our major concern about income inequality. The alleged "economic development" or growth that is said to have occurred can obscure the real issue: how fairly is the pie divided? As we shall see in the next chapter, differences between countries are monumental. Given the fact that income inequality is so widespread in our world, is it possible to cut the pie any differently?

In the end, we are still left with the age-old question of who is essentially correct in the description of income inequality. All of the theories described have a semblance of validity—being composed of both strong and weak points. And as any wise person knows, no one has a complete corner on the truth. In the next few chapters, a number of studies concerning the validity of world system and dependency theories versus development ideas will be looked at. But controversy rages here as well. Criticisms over the nature of the data, methodological bias, misspecification of concepts, and fuzzy hypotheses abound. In the end, there are no easy answers.

It seems that economists largely tend to opt for the development perspective, probably because they take as their starting point the existence of a functioning, free market system. My discipline of sociology nearly always looks at the types of inequality found within and among societies. Power and position become major causal variables determining who gets how much of what. These are the lenses in my theoretical glasses. It is a focus which leads me to see world system theory as yielding a more valid picture of contemporary American economic life and the global division of labor. Hopefully, the studies to be discussed plus analysis of my own data will convince you that this image of America and the world is more accurate than the market-forces image.

I must admit to a strong reaction to one important plank of development theory— the axiom that income inequality must get even worse before it gets better. Although some evidence indicates that there is a link between growth of income and reduction in income inequality, data measurements have not been taken at different points in time. In short, we are asked to accept on blind faith that currently developing nations, which show greater income inequality today, will have less of it in the future. This is based mainly on evidence which states that such a process has already happened in advanced industrial nations. Whether this has actually been the case or not is another question. At least one researcher has found no change in the rate of income inequality in the state of Wisconsin over the past one hundred years.[98] But even if this were true, world system theorists would be correct to argue that today's core nations were able to develop their economic lead in a bygone age with much less challenge from other countries. This was at a time when the earth was not so finite, when frontiers beckoned, lands were open to exploration and settlement, entirely new trade routes were established, and so forth. The peripheral nations of today live in a much more constrained world than that of the United States two hundred years ago. Presently, possibilities are much more limited. Free choice has become more the exception than the rule.

The idea that periphery countries must experience more inequality before true economic development can take place is belied by at least two examples. One model is that of modern-day Taiwan. After World War II this country began its long economic ascendancy from a base of poverty. But as general affluence went up, income inequality steadily went down.[99] A similar process has been documented for South Korea. Both Taiwan and South Korea actually tripled their gross domestic product (GDP) per capita in the last decade.[100] The fact that both countries were so successful in achieving a middle-class status among nations, however, may not be due mainly to their export success.[101] South Korea is unique in the strong state control used to regulate and control direct foreign investment within its boundaries.[102] Both Taiwan and South Korea also started with much reduced income inequality, which provided the spark to run the engine of economic growth.[103] Even if Taiwan, South Korea, and a few other similar nations are dismissed as mere anomalies, the idea that LDCs must somehow "bite the bullet" of growing income inequality is highly suspect. No one ever talks about how much inequality there needs to be, or for how long. Ultimately, we get back to the same question raised in chapter 1: "How much is enough?" At what level are pain and suffering, starvation, and unmet basic human needs permissible in a world where some nations are affluent but where the majority of humans experience true want? In the end, the Kuznets hypothesis and development theory reflect a conservative justification and rationalization of the global income-inequality status quo. The implication of the theory is self-serving for the rich, while it carries a perennial message to the poor. You too, the theory seems to say, can be wealthy if you only grit your teeth, sacrifice, save your pennies, and work hard at subsistence wages. In the meantime, the rich core countries reap profit from the "sweat of the brow" of poor countries. LDCs are told that if they only keep their nose to the grindstone, some financial nirvana awaits them in a vague and distant future.

The conservative bias of development theory is the same as the right-wing bias of the functional, human capital, and supply-side economic arguments. All of these explanations essentially accuse those actors who do not do well of ineptitude. The operation of the system they compete in is seen as fair and without many problems. The rational marketplace is claimed to be an impartial arbiter which rewards those who prove the most talented and/or hardworking. All the while, the weak and incompetent are being weeded out through the inexorable force of competition. None of the theories give serious attention to a fact of life we all know to exist from common, everyday experience: those who have, get. It is a wealthy and privileged economic elite which gets special treatment, benefits, and help. Favoritism also allows them to perpetuate and enhance their high position. This is true whether the economic actor is a rich person or a wealthy social class within a given nation. It is also the case for core countries within the modern world system.

The end result is that system variables are the most important forces which shape income inequality. Income does not grow because individual efforts at improvement are made by a factory worker here or an impoverished Third World country there. This is not to say that we are completely powerless as individuals. Joining together for collective action still remains a personal choice. Nor can we afford to be complacent about honing our job skills along human capital lines. Moreover, nations that completely ignore market forces do so at great peril. But these realities should not obscure the tremendous unfairness built into the system at the start. Our society acts to reward those with an initial power advantage in much greater proportion than the unfortunate poor and downtrodden. This reality is often glossed over, ignored, or forgotten when opposing ideas about income inequality are looked at. Yet system bias is the driving force behind the allocation of riches or rags to participants in our domestic and global economies.

Chapter

3

ECONOMIC INEQUALITY AROUND THE WORLD

The overall degree of global income inequality is extremely lopsided. We live in a time when the world's 358 billionaires have a combined net worth of $760 billion, which equals the net worth of the poorest 45 percent of the earth's people.[1] The top 1 percent of income recipients in the world receives about 15 percent of worldwide income, while the top 5 percent of recipients gets 40 percent of all income. At the other extreme, the poorest 20 percent of the world's population gets only 1 percent of global income.[2] Given this incredible degree of inequality as a starting point, fierce debate continues on whether today's poor nations can ever have any real economic growth. Yet there is almost complete agreement that income differences among countries are currently way too high. Nearly all those who review the international financial scene call for an end to the gross inequities in wealth that are now frozen among the nations of the world. The answers needed to fix such an unfair income schism are less easy to come by. To begin this feat, we need to see how such huge differences among the economies of modern nations came about in the first place.

A good starting place would be the Bretton Woods (New Hampshire) conference held by America and its European allies in 1944, shortly before the end of World War II. Before the killing even ended, it was felt by all Allied nations that America had the only healthy and intact economy left. War torn Europe had been laid to waste and its industrial might was in smoldering ruins. It looked as if old trade patterns were gone for good. Fears of a postwar decline were very high. There was even deep worry that Europe would no longer be able to feed itself:

It was hard to procure raw materials and even harder to find machinery or parts. Railways were destroyed, bridges bombed, harbors blocked. In 1946 there was only one passable bridge over the Rhine, and one over the Elbe. The French ports of La Rochelle, Calais, Boulogne, Dunkirk, and Toulon were virtually unusable. Much of Holland was flooded with salt water; canal systems everywhere had been closed or diverted. A flourishing black market in every imaginable product from ball bearings to synthetic fibers vastly complicated the task of restoring production. The sole source for most goods was the U.S., the only major country to have escaped the war's devastation. Machines and agricultural products that once had come from Germany now came from the U.S. . . . The devastation was so overwhelming, the problems of reconstruction so difficult, that it was not certain for several years that the world would recover.[3]

Such was the economic power vacuum that America found itself in at the end of the war. The major point of agreement among all countries signing the Bretton Woods accord was a desire for economic growth. Nations badly wanted to avoid the stagnant trade, declines in business, and high unemployment which had cursed the Depression decade of the 1930s. In order for recovery to come about, the ravaged countries of Europe (including Germany) had to have the means to buy American goods. The United States was to supply the dollars needed for rebuilding through such efforts as the Marshall Plan. In short, America was to spend whatever money was needed to foster growth in the world economy.

The dollar became the global currency upon which the value of all other types of money was pegged. In this successful quest for world economic order, each nation agreed to keep the value of its own money within 1 percent of its par value to the dollar at the time of the agreement. European countries had to buy extra dollars, but only if the United States ran a trade deficit in its effort to buy goods from the Allies. In this system of fixed currency exchange, the U.S. Federal Reserve became the world's banker and the dollar became the bedrock of the international monetary system—at least for awhile.

A second major area of consensus at Bretton Woods was a "yes" vote for free world trade. The protectionism in vogue before World War II had all but destroyed trade among countries. Nation after nation had been dragged down by the others as their economies sank. A common ploy was to stop buying foreign goods when times were bad. Trade barriers were built for the protection of home markets during the Depression. Yet all suffered as a result.

The new strategy was to prevent this from happening again by means of the General Agreement on Trade and Tariffs (GATT). Before GATT, nations usually dealt with each other on a one-to-one basis. Their aim was to reduce or remove trade barriers. As an example, Canada might agree not to attach a duty fee to American computer chips if the United States did the same with imports of Canadian wood products. Yet such bilateral pacts are impractical. Since a huge number of goods are being traded by a large number of countries, the deals get to be very complex. GATT imposed some order on this chaos by using these same tariff cuts for all other nations. In a word, the system brought strong pressure to bear for open, free trade between all nations. Its effect was to improve the running of the world economy via the marketplace.

A third popular idea in the Bretton Woods pact was multilateralism. Global institutions such as the International Monetary Fund (IMF) and the International Bank for Reconstruction and Development (World Bank) were created to get as many nations as possible involved in world trade. The first goal of the IMF was to help bankroll the recovery of European nations. Today it continues to govern balance-of-payment problems, but it is also now deeply involved with Third World growth. By contrast, the intent of the World Bank from the start was to help LDCs by funding investment projects with low interest loans.

For at least two decades, the system worked well with few major problems for both America and its allies. Western European growth spurted to new heights. Between 1949 and 1963, for example, industrial production tripled in Austria, Italy, Germany, and Greece while doubling in Denmark, Finland, France, the Netherlands, and Norway.[4] The United States had over half of all the productive capacity in the world at the end of the war. With this base and the privilege of having the dollar as the only legal path of exchange, our country was able to invest in other nations with great ease and under immensely favorable conditions. American business funded a massive buildup in new foreign plants which produced goods for later United States imports.

One corporation alone, General Electric, increased its overseas capacity fourfold, from twenty-one foreign plants in 1949 to eighty-two in 1969. The proportion of total plant and equipment investment located outside the United States doubled in the metal and machinery industries. . . . By the early 1970s, nearly one-third of annual U.S. automobile company investment was being placed abroad. The widespread plants, mines, distribution centers and offices of the multinational corporations made up entire production systems linked on a global scale. . . . During the 1960s, the productive capacity of the American economy nearly tripled, even after accounting for inflation. This meant uninterrupted, unparalleled, and unprecedented economic expansion from the end of the 1961 (Eisenhower) recession to the 1969–1970 (Nixon) crash. It was a period in which economists declared the business cycle obsolete and families saw their real income grow by a third. Exports to overseas markets and production abroad were more than matched by an enormous burst of growth in the home market.

> . . . Growth of . . . discretionary incomes of working families provided an opportunity for business to develop and market a wide range of new consumer goods and services. The postwar suburbanization of middle-class households, itself in part an aspect of this explosive consumerism, set the scene for the proliferation of shopping centers, tract-housing projects, and a seemingly endless array of services related to the automobile, from drive-in restaurants to drive-in movies.[5]

America raked in money. Such a scale of affluence and wealth had never been seen before in the world. We became the envy of nearly every other country. Although the United States has since met severe economic trials, huge income gaps will remain between our nation and others if only because of this postwar miracle. Due to postwar growth, core countries of the modern world system will continue to earn higher income than nations of the periphery and semiperiphery.

But what of the Third World countries? What has become of the poor nations that are at times euphemistically called less developed countries (LDCs)? How did they fare under the international new deal which sprang from the Bretton Woods agreement? In a word, growth for some of these countries proved just as healthy. For a short time it was as phenomenal for a few LDCs and equal to the high pace of increase in Europe, America, and its vanquished enemies Japan and Germany.[6] In this hemisphere, countries such as Mexico and Brazil chased explosive, outward-oriented growth rates. During the past two decades, they were at first successful in their attempt to gain large profits from exports. The Newly Industrialized Countries (NICs) of East Asia, especially the "Four Tigers" of Hong Kong, Singapore, South Korea, and Taiwan, had phenomenal growth. Much of this has been due to their dramatic penetration of the U.S. market. Fully one-ninth of today's American imports come from these four countries.[7] Mainland China provides further testimony to the power of careful development. This nation's share of world trade grew from 0.6 percent in 1977 to 2.5 percent in 1993, which ranks as the world's eleventh largest exporter. As of 1993, in fact, the United States was running a $31 billion trade deficit with China.[8] The astounding success of Japan is also the Horatio Alger story of the late twentieth century.

The average gross national product (GNP)[9] per person did, in fact, grow faster in LDCs in the years immediately following World War II than in countries of the core.[10] In the 1945–1955 period, the total GNP for all Latin American countries went up nearly 5 percent per year while GNP per person increased 2.4 percent annually.[11] Both figures were much larger than those for the United States during the same period.

There is another side to this coin. Despite higher growth rates in GNP totals, the actual gap in real dollars between rich and poor lands in

GNP *per person* has actually gone up. The latest World Bank figures (table 3.1) are misleading in this way due to the use of a weighted average to account for population size within each country. The inclusion of only one nation—China—with over one billion people sharply skews the inequality picture. With China's entire population added in, it seems as if low-income countries are actually doing better than industrial nations. Yet it could be argued that a better unit of study should be the state. Some may also feel that China should be kept out of any discussion of development to begin with. It has been removed from any Western expansion until only recently. Once it did start to allow foreign investment, it did so on its own terms in an effort to protect the tenets of communism from the spread of capitalist values. Thus, China is a major exception to begin with.[12]

TABLE 3.1   Average GNP per Capita and Average Annual Growth Rates for Countries by Level of Income: 1992

| | GNP per Person | | | |
| | 1992 Dollars | | Average Annual Percent Growth, 1980–1992 | |
Income Level of Country	By Population	By Country	By Population	By Country
Low-income economies[a] ($675 or less)	$390	$337	3.9	0.32
Lower-middle-income economies[b] ($676–$2,695)	Not Available	1,395	Not Available	−0.25
Upper-middle-income economies[c] ($2,696–$8,355)	4,020	4,533	0.8	0.97
Upper-income economies[d] ($8,356–$36,080)	22,160	20,799	2.3	2.07

a. Mozambique, Ethiopia, Tanzania, Sierra Leone, Nepal, Uganda, Bhutan, Burundi, Malawi, Bangladesh, Chad, Guinea-Bissau, Madagascar, Lao PDR, Rwanda, Niger, Burkina Faso, India, Kenya, Mali, Nigeria, Nicaragua, Togo, Benin, Central African Republic, Pakistan, Ghana, China, Tajikistan, Guinea, Mauritania, Sri Lanka, Zimbabwe, Honduras, Lesotho, Egypt, Indonesia.

b. Ivory Coast, Bolivia, Azerbaijan, Philippines, Armenia, Senegal, Cameroon, Kyrgyz Republic, Georgia, Uzbekistan, Papua New Guinea, Peru, Guatemala, Congo, Morocco, Dominican Republic, Ecuador, Jordan, Romania, El Salvador, Turkmenistan, Moldova, Lithuania, Bulgaria, Colombia, Jamaica, Paraguay, Namibia, Kazakhstan, Tunisia, Ukraine, Algeria, Thailand, Poland, Latvia, Slovak Republic, Costa Rica, Turkey, Iran, Panama, Czech Republic, Russian Federation, Chile.

c. South Africa, Mauritius, Estonia, Brazil, Botswana, Malaysia, Venezuela, Belarus, Hungary, Uruguay, Mexico, Trinidad, Gabon, Argentina, Oman, Slovenia, Puerto Rico, South Korea, Greece, Portugal, Saudi Arabia.

d. Ireland, New Zealand, Israel, Spain, Hong Kong, Singapore, Australia, United Kingdom, Italy, Netherlands, Canada, Belgium, Finland, United Arab Emirates, France, Austria, Germany, United States, Norway, Denmark, Sweden, Japan, Switzerland.

Source: World Bank, 1994, table 1.

A fundamental contrast turns up even when China is included in the data as one unweighted case among the thirty-eight countries with low income (GNPs per person of $675 or less in 1992). The early conclusion is reversed. The average annual growth rate in GNP per person between 1980 and 1992 declines from 3.9 percent to .6 percent for low-income countries. While there is a slight fall-off in upper-income economies (mostly industrial) in average increase when removing the effect of population (from 2.3 percent to 2.07 percent), the change is essentially insignificant. In essence, the developed industrial countries of the First World largely outperformed LDCs both in GNP per person and in the percent increase of GNP per person between 1980 and 1992.

It is true that some LDCs have shown great promise. Yet most countries in the periphery have not shared such success equally. There are over one hundred countries in what could be called the Third World. All are arrayed along an income spectrum running from dire poverty in sub-Saharan Africa to levels nearing relative affluence. Some GNP increases are not likely to last very long. Many are the product of boom-and-bust business cycles. For example, Mexico expanded the drilling and export of oil to meet a thirsty demand during a lucrative market caused by the OPEC embargo. The market came crashing down a few years later as the world oil glut developed during the 1980s. Currently, there is also a severe drop in the world economy. This threat may still bring on financial collapse for all countries, whether core or periphery. The global economic recession has already had a disastrous effect on many nations during the 1980s. (We will explore this more fully in a section on debt among LDCs in the next chapter.) Although many incomes in the United States have been changed as a result of the decline, most Americans are still unaware of any ill effects brought on by the world recession.

The gap between LDCs and developed countries is quite complex. Yet it is a safe bet that absolute income differences have become worse. During the past decade, the share of world trade for LDCs dropped from 28 percent to 19 percent within only a few years.[13] Even before this period, the average per capita income gap between low-income countries and industrialized countries was two-and-one-half times larger at the end of a thirty-year period.[14] The poor saw average income climb only 150 percent in these three decades. Thus, in 1950 LDCs earned 4.3 percent of the world's total income, but this had dropped to 2.5 percent in 1980.

Erosion of income for poor countries can be seen in more recent data as well. The World Bank meticulously calculates what are called "purchasing power parities" (PPP).[15] These are prices computed for a fixed basket of goods, which take into account costs in each country and the performance of their gross domestic product over time after inflation is removed. The PPP figures are expressed as percentages of a base equal to 100 in the United States. As an example, in 1992 the PPP figure for Germany was 89.1, meaning that the German economy could purchase—on average—89.1 percent of the complete basket of goods of which the U.S. economy could purchase 100 percent. This was a healthy improvement over 1987, however, when Germany's PPP figure stood at 80.7. In short, because of real growth in the gross domestic product of Germany between 1987 and 1992, this nation's economy increased its purchasing power by 10.4 percent. If countries are separately analyzed by the modern world system position, large differences emerge. In 1992, the average country in the periphery could only afford 16.8 percent of the basket of goods, compared to the average semiperiphery country at 43.5 percent and the average core country at 81.4 percent. Low purchasing power among poor nations is not a surprise, but it is the direction of the change that is especially meaningful. Periphery countries saw a reduction in their purchasing power of −0.11 percent (in essence, stagnation of their economies) while semiperiphery countries experienced an average loss of −3.7 percent in their purchasing power. Core economies flourished, adding 4.6 percent to their average purchasing power over the past five years.

Insight into why this has happened can be seen in table 3.2. The table lists GNP per person in 1992 dollars and its increase from 1980 to 1992 for nations by position in the modern world system. A clear hierarchy can be seen between the three groups of nations in average 1992 GNP per person. The thirteen countries of the core lead with a mean GNP per person of $23,085, followed by sixteen countries of the semiperiphery ($9,562), and forty-six nations of the periphery ($1,775). There are also large gaps in growth of GNP per person that follow theoretical expectations. Core nations show the highest increase per year (2.11 percent), followed by the semiperiphery (1.75 percent) and the periphery (−0.16 percent).

The world map in figure 3.1 depicts GNP per person for the World Bank typology of economies: low-income, lower-middle income, upper-middle-income, and upper-income countries. It is apparent that most of the low-income nations are concentrated in Africa. Of the thirty-eight low-income nations, twenty-four are situated in Africa (nearly two-thirds of all poor countries). Put differently, well over half of all nations in Africa are among the poorest of the poor. At the other end, Western Europe captures the regional honor for number of countries with high GNPs ($8,356 per year or more). Other places also qualify for this coveted position, including Asian entities such as Japan, Hong Kong, and Singapore—together with the United States and Canada. We shall see later, however, that low income among nations is more than just a regional problem. GNP per person is only a gross average. This crude figure can hide large numbers of poor people, even in rich countries.

TABLE 3.2   Average GNP per Person and Average Annual Growth Rate
(1980–1992) for Countries by Modern-World-System
Position: 1992

World Position	GNP per Capita	Average Annual Percent Growth
Core		
Spain	13,970	2.9
United Kingdom	17,790	2.4
Italy	20,460	2.2
Netherlands	20,480	1.7
Canada	20,710	1.8
Belgium	20,880	2
France	22,260	1.7
Germany	23,030	2.4
United States	23,240	1.7
Denmark	26,000	2.1
Sweden	27,010	1.5
Japan	28,190	3.6
Switzerland	36,080	1.4
Group Average for Core	$23,085	2.11%
Semiperiphery		
China	470	7.6
Romania	1,130	−1.1
Bulgaria	1,330	1.2
Poland	1,910	.1
Russian Federation	2,510	Not Available
Brazil	2,770	.4
Benin	410	−0.7
Central African Republic	410	−1.5
Pakistan	420	3.1
Ghana	450	−0.1
Guinea	510	Not Available
Mauritania	530	−0.8
Honduras	580	−0.3
Egypt	640	1.8
Ivory Coast	670	−4.7
Indonesia	670	4
Bolivia	680	−1.5
Philippines	770	−1.
Senegal	780	0.1
Cameroon	820	−1.5
Peru	950	−2.8
Guatemala	980	−1.5
Morocco	1,030	1.4

TABLE 3.2: (*Continued*)

World Position	GNP per Capita	Average Annual Percent Growth
Dominican Republic	1,050	−0.5
Ecuador	1,070	−0.3
Jordan	1,120	−5.4
El Salvador	1,170	0
Colombia	1,330	1.4
Jamaica	1,340	0.2
Paraguay	1,380	−0.7
Tunisia	1,720	1.3
Algeria	1,840	−0.5
Thailand	1,840	6
Costa Rica	1,960	0.8
Hungary	2,970	.2
Greece	7,290	1.0
Portugal	7,450	3.1
Ireland	12,210	3.4
New Zealand	12,300	.6
Israel	13,220	1.9
Australia	17,260	1.6
Finland	21,970	2.0
Austria	22,380	2.0
Norway	25,820	2.2
Group Average for Semiperiphery	$9,562	1.75%
Periphery		
Ethiopia	110	−1.9
Tanzania	110	0
Sierra Leone	160	−1.4
Malawi	210	−0.1
Chad	220	3.4
Madagascar	230	−2.4
Niger	280	−4.3
India	310	3.1
Kenya	310	0.2
Mali	310	−2.7
Nigeria	320	−0.4
Nicaragua	340	−5.3
Togo	390	−1.8
Turkey	1,980	2.9
Iran	2,200	−1.4
Panama	2,420	−1.2
South Africa	2,670	0.1
Chile	2,730	3.7
Group Average for Periphery	$1,775	−.16%

Source: World Bank, 1994.

FIGURE 3.1   Global GNP per Person in 1992

$8,356 to $36,100
$2,696 to $8,355
$676 to $2,695
$60 to $675
Data Not Available

Source: World Bank, 1994.

In summary, sharp income differences between nations continue at very high levels and seem to be increasing. This is aside from the fact that some growth has taken place in LDCs with respect to their overall GNP levels. In making comparisons through time and between countries, the presence or absence of only a few countries can greatly affect the outcome. The use of formerly socialist countries in comparing GNP per person can also produce results that are atypical from previous analyses. Many have undergone an economic meltdown as their economies struggle to convert from communism to capitalism. Whether a significant number of these countries will be able to make the transition to higher income as they unfold their lands to development is an open question. But in the meantime, their new freedom has come at a cost of economic deterioration—and that in itself is starting to show up in the data and skew it toward more bleak estimates.

Although countries such as India or China are fairly poor, the size of their populations weighs heavily in making comparisons. Any big change in China may end with dramatically shifting cycles. If China were included, for example, there would be no change in world income inequality for almost thirty years.[16] Theoretically and practically, communist countries before the fall of the Iron Curtain were only a minimal part of the emergent world economy But for noncommunist countries, there has been an unmistakable increase in income inequality during the quarter century after World War II:

> For the period as a whole distribution seems to have worsened somewhat. . . . The share of the upper-middle deciles rose by 1.8 percent of total income, with 1.3 percent coming from the bottom six deciles. . . . The decline in share of world GNP . . . was most marked for the lowest deciles, reaching almost 20 percent for the first decile.[17]

The relative incomes of most people in the free-market world have become much worse during this postwar economic expansion. Only when communist China is included among the statistics does this picture change. Yet even here the income gap gets better only in the sense that today's inequality is no worse than it was in the 1950s. The average person did not become worse off, but neither was there improvement. The income shifts that did occur acted to impoverish the poor even more. A decline in real income of 20 percent set in for the poorest 10 percent of income recipients.[18]

Albert Berry and his associates were able to break down the sources for the world's decline in relative income. Faster population growth among the poorest LDCs slowed emerging equality. So did a lag in India's economic growth. Since India is the second highest in sheer numbers and one of the poorest nations, it is a severe drag on overall global growth. United States growth in per person GNP was below that of the entire world. When this is added to the increase in real incomes for the Organization for Petroleum Exporting Countries (OPEC) and newly industrialized countries, the trend to a larger gap in world income is eased. All told, U.S. growth in real GNP per person during the thirty-year period from 1950 was 6.7 percent *below* the worldwide norm of 2.8 percent increase per year. The only areas of the world with a worse record than the United States were: the subcontinent made up of India, Pakistan, Sri Lanka, and Bangladesh (–20.5%); poor African countries (–25.2%); and Argentina, South Africa, and Uruguay (–18.5%). Communist countries such as China (63.8%), the former U.S.S.R. (40.7%), and former East European socialist countries (50.3%) did much better. Not surprisingly, newly industrialized countries had per person GNP increases 50 percent above the norm, while Japan led all national groups with an increase of almost three times the average. Even our closest allies of Canada, France, Italy, the United Kingdom, and West Germany managed to ring in real GNP increases per head that were 23 percent above average. One fact stands out. The stark reality is that the United States fell behind nearly all other industrialized countries in its per person GNP growth rate. Had our GNP level not been so high to begin with, this situation would have been even more extremely urgent than it actually is.

Looking at the most recent data, one study analyzes growth rates of real Gross Domestic Product (GDP) from 1961 to 1990, removing the effects of inflation. The facts reveal that economic growth for all countries slowed considerably in the 1980s. Thirty-seven out of 110 countries actually lost ground in the last decade (22 were in Africa).[19] Overall, the gap between the poorest and richest countries appears to have increased during the last three decades. It also appears that there has been an increasing divergence among low-income countries during the past thirty years.[20] This means that there is increasing heterogeneity among poor countries with respect to growth rates in real GDP. While some have lost ground, others stayed the same or did well.

Even among the 21 wealthy market economy countries, 9 countries performed poorly and did not keep up with average GDP growth (the United States is one of them). In fact, from table 3.2 it can be seen that the GNP growth rate for the United States was 19 percent below the average for core countries. Underperformance among core countries, however, means something vastly different than it does among poor countries. An annual growth rate of 1.7 percent, applied to a GNP per capita of $23,240, increases the per capita income of the United States by $400 per year. The Swiss, who are even worse underperformers (1.4%), increase their per capita income by $505 per year while the Japanese (3.6%) grow by over $1,000 per year. From a Third World perspective, this is dumb-founding. The entire GNP per capita for China is $470, and for India it is $310. Even accounting for the smaller population of the United States, our growth rate in GNP over four years is enough to support the entire population of China for one year.

THE DEBT CRISIS AND WORLD RECESSION

The worsening of the world's economic scenario sped up dramati-cally during the 1980s and continued into the 1990s. Today it is often referred to as the debt crisis. This can be a loaded term, however, suggest-ing that LDC borrowing is out of control and conjuring an image of wan-ton poor nations seeking to recklessly expand their economies overnight. This single leap from marginal subsistence to modern industrialism is to be helped by loans from the core. While some LDCs may have such a utopia in mind, the depiction smacks of blaming the victim. Certainly the World Bank and its sanction-wielding IMF twin are not hesitant to cry for austerity. Such programs are viewed as an indispensable aid to increasing national income through more exports. The World Bank has continued to push for cuts in public spending and subsidies as a means of deficit reduc-tion.[21] The message is clear and unmistakable: most of today's global fis-cal crisis is laid at the feet of poor countries who simply do not have the discipline to balance their own budgets.

Notice that officials, bankers, politicians, and economists in core nations never urge LDCs to simply stop borrowing money. The crux of the problem may lie with prolific and careless lending on the part of core nations. More than half of the World Bank's lending capital is supplied by core countries.[22] It is these very countries, especially the United States, that decide who gets how much of what. Such largesse is in turn always awarded to the LDCs who conform most closely to IMF mandates. The impact of these development loans is very questionable, since many of the projects are geared toward high technology uses which need an elaborate support structure. The necessary supplies and technical expertise for many of these projects are unavailable in the host country. Of course, such crucial imports are ready for purchase from the core nations who make the loans. An example can be found in the building of Mexico's Laguna Verde nuclear power plant in Vera Cruz. The plant was constructed despite the fact that it is situated between two geological faults and within ten miles of an active volcano. The General Electric reactors it uses may have serious design flaws. G.E. is being sued in the United States for such alleged flaws.[23]

When oil prices quadrupled during the mid-1970s, LDCs were urged by core banks and OPEC depositors to borrow heavily. The goal was to fuel a buildup of industry and business to feed exports. In this way, the heavier cost of oil imports could be more easily paid by increased profits. Such plans assumed that core countries like the United States would be willing to import more goods from the periphery. All parties to the mutual-aid game also took for granted that the variable rate interest on their loans would remain the same. Neither assumption was met. The second oil price increase of 1979 wreaked havoc upon Third World economies. Input costs for manufacturing and agribusiness went wild, spinning out of control. For the 1974–1982 period, non-oil LDCs paid for $345 billion worth of oil (the bill would have been $85 billion before the price hike).[24] By then, many LDCs were trapped in the vicious circle of hopeless debt. They had to borrow even more money to pay for higher oil imports in order to produce more goods for export to pay off their debts.

The second blow came from a sharp drop in the demand for their goods. A glut of products on the world market rapidly developed, forcing panic sales at bargain basement prices. Many more factors add to the vulnerability of poor countries on the export treadmill:

> Compounding the problem is the limited range of goods that the debtors can offer, which pits one against the other. African coffee producers compete not only against other Africans but against Latin Americans as well. The least developed economies are dependent on one, two or three agricultural or mineral raw materials, and it is folly to tell them to diversify in their present capital starved condition. . . . Third world commodities are further undermined by substitution. Every time the industrial countries think that the price of some raw material is out of line, they introduce a substitute that they can produce without recourse to outside suppliers. Sugar is a flagrant case. . . . As for non-agricultural raw materials, more efficient Western industries are using fewer mined metals, employing more synthetics. . . . Glass fibre as a substitute for copper wire in the telecommunications industry is one example. When Third World countries struggle to sell a limited range of goods in the face of shrinking demand, over-supply and plunging prices are the predictable results. Markets . . . simply cannot swallow unlimited quantities of foods or textiles or transistors.[25]

With so much of the national budget going to increased energy costs, less was left over for other consumption in many parts of the world. To make matters worse, the United States began to sharply raise its interest rates. This acted to siphon off from the Third World much capital that would otherwise have gone for investment and local consumption. The high interest rates meant that more now had to be paid on development loans taken out at initially lower rates. The end result has been a hemorrhaging of money from LDCs to the advanced industrial countries. For Latin America alone, new capital inflow in terms of aid and investment came to $38 billion in a four-year interval during the 1980s. Yet during the same period, the region paid back $144 billion in debt servicing.[26] In fact, for most of the decade approximately $20 billion to $30 billion was being wrested every year from the region in a net transfer of wealth to core countries. Rates of economic growth for nearly all Latin American countries became negative for the first time in half a century during the 1980s, skidding by –16 percent in Venezuela, –22 percent in El Salvador, and –25 percent in Bolivia. For the entire region, per person economic output was cut by 10 percent from 1980 to 1990. In total, output, real wages, consumption, and profits fell throughout Latin America, while unemployment grew by 30 percent during a five-year period of the 1980s.[27]

While effects have varied from one country to another, they are all variations along a continuum from bad to worse. The situation was even more desperate in Africa, where per capita income fell by one-fourth over the decade.[28] One of the most catastrophic cases occurred in Brazil. While per person economic output in Brazil fell 12 percent in four years, the inflation rate increased at an annual level of 200 percent. Brazil still owes $120 billion of debt to foreign banks. Brazil's 40 percent unemployment rate has set the stage for eventual default. With nearly two-thirds of its people undernourished, the cumulative effect is utter disaster.[29]

In total, more than seventy poor countries have been subjected to over 566 IMF structural adjustment programs in the last fourteen years.[30] Despite their proliferation, all the austerity programs have done is lock more poor countries up in debtor's prison for longer periods. Under structural adjustment, the Third World's debt burden actually doubled—rising to $1.3 trillion in 1993. Thirty-six of Africa's forty-seven countries have undergone structural adjustment, yet total external debt for the continent in 1994 was 110 percent of its GDP, compared to 28 percent in 1980. This horrendous debt, and the interest rate that accompanies it, effectively hinders these LDCs from investing in productive areas because money must first be used for repayment.

The natural response to these conditions on the part of LDCs is to cut back on imports from other countries. This acts to improve the balance of trade so that they can pay back their loans. Ultimately, however, this destroys trade. Within a quarter of a century (1960–1985), the Latin American share of total world imports dropped from 10 percent to 4 percent.[31] The tragedy is that many of these imported goods are also essential for Third World economic and physical well-being. Orders for medicine, food, books, machinery, spare parts, fertilizer, and the like have been cut to the bone. Quality of life has meaning only for rich nations in the core.

Export profits of core countries who ship to LDCs also dried to a trickle during the 1980s. During the first half of the 1980s, total exports from the United States to Latin America fell by over $11 billion. This made up 55 percent of the total drop in all our exports. Import cutbacks by Mexico alone accounted for a full one-fourth of the drop.[32] Although the volume of trade between Latin America and the United States has improved during the 1990s, poor countries remain seriously weakened by debt and low prices for their exports. The distinct message is that the United States and other core countries do not live in a vacuum. We are tied to the fate of Third World countries. Ultimately, when they fail to prosper, so do we.

The idea may still be held that the United States has benefited from the problems of our poor neighbors. To a small extent this is true. Yet the costs far outweigh the advantages in the grand scheme of things. To put it more bluntly and in different terms, we are living on borrowed time. At the start of the 1980s the United States was in a strong position as a lender nation when our country held over $140 billion in foreign assets (more than any other country). Incredibly, the United States had slipped deeply into red ink by as early as 1985. By that year we had become a debtor nation to the tune of $107 billion.[33]

> The United States had become the world's largest debtor . . . the United States was borrowing approximately $100–120 billion net each year and foreign holdings of American government securities soared. Projections of future borrowing indicated that by the end of the decade, the American foreign debt could reach $1 trillion. The world's richest country in less than five years had reversed a century-long trend and become the world's most indebted nation. . . . By the mid-1980s, the evidence supporting the relative decline of the American economy had become overwhelming. In the early 1950s, the United States, with 6 percent of total world population, accounted for approximately 40 percent of the gross world product; by 1980, the American share had dropped by half to approximately 22 percent. Whereas the United States in the early postwar period produced 30 percent of world manufacturing exports, by 1986 its share had dropped to a mere 13 percent. American productivity growth, which had outpaced the rest of the world for decades, declined dramatically from a growth rate of 3 percent annually in the early postwar years to an incredible low of 0.8 percent in the 1970s. As American productivity lagged behind that of other advanced economies, particularly Japan, West Germany, and the NICs, the result was a less competitive economy and a substantial lowering of the American standard of living. In capital formation, technological leadership, and the quality of the labor force (human capital), the United States was falling behind in a growing field of industrial competitors.[34]

By the beginning of the 1990s, the U.S. federal budget deficit had tripled from its $1 trillion base in 1980.[35] Our gross federal debt as a percent of our Gross Domestic Product has risen steadily every year since 1981. By 1994, federal debt had doubled—accounting for 70.4 percent of GDP in that year.[36] This massive flow of red ink, which not only continued but actually got worse throughout the past decade, showed only a small sign of levelling off by 1992. Yet there is no cause to celebrate. As figure 3.2 discloses, today over one-fifth of the federal budget goes directly to paying the interest alone on this enormous debt (in 1980, it absorbed less than 13 percent of the budget). The bottom line: just to stay even—without paying back the principal—requires twenty cents out of every tax dollar. In 1994, the interest payment came to about $300 billion, which was about $20 billion more than our annual defense budget. This amounts to well over $1,000 that each child, woman, and man must fork over each year to pay *only* the interest on the deficit monster.

There is insult added to this injury. Who, might we ask, benefits from the massive run-up in debt? According to one study, debt financing shifts income from the middle class to the richest 10 percent of our population.[37] In financing the debt, the rich pay higher taxes—but they also own more of the debt in terms of government bonds. They get paid more in interest than the rest of us. For every dollar of interest payment on the federal debt, the wealthiest 10 percent of households receive 75 cents while the poorest 30 percent of households get one penny. Even when the elevated taxes for higher income households are subtracted, the richest 10 percent still get a net transfer of 16 cents for every dollar of tax-financed interest payments.

Appalling as this explosion in government debt may seem, the story actually gets worse. Our national debt situation is the same as debt on all other levels. Total household, corporate, and consumer debt rose by 60 percent within five years during the last decade.[38] Families were increasingly forced to the wall. By the mid-1980s the proportion of total farm debt to net farm income had hit a postwar record, forcing painful foreclosures and the loss of farms that had been in families for generations. By the mid-1980s, household installment credit debt had reached 16.5 percent of total personal income—a postwar record.[39] But the relentless march toward further indebtedness continued well into the 1990s. By 1994, the ratio of household debt to disposable personal income had reached a record 81 percent (it was below 60 percent in the mid-1980s).[40] The only good news was that interest rates had declined, so that the impact of the debt service relative to income was lighter than it would have been in 1989, at the peak of the load. But by the mid-1990s, credit card debt had once again started to climb sharply while personal bankruptcies continued to run at record rates. Even home mortgage delinquencies had reversed a nearly decade-long downward trend.[41] Businesses suffered from mounting liability during the past few decades as well. Corporate business debt took $1 out of every $7 earned during the 1960s. It rose to $1 for every $3 in the 1970s, and soon reached a crescendo of half of all earnings in the last decade.

FIGURE 3.2   Interest on U.S. Public Debt

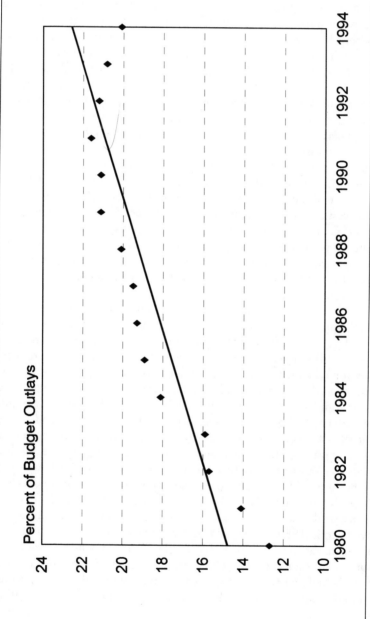

Source: U.S. Statistical Abstract 1994, CD-ROM, table 509.

To make matters worse, much of this money is not owed to U.S. banks or is not in the form of Treasury notes owned by American citizens. The debt payoff is increasingly going to foreign powers. Annual interest paid out to foreigners on the U.S. debt doubled in the 1980s.[42] The Japanese during this period bankrolled the massive increase in U.S. debt for a number of reasons. To begin with, they had the money to do so. The economic success of Japanese industrial development since the end of World War II has been breathtaking to see. Japan has used different means to subsidize heavy manufacturing (steel and autos) and high-tech industry (especially electronics) to develop an export economy. The Japanese have also gone to great lengths to protect local business markets from foreign penetration. They have encouraged personal savings at very high rates while aggressively marketing their products around the globe. Japan had undeniably usurped the American position of top world creditor by the mid-1980s. At a time when the net asset value of the United States was plunging into the red, Japanese trade surplus was the largest in the world. In 1970, seven American banks ranked among the globe's top twenty-five in terms of assets. By the end of the 1980s, Citicorp was the only United States bank in this chosen circle (tenth place). Today, the ten largest banks in the world in terms of deposits are Japanese, with no U.S. banks in the top twenty-five.[43] Much of the investment coming from these banks has been in the form of U.S. Treasury bond purchases.[44]

One other major reason led Japan to feed America's spendaholic deficit binge. We are without doubt Japan's best customer. In a five-year period during the 1980s, Japanese exports to the United States more than doubled.[45] From 1989 to 1993, Japan managed to increase its exports to the United States by 15 percent, while our country was undergoing its most severe recession since the Great Depression of the 1930s. Today, nearly one-fifth of the value of goods imported by the United States comes from Japan.[46] Preserving our financial health keeps their labor force fully at work.

The other magnet which drew foreign money and investment to the United States was the decision by the Federal Reserve to increase interest rates to all-time highs. This was part of an effort to curb the massive inflation endemic to the 1970s. While the result was to produce the severest U.S. and world economic tailspin since the Depression, it did work to stabilize prices. Money from rich and poor countries alike gushed in from around the world. Our interest rates were off the charts in comparison to those of other countries. Since the U.S. political climate was both conservative and stable, the mix formed an attractive investment package.

The massive inflow of capital to America also took away needed cash from LDCs for their own development. Profits from local firms in poor nations ended up in U.S. banks, rather than being recycled into their own economies. While money flowed to the United States, it was less available in other lands to purchase goods produced in LDCs. This made the recessionary problems caused by the OPEC oil price hikes much worse. Even within the United States, however, our new-found wealth failed to be put to productive use. Rather than using the money to reinvest in our own aging and badly outmoded industries, recipients frittered it away. Much of it was spent for nonproductive mega-mergers between avaricious U.S. corporations. This has been the major reason behind the long-term corporate debt alluded to earlier.

Most of the money was also spent on the largest peacetime military buildup the world has ever witnessed:

> Spending for defense more than doubled, from $134 billion in 1980 to $282 billion in 1987, in the process coming to consume nearly 60 cents out of every personal and corporate dollar raised by the federal government through income taxes. Even after accounting for inflation, this amounted to an average growth rate of over 7 percent a year—nearly three-and-a-half times the real growth rate of GNP. . . . As a result of all of this defense activity, more and more American manufacturers were becoming increasingly dependent on the Pentagon as a customer for their wares.[47]

The result of this massive arms buildup was a short-term growth spurt in our GNP. It is true that some jobs were created within the United States because more defense dollars were being spent. While employment in defense-related work grew by less than 4 percent per year just prior to the buildup, this expanded to 20 percent under the Reagan presidency. There was a 45 percent increase in the private sector as well, which could be laid at the feet of the newly fattened military calf. During the first five years of President Reagan's tenure, 1.2 million new jobs were added to the defense-related payroll.[48] The result of such an incredible military spending spree is cancerous. Like so-called banana republics, many American industries have become almost helpless in their dependence upon the defense budget. They are very vulnerable, along with the entire American economy, to any major defense cuts. This explains why it is so difficult to pare our military expenditures, even as the Cold War has ended and no military threat of any significance for the United States remains. A reduction in our defense budget means a loss of jobs (especially in a few concentrated areas, such as southern California and Texas). The loss of jobs translates into voter pressure by defense workers and their families, together with financial pressure and campaign contributions from huge corporate defense contractors.

Because of overreliance upon military spending, much of America's research and development effort is now aimed at esoteric and impractical ends. Products are made that we all fervently hope will never be used. While the United States develops stealth bombers and creates reams of SDI (Star Wars) software that critics claim will never work, our competitor nations develop computers, VCRs, high-resolution color TV sets, and more such goods for real people who want to buy these products. The United States has cut its own throat by weakening its ability to trade and to produce consumer goods. Most experts feel that the direct application of this exotic military hardware to the commercial market is quite limited or nonexistent.

Since the United States was forced to borrow heavily in order to finance the military buildup, interest rates were driven up around the world. This depleted investment capital of LDCs while also forcing LDCs to pay more on their variable interest rates. The worldwide recession was made worse. An already anemic consumer market in the developing world was weakened even more. Lastly, some believe that the results of all of these changes enabled the United States to cunningly levy a hidden military tax upon the Third World.

The other side of the imposing U.S. debt crisis was the effect of tax cuts instigated by conservative politicians in the 1980s. Not only did the cuts fail to stimulate the economy (other than in military areas), even more red ink was spilled. Rather than saving, the public went into debt at even higher rates. Total spending on new manufacturing plants and equipment actually declined as did spending for firms outside of primary industry. Labor productivity, which was supposed to increase as a result of greater investment in new technology, took a deep dive and was actually negative by the mid-1980s.[49]

The last factor underlying American debt has been the rapidly developing trade gap between the United States and the rest of the industrialized world. By now there can be no doubt that a global economy is a reality. The United States, Japan, Germany, and a few other core nations form its decorative centerpiece. Yet, the economic performance of the United States during the last decade was at best questionable. At worst it could be seen as fiscal irresponsibility. Our nation has shown a compulsion to borrow beyond its means. The Japanese have enabled our addiction to debt by lending us more money than is healthy for either country. Given the global recession and the resulting glut of goods on the world market, the United States has emerged as the buyer of last resort among all nations. We are the only country with the means and the desire to bury ourselves in a deluge of imported goods. Indeed, our free market rhetoric almost demands that we abandon any strategy of nationalistic consumption ("Buy American"). Any planned industrial development is seen as contrary to the American way of life. Instead, as a nation we have favored a laissez-faire strategy that leaves cheaper imported goods on our doorstep, despite the long-term damage this may do to our economy.

The high value of the dollar, sparked by the Federal Reserve's unheard of increase to a 21 percent prime lending rate, made American exports far too costly for the rest of the world. This killed any chance of selling our goods abroad. The fall in oil prices lessened the ability of OPEC nations to buy products. The rising debt and interest payments of Third World countries destroyed their purchasing ability. As a result, demand for imports in the developing world actually fell by $100 billion in four years. The strategy pursued by nearly every country except the United States during the last decade was to push exports while cutting imports. In a time of recession, this is known as "putting your house in order." America went the other way. For the first half of the decade, United States imports rose 35 percent while our exports fell 3 percent. Japan during the same period increased exports by 35 percent while slashing imports by 8 percent. The result was massive and immediate: a burgeoning trade debt that has shown no sign of ending. It is still too high, even with our government's success in reducing the value of the dollar. The trade deficit has been on a stubborn course of increase since 1981. The latest data estimate shows the deficit to be a negative $164 billion in 1994.[50] According to a study by New York University's Institute for Economic Analysis, the size of the deficits during the 1980s ended up costing the United States 5.1 million jobs in manufacturing.[51]

The point of all of this comes down to one basic assertion: the so-called economic recovery of President Reagan's administration was false. Although there were cosmetic repercussions, such as the creation of jobs in defense-related industries, what little growth there was during the decade was due to the onset of ruinous debt. The fact that the United States was able to get away with this for such a prolonged period of time was due to its past position as chief economic power in the world. We have been badly abusing this privilege since 1971 when President Nixon declared that our country would no longer pay in gold for U.S. dollars held by other nations. Despite the loss of real productivity and the sickness of many industries, the United States embarked on a spending binge during the last decade while the rest of the world was prudently dieting or being forced into starvation. The American public has not even begun to perceive the long-range injurious results of this fiscal irresponsibility, which have even deepened in the 1990s. Such facts were carefully kept from voters during the various presidential elections until 1992. Candidate Bill Clinton gained some mileage from the bleak economic scenario (his campaign note to himself became famous: "It's the economy, stupid!"). To some degree, he managed to tap into the bleak economic malaise of voters. But it was the emergence of Ross Perot as a viable third-party candidate which finally forced attention, although all too briefly, upon what the consequences of living beyond our means would be for all Americans.

Harvard economist Benjamin Friedman points out that the net federal debt after eight years of the Reagan presidency amounted to $2.1 trillion, or 43 cents for every dollar of national income. Prior to Reagan's election, this debt was 26 cents for every dollar—a rate that was two-thirds lower. Friedman concludes that unless the nation acts to reverse the accumulating debt, "today's voters and their children will pay the consequences in the form of a diminished standard of living and a far different role for America in world affairs."[52] Princeton University professor Robert Gilpin also warns us of the dire consequences:

> First, the competitive position of important sectors of the American economy has been permanently damaged and the structure of the entire economy has been distorted. Second, repayment of the immense external debt and the associated interest payments will absorb a large share of America's productive resources for many years to come; these costs will substantially lower the standard of living for a considerable period, even if defense expenditures are considerably curtailed. And third, the newly acquired preference of Americans for foreign goods and the expansion of productive capacity abroad have decimated many industries in which the United States once had a strong comparative advantage; America will be required to develop new products and industries if it is to regain even part of its former competitive position in world markets. The task of reversing the trends toward deindustrialization will be difficult and very costly . . . the United States had indulged itself in overconsumption and underinvestment for too long. Americans were consuming the source of their national wealth and that of other societies as well rather than putting it into productive investments. . . . For a time, the United States was able to mask its decline . . . . through foreign borrowing, especially with the financial assistance of the Japanese. . . . Despite the cries of a few Cassandras, the false prosperity of the Reagan "economic miracle" hid from the American people the reality of their true situation and the fact that they were prospering only on other people's money. The country as a whole failed to appreciate the historic meaning of the budget deficit and its long-term implications for the society.[53]

AUSTERITY, INCOME INEQUALITY, AND THE THIRD WORLD

Those within the Third World are aware of all of the trends discussed so far. Many of their intellectuals and leaders have earned Harvard Ph.D.s in economics or M.B.A.'s to better plan the modernization of their own nations. One thing our contemporary world does not lack in this age of telecommunications, satellites, cheap jet travel, fax machines, and computer modems tapped into the Internet and World Wide Web is instantaneous, up-to-date information. Third World leaders cannot help but know of the disarray in the American house of finance. There is thus deep resentment when the United States demands fiscal austerity from LDCs while turning a blind eye toward its own economic chaos. Such a position is often seen as hypocritical from the perspective of developing countries. Yet the message from the World Bank and the IMF, almost wholly under U.S. tutelage, continues to be ground out. Developing countries are admonished to stop their outrageous public spending on "questionable" programs designed to meet social needs. They are sanctimoniously told to trim their budgets in the provision of human services. They are lectured without mercy to make goods for export to the world market. Pressure to cut local consumption by ending food price subsidies, barring unions, lowering and/or putting caps on wages are usual parts of the IMF package. Removal of price controls and letting costs rise for electricity or transportation form still another part of the typical adjustment scheme. The plan calls for higher taxes. Interest rates go up in an effort to end runaway inflation. The big picture we are left with in this group portrait of LDCs is one of sacrifice and poverty. Can there be any question that a shift takes place in income from the poor toward the wealthy? Bankers now come right out and say they want to do this in order to encourage savings and to drive down wasteful local consumption.[54]

What is never discussed is the dwindling market for all these goods. No one talks about what happens when the economic pie actually gets smaller. Conflict will rear up in a nation when its piece of the economic action diminishes. The spread of income within any society is marked by rancor, hostility, and intense bitterness even during the best of times. Unfortunately, during epochs of decline there is an even greater tendency for the strong to prey upon the weak as they try to keep their expected income coming in. As inequality gets worse *between* nations, income inequality *within* nations also tends to grow.

Susan George cites the case study of Jamaica as a typical example of how an IMF austerity program works on its target population.[55] Jamaica's economy is dominated by bauxite and sugar exports as well as tourism. Foreign investment by aluminum companies in this nation was heavy in the 1950s and 1960s. By the 1970s this industrial initiative had dried up. Jamaica was left to wrestle with oil price hikes and world recession at the same time. A brief fling with a socialist government brought further decline as foreign firms began to flee the country. To meet rising costs while its export revenues declined, Jamaica increased its borrowing. It raised its foreign debt from $150 million in 1971 to $813 million by 1976. By that year, Jamaica was forced to restructure its debt along IMF adjustment policies. By 1980 the Fund demanded a $300 million cut in state spending (26 percent of the previous year's budget) which resulted in the layoff of eleven thousand public sector employees. Because it was so compliant and as a reward for voting its socialist government out of office in 1980, Jamaica was chosen by the Reagan administration as a showcase example of free-market, conservative economic development. Financial aid went through the roof. Between 1981 and 1984 Jamaica borrowed $495 million from the United States, more than twice the amount of aid it had received in the previous twenty-four years. Yet the reinvestment still failed to ignite the economy. Since by then the IMF/World Bank had allocated $900 million to Jamaica, it demanded its pound of flesh to service the new loans. Further austerity cut all food subsidies, public investment fell by 30 percent, and real income was slashed by nearly half in a period of two years. While water, telephone, and electricity rates all went up by 100 percent in 1984, local food production fell 13 percent in the same year. This was at a time when the minimum wage was not allowed to rise above $8.95 per week. Susan George correctly pinpoints the major result as the destruction of Jamaica's ability to meet basic human needs:

> The shrinkage of the "social wage" . . . had a further regressive effect on income distribution, dumping most of the welfare costs of adjustment on the doorstep of the worst off in the society. But this is exactly what the IMF *wants*. According to Fund doctrine, "redistribution of income" (read "more to the rich, less to the poor") will result in higher profits, which in turn will result in higher investment and thus create jobs so that people can earn more money, etc., etc. That is the theory. In practice, though there certainly has been "redistribution," the rich have invested their windfall not in job creation but simply in speculation (a lot in real estate) or even more simply in foreign bank accounts. Here, once more, is a classic case of the banks getting their money back twice—in payments on the debt made at the expense of the poor and, simultaneously, in the form of cash deposits that they can then reloan ad infinitum, or at least until something cracks. Jamaica is unlikely to struggle back to . . . prosperity through IMF measures.[56]

The economic havoc has continued into the 1990s, despite the protest election of socialist Michael Manly as prime minister. Pay raises have been all but impossible in this country under austerity. During the last decade, the average wage of Jamaicans declined by 82 percent while the inflation rate was running at nearly 200 percent.[57] Even in 1990, before sugar cane workers literally revolted, the average wage was $3 per day. The result has been described as "going back into slavery" for most Jamaicans, although the new masters have changed from plantation owners to regulators of the International Monetary Fund. To pay back the $4 billion in debt, each man, woman, and child now owes $1,839. To service this crushing load, 40 cents must be paid from every dollar earned in exports. In short, the situation seems like a never-ending treadmill for citizens of this tiny Caribbean country.

Horror stories such as these and others are common. Yet, we are still dealing with income inequality at the level of lone nations with only a few selected case studies. Although such accounts tend to be rich in detail and suggest paths for more research, they may still not be representative of the majority of less developed countries. The experience of Jamaica suggests that dark forces are at work in LDC economies, but it is hard to generalize from only a few examples. While Jamaica's story may be typical, we cannot be sure until all nations are examined together. In such research, the focus must be on their income inequality and how this varies with development and their position within the modern world system.

Such an analysis was conducted at about the same time Jamaica's economy came under the scalpel of IMF austerity. A study of thirty-two countries found strong evidence that government spending reduces income inequality.[58] This is particularly true for funds spent on such "social wage" items as social security, government-owned enterprises, and expenditures meant to prime the economy. For every 10 percent increase in direct government spending, there is an associated decrease of 3.6 percent in income inequality. The United States, with the highest unemployment rate of 6.7 percent among these nations, had one of the lowest government expenditure rates at 21 percent of GNP. Sweden, with one of the highest government spending rates (39.1 percent of GNP), had an unemployment rate of only 1.5 percent. Put differently, while government spending was twice as high in Sweden, unemployment was four times as low. The conclusion is obvious. If you slash government spending and its safety net of ways to meet basic human needs, then income inequality will skyrocket. What is most compelling about this study is an added finding that sharply challenges IMF and World Bank wisdom. All other things being equal (GNP per person and direct government spending), an increase in economic growth does not lessen the degree of income inequality in these countries.

Austerity programs do not work. It is a faulty strategy perpetrated on poor countries to justify enriching already corpulently wealthy elites. All the while, the majority in poor countries are deprived of even the food needed to keep alive. There is conclusive evidence that local consumption plummets when income inequality goes up within LDCs. A study of twenty poor nations found that more income inequality was associated with less money spent on personal needs.[59] Austerity leads to suffering, deprivation, and the destruction of local consumer markets at home.

The impact of austerity programs upon women and children is devastating. Empirical evidence now exists to document the profound hardship such policies create in many poor countries.[60] Among a number of predictors examined, including debt ratio, dependency on foreign investment, and poor economic growth performance, IMF pressure upon LDCs was the single most pervasive variable producing deterioration in health and well-being. The study found that the austerity measures typically included government spending cuts and wage freezes. This led to the allocation of fewer resources for immunization, general health maintenance, prenatal care, adequate nutrition, and controlled urban development. The result is much higher mortality rates among children under five years of age.

Some research in eighteen Latin American countries does show a better balance of payments resulting from the introduction of austerity programs. Yet scholars found the improvement was more artificial than real. Conditions got better despite lack of improvement in the trade balance. They also found outright declines in the balance of goods, services, and income available within these nations. Signing up for an IMF austerity package served as a seal of approval which allowed docile countries to obtain further loans from private banks. It was a fresh influx of money rather than real recovery that led to healthier looking ledgers for poor nations. All of this took place while inflation ran wild as a by-product of IMF austerity. The most important finding was a consistent weakening of income for laborers and workers, that is, a relentless, growing poverty. The decline in the social wage and real income of workers results because the IMF helps core nations dominate the periphery by use of judicious loans:

> This is partly because, as argued previously, the cooperation of local elites can be obtained by sparing them the burdens of adjustment. This rise in surplus—given no general effects on growth rates—should lead to an increase in profitability sufficient to attract the private capital inflows demonstrated. This inflow may allow a rosy picture of balance of payments improvement, but beneath it lies worsening income distribution, exacerbated social tension, and little or no improvement in the inflation, current account, and growth fundamentals. . . . After riots induced by the IMF-recommended price increases left 60 dead, 200 wounded, and 4,300 arrested, the planning minister of the Dominican Republic rejected (temporarily) the IMF pact commenting that "It is not that we are unwilling to put our own house in order. It is that we want to keep our house and not let it go up in flames." Economic policymakers both inside and outside the Fund should recognize that the [demonstrated] inequitable apportionment of adjustment burdens . . . cannot persist without letting societies "go up in flames." The design of stabilization policies to replace the unfair and often ineffective Fund policies should be the object of new research.[61]

Another example of violence sparked by an IMF austerity plan in a Third World country happened in Venezuela It was due largely to adjustments suggested by the IMF, such as a currency devaluation, an increase in loan interest rates, and a series of price increases. Because of heavy payments to service Venezuela's external debt of $33 billion—the fourth largest behind Brazil, Mexico, and Argentina of all Latin American countries—the average real wage in this nation has fallen by 38 percent since 1983.[62] Venezuelans saw the price for fuel increase by 80 percent and for public transportation by 50 percent almost overnight. These price hikes sparked rioting among the country's already destitute population, eventually leaving 347 officially listed as dead (Caracas newspapers estimated the dead at over six hundred), and almost one thousand injured and two thousand arrested. The severity of the outbreak led to unusual candor by Venezuela's then newly elected President Carlos Andres Perez. Although seen by many as being in the pocket of international bankers, Perez showed no hesitation in placing the blame: "The crisis Latin American nations are undergoing has a name written with capital letters—foreign debt."[63] He went on to fault the industrialized nations for giving little or no help to support democratic governments. His claim is true. Venezuela saw the proportion of money going to service its debt increase as a percentage of its export earnings from 4.2 percent in 1970 to 37.4 percent in 1986 (nearly an 800 percent increase). Adding to this picture have been debt service increases of 234 percent for Brazil, 24 percent for Argentina, 52 percent for Chile, 66 percent for Columbia, and 140 percent for both Ecuador and Bolivia.[64] In a word, the drain has been horrendous.

The costs have been enormous as well. In almost half of all debtor countries (thirty-nine of eighty) protests caused by the use of austerity programs were experienced during the 1976 to 1992 period, involving 146 separate incidents.[65] These were mass actions that were specifically in reaction to austerity policies (cuts in food subsidies, price increases as a result of devaluation, etc.) introduced by Third World governments who were pressured by the IMF to adopt such policies. The severity of the protests varied from one country to the next. Researchers John Walton and Charles Ragin used four indicators to measure the degree of violence involved: reported number of deaths and arrests; number of distinct protests; the specific presence of rioting; number of cities reporting protest events.[66] In comparison to a host of other possible independent variables that could be expected to be associated with the presence and severity of these protests, two stand out. The correlation with a nation's severity of debt (measured by average debt service as a percentage of the value of exports) and the presence of protest is significant at .368. A similar level of correlation also exists between severity of debt and the severity of protests. Lastly, International Monetary Fund pressure on a country (the number of renegotiations, restructuring agreements, and extent of debt owed to the IMF) had a very high correlation of .448 with the presence of austerity protests. This essentially means that one-fifth of the variation in protest activity can be explained by the rise of IMF pressure. The results of this study support the common sense idea that austerity protests tend to occur where debt and IMF pressure are the highest. When governments of poor countries impose austerity to pay off onerous loans to the IMF, protest and revolt normally follow—especially in nations with large, impoverished urban populations. The dollar costs of these riots are never entered into the ledgers of the World Bank and IMF, but they are very real. Typically, they are borne by the citizens and governments themselves. There are human costs as well. People are killed, beaten, jailed, and sometimes tortured. The social cost is bitterness, hopelessness, and hatred of a nation's people toward their own government, world lending institutions, and the United States itself.

INCOME DISTRIBUTION AND AVAILABLE DATA

On a general level, such concepts of economic activity as Gross Domestic Product and Gross National Product leave a lot to be desired. To begin with, there is an inherent prejudice in comparing poor countries to wealthy ones. In order for economic activity and goods production to be measured, the service or commodity must have a price tag, that is, be valuated in a market economy. Goods exchanged by barter or gift, such as in a subsistence economy, are simply not counted. The informal economy of less developed countries can be vibrant and thriving, yet never show up in formal measurements of GDP. If policymakers truly believe that the only real economy is the measured one, there will be increasing bias and distortion in their conclusions.[67] Women's work and unremunerated work done on a voluntary basis never get entered into a GNP or GDP compilation. Especially in Third World countries, the domestic products of women (often consumed in the home) are never entered in the ledgers by the World Bank, although the impact of their contribution to societal well-being is massive.

Various organizations have engaged in many attempts over the years to broaden the conception of economic activity to meet many of these defects, with varying degrees of success. The first major effort came from international organizations, such as UNESCO.[68] The United Nations Children's Fund (UNICEF) has maintained for quite some time that GNP per capita is an inadequate measure of a country's well-being, and that it fails to capture real development.[69] The United Nations Research Institute for Social Development (UNRISD) constructed a standard of living index that started out with over one hundred indicators.[70] Because of defective or missing data, the number of indicators was eventually reduced to nine—only one of which was GDP per capita.

At bottom, many critics emphasize that nations should be judged on how they meet their citizen's basic human needs. This assumption is at odds with the traditional dogma in economics that rising production of goods ultimately translates into a higher standard of living for everyone. Thus, analysts have worked out varying measurements designed to monitor either one of the concepts or a mixture of both. Predictably, the results vary widely. One "Index of Social Progress" based upon forty-four social indicators tapping welfare issues ranks the United Kingdom and the United States lower than some countries in the former East European bloc of nations or Costa Rica.[71] When four different indexes for a large sample of countries are compared:

> The two based mainly on economic indicators tend to rank the United States very high (first and second) while the other two, more widely based, classifications rank it lower (sixth and twenty-fourth). This is clearly a very controversial question. . . . Basic needs theorists argue that it is more fruitful to stress results rather than inputs in order to measure the adequacy of development policy. For example, life expectancy is a better measure of health services than numbers of doctors per person, and calorie supply per capita is a better measure of nutrition than total production of food.[72]

Further impetus for questioning and redefining the basic philsophy behind the GDP measurement comes from the directors of a nonprofit public-policy organization called Redefining Progress.[73] The impetus for change from their direction stems from the failure of the continuously rising U.S. GDP to reflect the deterioration in the lives of the great majority of Americans. Specifically, the Gross Domestic Product tallies the value of all money transactions, whether for good or ill. It does not distinguish between costs and benefits, or between productive and destructive activities. A terminal cancer patient going through a costly divorce adds to the GDP, as does the cost of repairing the devastation of earthquakes and hurricanes. This group proposes some subtractions from the GDP tally that most people would see as costs and as undesirable: the costs of crime ($65 billion per year in prevention alone), divorce (lawyers fees, new household costs, counselling expenses), resource depletion and degradation of the habitat, loss of leisure, and the jump in income inequality. In addition, it adds in the value of household and volunteer work, which is not now counted in the official GDP. This group's new barometer of economic and social well-being is called the Genuine Progress Indicator (GPI):

> The GDP would tell us that life has gotten progressively better since the early 1950s—that young adults today are entering a better economic world than their parents did. GDP per American has more than doubled over that time. The GPI shows a very different picture: an upward curve from the early fifties until about 1970, but a gradual decline of roughly 45 percent since then. This strongly suggests that the costs of increased economic activity—at least the kind we are locked into now—have begun to outweigh

the benefits, resulting in growth that is actually uneconomic. Specifically, the GPI reveals that much of what we now call growth or GDP is really just one of three things in disguise: fixing blunders and social decay from the past, borrowing resources from the future, or shifting functions from the traditional realm of household and community to the realm of the monetized economy.[74]

There is an even greater need to be cautious when comparing countries in terms of GDP. In 1989, then-president of the World Bank Barber Conable admitted GDP did not adequately reflect the importance of environmental issues. There are also profound and hidden dimensions to the change from GNP to GDP by the United States in 1991. While the amounts are not greatly different in our own country, using GDP paints a false, rosy picture of life in the Third World. Output from Japanese, American, British and other multinational corporations located in poor Third World countries is now counted as the product of the country where it occurs, despite the fact that profits, tax breaks, and loan development costs all return to rich, core countries. The Redefining Progress group terms this an accounting shift that creates statistical boomtowns out of many struggling nations. While aiding the push for a global economy, it hides the basic fact that nations of the North are skimming off the South's resources. Instead, focusing only upon GDP figures labels this operation as progress, as a gain to Third World countries.[75]

There is some indication that the World Bank may finally be confronting at least some of these criticisms.[76] It has lately constructed a new economic yardstick that is in the experimental stage. The measurement breaks down national wealth into three major categories. "Produced capital" is the economic value of machinery, factories, roads, and the rest of the nation's infrastructure. "Natural capital" consists of the value of natural resources, such as timber, oil, mineral deposits and the like. The third element is "human resources," such as the education level and nutritional standing of a population. "Produced assets" most closely parallel the traditional concept of GDP, but under the new system it accounts for only 20 percent of a nation's real wealth. The major wealth of nations appears to be on the social, human-resources side of the ledger. Richer countries stay that way by providing adequate nutrition, health care, and education to their populations. The point is driven home by the fact that Madagascar and the United States are dead even, deriving about 16 percent of their wealth from produced assets.

The major point is that, in evaluating the economic well-being of countries, other indicators—in addition to GNP or GDP—are needed. A high GDP or GNP per person, as we have seen, is often falsely viewed as a high level of wealth. Yet it is also possible for a country to have a high level of GNP and a very extreme degree of internal income inequality. It is not enough to measure the average level of wealth without also asking how a country's economic pie is divided. How a nation's wealth is distributed among its people is of paramount interest. We can ask whether income is shared relatively equally, so that all may reasonably benefit, or whether rewards are apportioned unjustly, so that only a few benefit while many suffer in poverty. The answer tells us much about the internal dynamics of any country.

For example, Brazil is often held up as a model of economic development. Its GNP per person in 1992 was $2,770—enviable by Third World standards. Just over half of this income, however, went to the highest 10 percent of all households. Only 2 percent of all income went to the poorest fifth of all households. Hungary is an equally successful LDC nation with a GNP per person ($2,970) roughly equivalent to Brazil's. Yet Hungary shows only 21 percent of its income going to the top 10 percent of all households, while 11 percent of all income ends up in the poorest fifth of all households.[77] In essence, economic development is very questionable if only a small minority of the population benefits. Much of the concern for income inequality must be looked at in terms of how it is spread out within countries. Without this, average GNP or GDP is very misleading.

Unfortunately, the quality of existing data which looks at the internal distribution of income in nations is even weaker and less complete than GNP per person figures. The main sources for the income distribution data have been gained from sample surveys. Even within developed countries there is a risk of bias resulting from the way the sample is selected. There is always a chance that the sample will not be a good proxy for the entire population. For example, Taiwan is often held up as a shining example of how export-driven development policies can enrich a nation and reduce income inequality. Yet this conclusion is highly suspect. It is based upon shaky inequality figures derived from a small, nonrandom sample of three hundred households (representing .02 percent of Taiwan's 9 million people in 1953).[78] The treatment of self-employment, property income, small-scale agriculture, and the informal sector can diverge from one country to the next. Within LDCs the problem of inadequate coverage and nonrepresentativeness can be more severe due to a number of factors: nomadic groups, hard to reach regions, fear of government officials using such information to increase a household's tax burden, failure to include non-money income such as food, and so on.[79] Comparing the inequality of family incomes in Third World nations to such inequality in developed countries may not yield valid results because even the very definition of what a family is can vary widely from one country to the next.[80] How "income" is defined can bias and distort international comparisons as well.

Especially in developing countries, in-kind payments (such as with food) need to be counted as income to dispel the notion that such countries are even poorer than they actually are. The issue of whether to measure pretax or posttax income should be kept in mind as well, although it may be of less importance in comparing Third World countries (where tax structures are undeveloped) to core countries.[81]

Nearly all analysts wish to use estimates that are close together in time, are based on large and representative national samples, and tap the same unit (most frequently, the household). Figures released by the World Bank in its latest report come closest to this ideal.[82] The data taken from the *World Development Report 1994* are fairly comparable in quality, represent a posttax distribution, count cash and in-kind payments as income, and are for households. The World Bank itself admits there are issues of weakness in their data sets, which are all drawn from nationally representative household surveys between 1978 and 1992. Nonetheless, although comparability problems are still present, the World Bank believes they are diminishing over time. This may be due precisely to the United Nations Household Survey Capability Program and two of the World Bank's projects (Living Standard Measurement Study and the Social Dimensions of Adjustment Project for Sub-Saharan Africa).[83]

All World Bank estimates of income distributions for nations categorized by modern world system rankings are based on data compiled within seven years of a 1985 benchmark (only four of sixty-four countries in this data set were analyzed earlier than 1980). Over 60 percent of these nations conducted their household income surveys in the most recent four-year period (1989–1992). Muller feels that data sets gathered within eleven years of each other are roughly comparable.[84] Thus, the income distribution results in table 3.3 should be the best and most up-to-date figures available for households in the world. An obvious flaw in these data, however, is the underrepresentation of LDCs in comparison to core nations. This is often because small, poor nations cannot afford to conduct expensive surveys and/or simply lack the expertise.

Table 3.3 can be more easily understood if we think of all households that receive income as occupying a continuum from the lowest income to the highest income. The fourth column in table 3.3 thus shows that for the Netherlands, the lowest 20 percent of all households receive only 8.2 percent of all the income distributed in that country. The fifth column states that 36.9 percent of all income goes to the households in the top 20 percent of all households. The sixth column says that 21.9 percent of all income goes to the top 10 percent of all households. There are three quintiles (percent of income going to households at the fortieth, sixtieth, and eightieth percentiles) that have been left out of the table to improve simplicity, but which would give a more complete picture of the

TABLE 3.3   Household Income Distribution within Countries by Modern
              World System Position

Country and World Position	Household Gini	80/20 Ratio	Percent of Income Going to Households		
			Lowest 20 Percent	Top 20 Percent	Top 10 Percent
Core					
Spain	0.271	4.41	8.3	36.3	21.8
Belgium	0.272	4.56	7.9	36	21.5
Japan	0.277	4.31	8.7	37.5	22.4
Netherlands	0.279	4.50	8.2	36.9	21.9
Sweden	0.281	4.61	8	36.9	20.8
Germany	0.322	5.76	7	40.3	24.4
Denmark	0.326	7.15	5.4	38.6	22.3
Italy	0.329	6.03	6.8	41	25.3
Canada	0.335	7.05	5.7	40.2	24.1
France	0.347	7.48	5.6	41.9	26.1
United States	0.362	8.91	4.7	41.9	25
Switzerland	0.372	8.58	5.2	44.6	29.8
United Kingdom	0.386	9.63	4.6	44.3	27.8
Average	0.32	6.38	6.62	39.75	24.09
Semiperiphery					
Hungary	0.223	3.16	10.9	34.4	20.8
Bulgaria	0.247	3.48	10.4	36.2	21.9
Poland	0.259	3.92	9.2	36.1	21.6
Norway	0.301	5.92	6.2	36.7	21.2
Finland	0.311	5.97	6.3	37.6	21.7
Israel	0.326	6.60	6	39.6	23.5
China	0.344	6.53	6.4	41.8	24.6
Australia	0.367	9.59	4.4	42.2	25.8
New Zealand	0.379	8.76	5.1	44.7	28.7
Brazil	0.604	32.14	2.1	67.5	51.3
Average	0.336	8.61	6.7	41.68	26.11
Periphery					
Pakistan	0.297	4.73	8.4	39.7	25.2
India	0.307	4.69	8.8	41.3	27.1
Ethiopia	0.308	4.80	8.6	41.3	27.5
Indonesia	0.317	4.86	8.7	42.3	27.9
South Korea	0.329	5.70	7.4	42.2	27.6
Ghana	0.353	6.30	7	44.1	29
Algeria	0.373	6.74	6.9	46.5	31.7
Morocco	0.376	7.02	6.6	46.3	30.5
Tunisia	0.385	7.85	5.9	46.3	30.7
Jordan	0.389	7.34	6.5	47.7	32.6
Philippines	0.391	7.35	6.5	47.8	32.1
Jamaica	0.401	8.07	6	48.4	32.6
Bolivia	0.405	8.61	5.6	48.2	31.7
Mauritania	0.407	13.23	3.5	46.3	30.2

TABLE 3.3   (*Continued*)

Country and World Position	Household Gini	80/20 Ratio	Percent of Income Going to Households		
			Lowest 20 Percent	Top 20 Percent	Top 10 Percent
Singapore	0.415	9.59	5.1	48.9	33.5
Zambia	0.416	8.88	5.6	49.7	34.2
Thailand	0.42	8.31	6.1	50.7	35.3
Venezuela	0.423	10.31	4.8	49.5	33.2
Peru	0.437	10.49	4.9	51.4	35.4
Costa Rica	0.442	12.70	4	50.8	34.1
Malaysia	0.463	11.67	4.6	53.7	37.9
Dom. Republic	0.483	13.24	4.2	55.6	39.6
Mexico	0.486	13.63	4.1	55.9	39.5
Colombia	0.492	15.50	3.6	55.8	39.5
Senegal	0.517	16.74	3.5	58.6	42.8
Panama	0.543	29.90	2	59.8	42.1
Kenya	0.545	18.18	3.4	61.8	47.9
Chile	0.547	17.00	3.7	62.9	48.9
Honduras	0.565	23.52	2.7	63.5	47.9
Tanzania	0.566	26.12	2.4	62.7	46.5
Guatemala	0.569	30.00	2.1	63	46.6
Average	0.431	12.03	5.26	51.05	35.53
All Nation Average	0.387	10.04	5.86	46.6	31.03

income distribution. It is not really necessary to look at these other three quintiles to get a clear idea of the skewed nature of income data. It is apparent that the poorest 20 percent of all households receive much less than 20 percent of all the income. At the other end, the richest 20 percent of all households receive much more than 20 percent of all income distributed. Japan, one of the most equal of all countries, shows only 8.7 percent of all its national income going to the poorest 20 percent of households. Almost 40 percent of all its income ends up in the wealthiest 20 percent of all households.

Such patterns are typical of all countries to some degree. If the data were to be divided into even finer groups of 1 percent each, the resulting graph would show a large hump among the lower-middle-income range with a long tail trailing off to the right. This skewing vividly shows that high incomes go to only a small proportion of the population. A more graphic picture than this has been used to describe the typical income distribution:

Jan Pen (1971) paints a fascinating "word picture" of the income distribution of individuals when he likens it to a parade of people whose height is proportional to their income and who all must pass a certain point in an hour. It takes 48 minutes before one sees marchers of average heights (income) and the parade grows with agonizing slowness until giants of 27 feet loom up at 1 minute to go. From then on, their height increases with dizzying rapidity—in the last few seconds of the march come businessmen and executives 100 feet tall, while the final marcher (a multimillionare) is some thousands of feet high.[85]

The data show that world position has much to do with the degree of income inequality within a nation. Among core countries, 6.6 percent of all income goes to the poorest 20 percent of all households—compared to 6.7 percent in the semiperiphery and 5.3 percent in the periphery. *Within the poorest of countries, the poorest of households receive less income.* Hope for persons of the Third World dims with this fact. As a nation becomes less wealthy, the poorest segment of the population suffers even more than the poorest in wealthier countries. The opposite is true for the richest 10 percent of all households. In core countries they get 24.1 percent of all income compared to 26.1 percent in the semiperiphery and 35.5 percent in the periphery. Again, as nations become less well-off, the rich actually receive more income than do the rich in the wealthiest of countries. Apparently there is some truth to the old adage that "the rich get richer, and the poor get children."[86]

What is a *fair* distribution of income? Harking back to the original question raised in the first chapter, we again ask, "How much is enough?" Obviously, comparing income distributions puts us into a more subjective, relative range. It is plain that all national income distributions are somewhat unfair. Not all share equally. That some of this may be due to intrinsic factors goes without saying. Hard work, motivation, risk taking, getting more education, and finding skilled employment are all crucial in the race for income. Yet major doubts about human capital must remain in the face of such huge national differences in income inquality.

Even among core nations, the large differences between countries suggest that other forces are at work to produce greater or lesser income inequality. The United States, for example, shows the lowest 20 percent of households to be among the poorest of all core nations. Only 4.7 percent of all our income goes to this group. Japan has the most well-off bottom quintile at almost twice this percentage—with 8.7 percent of its household income being claimed by the bottom fifth. At the other extreme, however, the United States actually shows a "modest" 25.0 percent of all income going to the richest 10 percent of all households. This is in contrast to 29.8 percent for Switzerland. All of these comparisons in the core pale into insignificance, however, when distributions in the periphery are examined. Without doubt, Brazil has the worst recorded picture of income inequality in the world. While only 2 percent of all income goes to the poorest 20 percent of its households, over one-half of all income ends up in the hands of the richest 10 percent of its households.

Another common short-hand way of looking at inequality is to divide the percent of all income going to the the richest fifth of the population (the eightieth percentile and above) by the income that goes to the poorest fifth (twentieth percentile and below). Based on this eighty to twenty ratio, Japan once again emerges as the most egalitarian of nations. Its richest fifth of households gets 4.31 times the amount of money the poorest quintile receives. At the high end of the extreme is the United States and United Kingdom, where the ratio is nine or ten times as much. Looking at each category of the modern world system, nations in the core on average have a more equal distribution of income (6.38) compared to nations in the semiperiphery (8.61) or the periphery (12.03). Indeed, the correlation between location in the modern world system and the size of the eighty to twenty ratio is .35, statistically significant and of moderate size. Although there is extreme variation within each of these three groupings, it is notable that the U.S. ratio of 8.91 compares poorly to most nations regardless of world position. We are the second most unequal nation among those in the core, and the third most unequal in comparison to the ten countries in the semiperiphery. Even if we were in the periphery, our eighty to twenty ratio would put us right in the middle of the thirty-one nations of the Third World. A separate report, prepared by the United Nations Development Programme (UNDP), compared twenty-one top industrial countries on the eighty to twenty index as well.[87] Our country had the second highest degree of income inequality by this measure and the second lowest amount of income going to the poorest 40 percent of our population (edged out only by Australia).

But what is the right statistic to look at? Should we be concerned with the bottom 10 percent of all households? The richest 20 percent? The elite top 5 percent? The eighty to twenty ratio? What about the quintiles left out of the picture? A society with a high percent of income going to the bottom quintile and a low percent to the top 20 percent can still be greatly unequal. This will occur if a high percent of income is found in the fourth quintile (60 to 79 percent range), a low percent in the second quintile (20 to 39 percent range), or both. In short, the picture can become very muddy if all five quintiles are compared against all fifty-four countries. Even if we were to follow this analytical avenue, it might not lead us to any clear conclusions. We would still be left with the task of sorting out and deciding which quintile or portion thereof (e.g., top 10 percent) we felt was the most important.

Gini Ratios: A Single Measurement
of Income Inequality

While it is acceptable to look at both ends of the income distribution scale in comparing countries, there is an easier way to describe all of a nation's income inequality at one time. We can draw a graph that theoretically shows what a perfectly equal income distribution for a population would look like. Figure 3.3 looks at family income inequality in the United States as of 1993 (the most recent data available). Perfect equality would here be seen in the 45-degree diagonal line A to C. What this shows is that 20 percent of all families receive 20 percent of all income, 40 percent of families receive 40 percent of all income, and so on, up to 100 percent. A distribution such as this, of course, does not exist in reality. It is the departure from this perfectly equal state of income distribution we are interested in. This can also be plotted in a curve using quintiles or deciles from data used to construct table 3.3. The result is what is called a Lorenz curve, which is represented by the curved line in figure 3.3.[88] The further the curve is from the diagonal line, the more income inequality there is.

Although this graph was quite an advancement in showing income inequality in its entirety, it was still akward. It was simply too difficult to compare many nations on the basis of numerous Lorenz curves. Italian economist Corrado Gini proposed that a single ratio be used instead. It divides the area of the curve which departs from the line of perfect equality (shaded in figure 3.3) by the entire possible area (represented by the triangle ABC). The degree of income inequality can then be summed up in one number, and populations can easily be compared. Calculating household Gini ratios using the most recent data from the World Bank and International Labour Office yields the figures in column two of table 3.3.[89]

The figures indicate a Gini ratio of .362 for the United States and one of .372 for Switzerland (the wealthiest among all nations by GNP per person). The lower the ratio, the less income inequality. On this basis, the United States has more income inequality than most European countries. We are a shade *more unequal* than our neighbor to the north (the Canadian Gini is .335) and dramatically *more equal* than Mexico (.486). Overall, the U.S. Gini ratio is close to the median score of all countries. We are at the forty-fifth percentile, meaning that 44 percent of all countries are more egalitarian in distribution of income than our nation. Our Gini ratio is still 25 percent above the world's lowest score. It should also be noted that the poorest 20 percent of our households get the lowest percent of income among all core countries, with the exception of the United Kingdom. Lastly, when the eighty to twenty ratio is used, the United States has an unenviable record of very high inequality. We have the twelfth highest score out of thirteen core countries, show more

FIGURE 3.3   Lorenz Curve (U.S. Family Income in 1993)

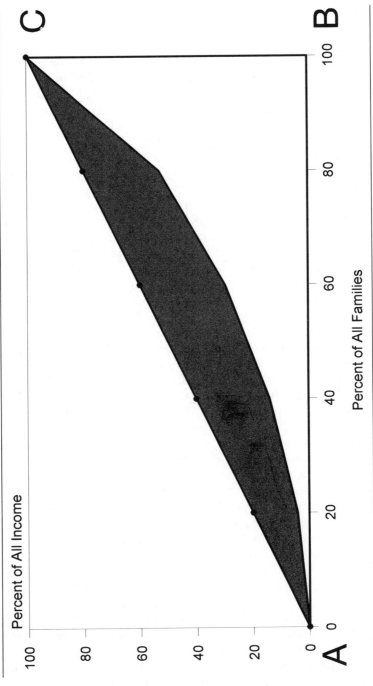

Percent of All Income

Percent of All Families

Source: Data from U.S. Census Bureau, *Income and Poverty, 1993*, CD-ROM, table F–2.

inequality than nine of the ten semiperiphery countries, and are about in the middle of the pack if we were placed in the periphery (fifteenth of thirty-one nations)

Also obvious are the large differences in Gini scores by world position. Core countries have an average of .320, compared to .336 for the semiperiphery and .431 for the periphery. Such large differences are also quite profound, despite what apprears to be a small size. The real width of the inequality gap again becomes more obvious when we look at the ratio of the percent of all income going to the richest fifth of all households divided by the percent going to the poorest fifth (the eighty to twenty ratio). Here, the mean ratio for core countries is 6.38—meaning that the richest fifth of all households has over six times the percentage of national income that the poorest fifth has. In the semiperiphery this ratio is 8.61 while in the periphery it is twice as large as in the core—at 12.03. Thus, whether using an indicator that captures the entire spectrum of income inequality (the Gini ratio) or the important contrast between the top and bottom of the distributions (eighty to twenty ratio), the conclusion is inescapable. Less developed countries suffer from relative income deprivation when compared to core industrial countries. LDCs are not only poorer in terms of absolute income, but *the poorest parts of their populations are relatively worse off* than the poor in wealthy nations.

The Gini ratio does have a number of weaknesses as a summary measure of income inequality.[90] Yet this seems inevitable with any figure that tries to show a complex process with one number. The Dow-Jones industrial average, although widely used and almost religiously followed, is only a composite of a few selected stocks among the many traded on the New York Stock Exchange. At times it has also been criticized for not being representative of activity on the exchange because of its makeup. Yet it continues to provide a useful function for analysts. So, too, does the Gini ratio. This technique is the oldest and most reliable of income inequality measurements. It has now been in use for three-quarters of a century. Its utility also depends greatly on how we wish to use the Gini ratio. In an analysis comparing Gini scores with seven different income inequality measurements, the choice of measure did not really matter in looking at the factors we wish to pursue in this book.[91] Other studies show that the Gini ratio is an ideal way to measure income change within the middle class, which is of great concern to us when we look at trends in the United States[92]

In the end, income inequality has both absolute and relative dimensions. Both aspects are looked at in this book, since the picture we get of income inequality can shift somewhat depending upon what perspective we use. This does not mean that there will always be disagreement. It is already plain that poor countries are cursed with both absolute and relative income deprivation. Yet there are times throughout the book when the distinction will become more important. Core countries tend to be high on absolute measures such as GNP per person. They are also low on relative measures such as Gini, the proportion of income going to the top 10 percent of households, the eighty to twenty ratio, and so on. This is true among American states. Wealthy states ranked by mean family income also tend to have low Gini ratios. The difference is important and must be kept in mind when we look at income inequality.

HOW INCOME INEQUALITY HAS CHANGED: A CROSS–NATIONAL COMPARISON

The World Bank has been monitoring income distribution for many years. Although there are problems in comparison, the quality, extent, and coverage of the data sets have improved over time in the great majority of countries. Thus, an approximate estimate can be made regarding the direction and degree of change income inequality has undergone in the immediate past. Using country Gini scores which were constructed from World Bank data in the first edition of this book, it is possible to make a rough comparison with contemporary income distribution and current Gini scores. Again, all figures are derived either from the *World Development Report 1988* (time 1 Gini scores) or the *World Development Report 1994* (time 2 Gini scores). All have been calculated using the percent of income accruing to the bottom four quinitiles, the eightieth–eighty-ninth percentile, and the top 10 percent of all households ranked by income.

At this juncture, however, the comparisons are quite crude—and should be interpreted in that light. Some nations did not have another income distribution survey between time 1 and time 2, and thus had to be excluded. Among those that did, a few countries were dropped because the Modern World System categories changed somewhat between the first and second editions of this book. Since semiperiphery countries are few in number (by definition) and relatively poor, most do not take income distribution surveys very frequently. Only three nations in this group could be compared, so the entire semiperiphery category was dropped from the analysis. While periphery countries are even more destitute, they are also more numerous—so twelve nations in this category could ultimately be compared to nine nations in the core group.

The dated nature of the figures is a perennial problem in comparisons of this sort. The most recent income distribution surveys in the 1994 report were for 1992, while the most recent surveys in the 1988 report were for 1986. Unfortunately, there is also wide variation in the years surveys were taken in both reports. Among core countries, there

was an average of 7.89 years between time 1 and time 2 inequality estimates (the average survey year for the *World Development Report 1988*
was 1979 for core countries, and 1987 for data from the *World
Development Report 1994*). Among periphery countries, there was more
variation. The average survey year for time 1 was 1975, and for time 2 the
average was 1988. Typically, then, the gap between readings was 13.38
years among periphery nations. This could bias the results because of the
longer lapse of time in the periphery, so the change in Gini scores was
adjusted to give an annual rate of change.[93]

The overall changes were very surprising, at least when seen
through the prism of traditional development economics. Among the rich
core countries, there was a 3.7 percent increase in the Gini ratio between
time 1 and time 2 (0.86 percent per year). The average core Gini ratio was
.3218 in 1988 data, and .3338 in 1994 data. Among the poor countries of
the periphery, there was a −10.8 percent decrease in the Gini ratio (−1.07
percent per year). The average periphery Gini ratio was .4759 in the 1988
data, and .4246 in the 1994 data. Thus, income inequality has increased
in wealthy industrial nations of the core, while it is actually going down in
less developed countries. The inequality reduction in the periphery is not
due to any development, at least not when measured in terms of GNP per
capita gains (table 3.2 indicates a slight drop for the periphery in the
1980–1992 period). More analysis of Third World countries and their
economic changes will come in the next chapter.

Development theory also postulates that in technologically
advanced, rich nations there should be an ongoing slide of inequality. With
greater riches, the pie is supposedly cut more equally, since there is more to
share and less to argue about. This is clearly not happening, at least from
data reflecting the experience of the 1980s. Confirmation of the hike in
income inequality among core nations is not hard to find. In fact, the data
that support this conclusion are much firmer and of higher quality.

The Luxembourg Income Study (LIS) began in 1983 under the joint
sponsorship of the government of Luxembourg and the Center for
Population, Poverty, and Policy Studies in this country. Funding has since
expanded to include the national science and social science research foundations of each country that now participates in this ongoing research
project. The major goal of the LIS Project is to encourage and promote
comparative research on the economic status of populations in different
countries. This is accomplished by making the data sets readily available
via electronic, on-line access in a reorganized format to improve comparability. Flexibility in study design is enhanced because the data sets consist of original, individual, or household records. In effect, researchers are
then able to choose their own definitions of income, unit of analysis, time
period, age groups, and so on. The membership roster of participating

countries that conduct similar income surveys now consists of more than twenty nations, including Australia and most countries in Europe, North America, and the Far East. A plan to include data and surveys from other nations is also being pursued, especially the addition of South Korea, Russia, Portugal, and Mexico. There are now well over fifty data sets available for analysis in the LIS covering the period from 1968 to 1992. Until recently, there have been two waves of data available to perform not only comparative, cross-national analysis, but also a rough time-series appraisal. Wave 1 surveys include data gathered in the 1979–1982 period, while wave 2 studies were mostly done in the 1984–1987 period. A third wave of studies is just now being added to the database, comprising surveys done mostly in the 1990–1992 time period. For the most part, it was too early to analyze changes between wave 2 and wave 3 as of this printing. Nonetheless, a good idea of what shifts have been taking place among and between developed countries can be garnered from studies dealing with waves 1 and 2.

Johan Fritzell compared recent developments in income inequality in Canada, Germany, Sweden, the United Kingdom, and the United States between the early and late 1980s using LIS data.[94] He studied total disposable income (after cash transfers such as AFDC or child support are added in and taxes subtracted). To more adequately compare the income well-being among households of different sizes and compositions, all income figures were divided by an equivalence scale. The scale is set to 1.0 for a one-person household, and adds 0.7 for each additonal adult and 0.5 for each additional child (under eighteen) in the household. Thus, size of family is taken into account while recognizing the economies of scale present in larger households. In essence, two can live more cheaply than one in the sense they need 1.7 times as much income to be as economically well-off as a single person (not twice as much). Fritzell found huge differences among these five countries on such factors as poverty, overall inequality, and the degree of welfare distribution. Using a variety of inequality measures, including the Gini ratio, the United States is the most unequal of the five countries. Between the early and late 1980s, inequality also increased markedly in the United States and the United Kingdom while growing only slightly in Canada and Germany. Sweden had an inequality growth rate comparable to that of the United States, but since it started at such a low figure to begin with (its Gini ratio was 0.196 in 1981), absolute changes in inequality were fairly small. When demographic factors were examined, it was found that changing age structure (more postwar baby boomers entering the labor market) and shifts in family composition (more female-headed, single-parent families) had almost nothing to do with the expansion of income inequality in any of the nations studied.[95]

The 1980s were a period during which significant changes in tax and transfer policies were introduced in these countries. A parallel movement toward less progressivity in taxes took place—for example, the "bite" lessened for higher income households, while swelling for the middle class. In all countries, welfare and social transfer programs came under increasing attack and were pared down. The new policies were a contributing force to growing income inequality in a number of countries, but did not account for the major segment of the disintegration of income equality. Fritzell divided the national samples into four wage-earning groups to isolate the effect of market forces (recession, growth, unemployment, etc.): earnings less than 50 percent of the median; earnings from 51 to 149 percent of the median; earnings 150 to 199 percent of the median; earnings more than 200 percent of the median. In short, this produces one low-income, one middle-income, and two high-income categories. In the United States and the United Kingdom, there were large increases in the extreme groups (Canada displayed a similar but much less pronounced pattern, while Sweden and Germany showed no growth of their income gaps). Low-income earners increased substantially between the two time periods as did high earners, while the middle class contracted. The bottom line was an unmistakable trend toward polarization in the United States and the United Kingdom. Because of these changes, the differences in inequality among the five countries was actually larger at the end of the 1980s than at the beginning. Although growth of overall inequality had occurred in four of the five countries (Canada excepted), it was almost nonexistent in Germany and minimal in Sweden.

Peter Gottschalk has focused more upon the earnings-inequality experience in his comparative study of seven developed countries (Australia, Canada, France, the Netherlands, Sweden, the United Kingdom, and the United States).[96] He initially limited his samples to male heads of households twenty-five to fifty-five years old employed full-time year-round (traditionally regarded as prime earners), later adding total family income which included other earners. Income was adjusted on an equivalency scale and measured at several different points (earnings, other sources of income, private transfer, government transfers). His chosen measure of inequality was a ninety to ten ratio (the income of the person at the ninetieth percentile divided by the income of the person at the tenth percentile). It was immediately apparent to Gottschalk that the wages of prime-earning males had become more unequal between the early and late 1980s in all countries. Thus, the United States was not alone in its exposure to whatever market forces were driving up inequality. We did, unfortunately, experience the largest growth in inequality of earnings. When other family earnings are combined, an equivalency introduced to control for variation in size and composition, and posttransfer income is measured (adding AFDC, child support, family allowance, etc.), the United States emerges as the most unequal of all nations. Our ninety to ten ratio of 6.11 places us far above all other nations in this study: Canada (4.65), Australia (4.04), France (3.69), Sweden (3.65), the United Kingdom (3.60), and the Netherlands (3.56). While earnings inequality increased in all nations, the welfare system in the United States was largely ineffective in blunting the severity of lessened income for the poor. While the United States underwent massive cuts in social and welfare expenditures during the 1980s, Canada and most European countries buttressed their transfer programs and increased social spending. These moves helped to counteract the ill effects of this conspicuous surge of income inequality in other countries, but they were not made in the United States.

Timothy Smeeding and John Coder adopt a research design similar to Gottschalk's, studying changes in income inequality for male prime-age earners and for families as a whole.[97] Again, two time periods in the 1980s were used for the United States, the United Kingdom, Canada, the Netherlands, Australia, and Sweden. Although no equivalency scale was used, the data were weighted to account for differences in family size. Three income concepts were employed: earnings, market income (earnings plus pensions, interest, rent, dividends, alimony, child support, and other private income), and disposable personal income (market income plus government cash transfers minus taxes). Three measures of inequality were also employed: the income at the twentieth percentile divided into the median income (income at the fiftieth percentile, or twenty to fifty ratio); the income at the eightieth percentile divided by the median income (eighty to fifty ratio), and income at the eightieth percentile divided by income at the twentieth percentile (eighty to twenty ratio). In this way, changes at the top and bottom of the income distribution were captured, as well as the changes in the spread between the two extremes. Among earners, U.S. householders at the twentieth percentile are worse off than those in all other nations except for the United Kingdom, while our richest eightieth percentile is better off than that percentile in all other nations. Incomes are more equal in all countries when disposable income (after taxes and transfers are included) is compared to market income. In fact, the top to bottom (eighty to twenty) inequality is cut by one-fourth for all six countries taken together. The tax and transfer policies are least effective, however, in the United States—leaving our country with the highest inequality readings on all three measurements. While tax and transfer policies grew less effective in equalizing income in all countries (aside from the Netherlands) during the 1980s, the erosion was worse in the United States. The weakness of the U.S. redistributive system consists of an anemic welfare system (especially compared to our

European counterparts) and a tax system that falls much more lightly on higher income citizens than do taxes in other developed countries. Because little government help was given to the poorest fifth of America's households, their disposable personal income decreased during the 1980s. By contrast, the wealthiest fifth of our population actually gained a considerable advantage—which increased the spread between the highest and lowest households by 11 percent (ahead of all other nations).

Comparative income inequality research within wealthy nations has expanded greatly during the 1990s. An exhaustive, in-depth monograph analyzing the details, nuances, and limitations of such research among nations in the Organization for Economic Cooperation and Development (OECD) has just been released.[98] This study has greatly expanded the methodological advances made by the Luxembourg Income Study Project, allowing comparison of housheold surveys done in Australia, Belgium, Canada, Finland, France, the Netherlands, New Zealand, Norway, Sweden, the United Kingom, and the United States. Looking at disposable income (after taxes are subtracted and transfer payments are added) among persons, adjusted for household size, the study found a majority of wealthy countries experienced growing inequality during the last decade. Individual nations had greatly varying experiences. Looking at the ninety-ten ratio (the income of persons who receive more than 90 percent of all income, divided by the income of persons who receive more than 10 percent of all income) is instructive.[99] The comparison shows how much richer the top 10 percent of income recipients are in comparison to the lowest 10 percent. The United States leads the pack with an inequality ratio (5.94) nearly one-half times higher than the next most unequal country, Canada (4.02), followed closely by Australia (4.01), the United Kingdom (3.79), New Zealand and France (3.48). Norway, the Netherlands, Belgium, Finland, and Sweden all have low inequality ratios (under 3.00) that if doubled would still not match the degree of inequality in the United States

Where two time points were available for comparison, data show there was actually a slight reduction of inequality by this measure in Australia, Canada, and New Zealand during the 1980s— with no change in France and a barely perceptible increase in Belgium (figure 3.4). Large increases took place in Norway and Sweden, but these countries had a low ratio even at their last inequality measurement (2.93 and 2.72, respectively). The combination of a great degree of income inequality at the start plus a large yearly growth rate in percentages for the United States is lethal to our middle-class way of life. Among all advanced countries where data are available, we show the most significant expansion of inequality. The United States, in comparison to fourteen other OECD countries, also has the highest percentage of low-income persons. The OECD study calculates that 17.8 percent of U.S. citizens under age sixty

FIGURE 3.4   Annual Percent Change in Rich-to-Poor Ratio

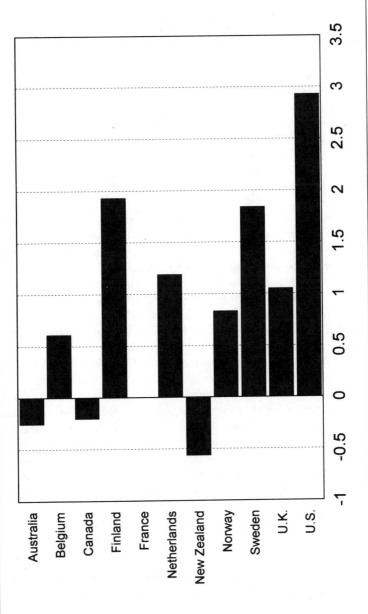

Source: Calculated from Atkinson, Rainwater, and Smeeding, 1995a, table 5.

can be regarded as low income.[100] This percentage is one-third higher than that of the second worst performer, Ireland, and five times higher than Belgium's—which turned in the best low-income figure (only 3.5 percent of its population).

Comparing the change in Gini ratios for the ten OECD countries where data are available is also instructive. As is obvious from figure 3.5, there were increases in this summary measure of inequality for all nations (with the exception of France). The increase for Canada between 1981 and 1987 was almost nonexistent, however, going from .286 to .289. The length of the bars in figure 3.5 gives some indication of the amount of change over time, marking both the starting and ending point of the Gini ratios. It can be seen that while nearly all OECD countries witnessed increasing inequality, the change was more dramatic in Sweden, the Netherlands, the United Kingdom, and the United States. The horizontal line across each bar represents the equivalent of a year's change in each country (reading from the bottom of the bar). This gives some idea of the rate of change, since the time gap between readings is not similar from one country to the next. For example, Finland shows little change in the Gini ratio since its bar is relatively short. The shorter bar represents a three-year gap between the first Gini score in 1987 and the second reading in 1990. The horizontal line, however, is higher from the bottom of Finland's bar than from the bottom of Norway's. Finland's inequality index is climbing at a faster rate than Norway's.

Finally, the graph also shows where each nation currently stands in comparison to the other countries. The top of the filled bar indicates the value of the Gini score when last measured. The countries can be roughly grouped. Even though Scandinavian countries have had serious inequality surges, they are still considerably more egalitarian than other OECD nations. Together with France, the English-speaking countries of Australia, Canada, and the United Kingdom are comparably high in inequality. In a class by itself, at the top of the inequality pyramid, is the United States. The Gini ratio for our country at its start (the year of the first measurement) was higher than for any other OECD country at its second measurement time. Moreover, the bar shows substantial change in the United States between its two readings (1979 and 1986). Even on a yearly basis, the increase in household-income inequality in America has been incredible in comparison to similar increases in our industrial peer nations.

A CASE ANALYSIS OF ABSOLUTE VERSUS RELATIVE INCOME INEQUALITY: THE ECONOMIC "MIRACLE" OF BRAZIL

There is no shortage of studies examining the impact of development upon Third World countries. This is very true in looking at individual

FIGURE 3.5   Change in Gini Ratios in OECD Households

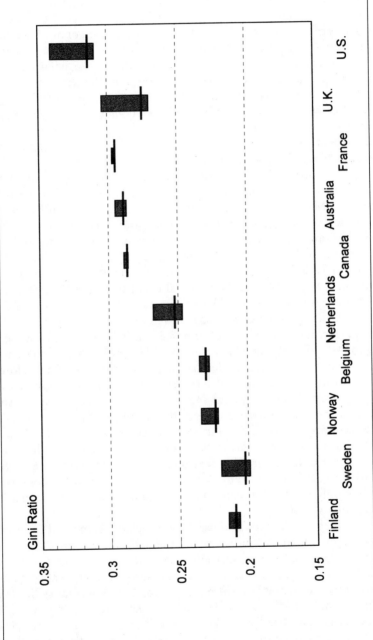

Source: Calculated from Atkinson, Rainwater, and Smeeding, 1995b, table 4.8.

nations (the case-study approach). Although longitudinal data measured over a long period of time are scarce for nearly all poor nations, there is some evidence of trends in the Third World.[101] Looking at figures kept over the past ninety years, for example, shows that the growth in Mexico's relative income inequality has shown no sign of turning around. The poorest 40 percent of its families have not shown any material improvement since the Revolution of 1910. Meanwhile, while the richest fifth of Mexican families received ten times as much income as the poorest fifth in 1950, this ratio had risen to twenty to one by 1977. By this year, Mexico's distribution of income was among the most uneven in the world, but it was to get even worse. During five short years in the 1980s, the share of Mexico's gross domestic product going to wages and salaries declined from 37.2 percent to 24.9 percent.[102]

This painful slide was largely a result of IMF restructuring which was planned in collusion with Mexican business elites. In the six years of the de la Madrid regime during the 1980s, real wages fell by one-half while debt-service payments from Mexico to its creditors rose to nearly two-thirds of the national budget. The inflation rate rose to nearly 25 percent a year in Mexico, while it has had to import over half of its basic grains in recent years. What this means is that people are starving. Between 1981 and 1988, the personal consumption of rice, beans, and corn was cut nearly in half. All of this occurred against the backdrop of nominal success as Mexico managed to increase its industrial exports by 130 percent in six years. This is supposed to bring in more money, according to IMF wisdom, but what profits were generated also left the country in a massive capital flight. Mexico is actually worse off than before its restructuring, and now has a government that was forced to resort to massive voter fraud in order to stay in power.[103] After thirteen years of structural adjustment, involving nineteen adjustment programs, Mexico's debt had increased to $100 billion in 1992 from $85 billion in 1982. Yet its GDP growth rate remained at a standstill.[104] As if to rub salt in the wound, the number of billionaires in Mexico increased from two to twenty-eight in the first half of the 1990s. Mexico's elite now "hail the country for trailing only the United States, Germany, and Japan as an incubator of ostentatious wealth."[105]

The case of Mexico underscores the lack of meaningful connection between development and help for the suffering majority of citizens within a country. Yet it is especially with Brazil that we can gain an even sharper idea of the dynamics between absolute and relative income inequality. This nation's performance since 1960 as a country brimming with growth earned it accolades such as "the miracle economy." From 1960–1981, the Brazilian economy racked up great income-per-person growth at an annual rate of 5 percent, year after year. For a few years in the late 1960s and early 1970s, average annual growth was approximately 10 percent. Even after the first shock of oil price increases during 1973–1974, it managed to achieve healthy industrial and agricultural growth rates. Much of its economic plan was based upon import substitution. In an effort to free itself from oil imports, Brazil introduced a massive sugar cane industry which produces alcohol fuel. It has also built up its industry and its agribusiness (soybeans) to increase exports. Gigantic internal projects meant to stimulate economic growth have also been mounted, such as development in the Amazon.[106] Much evidence also shows that promoting exports over industries catering to a domestic market and/or substitutes for imports actually produces more employment in Brazil.[107] In this manner, more income is redistributed downward to the poor because unskilled jobs are provided. Thus, Brazil is a fertile testing ground of the relationship between real economic growth and relative income inequality.

Did the majority of Brazil's population share in the run-up of wealth and economic growth during these two decades? In terms of absolute income, the answer is a resounding "maybe." Despite high initial income inequality, Brazil's own census figures for 1960 and 1970 show that all income groups had very large increases in real income. Each 10 percent (decile) of households had nearly 50 percent growth over the decade.[108] Persons in the bottom 10 percent, 20 percent, and 30 percent brackets showed the same degree of increase as persons in the top decile brackets of 70 percent to 100 percent. It was the middle classes that showed slightly smaller increases in real income, averaging about 35 percent over the decade. Yet Professor José Camargo, a Brazilian economist, disputes this interpretation. If persons who have not earned any income (e.g., those unable to find jobs) are included in both the officially designated "poor" and "nonpoor" groups, a sharply different picture emerges. The relative income gap between poor and nonpoor persons actually increased from 1960–1970 by 44 percent.[109]

Noneconomic indicators of poverty and social well-being are less equivocal and do show progress during the 1970–1980 period. Literacy rates increased, as did school enrollment percentages. Piped water and sewer connections went up dramatically. Households with electricity climbed from 48 percent to 67 percent. By 1980, over half of all households had a refrigerator and/or a TV, three-fourths had a radio, and nearly one-fourth owned a car.[110]

The growth statistics are quite breathtaking until 1980. Yet the expansionist economic policies followed by the military junta which ruled Brazil from 1964 to 1985 ultimately led to disaster. Import substitution worked in only a few select industries such as iron and steel. For aluminum, fertilizers, and petrochemicals, real imports have continued to grow, as they have for capital goods. The crucial area of oil substitution, which makes Brazil highly vulnerable to world price flux, showed no change at all during the 1970s.[111] To finance further growth at continued high levels, Brazil was willing to add even more external debt to fuel its booming economy. Credit subsidies and incentives were heavily used to favor export industries, while higher taxes to pay for the growth were avoided like the plague. Such debt increases came to about 28 percent per year between 1973 and 1978.[112] As the ensuing world recession developed, Brazil's export performance slowed. It became more and more difficult to pay off the mountain of debt it had accumulated from international banks. By 1982 Brazil was forced into an austerity program dictated by the IMF. The result for its economy exceeds the usual painful decline, however, ending in an annual rate of inflation for 1988 of 942 percent. As late as May 1994, the inflation rate was still running at 50 percent per month—which means prices double every two months.[113]

Some scholars argue that because of its vast size and richly endowed resources, Brazil is in more of a position to be independent of the world system. It may have the ability, if we consider it a semiperiphery state with *some* inherent power, of weathering the periodic storms sweeping through the world economy. Sociologist Peter Evans points out that it is too simplistic to think of emerging industrial powers in terms of the old Dependency school, that is, to see them as operating only at the whim of core countries.[114] In fact, in some countries like Brazil there may often be a triple alliance among the state government, large native capitalist interests, and multinational corporations from core countries eager to invest in LDCs. The state acts to mediate the alliance in such a way as to encourage multinational investment. At times it may combine with large local firms; or it may at times choose to set up its own enterprises in conjunction with foreign firms. The point of his detailed analysis of Brazilian industry is that Brazil is not at all powerless in the worldwide economic pecking order. The government will insist that certain conditions be met before multinationals are allowed to come in and set up shop. Many of these conditions require coalitions with the state and/or local capital, adherence to import restrictions, observance of quotas for the hiring of Brazilian labor or contracting, and so on.

Yet the bottom line of the triple alliance still leaves Brazil under the thumb of countries in the core. The success of such industrial growth ultimately lies with the willingness of multinationals to invest, international bankers to make huge loans, and core countries to consume exports from Brazil. The government of Brazil and its largest firms may wrangle among themselves in the quest for greater accumulation and profit, but they must ultimately satisfy foreign investors with enough sugar coating on their deals. Since this can threaten the profit margin of Brazilian big business and the state, they always respond by squeezing the wages of workers:

> Whether products were developed locally or abroad by multinationals or local firms, they were developed in response to perceived opportunities for profits, which given the skew of Brazilian income distribution, is hardly synonomous with the satisfaction of needs. . . . The maintenance of the delicate balance among the three partners militates against any possibility of dealing seriously with questions of income distribution, even if members of the elite express support for income distribution in principle. . . . Nothing indicates the divergence between the benefits expected from industrialization and the actual results better than the changes in Brazilian income distribution between 1960 and 1970. In a decade during which the production of television sets tripled and the top 5 percent of the population almost doubled their average income, the 80 million on the bottom remained stagnated at incomes averaging below $200 a year. . . . The rate of infant mortality in Sao Paulo, one of the most modern and heavily industrialized cities in the world, was 84 per 1,000 live births in 1969, about a third higher than the rates for the entire nation of Argentina. The rate for Recife, the largest and most advanced city in the northeast, was 263.5 per 1,000 in 1971, about 60 percent higher than the rate for the rural population of Chad, which was the highest national rate recorded by the United Nations. Even more discouraging is the fact that Recife's rate of infant mortality is apparently no lower than it was a generation ago.[115]

In other words, Brazil as a nation may remain strong despite economic penetration by foreign firms. Perhaps some of its sectors will even be better off and more developed as a result of continued investment from the outside. A few of its corporate elites may also benefit greatly from such growth. Yet the vast majority of its citizens still remain in abject poverty. An ever escalating GNP per capita has no meaning to hungry children in the favelas overlooking Rio, or to the rural poor in the northeast section of the country.

Yet even within the affluent major cities of the southeast, such as São Paulo and Rio de Janeiro, there is a major question whether Brazil as a nation is better off from having followed the golden path of development. São Paulo is Brazil's richest city, where most of the nation's manufacturing is concentrated. Local jobs pay relatively high wages. Yet after the "miracle" had started pumping out industrial products for export, infant mortality actually increased from the already high level of eighty deaths per one thousand live births. Even in rich São Paulo, one-third of households lacked running water during the 1970s and 60 percent had no sewer connections.[116] Despite the healthy growth in manufacturing exports, the proportion of the labor force in the informal sector (24 percent) did not budge between 1969 and 1983.[117] The informal sector is made up of people doing anything to survive, that is, people who do not hold regular jobs or earn a wage. In Rio they are the street vendors selling T-shirts, drugs, sex, a gamut of personal services, and the like. Since unemployment nearly doubled in the same period, this means that jobs in the formal sector have slid by 5 percent as well. At the bottom line, export growth has not created enough jobs to even make a dent in bettering the distribution of income.

In Berkeley economist Albert Fishlow's opinion, the state definitely dropped the ball.[118] By heedlessly chasing growth through unwise borrowing, the government failed to introduce needed change in order to maintain the health of its economy. When the crunch came, the common people had to pay the final reckoning. Ironically, this led to the downfall of the military junta and the return of civilian government in Brazil—which was to be the only ray of sunshine in this debt calamity.

On a relative basis there is much poverty. While Brazil has the highest per capita income in Latin America, 40 percent of its inhabitants now live below the poverty line, because the lower half of the population receives only 13 percent of all income.[119] Between 1960 and 1990 the richest 10 percent increased their share of income from 40 percent to 68 percent. At the same time, the share of income received by the lowest 80 percent actually fell. The richest one-fifth of all households in the United States averages twelve times the income of the poorest fifth. The same ratio in Brazil is thirty-three to one.[120] The Gini ratio of income inequality actually increased from .52 in 1960 to .565 in 1970 to .590 in 1980 among the economically active population.[121] By the beginning of 1990, Brazil still held the world's inequality record with a Gini ratio of .6039.

Wherever one looks, economic inequality in Brazil has increased sharply during the past three decades of development. During the 1960–1970 decade, the upper 3.2 percent of the population enlarged its share of Brazil's total income from 27 percent to 32 percent. The increase in real income during the same period did not reduce the percent of those under the poverty line. The absolute gap of real income between the poor and nonpoor instead widened by 25 percent during those ten years.[122] Helga Hoffman, an analyst at the United Nations, also reports that income of individuals, income of the economically active population, and family income all became more unequal during the 1960s. This growth in inequality slowed somewhat during the 1970s (even slightly reversing for families who were forced to put children to work).[123] But by 1980 the income share of the poorest half of Brazil's population had actually gone down from 15.6 percent to 13.4 percent while the richest 10 percent of the population laid claim to one-half of all income in the country. By 1990, the richest fifth of the population gleaned over two-thirds of the nation's income—leaving scant resources for the bottom 80 percent.[124] Industrial concentration, already massive, actually went up during the same decade. The top half of 1 percent of firms now accounts for one-half of all total sales in the country, while the top 3 percent of firms accounts for 70 percent of all industrial capital.

As business firms were growing larger and ever richer, an economic nose-dive took place for the Brazilian working class. Professor José Camargo of the Department of Economics, Catholic University of Rio de Janeiro, presents evidence that this slide in wages was deliberately planned:

> The period 1964–1974 was one of strong political repression. After the military coup of March 1964, unions were dissolved and strikes repressed by the military force and, as a result, the position of the workers was drastically weakened. Unable to press their demands, the workers had no choice but to accept the wage adjustment imposed by the government. On the other hand, government priority was to reduce the rate of inflation and a wage squeeze was a powerful instrument by which to attain this objective. . . . As a direct consequence of this policy, those wages whose adjustments followed the formula grew at rates much lower than the inflation rates. The real wage declined sharply . . . there is a decline of 20 per cent for the real minimum wage between 1964 and 1971 . . . in 1978 it was still 14 per cent lower than in 1963.[125]

There were other costs to the rule by military junta as well. As is true whenever the military dominates a country, spending for defense tripled in the first decade of their rule. Jan Knippers Black details many cases of jailing, killing, torture, and disappearance of politicians, scientists, professors, educators, artists, and others who dared oppose the new order.[126] In her book *United States Penetration of Brazil,* she presents evidence of U.S. complicity through military channels in the 1964 coup which brought the Brazilian army to power. Her analysis also tracks multinational corporations in Brazil and their role in bringing about the downfall of this nation's constitutionally elected democratic government.

There was much to lose in the eyes of these corporations. The rake-off for U.S. firms in Brazil has been astronomical in comparison to the share they receive in other countries. In 1969, for example, many U.S. businesses recorded the following high profit levels for their Brazilian subsidiaries: Exxon (15.7 percent), Johnson & Johnson (16 percent), GM and RCA (20 percent), Squibb (20.5 percent), Union Carbide (21.8 percent), GE (22.6 percent), Goodyear and Atlantic Richfield (23 percent), Texaco (28 percent), Xerox (64 percent).[127] After the coup, U.S. dollars gushed into Brazil in the form of government aid and new business start-ups, climbing 71 percent during the 1960s alone.

The major result, according to Black and other scholars, has been denationalization of Brazil's most profitable industries. Lavish concessions flowed to U.S. firms from the generals, including full repatriation of corporate profits to the core nations, much lower taxes, more favorable bank credits, and so on. While foreign firms could borrow money at 7 to 8 percent, Brazilian domestic firms had to pay as much as 48 percent. Most of them went belly-up within a few years or were devoured by the very same multinational piranhas that were being so unfairly subsidized by the junta. In this manner, foreign capital came to control 70 percent of the 679 largest firms within Brazil by 1970. From the coup through 1976, Brazil received more U.S. economic and military aid than any other Latin American country. Given the high profit levels and the very favorable multinational investment climate that Brazil offers, this is not surprising.

A more insidious result than the bleeding of capital from Brazil has been the destruction of anything unique to its own native culture. Deep penetration has been economic, but it has also been strong in social, cultural, political, and military channels. Brazil has come to resemble a chameleon, taking on the hues and colors of the nations it hosts through their corporate subsidiaries. Jan Black provides an eloquent quote by former Brazilian congressman Marcio Moreira Alves that distills the essence of resentment such institutional domination causes in the people who are forced to live under the deep shadow of U.S. penetration:

> Everything we consume in our daily lives betrays a foreign presence. We are bandaged at birth by Johnson & Johnson. We survive on Nestle or Gloria Milk. We dress in synthetic clothes produced by French, British, or American firms. Our teeth are kept clean by Colgate toothpaste and Tek brushes. We wash with Lever Brothers' and Palmolive soaps, shave with Williams and Gillette. Resting in the sun we drink Coca-Cola—and now even the largest producer of *cachaca,* the white rum national drink, is owned by Coca-Cola. We ride Otis elevators, drive Volkswagens and Fords and ship our goods on Mercedes-Benz trucks fueled by Esso and Shell, our rubber is Pirelli, we talk with Ericsson telephones, communicate through Siemens telex, type on Olivetti machines and receive IBM-processed bills. We eat out of American and Canadian made cans packed by Armour, Swift, and Wilson. The Beatles' beat comes out of Phillips radios, and we dance to RCA records. Our General Electric TV sets are connected to ITT satellites. We can rely on old Bayer for aspirin, or, if trouble develops, on Squibb for antibiotics. From the comfortable Goodyear mattresses of our American hospital beds we can look through Saint-Gobain windows on gardens tended by Japanese lawnmowers. If we die (say, from lung cancer puffed from British or American cigarettes) we may finally have a chance of entering into our 94 percent share of the economy—graveyards are owned by the Santa Casas de Misericordia, an old Brazilian institution. But the family must pay the electricity bills to Canadian Light and Power—with money

manufactured by Thomas de la Rue or by the American Bank Note Company.[128]

Dependency scholars predict that income inequality cannot help but grow in Brazil and other countries like it.[129] This will always be true wherever there is industrial development and growth in a modern sector going on next door to traditional, poor sectors surviving on subsistence agriculture. The contrast in Brazil is quite vivid between Rio de Janeiro and São Paulo versus its northeast region. Large gaps in both absolute and relative income remain between Brazil's states, as do marked urban and rural contrasts. The disparities between Brazil's south and its northeast are so enormous that the United Nations has issued a warning that disastrous social upheavals and explosions could result. In the northeast, life expectancy is seventeen years shorter, adult literacy 33 percent lower, and average income 40 percent lower.[130]

Still, the majority of growth advocates point out that real income will nonetheless advance for most people. They argue that for a relatively affluent LDC such as Brazil, the increase in real absolute income can amount to a tidy sum. Even if this were true, however, we must never forget that we are dealing with incredibly low incomes at the bottom of the Brazilian income hierarchy. During the economic boom period of the early 1970s, one expert estimated that it would take twenty to thirty years of growth at current rates to raise the poorest 10 percent of the population up to an annual $100 per person income.[131] Even during the high-growth period of the late 1970s, 40 percent of Brazil's export earnings had to be spent on servicing its foreign debt.[132] This was before Brazil, along with other periphery countries, ran into serious trouble during the early 1980s because of the second oil-price increase and world recession.

The downturn in its economy was severe. Within a few years at the start of the 1980s, industrial production declined 11 percent and capital goods production slipped by 45 percent. Within four years, Brazil's standard of living declined 17 percent—and was projected to drop another 20 percent by the start of the 1990s.[133] In order to pay off its massive debt, Brazil submitted to an IMF austerity program in 1983, leading to further cuts in domestic spending, financial speculation, out of control inflation, and a brutal recession. Put bluntly, Brazil's economy ended up in the gutter because of this debt. For the rest of the decade, Brazil's political and industrial elites managed to spin off these shaky, mostly private loans to the public sector. By the beginning of the 1990s, 90 percent of Brazil's debt had been assumed by the government. By means of renegotiation and a slight 7.6 percent debt forgiveness, Brazil put together a new deal to pay its bills through the Brady Plan in 1994.[134]

Despite a temporary respite, it is very likely that even more doom and gloom lie waiting in the future for Brazil. Because of the decline in prices of its exports and the increase of interest payments on its mountain of debt, the vital force of the country's economy is being bled off. When the military took power in 1964, Brazil owed only $3.5 billion. By the time it stepped down in 1985, over $100 billion of external debt had accumulated. By the end of the 1980s, Brazil owed a full 10 percent of all Third World debt. Thus, in a period of two decades, debt owed to foreign banks increased twenty-nine-fold while per capita income went up by only 130 percent.[135] By the mid-1990s, Brazil's total debt had topped $145 billion, with annual payments of $10 billion. Within the past decade, Brazil sent $123 billion out of the country as debt repayment—an amount roughly equal to what it now owes. Despite the country's prior history of shaky loans and threats to default, new loans keep getting approved. In fact, the cost of restructuring nearly always includes a new debt load. As a result, Brazil's arrears actually increased during the last decade from $98.4 billion in 1983 to $120.8 billion in 1992. From 1988 through 1993, payments to the World Bank, IMF, and Inter-American Development Bank (IDB) increased by 55 percent.[136]

Paying off this huge debt has made Brazil the world's largest capital exporter after Japan. It has also caused endless turmoil and bickering among political and economic elites within its own borders. The acrimony has stopped any coherent, workable plan from emerging despite the return of democracy. Instead, a variety of economic policies have alternated between expansion and contraction in the 1980s. Such a stop-and-go mentality has virtually destroyed stability and the credibility needed for the economy to work normally.

While all the dramatic wheeling and dealing between bankers was going on, actual people were suffering. Many experts continue to attest to a loss of real income and an increase in grinding poverty among the masses. Since 1980 the poor have born the brunt of austerity measures. There can be no doubt by now that even the paltry gains in absolute income eked out during the 1960s and 1970s were lost during the 1980s. Although the government has set a minimum wage, this has done nothing to help stave off the slip in Brazil's living standard. Yet it does serve as a stark yardstick of how bad the situation really is. Data for the 1980s reveal the minimum wage to be one-fifth to one-seventh the amount needed to cover a family's basic needs. This was at a time when only 10 percent of the work force earned the equivalent of five minimum wages. Put differently, nearly two-thirds of the Brazilian population lived at a level between misery and extreme poverty. The purchasing power of a minimum salary had declined by 50 percent.[137] Such trends can mean life or death to poor Brazilians, at times through slow starvation. The number of persons living there on less than $1 per day increased by 40 percent to 35 million during the 1980s. While Brazil is the world's second largest market for corporate jets, at least one-fourth of its people go to bed hungry every night.[138] To put food into the mouths of their family, some workers have ended up in virtual slavery. The number of documented cases of Brazilians enslaved because of debts owed to employers rose from 592 in 1989 to 16,442 in 1992. Only a fraction of these cases ever become known, but the scenario is always the same. Workers are recruited under false pretenses, promised a living wage, and then charged for such expenses as food, tools, transportation, and the shanty they sleep in. When they cannot pay, they are prevented from leaving and paid only in food:

> "Workers who try to escape are pursued by gunmen and returned to the estate, where they can be beaten, whipped, or subject to mutilation or sexual abuse," the International Labor Organization asserted in its March report. "In 1991, 53 victims of forced labor were murdered in Brazil, but few of the accused were brought to trial or received the statutory punishment."[139]

The current situation is neither fair nor just. This is especially true when we realize that productivity in Brazil has quadrupled since 1940. It is cruel to expect people to undergo such want and deprivation while they are this hardworking and industrious. Yet Brazil must labor at an exhausting pace just to stay even. Aside from the cost of mounting debt, its population continues to multiply at a much faster pace than the growth of jobs. The poor in Brazil have a rate of natural increase three times that of the rich. Over time, the fact that women in poor Brazilian families have an average of 7.6 children compared to 3.3 children for high-income women can mean only one thing. The larger absolute and relative numbers of needy persons will increasingly twist the maldistribution of income in Brazil. Income inequality in this country will thus be even more grotesque at the beginning of the next century.[140] According to projections, the ranks of the poor and deprived will swell larger even if the miracle growth rates briefly evident in the 1970s stage an unlikely comeback.[141]

DESPERATION AND THE FIGHT TO STAY ALIVE:
THE HUMAN COST OF INCOME INEQUALITY IN BRAZIL

To find out more about the impact that such gross inequality has upon its people, my wife and I visited Brazil. It was an opportunity to meet with economists, scholars, and government officials who graciously donated their time and services to my study. By interviewing Brazilians closely connected with the development of their nation, it was possible to get first-hand opinions about whether their country was dominated by foreign interests. Almost every expert said "no" to this question. The intellectuals we talked with were unanimous in their view that Brazil was powerful enough in its own right to resist or manage corporate domination from abroad. Nationalism in Brazil is rampant, extending far beyond the fanaticism of its soccer fans. One authority proudly told me that his nation had managed to build the world's first and only Apple MacIntosh computer clone. Ironically, although he felt foreign powers were relatively weak in Brazil, he said in the next breath that the government had yielded to pressure from Apple Computers and had decided not to sell the clones.

Brazilian economists have a touching faith in growth-oriented policies, even though these procedures have brought them so much current distress. One official jokingly told me that "Brazil was doomed to grow." We could see he was pleased with the prospect. The general message from all economists was that although Brazil had serious problems, it would once again surge forth among the world's economies with explosive growth. In my opinion, this is a tenuous assumption. It is accepted on good faith that free markets will reign supreme long into the future. Reality seems to point toward the other extreme. Many contemporary economists believe a once-unified world economy will split into at least three trading blocs (Europe, North America, Japan/Asia). Tariffs and trade barriers will prevail in the future as exports dry up.[142]

But whether this scenario is accurate or not really begs the question. By focusing on growth, Brazilian officials have turned the spotlight away from the enormous inequities that are so painfully obvious in their country. The implicit view is that growth will take care of poverty more or less automatically. As real income goes up, the need to be concerned about unfair income distribution will go down. Perhaps as foreigners unused to the country, my wife and I saw things that Brazilians may have long taken for granted. Luxury hotels line Avenue Atlantica beside Copacabana beach in Rio de Janeiro, "glamour capital of the world." Outside their doors, literally only a few feet away, is the most dire poverty I have ever encountered in my life. Homeless families live on mattresses on the sidewalk or beneath highway viaducts. Beggars accost tourists wherever they go. Returning to our room one morning at 2:00 A.M., we counted nearly a dozen prostitutes on one corner.

These are the seamier sides of life without money, of being forced to do anything to feed yourself. If I were asked to describe Rio in one sentence, it would be "Two blocks of Paris along the beach with Ethiopia surrounding it." In most areas, the city of Rio de Janeiro is only four to six blocks wide before steep hills are encountered. On these hills are the favellas, slums where the poorest of the poor live. Two totally different classes exist side by side and the streets of Rio reflect this. Graffiti is scrawled everywhere, even on the churches. Rich Brazilians live in condominiums along Ipanema Beach or one block in, only three blocks away from devastating poverty. Armed guards continuously patrol the entrances to these condos, while all ground floor windows have iron bars or steel shutters.

It is unsafe to walk in most areas of the city, whether you are a tourist or a *carioca* (native of Rio). It is mandatory not to wear a watch, carry a camera, have much money on your person, speak English, or even look as if you do not know where you are going. Tourist books give some warnings of these facts of life for vacationers headed for Rio. Yet the warnings simply cannot convey the reality of the scene, which is akin to entering a war zone. A tourist is continuously watched by many street people, vendors, beggars, taxi drivers, thieves, pickpockets, pimps, armed police, and assorted persons seeking favors. Nearly all of them are desperately trying to get money in any way possible. Although we were told it was safe to walk along Avenue Atlantica (Copacabana Beach) at night, since manned police kiosks alternate every few blocks, it proved to be poor advice. One hour after dusk on our third night in Rio, we were jumped by eight *favella* teenagers on this busy avenue. They jostled and pushed us. As fear and confusion quickly set in, one youth reached for my shirt pocket and ripped it open. Money scattered on the ground, and we were forgotten in the scramble to pick it up. Then the gang melted into the shadows. The attack lasted only thirty seconds at the most!

With my shirt in shreds, we returned to the safety of our hotel brimming with anger and a deep sense of violation. This soon gave way to fear, as we began to realize what could have happened beyond the loss of a few dollars. In the remaining week in Rio, fear was our constant companion. A deep wariness of everyone we came in contact with set in. We began to suspect that we had overreacted, that we were being paranoiac. When we told *cariocas* about our mugging, however, it was as if a dam had broken loose. Their own fear came gushing out. Native Brazilians who had lived all their lives in the city began to tell us about the danger they felt and how unsafe things had become. On our plane ride back, we met a young Brazilian man who had emigrated to New York City eight years before. He said New York was safer than Rio because he could actually walk to work without fear.

One native told us there is at least one murder per night along Copacabana during Carnaval. According to official statistics, this may be an understatement. While New York City is about the same size as Rio, New York had 1,896 murders during all of 1988. In contrast, Rio had a virtual avalanche of homicides—500 killings in April 1989 alone.[143] Traffic from tourism has plunged as a result of this mushrooming violence, which has culminated in a rising homicide rate (a one-fourth increase in two years). Ten people were shot through the head in one twenty-four-hour period, armed robberies occur at funeral processions, and death squads are made up of moonlighting police.[144] The everyday routineness of such terminal violence has led to a culture of institutionalized indifference where grief over death becomes almost completely unknown.[145]

Brazilian sociologists are reporting that the violence is increasingly based on social class differences, and that many of the poor are now seeing the killings as acts of social justice. It has become routine to rob and kill the unwary rich who are foolish enough to drive through poor neighborhoods. The slums of Rio bristle with automatic weapons which outgun the police. Drug-dealing teenagers are behind much of the violence, but the angry, resentful, and desperate residents of the *favela* tend to support the killings. They see the murders as a blow for social justice, as helping to right many of the wrongs inflicted upon them because of the gross income inequality in Brazil.

There are now also at least 200,000 homeless children living on the streets of Rio. Their presence, like gasoline poured on a fire, fuels the crisis of robbery, mayhem, and violence as they fight like animals merely to stay alive. Tragically, they are also frequent victims of outright murder by gangs of moonlighting policemen. Among the 7 million Brazilian children and adolescents forced into homelessness, four per day are killed, while a half-million girls resort to prostitution to feed themselves.[146] Roger Cohen, in a *Wall Street Journal* article, gives us a telling glimpse of just how brutal life can be in the slums of Rio:

> Luiz Pereira da Silva looks out over the slum to the middle-class district of Santa Teresa only a few hundred yards away. Beyond it rises Sugar Loaf mountain, with its cable cars filled with tourists. Below it is the blue of Botafogo Bay and its yacht club, where membership costs several thousands of dollars a year. Luiz Perreira da Silva earns $80 a month as a laborer. He keeps his wife and five children in the one-room mud-and-timber hut he bought seven months ago for about $100. . . . He says the neighborhood is full of drug dealers. "But it's best not to look, best to ignore it," he says. Below his home chickens feed on garbage, and open sewers gurgle down the hillside. A boy is urinating against a wall. Another, standing nearby, has a pistol tucked into his Bermudas. Such children often pour down into the suburban streets to rob rich children on their way to private schools. . . . Some of Rio's homeless children spend their nights beneath the bridges of an expressway that snakes along the glittering shore past beaches, yachts, and luxury apartments. . . . The grassy shoulder of the expressway that whisks commuters past the scantily clad beach crowd to downtown offices has become a dumping ground for corpses. "Killers seem to favor it because they can make a quick getaway," says Military Police Sgt. Alberto Gatti. A

corpse found last month was identified as a lawyer's. It lay half-covered by a black plastic garbage bag.[147]

The experiences my wife and I had in Brazil are totally subjective and profoundly emotional, as are those of this journalist. Yet they add conviction to the statistics on poverty and low income already given. A very unfair income distribution, especially in a country that is poor to begin with, is a recipe for disaster. It guarantees the growth of various social pathologies, death, disease, inhumanity, and a smorgasbord of vices. It is cemented through feelings of hopelessness and despair. A tortured anguish sets in when such desperate people are forced to watch rich tourists and *cariocas* live in what they see as opulence. While the children of the poor go hungry, a few blocks away the children of the rich cavort in Rio's nightclubs.

Things may be getting this desperate in the United States. Income has begun to skew more to the wealthy while the poor get even less. Chapter 5 shows how the middle class in America is eroding and documents the parallel rise in poverty. With an obvious bourgeoning of homelessness, climbing crime rates, and constant media hyping of "lifestyles of the rich and famous," it can well be asked if the Brazil model has already come home to roost in the United States. The point is certainly worthy of debate, although it is unlikely to be broached on the mainstream television networks owned by conglomerate corporations intent on preserving "the American way of life" for the rich.[148] It is certainly too early to ring a death knell of hoplessness. Yet it is time to sound the alarm bell. It is only by a stroke of fate that we were not born in the *favellas* of Rio, a hut in Ethiopia, or a hovel in Bangladesh. The station in life we start at is beyond our control. We still have some say, nevertheless, in how the economic pie is cut in our own society. Therein lies hope for the future.

4

MULTINATIONAL CORPORATIONS: INCOME INEQUALITY AND BASIC NEEDS

The evidence which points to large gaps in world income is impossible to deny. In terms of real dollars, poor countries are actually worse off in comparison to industrial countries today than just after World War II ended. This is also true when measured by relative wealth. Less Developed Countries (LDCs) have seen their economic fortunes plunge in the past few decades. How has all this come about? Why is it permitted to continue? How can politicians, decision-makers, and scholars who have witnessed this slide remain unconcerned?

One major reason is that such inequity is mistakenly viewed as progress by bankers, corporation heads, state leaders, and others who stand to greatly benefit by development and investment. Another persuasive reason comes from the realm of social science, which has been used to justify such obvious deprivations of the poor. In making global comparisons, economist Simon Kuznets noted that most nations tend to follow an inverse U-shaped curve. Countries will typically go from small income inequality at the start to greater inequality as they develop. Growth ends with much more equality during the final stage of evolution.[1] At this point, so the scenario goes, a greatly enriched middle class will share the wealth. The poor and the rich will be only small parts of the population.

There are a number of good reasons why we should expect income inequality to get worse in countries as they develop. Real growth is nearly always a result of getting rid of subsistence agriculture. The other route to more wealth is to bring in advanced industry. This growth of good-paying jobs and economic change takes place first in a nation's major cities. Thus, the already yawning chasm between rural and urban incomes becomes still larger. Even in the giant cities of LDCs, an underclass informal sector is born which serves as a classic surplus proletariat to keep wages low for much of the working class population.[2] There emerges a very modern, corporate group of well-educated executives and highly skilled labor who command high wages. Opposed to this well-off minority is a much larger traditional sector of unskilled service workers or low-paid factory assemblers. In this bizarre hodgepodge, it is not unusual to see street vendors barely surviving by selling single cigarettes to pedestrian traffic. Such life and death drama occurs in the shadow of monolithic, modern skyscrapers which house the headquarters of the new economy. Yet this will eventually change, the poor are told, as the productive power of their nation reaches a blastoff point of self-sustaining, continuous growth. At this point, an increase in education, unions, and productivity will lead to a more equal sharing of the new wealth among all citizens.

The evidence for this rosy picture does seem overwhelming—at least at first glance. Kuznets' idea led to a great deal of research that showed a curved relationship between economic development and income inequality. The first systematic study was done by the International Labour Organization (ILO).[3] Its analysis of fifty-six countries found that as gross domestic product (GDP) per person went up— especially among countries in the $300 to $500 range—income inequality also grew. When a fairly high level of income of GDP per person ($2,000 or more) was reached, inequality took a steep nose dive. The findings were the same whether income inequality was measured using Gini ratios or percent of income going to the top 5 percent, top 20 percent, or bottom 20 percent of recipients. Research by the World Bank followed close on the heels of the ILO study. The World Bank verified that LDCs show much greater income inequality than developed nations. Income inequality was substantially higher among the lowest- and middle-income groups in the sixty-six nations in this study, because of their smaller shares of income in comparison to that of the rich.[4] As GDP per person increased in countries from $250 to $500, the percent of income going to the poorest 40 percent of the population actually went down from 12 percent to 10 percent. Although this may not appear to involve much money, at close to subsistence levels the difference can mean life or death.

As the base of countries expanded to seventy-one and research became more refined, scholars were able to estimate a turning point of $243 (in 1971 dollars). At this point the Gini ratio peaks for developing countries. Thereafter, as GDP per person increases in countries, income inequality begins to fall.[5] Since the Gini ratio is an average number for the entire population, not all segments will necessarily become more equal at the same rate. The share of those at the top 5 percent of income actually starts to fall at the $200 level, while the share of the bottom 20 percent does not begin to increase until GDP per head reaches $500. A more detailed analysis by the ILO supports what seems to be an indestructible Kuznets curve. Gini ratios climbed from an average of .437 for countries with GDPs per person under $200 to .545 ($331–$700). Thereafter Gini scores decline to .487 ($700–$1,650) and bottom out among the developed countries at .381 ($1,651–$4,760).[6] Yet this study used more dated calculations of Gini ratios than desirable, as well as GDP figures for 1971. Table 4.1 contains the latest available figures from the World Bank. These results also confirm prior studies.

Virtually the same results uncovered in all prior research obtain in 1994 World Bank data. The industrialized, wealthier developed countries found in the richest group—with GDPs per person ranging from $10,000 to $35,000—are more equal in their income distributions than are LDCs. This is true no matter which of the three measurements is used. Core countries have the lowest household Gini score (.33), lowest percent of income going to the top 10 percent of households (25 percent), and second highest percent of income going to the poorest 20 percent of households (6.2 percent). The poorest countries with GDPs of under $500 per person actually show less inequality than the peak middle group of countries ($1,500 to $2,999 per person). This fourth group can be equated with developing nations. The differences between core countries in Group 6 and developing countries in Group 4 are both large and significant.[7]

The usual warnings of potential bias in the sample of countries and possible problems with representativeness are even more crucial in this instance. One study indicates that the relationship between income per head and income inequality is much less strong among LDCs. This is mostly because of poor quality data in comparison to data from devel-

TABLE 4.1 Household Income Distribution by Country Level: Gross Domestic Product per Person

Country GDP Level	Average GDP per Person	Average Household Gini Ratio	Average Percent Income Going to:	
			Lowest 20 Percent	Top 10 Percent
Under $500	$ 238	.394	6.43	33.09
$500–$999	718	.416	5.54	34.15
$1,000–$1,499	1,227	.416	5.79	34.38
$1,500–$2,999	2,170	.453	4.70	35.75
$3,000–$9,999	5,383	.411	5.92	34.65
$10,000 and over	20,556	.333	6.23	24.95

Source: World Bank, 1994, tables 1, 3, 30.

oped nations.[8] There can also be great instability where the so-called turning point toward greater income inequality for LDCs is reached. The size
in the amount of GDP per person, where the curve starts to turn and
reverse direction, depends upon the number of countries used in the
analysis. The amount has ranged in samples from $371 (forty countries)
to $468 (sixty countries) to $800 (thirty countries).[9] At least one study
involving sixty-seven countries has also raised the question of whether
there actually is a turnaround. It found, instead, a tendency for the
income share of the poorest 20 percent to steadily decline in the course of
economic development.[10] The eighteen nations in Group 4 (the most inegalitarian developing countries) represent countries with GDPs per person in the $1,500–$2,999 range. This boils down to only seven countries:
Botswana, Panama, Brazil, Poland, Costa Rica, Thailand, and Tunisia. Just
a few nations, then, are held up as representative of LDCs around the
globe. Because of the low number of nations involved plus the other flaws
just mentioned, major questions remain. Can we, without qualification,
regard the seven nations in Group 4 (where income distribution data is
available) as representative of developing countries?

The existence of the inverted-U curve between development and
income inequality at a *prima facie* level seems beyond question to most
scholars familiar with the area. The Kuznets curve is also shown to be cast
in concrete in two recent studies.[11] Both studies measured development as
energy consumption per capita, logged to the base 10, rather than as GDP
per head. Prior research has shown this to be an acceptable proxy with
which to chart a nation's progress. François Nielsen and Arthur Alderson
also used competing variables which would help explain the inverted-U
curve of income inequality obvious from the 279 observations in 88
countries during the 1952–1988 period, especially the presence of political democracy discussed in previous sections. In the end, they find that
the democracy variable is relatively unimportant in predicting the
inverted U, but that a higher secondary-school enrollment, a larger natural rate of population increase, a higher percent of the labor force in
agriculture, energy consumption per capita (curvilinear), and the presence of a Marxist-Leninist political regime are all predictive of lower
income inequality. Especially important in contributing to the growth of
inequality is sector dualism, or the sharp inequality existing between the
traditional agrarian sector and the emerging modern industrial sector of
manufacturing and services. This is measured by the absolute value of
$p - L$, where p is the percent of the labor force in agriculture and L is agriculture's share of GDP.

The study by Nielsen and Alderson marks an important advance in
research on development as it affects income inequality—theoretically,
conceptually, and empirically. As always, however, questions remain and

more research is needed. It would seem important to test the predictive efficacy of GDP per capita (used in much of the prior literature) as a substitute for energy consumption. The authors also admit that subsequent research must eventually employ Modern World System position as an important predictive variable as well. As discussed previously, a nation's debt has a great deal to do with retarding development and promoting inequality. Prior research shows it to be a potent predictive variable as well.

The World Bank data used to construct table 4.1 was also amenable to change analysis. For countries where the information was available, time comparisons in household Gini scores were calculated and constructed to yield a decade change rate. Change in GDPs for the 1980 to 1992 period was also measured. There is a large (.69) correlation between a nation's gross domestic product per person and a country's increase in its household Gini score. This is the usual cross-representational finding. But when the decade change rate in household Gini scores is correlated with the change in gross domestic products for countries, there is virtually no correlation. Indeed, if nations are simply split into high and low income on the basis of their GDPs (essentially, above and below $4,900 GDP per person), fascinating things begin to happen. Among the fourteen low income nations where data is available, there has been a 13 percent drop in income inequality as measured by the household Gini ratio. For the wealthier twelve nations, there has been an average decade increase in inequality of 6 percent. Statistical experts will quickly point out that the lack of an initial correlation does not mean a variable cannot be an important causal predictor in such techniques as regression analysis. But the absence of a correlation between a rise in GDP levels and a rise of income inequality over time should challenge the idea that greater unfairness in the distribution of income is the inevitable byproduct of development for poor countries. In the end, the "development → income inequality" question is destined to remain open for some time to come.

It is also true that social and economic reality is very complex. What appears real on the surface may be illusory when all other factors are considered. For instance, a common (and true) statistic shows a clear linkage between the birth of a great many babies and the noticeable presence of storks in given areas. Are we to conclude from these facts based upon valid data that storks bring babies? Common sense leads us to search for more realistic answers. The stork and baby example is a good illustration of what is called a spurious relationship. The correlation falls apart when other variables that can cause fertility are examined. One of the most important factors which encourages large families is living in rural areas. Storks also tend to flourish in rural areas. Ruralness therefore turns out to be an important variable causing increased childbearing when all other factors (e.g., the presence of storks) are held equal.

The stork illustration is an extreme example of a variable shown to be entirely spurious. On the surface storks appear to cause high birth rates, but in reality storks have nothing to do with fertility. Unfortunately, social and economic life is never this simple. A number of forces contribute to changing income inequality with widely varying impacts. Although questions have been raised about the human capital perspective, it does play a part in lowering income inequality. So does a country's growth in GDP per capita. So do, however, a number of other factors that are never remotely considered by traditional economists, bankers, financiers, and politicians. Most of these forces are unmentionables in the view of economic elites. Yet they have much to do with the reality of power and how money is spread around the world. We come back to a country's position in the modern world system. A nation's rank is of great importance to sociologisists and political scientists, if not to IMF and World Bank officials. What is the effect of this crucial variable? How does it compare to many of the other more conventional economic factors that act to reduce income inequality?

THE RISE OF MULTINATIONAL CORPORATIONS

World System supporters strongly argue that a nation's economic welfare does not hinge upon its balance of payments, exports, debt owed, type of industry, and so on. The crucial force which sets its income is its location in the global economic pecking order. The main idea of this theory is that periphery countries are forced into never-ending subservience. The root cause is an economic system of world commerce favoring core nations. The power of core nations allows them to dictate the terms of trade. LDCs pay more for the manufactured products from the core. Third World countries get less for raw materials that they sell to the core. The system works very nicely for core countries. If they are the buyers, they get to set the price. If they are the sellers, they get to set the price.

A deal is cut between the leaders of major firms in the core and business elites of the periphery. Its target is always labor in the periphery. This power bloc suppresses the cost of wages. While it keeps production costs down, it also brings about more income inequality. Local consumer markets and domestic industries are further eroded due to the presence of foreign multinational firms from core countries. These firms are successful at manipulating weak LDC governments into giving them a variety of gifts. Often these concessions take the form of tax reductions and start-up loans. Any desire of periphery countries to start their own industries is actively discouraged by foreign multinationals because these core firms would then have to compete for native capital and local markets. In the end, the core firms wish to keep alive the import dependency of these countries on their own products. As a result, independent local growth stagnates or declines in LDCs even while core multinationals open new factories in these countries. Nearly all major business and trade in the periphery is largely controlled by rules laid down through foreign corporate policy.

A hint at what frequently takes place in developing countries can be gained by studying Africa. In the Ivory Coast, for example, European managers and heads of large corporations earn more than twenty times the salary of Africans. In Swaziland the ratio is about twelve to one. Nearly three-fourths of all income inequality has been attributed to being white and European in the United Republic of Cameroon.[12] The presence of multinational staff in these foreign countries, then, can greatly distort the local salary structure. Wage scales are set in terms of the core country that serves as corporate headquarters. It is often believed that incoming firms create new jobs and raise the pay scale of laborers. Yet this is true only in the short run.[13] There is much evidence to show that dominance by multinationals in LDCs is a recipe for financial disaster in the long run.

The simplest definition of a multinational corporation would be a firm that owns and manages economic units in two or more countries. What this means is that direct foreign buyouts and management control take place across foreign boundaries. When the figures are examined, a great variation in size and degree of global spread rapidly becomes apparent. The size of these firms can vary, but the end result is always the same. The new global corporation is made up of a complex pattern of subsidiaries, branches, and affiliates. This became most apparent as multinationals bought foreign firms that themselves had foreign subsidiaries. Business abroad ends up involving many nations stretching over a wide range of products, services, and manufacturing processes.[14]

It is difficult to exaggerate the immensity and impact of today's multinational corporations. *Fortune* magazine compiles a list of the largest of these worldwide firms called the Global 500.[15] The combined revenue of these firms was over $10 trillion—an amount triple that of the entire U.S. debt and equivalent to a GDP half again as large as our nation's. More than a quarter of the world's economic activity now stems from only two hundred corporations, while approximately one-third of world trade takes place among different units of a single global company.[16] Oligopoly and monopoly at this level are common. The companies are so concentrated and powerful that they completely dominate economic activities in various sectors or in entire countries. Consider trade in manufactured goods, for example: a mere three hundred fifty companies control 40 percent of all buying and selling. The situation is even worse in commodity training. The largest six multinationals control 90 percent of the wheat trade, 75 percent of the oil trade, and 95 percent of the iron ore trade.[17] Such a concentration of power is awe-inspiring. It also helps to explain how companies from core nations are able to set the prices of their transactions with great ease—whether buying or selling—wherever they are.

As these corporations have become massive, they have also come to represent a wide gamut of interests. Thus, they can no longer be viewed as necessarily loyal to or affiliated with the home country of their headquarters. An increasingly complex degree of interpenetration between corporations has set in. This is particularly the case in core countries. Antimonopoly laws simply do not apply against foreign companies who decide to align themselves with domestic firms in the same fields. Goodyear Tire and Rubber has cooperated with its French competitor Michelin in the co-production of synthetic rubber while sharing technology with its Japanese competitor, Bridgestone. While U.S. Steel complained about unfair competition from abroad, it invested in partial ownership of eleven foreign companies operating on five continents. Their foreign buy-up came to include four mining and mineral processing businesses in South Africa. Even as early as 1970, General Electric became the single largest owner of stock in Toshiba electronics. By that year, GE had licensing agreements with over sixty Japanese companies, while Westinghouse had become the principal shareholder of Mitsubishi.[18]

The growth in size and dominance of the 180 largest U.S. multinationals shows that in 1950 over three-fourths of them operated in six countries or fewer while none owned businesses in twenty or more countries. Only two decades later, fewer than 7 percent were involved with only six countries and nearly one-third owned subsidiaries in twenty or more countries.[19] The growth of American multinational firms worldwide has been astronomical:

> From an accumulated direct investment of only $11.8 billion in 1950, the book value of American direct investment abroad had risen to approximately $234.4 billion by 1984. . . . In 1981, American foreign direct investment was more than two-fifths of the world's total foreign direct investment . . . the largest fraction of postwar investment went into advanced manufacturing industries (particularly automobiles, chemicals, and electronics). . . . By the early 1970s . . . international production by American multinational corporations had surpassed trade as the main component of America's international economic exchange. Foreign production by the affiliates of U.S. corporations had grown nearly four times as large as American exports. . . . By 1969, the American multinationals alone produced approximately $140 billion worth of goods, more than any national economy except those of the United States and the Soviet Union. Many of America's largest corporations had placed more than half of their total assets abroad, and more than half of their total earnings came from overseas.

. . . Although the rate of growth of foreign investment declined by the 1980s, the United States remained heavily dependent on its multinationals for access to foreign markets and for the earnings they produce.[20]

At the start of 1994, the market value of U.S. private direct investment abroad stood at just under $1 trillion ($993.2 billion). This was nearly four-and-one-half times the 1982 total.[21] Direct foreign investment by U.S. firms increased sixteen times within thirty years, but the rate of increase for America's gross domestic product was only half as much. Incredibly, the pace of expansion picked up even more momentum in the 1990s. According to the United Nations 1994 World Investment Report, investment flowing to the developing world doubled to $80 billion within only a few years.[22] In 1994, outflows from U.S. multinationals hit a record $50 billion in foreign investment, with one-fourth of that going to LDCs. One major factor reinforcing the rush to expand overseas is a lucrative profit margin. The share of net income from foreign sources has almost doubled over the past decade for U.S. multinationals—climbing from 26 percent to 45 percent.[23]

By the mid-1990s, 81 percent of all U.S. exports were affiliated with multinational corporations while 58 percent of our imports came through these businesses. Of all U.S. exports, nearly one-fourth consist of trade between U.S. parents and their foreign affiliates (17 percent of all imports).[24] What this means in reality is that nearly three-fourths of the manufacturing cost for an IBM PC computer is spent overseas. Over 27 percent goes to branch plants of American multinationals while nearly half is spent in direct buying from foreign firms (46 percent).[25] The manufacturing of Ford's highly successful Escort involves parts assembled in sixteen countries on three continents. It translates to seven General Motors and two Chrysler plants that were opened between 1978 and 1982 along the northern border of Mexico.

A survey of the two hundred largest U.S. multinationals has shown them earning one-third of their revenues through off-shore subsidiaries.[26] In 1993 alone, America's one hundred largest multinationals earned a total of $703 billion in foreign sales.[27] The amount of profit can vary somewhat from one firm to another, but it is always large. Table 4.2 contains the most recent data for the forty largest American multinational corporations, ranked by absolute amount of foreign profit earned in 1994. To minimally qualify for the top one hundred, a firm has to earn over $1 billion per year from foreign sources. Even such Americanized corporations as McDonald's have increased their sales by foreign expansion. McDonald's (forty-eighth largest multinational) now operates in eighty foreign countries that generate half of its total business volume.[28] Seventeen of the top forty U.S. multi-national corporations earn over

TABLE 4.2   Revenue and Assets of the Forty Largest U.S. Multinational
               Corporations: 1994

Company	Foreign Revenue (Millions of Dollars)	Foreign Revenue as Percent of Total	Foreign Assets as Percent of Total
Exxon	$77,125	77.4	57.9
General Motors	44,041	28.4	25.6
Mobil	40,318	67.6	58.8
IBM	39,934	62.3	57.0
Ford Motor	38,075	29.6	27.6
Texaco	24,760	55.9	43.4
Citicorp	19,703	62.3	54.4
Chevron	16,533	42.9	42.0
Philip Morris	16,329	30.4	34.2
Proctor & Gamble	15,650	51.7	41.9
E.I. du Pont	14,322	42.1	38.2
Hewlett-Packard	13,522	54.1	46.2
General Electric	11,872	19.8	17.4
American Intl Group	11,636	51.8	37.2
CocaCola	11,408	68.3	49.9
Dow Chemical	10,073	50.3	44.2
Motorola	9,770	43.9	33.9
Xerox	9,678	47.8	31.3
United Technologies	8,300	39.2	30.4
Digital Equipment	8,274	61.5	57.1
PepsiCo	8,226	28.9	30.6
Johnson & Johnson	7,922	50.3	42.0
ITT	7,785	33.0	9.2
Minn Mining & Mfg	7,568	50.2	40.6
AT&T	7,325	9.8	11.8
Eastman Kodak	7,123	52.5	39.3
Chrysler	6,569	12.6	13.7
Amoco	5,857	21.7	29.1
Sara Lee	5,787	37.2	49.3
Intel	5,695	49.4	21.3
JP Morgan & Co	5,430	45.6	49.5
Compaq Computer	5,393	49.6	46.5
Colgate-Palmolive	5,188	68.4	49.9
Goodyear Tire	5,158	42.0	38.7
Aflac	5,149	84.3	90.3
Bristol-Myers Squibb	5,005	41.8	33.0
UAL	4,920	35.3	NA
Merrill Lynch	4,900	26.9	36.6
CPC International	4,780	64.4	68.0
RJR-Nabisco	4,750	30.9	15.6

Source: "The 100 Largest U.S. Multinationals," *Forbes,* July 17, 1995, pp. 274–76.

one-half of all their revenues from overseas sales. Among them are: Exxon (77 percent), Citicorp (62 percent), Digital Eqipment (62 percent), Mobil (68 percent), Colgate-Palmolive (68 percent), Coca-Cola (68 percent), and IBM (62 percent). CocaCola, of course, has long prided itself on advertising its product as a forerunner of U.S. presence overseas. It has become the international icon of Americanism abroad.

We have seen that much of the rise of American multinationals has been due to the unscathed condition of the United States after World War II. Having the dollar as the medium of exchange also helped to fuel U.S. expansion in other lands. There are many added factors, however, which led to the voracious growth of overseas corporations. These had less to do with our enviable position and more to do with the nature of free-market capitalism. To begin with, revolutionary developments in communication, travel, and computer technology created a new and unique possibility. For the first time ever, separate parts of actual production could be easily linked around the globe. Information has now been more centralized while control is less concentrated. Privatized systems—such as Satellite Business Systems owned by IBM, Comsat, and Aetna Life Insurance—are used heavily by corporations. Such sophisticated communication hardware allows firms to effectively bypass any dependence on governmental agencies. Jet travel and electronic media, which allow contact to be made in an instant, have smashed barriers to world commerce.

Among other reasons is the explosive growth of manufacturing technology and the stampede it created in the rapid development of products. Core countries have long been involved in the costly process of research to create and develop new products for the market. During the initial stage of development, corporations enjoy a monopoly over these new technologies, and they use it to raise their profits. Over time, the new inventions and manufacturing technology spread to other countries. Rising trade barriers and the growth of foreign competition make it necessary to locate production in foreign lands. Saturation of the core market occurs. A search to locate and expand new markets assumes more importance as the product becomes commonplace at home. Since many global firms dominate markets with only a few others (oligopoly), their competitors also tend to locate branches and subsidiaries in countries where one company pioneers. Expansion in given countries then tends to produce a follow-the-leader mentality as each firm tries to stabilize and protect its share of profits. Production is by now standardized to the point where manufacturing can be done anywhere, often resulting in the location of plants where costs are cheapest.

Location of plants in LDCs may also be caused by threats from periphery governments to erect trade barriers to imports from core countries. As part of the cost of doing business in some of these lands, firms have been forced to guarantee that large portions of assembly and manufacturing will be kept in the host country. This is a way to preserve jobs and to protect the economy. Avoidance of trade tariffs and protected markets in the host country can also have a lot to do with the decision of a firm to open a branch in a given nation.

Much expansion of transnational firms can also be laid at the feet of new types of financial groups. A stateless capital has been created by the flood of dollars abroad (Eurocurrency). This is now being made available to fund risky mergers or questionable development schemes that traditional and more conservative lending agencies have refused to bankroll.[29] Yet more stable banks have not been entirely immune to risky ventures in the Third World. Banks have also entered into races with each other for the profits that can be made from expansion loans. Huge consortia of banks made up of members from many different countries evolved in the 1970s and 1980s to amass the gigantic sums needed by the multinationals.[30] In short, a global banking network was created with almost no constraints. No protective shields exist to prevent harm to nations or individuals doing business with these cartels. Capital now moves from nation to nation at the blink of an eyelash—or more accurately, at the click of a computer key. Frequently, in this high level game of financial lending, LDCs which seek development funds actually find themselves in competition for scarce capital with multinational firms intent on growth through merger.[31] It is almost impossible to overestimate the financial clout of these business leviathans. In 1993, for example, General Motors led all of the world's multinational corporations with revenue of $133.6 billion.[32] In the preceding year, only 26 of 136 countries had a gross domestic product larger than this. If you were a world banker and were presented with numbers like these, would you lend to General Motors or to Guatemala?

There are a number of drawbacks to the ease with which capital is transferred around the world, not the least of which is the increasing power it gives multinational corporations over workers and governments both at home and abroad. Since capital has become so globalized, it is sometimes impossible even to define the national origin of various financial deals. Since investors are so mobile, they put constraints on governments that may wish to loosen monetary policy by lowering interest rates or expanding credit. If investors disapprove of a government's policies, they can pull up stakes and be gone overnight. Such capital flight can also lower the value of a nation's currency, which fuels inflation and cuts back on business expansion. Canada learned this lesson the hard way when its central bank attempted to stimulate its economy with a modest interest rate cut. In a matter of a few days, speculators sold off enough Canadian dollars to force a three cent drop in its value. This in turn caused an interest rate reversal by the bank that was so sharp it threw the country into a recession.[33]

One of the most destructive trends has been the rise in currency speculation. The world's currency market—where an estimated trillion dollars changes hands every working day—has been described as a "cacophonous, high-tech global bazaar" that is bigger than all of the world's stock, bond, and commodities markets combined.[34] The scary part is that no one is at the helm. Most market experts admit much of the activity is driven by frenzied speculation in efforts to turn a quick profit via currency exchange. As money gushes from one current hot investment to another, it produces a bandwagon effect that multiplies, rocking the boat of high finance. Can it sink once and for all? The Clinton Administration thinks so. Mexico went through a profound devaluation of its peso that may have resulted in a financial free fall and possible crash for investor nations, had not the President intervened at the eleventh hour with billions of dollars of pledged securities from the United States.[35] U.S. officials have held talks with Federal Reserve administrators, top Wall Street financiers, and G-7 representatives at recent meetings in Halifax—but to no avail. While everyone shares in the worry that such hot money could create a crisis of the first magnitude, they are unsure that anything can be done to reduce the risks.[36] For the foreseeable future, the global money market will remain in peril.

U.S. military and economic policy has been used to greatly encourage the overseas expansion of multinationals as well. The foreign presence of American firms is seen both as a demonstration and an alternative development model to communist and nonaligned LDCs. Sometimes this has led to the use of covert and occasionally ugly activities in the host country.

> Whatever their manifest military purposes, American troops, military advisors, offshore cruising naval vessels, strategic long-range bombers, and, eventually, long-range ballistic missles all helped at least indirectly to extend American interests abroad. During these years, the U.S. government made commitments to a whole network of antidemocratic dictatorships, whose leaders seemed dedicated, along with keeping themselves in power locally, to promoting the entry of American business enterprise into their economies. In this context, many of America's new Third World allies—South Korea, Taiwan, Brazil, and Argentina—courted U.S. firms by offering terms that were unbelievably tempting, especially low wages and prohibitions on free union activity. The modern equivalent of U.S. gunboats—much more subtle covert intelligence operations and secret funding of military and paramilitary operations—protected these regimes from internal dissent and external attack, while U.S. diplomats averted their eyes from the official government terrorist campaigns to crush free trade unions and, for that matter, free democratic popular elections of any kind.[37]

Economists Barry Bluestone and Bennett Harrison see a variety of other ways in which American corporate foreign investment has been expanded.[38] *U.S. tax and tariff policies* clearly fed such an expansion of firms abroad. American corporations can deduct the full dollar value of all of their foreign income taxes against the domestic taxes they owe here. The normal way to deduct costs of doing business is from earned revenue. These taxes also do not have to be paid in the United States until the profit from foreign branches actually comes into our country. Rather than repatriate, firms simply use profits from their subsidiaries for further foreign expansion in a steamroller effect. *Transfer pricing* also makes foreign expansion quite lucrative. This is the practice of overpricing products made by American companies and sold to their foreign subsidiaries. It reduces the profits of the subsidiaries, and thus the taxes due on them in the United States. The opposite pattern is also used for host countries with high tax rates: the subsidiary then overcharges the U.S. parent company. According to some estimates, the bottom line for all these incentives is a real tax rate of 5 percent on foreign earnings. Lastly, subsidies for foreign expansion are available through cuts in tariffs. The slashes are made by calculating import value on only a portion of manufacturing costs. This allows U.S. firms to ship components abroad for assembly in LDCs where labor is cheap. The goods are then imported back to our country. The tariffs are then due only for the value added by the foreign subsidiary. The industries deeply involved in this process manufacture a compendium of products that once came from the American industrial heartland. The import roster includes autos, stereos, televisions, computer chips—even capital goods such as machinery used in textile manufacturing.

To make matters worse, there is often *outright collusion* between our federal government and the multinationals to assist in plant relocations to foreign lands, which result in job losses at home. Throughout the 1980s and well into the 1990s, the U.S. Agency for International Development (AID) funded a program to persuade factories throughout the United States to relocate in low-wage havens in Central America or the Caribbean.[39] The avowed purpose was to create new private investment in these areas through special export processing zones. In only two years at the start of the 1990s, multinationals with plants in El Salvador, Guatemala, and Honduras closed fifty-eight plants in the United States and left twelve thousand American workers jobless. The layoffs also had a multiplier effect which further depressed wages, especially in the apparel industry. In El Salvador alone, AID spent $32 million to enable the construction of 129 new factory buildings in new industrial parks. U.S. multinationals that relocate to, or open a plant in, El Salvador can get a 50 percent subsidy to defray costs of training workers, as well as a line of credit to develop exports. In return, El Salvador has put in place a foreign investment package that includes a 100 percent exemption from import and export duties, with no corporate income, dividend, or equity taxes.

Cheap labor and low wages may be the major reason so many firms have gone the multinational route. Much of this outsourcing has been to Japan, Germany, Canada, Mexico, Malaysia, Singapore, the Philippines, Korea, Taiwan, Hong Kong, and Haiti (in descending order). Nearly all countries involved are LDCs which have actively sought the entry of American, Japanese, and European corporations. Their avowed goal is always to jump start their economies. They see the entry of such firms as a means to improve wages and provide jobs. They believe this will enlarge their modern sector and improve the overall standard of living. Most of the newly industrialized economies, for example, have created Export Processing Zones (EPZs), areas specifically set aside by the host country to build new branch plants for multinational firms. The zones typically include state-of-the-art infrastructure (seaports, roads, airports, generating plants, etc.). They also allow companies that locate within them large cuts or a complete holiday from taxes, weak or nonexistent pollution controls, and a docile, nonunionized labor force. The first such zone was established near Shannon Airport in Ireland in 1967. By 1980 there were more than eighty such zones in the Third World.[40]

In Mexico, this approach is called the Border Industrialization Program or the *maquiladora* plan. A 12.5 mile strip along that country's northern border has been set up to help U.S. firms who are looking for cheap labor in a virtual tax- and tariff-free environment. Seventy-two American plants set up shop within the first two years of its inception in 1969, and by 1974 the number had risen to 655.[41] Not all production is low-tech or simple. By the end of 1987, General Motors had twenty-three plants operating inside Mexico, most within the *maquiladora* region. Ford's Hermosillo plant, which contains state-of-the-art production equipment, was built here in the latter 1980s.[42] By the start of the 1990s, shops in this zone—mostly owned by U.S. Fortune 500 firms—employed nearly a half million workers.[43] The region's two thousand-plus plants now account for a full tenth of the Mexican labor force, and make up that nation's second most important source of foreign currency earnings (led only by crude oil exports).[44]

Almost 70 percent of *maquiladora* workers are women, engaged mostly in electronics fabrication, garment assembly, and other light manufacturing. From the beginning, *maquiladora* plants have targeted women as their preferred employees. Stereotyped notions such as women's greater manual ability and notions of domesticity serve as rationales for the gender bias. But above all, managers prefer young women because of their willingness to conform; bosses report they simply have fewer problems with young women than with men.[45] Docility becomes an especially important trait to these companies. Most of the *maquiladora* jobs do not offer living wages that can sustain a small family, despite the fact that these women workers are more often than not the primary breadwinners for their families.[46] In El Salvador, despite working grueling twelve-hour days in the prisonlike atmosphere of factories surrounded by barbed wire and patrolled by armed guards, these women usually do not earn enough even to feed their children:

> What I was finding from interviews and meetings with dozens of workers was that the pittance they earned in the plants was not even enough to supply adequate food for themselves and their children. Infants and toddlers, for example, are commonly fed rice water or coffee instead of milk because milk costs too much. The woman in the car spoke through an interpreter. She said she was 19 and had a 3-year-old daughter. I asked if the child had enough to eat. "Oh no," she said matter-of-factly. "We are very poor." I asked if her daughter drank milk. "No," she said. "We can't afford it. We give her coffee." She said her daughter might have an egg for breakfast and boiled or fried beans for dinner. That's it. A meal with meat or vegetables was extremely rare. "It afflicts me greatly," the woman said, "because it is necessary for her to have a good diet. My daughter is very thin and also weak. Sometimes she falls down. . . . When asked about her own diet, a resigned look crossed the young woman's face. "There is not much food," she said. "My head hurts and sometimes I feel dizzy. I suffer that."[47]

The woman interviewed in the above passage works in the Doall plant, which makes jackets for Liz Claiborne. While these jackets sell in the United States for $178, the garment workers who make them earn about $.77 per jacket and are paid $.56 an hour. There is a grim reality behind the glitz and glamour of fashion.

The building of Ford's Hermosillo plant in the Mexican *maquiladora* region also highlights an entirely separate problem for the core firms. They fret over escaping high technology, which eventually erodes their business competitiveness. Yet little doubt exists that "99 percent of all inputs in these assembly plants are imported, making them little more than sweatshops exploiting cheap labor."[48] The worldwide gaps in typical manufacturing wage levels have guaranteed global expansion due to all of the conditions listed above. Chief executives of transnational firms, salivating after greater and greater profits, have not hesitated to pull up stakes. They are eager and willing to move entire plants anywhere the job can be done more cheaply as long as there is a stable political environment. In 1992, for every $1.00 of average hourly wage earned by U.S. workers in manufacturing, workers in Taiwan earned $.32; in Singapore, $.31; in South Korea, $.30; in Hong Kong, $.24; in Mexico ,$.15; and in Sri Lanka, $.02.[49] Corporations have not ignored the huge profits inherent in such subsistence level wages.

Nor have these trends been constrained only to blue-collar assembly work. The global factory is now being joined by the global office as firms again take every opportunity to exploit cheaper wage scales abroad. It was once argued that service and office jobs would be created to take the place of manufacturing jobs lost to foreign countries. Today the signs indicate that global corporations make every conceivable effort to shave labor costs to the bone, even for allegedly sacrosanct white-collar jobs which have until now been viewed as "safe." New York Life Insurance has set up a claims-processing center in rural Ireland where there is a large pool of educated young people who need jobs. They are willing to work for wages lower than those in the United States. In the operation, claims are sent directly to a post office box at Kennedy International Airport in New York City. They are flown to Ireland each day, processed, and then sent back via a leased transatlantic line to the company's mainframe computer in New Jersey. From there, the claim is settled almost immediately by simply issuing a check to the beneficiary.[50] Another firm (Saztec International) employs three hundred people at a data-entry branch which has been operating in the Philippines for over ten years. During the 1980s, American Airlines moved most of its keyboard entry processing to Barbados and the Dominican Republic. Its spin-off company, Caribbean Data Services, employs over one thousand workers at a wage scale starkly lower than that prevailing in the United States.

Some of examples of job transfer are staggering. A *Business Week* article reports that International Data Solutions scans case and client files for U.S. law firms, transmitting them via satellite to the Philippines, and then has workers in this country organize and index the documents into a computer retrieval network.[51] The whole operation employs two people in Virginia and three thousand Filipinos. Yet clerical and data entry jobs are not the only ones being exported to Third World countries. It was once thought that the globalization outsourcing process would leave high-tech work alone. It was assumed that the more demanding, professionally skilled and well-educated workforce would grow rapidly in the United States, along with higher pay, thereby replacing these low-end clerical jobs. The same *Business Week* report cites a great deal of evidence to indicate that professional jobs may end up disappearing as fast as blue-collar and clerical jobs did in past decades:

- While a good computer circuit-board designer in California may earn $60,000 to $100,000 per year, Taiwan is glutted with similar talent paid $25,000 per year (in India or China, Ph.D.s in engineering command only $10,000 per year).
- The thirty professional employees at Bilingual Educational Computing in Beijing, China, can produce a new CD-ROM computer

title for one-tenth to one-fourth of the U.S. cost because their key-punchers earn $75 per month and computer artists make $400.

- Five years ago, most mother boards for personal computers were designed and produced in the United States; today, 60 percent are subcontracted to Taiwanese companies, which employ 150,000 information technology engineers.
- The Scriptor pager offered by Motorola was developed almost entirely in its Singapore Innovation Center by native industrial designers using Singaporean software.
- India has the world's second largest number of English-speaking sci-entists in the world (including 100,000 software engineers), so multinationals in India can employ an experienced engineer for $800 per month.

There is a growing tendency of multinational corporations to farm out technical tasks to the intelligentsia of poor nations, thereby eliminat-ing many high-wage professional jobs in America. At times, this policy is pursued vigorously and with great purpose. Ford Motor Company, for example, has adopted a plan called Ford 2000 in an attempt to make the world's second largest industrial corporation a company that knows no boundaries.[52] It is explicitly trying to remove centralized management control while distributing authority around the globe. Thus, the members of its design team span three continents—from Dearborn, Michigan, to Britain and Australia. Teams work on car designs by using video screens, three-dimensional simulation techniques, and a supercomputer.

More insidious is paying educated locals in foreign lands to do what well-trained professionals used to do in this country:

> India is fertile ground for these companies because it inherited a strong English-language school system from the days of British rule and has emphasized mathematical education since then. Yet, computer scientists trained at Indian universities come relatively cheap. Experienced program-mers command salaries of $1,200 to $1,500 a month, compared with $4,000 to $6,000—or even $10,000 for stars—in the United States. The result has been explosive growth in the number of Indians working on com-puter programs, mainly for the American market, from several thousand in the early 1980s to nearly 75,000 today.[53]

U.S. multinationals are not the only villains in this drama of labor exploitation, especially in view of what has happened in the past decade. Japanese and European firms have become much more active in foreign expansion. This is true both in core countries such as the United States and in LDCs. Although Japan still accounts for only 7 percent of total world direct foreign investment, it has strongly increased its overseas presence. During the 1980s, it reinvested the explosion of profits from its large trade surpluses. Much of its buying has been for manufacturing plants, businesses, and real estate in the United States. It has also spent vast amounts of money in developing countries, however, particularly in the Pacific Basin. Although Americans are in the habit of considering South Korea one of our nation's pet economic development projects, Japan is the largest direct investor in that country. It also owns the majority of foreign firms in Thailand, Malaysia, Indonesia, and Iran.[54]

Toward the end of the 1980s, total Japanese business investment in Asia more than doubled over the span of two short years, rising to nearly $5 billion per year.[55] There are now more than three thousand Japanese companies operating in the Pacific. In Thailand alone, 77 percent of incoming foreign investment funding stems from the Land of the Rising Sun. This translates to the start-up of one company per day. The "Four Tigers" of South Korea, Singapore, Hong Kong, and Taiwan credit Japanese investment as a major reason for their success. At the same time, Japan has surpassed the United States as the foremost trading partner in East Asia. China, South Korea, Indonesia, Thailand, Malaysia, and Singapore now import more from Japan than from any other country.

Lest there be any lingering doubt that Japan has become the economic king of the hill, at least in the Far East, that nation has now become the largest provider of development assistance to most of Asia. Over two-thirds of its foreign aid—$3.5 billion per year—now goes to Asian nations. Japan has also pledged $50 billion in public and private funds to developing countries through the World Bank, the IMF, and the Asian Development Fund. By the start of the 1990s, Japan's Official Development Assistance approached $8 billion annually. Not surprisingly, Japan has become the second largest provider of Official Development Assistance in the world.[56]

Much of the reason for Japanese expansion is the search for wages lower than those in its own country. The major point is that the expansion of multinationals stems from only a handful of advanced core countries and shows no sign of ending. In fact, there is every indication that the trend toward opening branch plants in LDCs is likely to rise in the future as the global-office movement gains momentum. Completing the circle of expansion, ironically, are such successful newly industrial countries as South Korea that are also looking for opportunities to raise their own world presence. It is the developing global work force that is the very center of the new world economy. The economies of core nations are the nexuses that mainly administer, finance, and develop business deals. Branch plants are then set up to manufacture goods in peripheral countries for export to the world market. At the pinnacle of the new coterie marching forth from developed core countries is the multinational corporation.

These firms, above all else, have provided the vehicle which was needed to achieve the new supranational economy. During the last decade, while corporations invested within America's borders at the rate of 12 percent a year, the comparable rate of foreign expansion by U.S. firms was 19 percent—over one-and-a-half times as much.[57] The magnitude and extent of multinational corporate dominance of LDCs is bound to be very pervasive—for better or worse.

MULTINATIONAL PENETRATION AND INCOME INEQUALITY: DO PERIPHERY COUNTRIES BENEFIT?

The search for lower wages on the part of multinationals has been described as exploitation of labor by many critics, especially union members in core countries. Yet it is only fair to ask whether these nations actually benefit from the presence of subsidiaries. Transnational firms are among the first to claim that they help their host countries by introducing jobs. They say their plants greatly raise the local wage scale. Spin-off business demand opens up new opportunities for local firms to feed into the needs of these branch plants. Thus, how can scholars and critics from either developing or core countries really be skeptical about whether this expansion is good for nations that are so obviously poor to begin with? It would seem that for a host country to question the benefits of multinational activity reeks of ingratitude, of biting the hand that feeds them.

Defenders of multinationals claim that their expansion is mankind's salvation. They feel such firms have a Midas-like ability to bring riches to all lands they touch. Such firms always seem to bring the latest technology, modernity, economic growth, and a better life to LDCs. Some believe they are the hope of peace in the future. Many point out that these firms have a large number of intersecting allegiances. They are thus likely to exert pressure on nations to settle their differences without war. At the least, they increase efficiency and lower production costs by locating plants in areas close to resources and markets. Multinationals are frequently portrayed as evil, all-powerful manipulators of weaker, victimized nations. Yet this may be far from the truth. LDC governments have seized corporate property. Revolutions and nationalistic movements with hostile overtones do take place in the real world. Many host countries now insist, as a condition of entry, that multinational firms must manufacture a certain percent of a product's worth inside the host's borders. Tax privileges for multinationals have been harmfully altered or revoked once expensive plants are built. Domestic industries are frequently protected from the ravages of imports from core countries. Some multinationals report considerable local pressure to share the latest productive technology. This can allow industries in the host country to acquire knowledge and techniques for which core firms have spent a great deal of money in research and development. Hence, the agreement to open shop in some countries can end up cutting the throats of transnationals by stealing their competitive edge.[58]

Most of the claims that multinational rule is beneficent or benign are open to question. These assertions cannot be answered here for lack of time and space. We *can* ask, however, whether the presence of such firms in LDCs actually does cause economic growth—and for whom. The view that real income tends to increase because branch plants are set up in these lands is debatable. It may actually enrich a select few, who seem always to be the economic/political/military ruling oligarchy. Yet it fails to benefit the masses, who instead remain in desperate poverty.

This becomes clear when we speak to the very basic human need of getting enough to eat. Frances Moore Lappé and Joseph Collins point out that export-driven development plans in Third World countries have gone hand in hand with worsening hunger and outright starvation.[59] Brazil again provides a good example. Its attempt to boost agribusiness exports led it to win second place among world agricultural exporters. Ironically, as food and soybeans have been shipped abroad in ever greater amounts, the percentage of hungry Brazilians also grew. Those who are undernourished went from one-third to two-thirds of the entire population within the space of two decades. While beef exports from Central America have increased sixfold in the past twenty years, nearly three-fourths of Salvadoran infants are underfed. While export earnings from Mexico increased twelvefold between 1970 and 1980, the portion of imports devoted to food went down from 12 percent to 9 percent. The authors report that such huge multinationals as Dole and Del Monte virtually abandoned Hawaii as a source of canned pineapple production because field workers had successfully organized and had managed to raise their minimum wage to $3.25 an hour. After pulling up stakes, these firms relocated in the Philippines where labor was available at a more profitable level of $1 a day. These below-subsistence wages are damaging to workers who are now also exposed to harmful pesticides. More land that was once used to grow food for local peasants is removed from production. Many reclaimed acres have often been forcibly taken from the poor, who are beaten and occasionally killed. At the same time, the Philippine government has offered land-lease inducements to foreign agribusiness monoliths at $18 per acre in order to attract their business.

In essence, a worldwide, profit-driven commercialization of agriculture has taken place in the last half century. Accompanying this trend, to no one's surprise, has been a rapid retreat from susbsistence farming. This in itself would not necessarily be a negative development if enough food were grown and if it could be distributed more efficiently to everyone.

Such is not the case. Multinational agribusiness firms, mostly from the United States and Europe and with the connivance of their governments, have erected a complex web of dependency in food production and consumption. Harriet Friedmann identifies three such complexes: wheat, durable foods, and livestock.[60] The growth of the wheat complex, dominated by only a few global corporations, has facilitated import dependency in the Third World through U.S. government subsidized trade with less developed countries. Despite the fact that many countries grew enough to feed themselves, they began to import wheat in the 1950s because of rock-bottom pricing. They eventually became hooked just before prices skyrocketed during the 1970s. Thus, poor nations ended up importing wheat that carried a per-ton price tag one-fourth higher than maize, six times higher than rice, and five times higher than petroleum.[61] The durable food complex emerged as transnational corporations deliberately introduced chemical and biological substitutes for such Third World exports as sugar and peanut, palm, and coconut oils. The impact on poor countries has been devastating: a 1 percent drop in the price of sugar, for example, results in a one-third export-earnings decline in six countries. Lastly, multinational firms expanded livestock production into Central and South America, typically by burning down rain forests and evicting poor *campesinos* from landholdings where they had at least been able to feed themselves.

In a study of eighty-two nations, Krahn and Gartrell discovered that the effect of world trade participation among poor nations is both large and damaging.[62] Simply put, in less developed agrarian nations participation in world trade (as a percentage of GDP) is the strongest predictor of income inequality. Fully 85 percent of the profit from exports ends up in the hands of multinationals, bankers, traders, distributors, and various stockholders rather than staying in the hands of the people in the host country. World prices for food and mineral commodities have fallen steadily, recently reaching their worst level ever in comparison to costs of imported manufactured goods. Real commodity prices toward the end of the 1980s were one-third below the average for prices in the beginning of the decade.[63] In Latin America, the terms of trade declined by over 28 percent from 1981 to 1992. At that point, they stood roughly equal to where they had been during the Great Depression in the 1930s.[64] Since prices have deteriorated so badly, *many LDCs are now punished for their very success.* A large number of Third World countries have ended up exporting more while actually getting less money in return. Much of this is due to the fact that certain multinational firms in the world market have a monopoly which permits them to set prices. Such corporations often start price wars among LDC nations competing for multinational purchasing dollars. In the end, according to the founders of the Institute for

Food and Development Policy (Food First), we cannot assume that exports are the answer. It is a very real question whether world trade can generate and distribute enough income to alleviate hunger and poverty:

> Simply put, third world exporters find themselves in a buyer's market. They lack bargaining power. Over half of the countries in the Third World obtain more than 50 percent of their export earnings from just one or two crops or minerals. When prices fall, many countries have no alternative source of foreign exchange earnings; they cannot hold out for better prices. In fact, they feel even more compelled to step up exports. . . . Of course, this response undercuts prices still further. . . . In most Third World societies, the poor are hurt by export-oriented agriculture:
>
> - It allows local economic elites to ignore the poverty all around them that limits the buying power of local people. By exporting to buyers in higher paying markets abroad, they can profit anyway.
> - It provides incentive to both local and foreign elites to increase their dominion over third world agriculture and fuels their determination to resist economic and social reforms that might shift production away from exports.
> - It mandates subsistence wages and miserable working conditions. Third World countries compete effectively in international markets only by crushing labor organizing and exploiting workers, especially women and children.
> - It throws the poor majority in third world countries into competition with foreign consumers for the products of their own land, thus making local staple foods more scarce and more costly.[65]

Third world workers are not completely helpless against multinationals, and occasional resistance to their plans does arise. In India, for example fifty thousand farmers who are members of the Karntaka State Farmers Union have organized a nationwide campaign to stop transnational corporations from entering Indian agriculture.[66] The farmers are especially concerned with regulations in the General Agreement on Tariffs and Trade (GATT) and with encroachments by Cargill—a Minneapolis-based corporation that is the world's largest grain producer, trader, and processor. Cargill is attempting to gain patent control over seeds developed in India. This would deprive farmers of their right to reproduce, save, and distribute seeds freely among themselves as they now do. The development by Cargill of the new seeds and their extension to a seller's market in India are very questionable. To begin with, huge profits will accrue to Cargill if it reaches its stated goal of capturing 25 percent of the Indian oil seed market. Unlike current seeds, which produce seeds for future years, Cargill seeds can be used for only one season and must be bought annually. Crops from these seeds require much more artificial, chemical input (spraying and fertilizer) and greater quantities of water. This further necessitates irrigation, frequently requiring expensive equipment and dams. In short, the scheme for the Cargill agricultural infrastructure is both ecologically harmful and too expensive for poor farmers, who will end up being forced from their lands.

Although mainly sympathetic to multinationals, Theodore Moran points out that it is those firms like Cargill—which have only a few competitors (oligopoly)—that are most likely to locate in foreign lands. This allows them to dictate prices, set wages, and keep down operating costs with impunity in the host country.[67] Moreover, research has shown that these firms do not typically bring in much outside capital. When setting up shop in foreign lands, they instead eat up local money through LDC loans and other concessions used to attract them. By such means, they may end up completely dominating the economy of the host country. The very term "banana republic" acknowledges the role of economic supremacy played by such firms as the United Fruit Company in Central America. The end result can create a small labor elite while driving the bulk of workers into the ranks of the unemployed. At the same time, profits are siphoned off from the host country for repatriation to core firms. Evidence has shown that an entrenched habit of foreign firms is to avoid purchasing inputs from host country sources. Such buying could generate local industry and cause economic growth. Yet multinationals, to no one's surprise, continue to buy goods from their own branches in other lands.

Multinationals have often run behind-the-scene schemes in LDCs to enrich their corporate coffers. They encourage projects in Third World countries that are unneeded or that have dubious or harmful effects. Westinghouse reportedly paid a relative of then-President Ferdinand Marcos nearly $80 million in bribes to get the Philippine government to build the Bataan Nuclear Power Plant, in spite of its location near a volcano and several earthquake faults. Although the plant was mothballed by President Aquino when she assumed office, the Philippine people are still paying off the $2.2 billion cost of this boondoggle. The payback includes an enormous interest payment which alone amounts to half a million dollars a day.[68] A United Nations study proves that the presence of multinational firms aggravates the balance of payments problem for Third World countries. Profits earned through the trade generated by the branch plants is more than offset by the borrowing LDCs must do to keep these firms in their countries.[69]

The evidence cited to prove or disprove any harmful effects by multinationals on LDCs has been largely taken from single case studies of particular countries. Such studies can be useful in highlighting the problems of particular poor nations. Yet they may not be at all typical or representative of the majority of LDCs. What is needed in the end is research addressing the impact of core firms upon the economic well-being of host countries in the periphery. As many nations as possible should be included in such a study. A large sample tends to increase accuracy. Only with such research can we decide whether money spent by the firms of core countries for foreign expansion is actually of some help to the majority of LDCs.

The first comprehensive effort to answer this question came in what is called a "meta-analysis." Researchers Bornschier, Chase-Dunn, and Rubinson looked at all prior studies. They pointed out that the effects of investment upon income inequality were of the same two kinds we are examining in this book. Both absolute income as measured by Gross National Product (GNP) or Gross Domestic Product (GDP) per person and relative income inequality (often Gini ratios) are affected.[70] The authors found that among all five of the earlier research studies, greater investment caused more relative inequality. Contradictory evidence was found on whether investment caused absolute income to go up. The reason for the disagreement stemmed from the way investment was measured. If it is measured by the current inflow of foreign capital (either through International Monetary Fund loans, other foreign aid, or multinational expansion), it causes an immediate period of economic growth in GNP per person. Looking at the stock owned by foreign firms in relation to total stock in a country's market gives a dramatically different and more long-range picture. The result of investment dependence as measured in this way works as a drag on economic growth. This occurs despite a short initial burst in GNP per person.

To test this in more detail, the authors performed a new analysis of seventy-six less developed countries. Their analysis used both money flow and stock ownership to measure investment dependence. This study found that both factors work to change real income within LDCs, but again in opposite directions. Ownership of local businesses by foreign capital (stocks) was most important in reducing the growth in GNP per person within these countries. It was more powerful than the inflow of foreign investment, which does tend to increase average GNP somewhat. Thus, the net effect of foreign investment in LDCs is negative. Poor nations lose more money than they gain. *The effect of foreign investment was even worse among the wealthier LDCs* than among the poorest. This surprising finding is explained by the fact that poor LDCs are typically engaged in extractive exports such as coffee, sugar, bananas, and minerals. The richer LDCs are frequently more involved with manufacturing. Part of the reason why multinationals lower real income may thus be due to the makeup of their industry.

A study looking at fifty-six nonsocialist nations based on their core, periphery, or semiperiphery status also looked at economic development and relative income inequality.[71] At first it appeared that as development progressed, the share of money going to the richest quintile declined but so did the amount going to the poorest 20 percent. The biggest winner was the middle class in LDCs, which rang up large increases in its share of income. An initial glance thus seems to show that development leads to less income inequality. Yet another view is seen when a nation's strength vis-á-vis other countries is measured by the size of its debt as a percentage of GDP. Income inequality skyrockets when debt goes up. Being plugged into the world market means that a nation's relative income inequality will climb. This is true for LDCs who have either huge exports or massive imports (as a percentage of GDP). The portion of income going to the poorest 20 percent of persons also went down as the role of importing and/or exporting went up. The structural position of countries is also associated with an increase in the concentration of income among the rich.[72] The results conclusively show that a growth in economic development had no major effect in reducing a nation's income inequality. It was being tied in with the export market, having high levels of imports, being located in either the periphery or the semiperiphery, and/or owing huge amounts of debt that proved most responsible for greater income inequality. Any good effects of economic development largely vanished when these factors were strong.

The thrust of these major studies was to seriously weaken the argument that free-market economic forces help make everyone more wealthy. The challenge did not go unanswered. One critic pointed out that modern world system research forgot to use other important variables that have an effect upon GNP per person. High population growth rates in LDCs work to drive more people into poverty. Poor nations simply cannot make enough to feed and clothe their rapidly growing numbers. Population size can affect efficiency. Large economic enterprises —such as steel production—are more expensive in smaller countries than in larger ones where an internal market may exist.[73] Among seventy-two Third World countries, it was found that savings were an important part of growth in GNP per capita. In fact, when researchers added size of population to the analysis, the effect of foreign stock ownership on slower GNP-per-person growth actually disappeared. Other skeptics have claimed that high rates of military participation tend to equalize income inequality as well. The degree of democracy in countries is a variable that can greatly affect equalisation of income. Electorates have become angry with blatant unfairness and have voted their leaders out of office.

In a study that included these factors, a curve was found much like that predicted by Kuznets.[74] Using several different income inequality measurements (Gini, top 20 percent, low 20 percent), it was found that development was strongly related to more equality. This was true over and above the effect of trade, dominance by foreign firms, and the like. Military participation and the role of strong democracy in a nation also played major roles in reducing inequality. On the whole, little or no evidence was found that relative inequality went up because of dependence on external markets.

Edward Muller sees Canada as an example which refutes the dependency/world system perspective. Of all nations, Canada is one of the most highly penetrated by transnational corporations. Yet it ranks as one of the most economically equal countries in the world. The great bulk of multinational penetration (mostly from the United States) did not occur until after World War II. This growth has failed to produce any increase in the distribution of income inequality.[75] Muller goes on to fault earlier findings that show negative effects of foreign-firm investment by questioning the adequacy of the data used in these studies. He objects to including what was then Yugoslavia since it was a communist country outside the pale of free-market influence. After making other corrections to the data, he found that neither penetration by multinational firms nor amount of debt increases inequality, as measured by the percent of income going to the top 20 percent of recipients.

The debate among scholars has often been sharp, rancorous, and ongoing. A study by Miles Simpson, for example, criticizes the detractors of modern world systems theory by pointing out their failure to use education and political democracy variables. If they had done so, the skeptics would have tripled their ability to explain income inequality.[76] His numbers have also failed to detect any effect of economic development upon inequality. The bottom line for Simpson: using the correct predictive variables will make the Kuznets curve disappear.

In truth, what much of the fight boils down to is which factors should be looked at as a cause of later income inequality. In the search to throw out spurious variables—such as the false effect of storks in our fertility example—close attention must be paid to the theory behind it all. We ask what influences are expected to cause change in absolute or relative income on the basis of the ideas we have before we do the study. This is particularly crucial since nearly all of the studies on inequality use multiple regression. In order for a study to be valid, there should be only one more variable for every ten nations in the sample.[77] In studies of seventy nations, therefore, only about seven variables should be used. Since the size of samples where income inequality data is available is quite small, we cannot control for everything at once. It is simply impossible to make all other factors equal while looking at the impact of multinationals upon income inequality within nations.

Given these constraints, Bornschier and Chase-Dunn to their credit attempted to correct the flaws in their early work by a more detailed research effort.[78] Again echoing their fear of multinational dominance, the authors point out that fully one-half of all world trade today passes through such firms. The free market is completely avoided. Transnational firms now have a major influence on foreign trade and in determining who does what, within both the core and the periphery. This is very true of the "new dependency" which is created by locating new manufacturing plants in LDCs to exploit cheap labor. It is in contrast to the classical dependency of LDCs, which export only a few agricultural products such as sugar, coffee, minerals, and so on. When firms enter countries in the periphery by opening new industrial facilities, there is an initial growth spurt. Yet long-run economic stagnation, unemployment, and increasing poverty all come in the end for the majority of the population. Evidence from the Bornschier–Chase-Dunn study shows that this new form of dependency has become more important during the 1970s and 1980s among LDCs.

In their review of thirty-nine previous studies, the authors once again found that when measures such as foreign flow of investment into LDCs were used to measure penetration, the results nearly always indicated an immediate increase in real income.[79] Yet, again using the better way to measure penetration (stock ownership by foreign firms), the authors found that corporate dominance by core firms led to a great decline in growth of GNP per capita between 1965 and 1977 for 103 countries. This was true even when other variables were held constant.

As we have entered the 1990s, research examining the "economic development → income inequality" nexus has grown ever more sophisticated and complex. In a meticulous study of sixty-three nations by Terry Boswell and William Dixon, the great importance of world system position and multinational penetration in predicting income inequality is confirmed.[80] The study is important for corroborating findings in the 1980s—including the failure of economic development to effectively reduce income inequality or to follow an inverted U-shaped curve. In fact, their statistics essentially replicate Bornschier and Chase-Dunn's major finding that corporate penetration dampens economic growth after controlling for level of development and investment. Their study also introduces more refined measures of various independent variables. Military dependency (as gauged by arms imports) is introduced as a component of economic growth, as well as agrarian inequality (an index combining land concentration and the proportion of the labor force in agricultural occupations) as a determinant of income inequality. Both prove important in their subsequent conclusions.

The major finding—that foreign investment via multinational penetration works against economic growth in poor countries—has been severely challenged in a seminal article by Glenn Firebaugh.[81] In his study of seventy-six less-developed countries, Firebaugh zeroes in on method-ological weaknesses in the way Chase-Dunn and Bornshier measured multinational penetration. His critique of this methodology is especially important since all studies have essentially replicated the original penetra-tion scores in subsequent research. To find the yardstick fundamentally flawed calls into question a whole genre of studies. Firebaugh's argu-ments are imaginative, well-constructed, and too technical to include here. The major finding, however, is unambiguous. While foreign capital investment may not produce the economic growth for poor countries that domestic capital does, it is not harmful. The assertion that multinational dominance through capital investment is dangerous for the Third World is basically false—at least when penetration is measured correctly.

The criticisms by Firebaugh have been freshly reexamined and par-tially refuted by Dixon and Boswell.[82] Their new analysis measures the economic impact on countries by comparing *rates* of domestic and for-eign investment while utilizing two new penetration measurements that correct for the earlier methodological weaknesses.[83] In essence, they argue that the capital investment rate and foreign capital penetration are conceptually distinct from one another. Their finding, that foreign as opposed to domestic investment produces much less economic growth in LDCs, agrees with Firebaugh's conclusion. Their new models clearly show that when investment rates hold constant, increasing penetration by foreign capital in Third World countries slows LDC rates of economic growth. The reanalysis also indicates that as foreign owners command a larger portion of total capital stock, income inequality grows. For every 4 percent rise in penetration, another 1 percent of income is put into the hands of the wealthy.

To gain a better idea of current multinational influence in various nations, I have measured penetration by dividing the amount of net for-eign direct investment (FDI) in each country in 1992 by the country's GDP in that year, expressed as a percentage.[84] This is a snapshot of what is currently taking place among nations via multinational investment. The FDI measure is different from some prior research, however, in not including the value of portfolio stock and in not looking at domestic stock investment. Many other cross-national studies do measure investment dependence, however, by taking the value of foreign direct investment as a percentage of total GNP.[85] In the end, this approach gives a relatively simple, more immediate measurement of up-to-date transnational activity in foreign countries. The impact of the penetration is measured against all economic activity occurring within the country (GDP).

Table 4.3 includes the multinational penetration scores of countries arranged in a rough hierarchy of severity. Countries with negative scores, mostly core nations, essentially show greater outflow than inflow of direct investment.[86] The net direction is more investment in business in other countries after subtracting for investment from foreign countries in the homeland.

It may not be obvious from the table where much of the multinational penetration is actually occurring. Figure 4.1 indicates that, globally, South America seems to take top honors among regions in degree of dominance by investor firms. North America approaches this level as well. The United States is the only nation in the region with a low penetration

TABLE 4.3   Net Foreign Direct Investment as Percent of Gross Domestic Product: 1992

Investor Countries	Light Penetration Countries (.01–1.0% of GDP)				Heavy Penetration (Over 1.0% of GDP)	
Netherlands (–2.22)	Myanmar	(.01)	Jordan	(1.00)	Venezuela	(1.03)
Switzerland (–2.18)	Bangladesh	(.02)	Uruguay	(.01)	Spain	(1.19)
Denmark (-.98)	India	(.07)	Algeria	(.03)	Australia	(1.25)
Germany (–.84)	Kenya	(.09)	Zimbabwe	(.08)	Zambia	(1.31)
Gabon (–.61)	Cameroon	(.10)	Ethiopia	(.10)	Egypt	(1.37)
United States (–.51)	Rwanda	(.13)	Uganda	(.10)	Sri Lanka	(1.40)
Austria (–.50)	Canada	(.18)	Nepal	(.14)	Indonesia	(1.40)
Japan (–.39)	South Korea	(.19)	Mauritania	(.19)	Morocco	(1.49)
United Kingdom	Lithuania	(.20)	El Salvador	(.19)	Portugal	(1.54)
(–.33)	Latvia	(.28)	Uzbekistan	(.27)	Colombia	(1.63)
Mali (–.28)	Benin	(.32)	Romania	(.32)	Mexico	(1.63)
France (–.24)	Somalia	(.34)	Ghana	(.33)	Botswana	(1.65)
Central African	Norway	(.37)	Kazakhstan	(.35)	New Zealand	(1.69)
Republic (–.24)	Syria	(.39)	Bulgaria	(.39)	Bolivia	(1.76)
Italy (–.22)	Brazil	(.40)	Chad	(.40)	Chile	(1.79)
Sweden (–.17)	Oman	(.51)	Philippines	(.43)	Argentina	(1.83)
Iran (–.15)	Ivory Coast	(.56)	Lesotho	(.56)	Thailand	(1.92)
Belgium (–.15)	Mauritius	(.58)	Peru	(.57)	Honduras	(2.13)
Panama (–.02)	Pakistan	(.66)	Paraguay	(.62)	China	(2.20)
	Laos	(.75)	Ecuador	(.67)	Dominican	
Finland (–.01)	Poland	(.81)	Madagascar	(.76)	Rep.	(2.32)
	Turkey	(.85)	Nicaragua	(.81)	Jamaica	(2.64)
			Guatemala	(.90)	Tunisia	(2.74)
					Nigeria	(3.02)
					Trinidad	(3.30)
					Costa Rica	(3.37)
					Hungary	(4.20)
					Sierra Leone	(5.84)
					Malaysia	(7.15)
					Papua New	
					Guinea	(9.46)
					Estonia	(13.52)

Source: Organization for Economic Cooperation and Development, 1995, tables 2, 3; World Bank, 1994, table 22.

FIGURE 4.1   Global Multinational Penetration

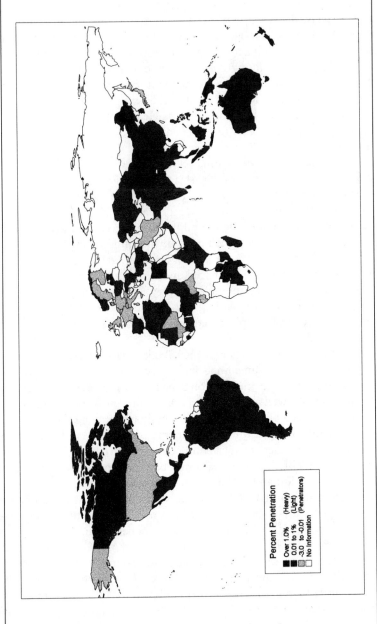

Source: Data from Organization for Economic Cooperation and Development, 1995; World Bank, 1994.

score (Central America is highly penetrated). The countries of Africa (figure 4.2) show most nations where information is available to be highly dominated by multinational firms. Europe (figure 4.3) is the region with, comparatively, the least penetration, although there are exceptions here as well. Estonia and Hungary, formerly East European communist bloc countries, are receiving large degrees of foreign direct investment. Spain is the only country among core nations that has heavy penetration. It remains to be determined in subsequent research, however, whether multinational penetration acts in the same negative manner in wealthy countries as in LDCs. In summary, it is clear that countries particularly vulnerable to dominance by foreign firms have been in Latin America and in Africa.

Multinational penetration, as measured by net foreign direct investments as a percent of Gross Domestic Product, becomes more important as we go from the core (–.52 percent) to the semiperiphery (1.05 percent) to the periphery (1.18 percent). The differences between core nations and semiperiphery and periphery countries are statistically significant. This is not surprising given the findings in the last chapter and the tendency of LDCs to actually encourage these firms to locate within their boundaries. The crucial question is whether their presence is actually good or bad for the host country. Since the impact of multinational entry may vary depending upon how developed a country is to begin with, core and periphery nations need to be separately analyzed in future research.

Using the latest available World Bank data (1994), countries were examined using development and inequality variables found important in earlier studies. Revised multinational penetration scores have been provided by Terry Boswell and William J. Dixon that answer the methodological criticisms raised by Glen Firebaugh. By simply ranking these scores and dividing them into equal thirds, we can glimpse the effect of multinational penetration on a variety of social and economic variables. Table 4.4 shows less difference (although there is some) between countries with high and medium penetration and nations with low penetration. On the important development indicator of GDP per person, there is a huge gap between countries with low penetration—which are significantly wealthier—as opposed to countries with high or medium penetration. This pattern also holds true for income inequality as measured by the Gini ratio, which is over one-fourth as large in higher penetration countries. Possibly the most salient measure of how penetration affects the economic well-being of nations can be seen in the GDP growth indicator. While the average annual change in GDP from 1980–1992 is 3.65 percent among low penetration countries, it is 2.38 percent and 1.99 percent among medium and high penetration nations.

FIGURE 4.2   Multinational Penetration in Africa

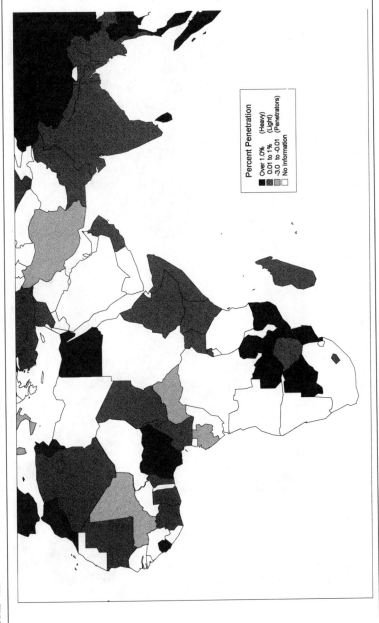

Percent Penetration

Over 1.0%   (Heavy)
0.01 to 1%   (Light)
-3.0 to -0.01 (Penetrators)
No Information

Source: Data from Organization for Economic Cooperation and Development, 1995; World Bank, 1994.

FIGURE 4.3   Multinational Penetration in Europe

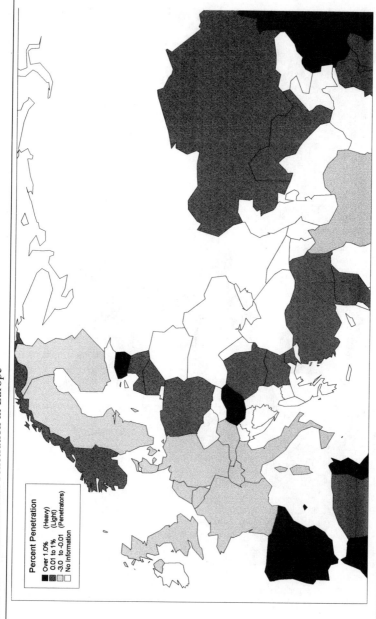

Percent Penetration

Over 1.0%   (Heavy)
0.01 to 1%   (Light)
-3.0 to -0.01 (Penetrators)
No Information

Source: Data from Organization for Economic Cooperation and Development, 1995; World Bank, 1994.

TABLE 4.4   Differences among Development Variables by Penetration: 1994

Indicator	Low Penetration	Medium Penetration	High Penetration
GDP per person	$4,824	$1,487	$1,660
Household Gini score	0.359	0.456	0.433
Average annual % change in GDP, 1980–92	3.65	2.38	1.99
Average annual % growth in Gross Domestic Investment (GDI), 1980–92	3.03	.93	–.54
GDI as % of GDP, 1992	22.05	19.04	18.4
Exports as % of GDP, 1992	18.8	25.7	30.8
Energy consumed per person (kilograms of oil), 1992	987	470	696
Av. annual % growth in energy consumption, 1980–92	4.0	2.4	2.1
Debt as % of GDP, 1992	53.4	79.6	81.0
Debt service as % of GDP, 1992	23.2	20.0	23.8
Debt per person, 1992	$845	$563	$884
Development assistance as % of GNP, 1991	7.7	8.4	8.4
Domestic savings as % of GDP, 1992	14.6	11.9	14.2
Number of countries in group	23	23	23

Source: World Bank, 1994, tables 1, 2, 3, 5, 9, 13, 19, 20, 23.

Another important mark of the healthiness and financial independence of an economy is its gross domestic investment. Gross domestic investment for the most part taps spending for fixed assets within an economy, such as machinery, new factories, a buildup of inventories, and so on. It is the degree to which a nation invests in itself. Not surprisingly, the capacity of Third World countries to do so has been cut severely since 1980 because of the worldwide recession. The effect of multinational entry serves to aggravate the problem for the Third World while it actually proves to be an advantage to core countries. Modern world system theorists declare that penetration destroys or dampens the ability of a nation to invest in itself, free from the constraints of foreign corporations. Among low penetration countries, there is an annual 3 percent growth in domestic investment, while the growth is less than 1 percent in medium penetration countries. In countries highly penetrated, there is an actual decline of half a percent per year in domestic investment. The slide in domestic investment as penetration goes up can also be seen in the proportion of the gross domestic product that is spent in this way. All of this transpires against a backdrop of ever-increasing exports as we go from low to high penetration countries. While medium and high penetration countries are successful at selling goods abroad, there is little reward at home. Debt, whether measured per person or as a percentage of GDP, remains higher in these penetrated countries. Because such nations are also poorer, the negative effects of these liabilities are worse. Lastly, a frequently used indicator of economic robustness is an increase in energy consumption. This indicator successfully gauges how modern and industrial a particular nation has become. In many studies, the log value of energy consumption has served as a proxy to GNP growth as an indicator of development. Again, the figures indicate much less growth among countries dominated by multinationals—although the difference between medium and high penetration countries is not great.

In an attempt to assess the differing impact multinational penetration may have upon the poorest of countries, these same development indicators were looked at separately after countries were divided into two groups based on median GDP per person in 1992 ($1,276). In the poor and rich nation groups, countries were further divided based upon the median penetration score (7.85). This score is measured by dividing foreign-owned capital stock by the total stock of invested capital. In the end, low and high penetration countries can also be compared in terms of poor and rich countries, for a total of four groups.

For low income LDCs there is one ray of sunshine. It seems that the twenty-two countries with high penetration actually have a higher gross domestic product per person ($545) than do low penetration countries ($445). The same is not true of wealthier Third World countries, however, where the trend follows the theoretical prediction of "high penetration → low GDP." The expected pattern also prevails in showing greater household income inequality in highly penetrated countries, whether rich or poor. There is a fall-off in the average annual percent growth in GDP in the 1980–1992 period as penetration increases, irrespective of how rich the country is. Among poor countries, greater penetration leads to a reduction in Gross Domestic Investment (GDI)—although no change occurs in richer LDCs on this variable. The same is true with energy consumption per person, which declines as penetration mounts among poor countries but climbs somewhat for wealthier countries. The growth rate of energy consumption shows penetration hampering development substantially in both rich and poor countries. Opposite trends on debt as a percent of GDP appear. In poor countries, the debt percentage decreases with penetration. In richer countries, it increases sharply as multinational dominance goes up. There is no substantial difference in debt service (the proportion of export sales used to pay off debt) by penetration in wealthier nations, but highly penetrated poor nations bear a heavier burden. While debt per person is roughly the same between high and low penetration poor countries, it climbs with penetration among richer LDCs. Lastly, penetration actually seems to generate a greater proportion of savings in both poor and rich countries.

TABLE 4.5   Difference among Development Variables by Penetration in Rich and Poor Nations: 1994

Indicator	Poorest (Under $1,276)		Richest ($1,276 and Over)	
	Low Penetration	High Penetration	Low Penetration	High Penetration
GDP per person	$445	$545	$7,827	$4,065
Household Gini score	.381	.438	.372	.457
Average annual % change in GDP, 1980–92	2.94	1.93	3.46	2.69
Average annual % growth in Gross Dom. Investment, 1980–92	1.46	.43	2.72	.48
GDI as % of GDP, 1992	19.0	16.7	23.5	23.9
Exports as % of GDP, 1992	18.3	23.7	18.7	47.0
Energy consumed per person (kgs of oil), 1992	164	140	1,609	1,631
Average annual % growth in energy consumption, 1980–92	2.53	1.53	4.39	3.99
Debt as % of GDP, 1992	96.1	69.7	32.1	62.9
Debt service as % of GDP, 1992	19.5	22.0	25.4	24.7
Debt per person, 1992	$491	$474	$1,208	$1,450
Development assistance as % of GNP, 1991	13.7	10.8	.59	1.25

Source: World Bank, 1994.

In summary, the picture is mixed, and it is obvious there is a need for further, more detailed and refined exploration. One can safely say that penetration is more harmful to poor countries when measuring development as the percent change in Gross Domestic Product, as growth in energy consumption, or as the ability of a nation to invest in itself (growth in GDI). Moreover, it is clear that penetration leads to increasing income inequality, regardless of how wealthy or poor a country is. Figures 4.4 and 4.5 depict these two trends for all nations. On other traditional development indicators, the picture is more mixed. There is ambiguity on some debt measures, which are discussed in greater detail below. Yet in light of these findings, it is clear that the net effect of penetration is at least as harmful to poor nations. In some cases, it may be more damaging because such countries have so few resources at the start to overcome obstacles to development.

A number of commentators and theorists have suggested that modern world system theory has not aged well—that there has been an erosion in its dynamism and explanatory power since the heady days of the 1970s and early 1980s. Indeed, there has been a great deal of discussion and worry about its legacy and support among members of the Political Economy of the World-System Section (PEWS) of the American Sociological Association. Newer explanations and models have currently been advanced to explain the growth of income inequality[87] or democracy,[88] neglecting or ignoring the role of modern world system position outright. In this sense, the intense controversy over the correct way to measure multinational penetration may be a good thing. Since we must now redo much of our research using new and better techniques, there may be a resurgence of creativity in the way this theoretical model is employed. Other competing variables will be added to the analytical mix—hopefully producing more rigorous findings. Supporters of modern world system theory are today poised to employ better measurement techniques for its major explanatory variables. Degree of coreness, as developed by Cornelis Peter Terlouw, is now a continuous measurement which can supplant the ordinal measurement inherent in the core, semi-periphery, periphery distinctions.[89] Penetration is now measured in a more straightforward and understandable fashion, devoid of earlier methodological biases. Yet there is still much to be done. There is a desperate need to update penetration scores, originally developed from data circa 1970, to reflect trends in the 1980s. Current penetration scores are also severely limited to a smaller number of nations and do not reflect the presence of many core countries. In essence, the measurement needs to be broadened to include new nations, more core countries, and more current economic data.

FIGURE 4.4   Multinational Penetration Decreases GDP Growth

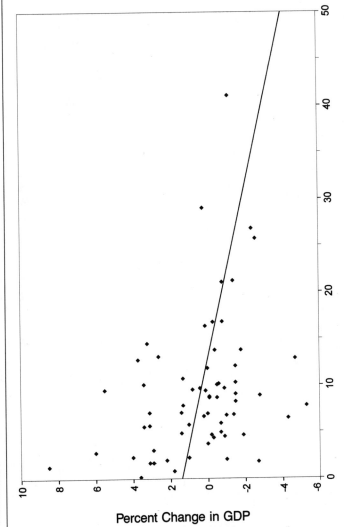

Source: World Bank, 1994, table 2.

FIGURE 4.5   Multinational Penetration Raises Income Inequality

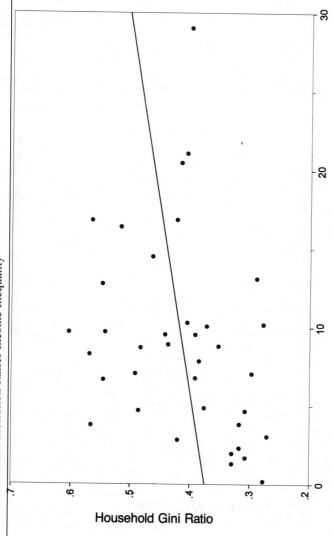

Household Gini Ratio

Foreign Capital Stock Penetration

Source: World Bank, 1994, table 30.

Yet it would be a mistake to write off modern world system theory as irrelevant on the basis of these weaknesses. Again, many of the measurements are more sound than those used in the past. Even with penetration scores based upon multinational foreign stock ownership around 1970, strong relationships can still be shown to exist. For example, the correlations between growth in GNP over the 1980–1992 period with multinational penetration (–.307) and with degree of coreness (.510) are large, statistically significant, and in the expected directions. The correlation with penetration and the Gini ratio of household income is .356, also significant and in line with theoretical expectations (greater penetration leads to greater income inequality). A regression analysis employing development indicators found important in past research[90] did yield the finding that increasing exports are predictive of growth in GDP. More important, however, is the degree of coreness a nation reflects, that is, how close a country is on a number of different properties typically found in core nations. Also statistically significant in the regression equation is penetration, which is significantly predictive of slower growth in GDP. In the end, a nation's position in the modern world system and the degree of penetration by multinational corporations are still strong determinants of its economic development.

Among competing variables that must also be looked at in the future as an explanation for lack of development and growing inequality is debt. What exactly is the role of the growth of debt, claimed by Susan George, York Bradshaw, and other critics of the development model as so important? It is alleged that the presence of foreign firms actually works to erect a mountain of debt for poor persons in the Third World while benefiting only a select few. Much of the debt is claimed to exist because corporations eat up scarce local capital. The debt is created either directly through loans or in the form of inducements such as forgone taxes, repressed wages, and the like. This in turn prevents local industry from forming, further retarding economic growth.

The World Bank provides a year-by-year accounting of the debt structure in LDCs, although the same data for core countries is unfortunately lacking.[91] Modern world system position is highly associated with greater debt in Third World countries as measured by a variety of data indicators. Among all periphery countries, the average long-term outstanding debt is larger than three-fourths their GDPs. The same figure for countries within the semiperiphery is only about one-half. Total debt service (percent of exported goods used to repay loans) is larger in the periphery—at 21.8 percent—than in the semiperiphery (14.4 percent). At first glance, the debt load looks lighter in the periphery—at $849 per person—compared to $1,204 in the semiperiphery. But the lower standard of living in the periphery must also be taken into account. The average

GNP per person in the semiperiphery is $9,562 compared to $1,775 in the periphery. Thus, in the periphery the average debt is almost one-half GNP per person, compared to 12.6 percent per person in the semiperiphery. Thus, a nation's ability to pay back loans becomes four times more difficult in periphery countries than in semiperiphery countries.

MEETING BASIC HUMAN NEEDS IN POOR NATIONS

By now, it is painfully obvious that the economic misfortunes of Third World citizens are bound up with the financial fortunes of those in the First World, mainly because of multinational dominance. This is because all nations are part of a world economic system that both causes and perpetuates income inequality. There may be some growth for all countries in this system. Yet it is always true that if a few rich nations in the core gain, many poor nations in the periphery will lose. Multinational entry leads to more relative income inequality while it cuts real GDP growth. The negative force of multinational impact is played out mostly in the Third World. A plan which relies on growth in export industry, led by foreign firms that seek to set up shop in a poor country, may therefore court ultimate disaster. In the end, the life-and-death needs of the bulk of the population will not be met by this strategy. It is even likely to backfire. Its effect is often a worsening of living standards, deterioration in wages, mounting debt, underemployment, ill health and even starvation.

Can the argument that living standards will improve as a country develops be a Trojan horse? Is there more to the story than just worry over the size of income shares? Most of us by now would agree that countries cannot be judged solely on the basis of their annual GNP growth rate. We know all too well that much of the supposed good effect of modernization does not seep down to the bulk of the masses. Side by side with futuristic skyscrapers, natives in LDCs often exist and barely survive in the most primitive poverty. Even if no major new shuffling of income shares has taken place in LDCs, however, some feel that living standards will automatically get better because of the development of a modern sector. Many argue that meeting basic human needs is bound to become easier as more income flows into an industrializing society through increased exports and participation in world trade patterns. Following economic rules laid down by core nations, then, will guarantee enough food to eat. Basic literacy will improve as foreign firms demand a more educated work force. It is possible that partial modernization may give large portions of the population many benefits even if they do not receive higher real income. Much attention has been devoted to widespread improvements in health through the introduction of easily imported, cheap Western medicine into Third World countries. For example, after World

War II the death rate in Ceylon was cut in half during one year simply because the island was sprayed with the insecticide DDT. This killed off malaria virtually overnight. And so the arguments go.

While there is some disagreement as to what basic human needs are, evidence indicates that measures that simply combine a large number of indicators are weak.[92] Yet even using a combination of infant mortality, life expectancy, and basic literacy reveals large gaps between rich and poor.[93] The higher the average GNP per person in a country, the higher is its quality of life. This is also true when each of the separate components is considered by itself. Richer countries have higher literacy rates and life expectancies and much lower infant death rates.

What other factors may be operating to improve or lessen the delivery of basic needs? Who benefits and who does not? What other forces, aside from a high GNP per person, may be operating to increase or decrease quality of life? As we have seen, penetration of poorer countries by multinational firms has had much to do with retarding real economic growth. It also drives up the gaps in relative income. Because of this, we might also reasonably suspect that penetration lowers standards of living to the point of threatening basic survival. Does a position in the periphery in the modern world system always lead to lower quality of life? Is this still true if GNP and penetration are the same? In the end, it is not enough to look only at economic performance. In a given nation, frequently there are major population segments that do not benefit regardless of how high the GNP may climb. The large number of homeless in the United States serves as a sobering reminder of this point.

A study by Bruce London and Bruce Williams looks at how dependency acts upon the provision of basic human needs in eighty-six LDCs.[94] A nation's GNP per person is the most predictive of all factors that work to meet basic human needs. The authors also found that multinational corporate dominance reduces provision for human needs in the Third World. This is true even when money inflow, Gross Domestic Investment, and initial GNP per person are high to begin with. The results are the same for every human need they looked at. Corporate dominance lowers the per-person amount of calories eaten daily. The presence of foreign firms also reduces the amount spent on welfare benefits in host countries. Multinational entry cuts the average length of life as well. Lastly, multinational presence tends to raise infant mortality and contribute to the lack of physicians in LDC populations.

Nonetheless, a more valid test of market forces versus world system variables was not present in this study. It left out the all-important effect of debt, exports, growth of GNP, and world system position. Structural adjustments that nearly always accompany loans to Third World countries actually harm children. There is an association between such debt processes and

lower immunization rates, poorer general health maintenance, lack of prenatal care, and inadequate nutrition that severely affects children.[95] Some
attention must also be given to the effect of relative income gaps as well as
to absolute income (GNP per person). We know for a fact that relative
income inequality is itself a cause of meeting or preventing the satisfaction
of basic human needs. This is true for such basics as death rates and population per physician.[96] The impact of income inequality can be seen in table
4.6. Nations have been split based on high or low inequality according to
the median for household Gini scores (.374). Looking at eight indicators
reflecting a variety of crucial human needs, it is easily seen that low income
inequality goes with higher levels of well-being. To begin with, egalitarian
countries are marked by much lower infant mortality rates. For every one
thousand live births, approximately thirty-six infants die before reaching
age one compared to forty-five infants in countries with high inequality.
The life-and-death struggle continues during early childhood in nations
plagued by inequality. While fifty-five children per thousand end up dead
before reaching age five in egalitarian countries, nearly seventy die in
unequal nations. Much of this is due to diarrhea, the number one killer of
children in Third World countries. Diarrhea is frequently caused by unsani-
tary drinking water. The table reveals that access to safe water is denied to
one-fourth of the population in highly unequal countries. Ironically, babies
die because of the dehydration associated with this disease that is itself
caused by toxic drinking water.

TABLE 4.6   Comparison of Provision for Basic Human Needs
             by Degree of Household Income Inequality: 1994

Basic Human Need Indicator	Low Household Gini Ratio	High Household Gini Ratio
Infant mortality rate	35.8 deaths	45.4 deaths
Male child mortality rate	55.5 deaths	69.8 deaths
Female life expectancy	71.5 years	67.4 years
Contraceptive usage	51.2%	38.9%
Access to safe water	83.7%	73.6%
Households with electricity	79.9%	56.4%
Amount of total government expenditures spent on housing, welfare, and social security	28.2%	17.3%
Children enrolled in secondary school	72.1%	47.7%

Source: World Bank, 1994, tables 10, 26, 27, 28, 29, 30, 32.

Surveys have shown that large proportions of women in less developed countries do not really want as many children as they give birth to. Studies also reveal that they would avail themselves of artifical contraception were the materials available to them. In many societies, access to this fundamental right to control reproduction is denied to many women. Consequently, nearly two-thirds of women in the childbearing age group in highly unequal countries do not use contraception, while nearly half of all women in countries that are more equal utilize such means to control conception. Countries with more equal income distributions also spend half again as much on housing, social security, and welfare than do inegalitarian countries. Finally, access to education has been associated with human development in study after study. While nearly three out of four elgible youngsters are enrolled in secondary school in egalitarian countries, fewer than half are able to attend in unequal countries. Ignorance perpetuates inequity.

These drab statistics are very real in their impact. They can come down to sheer survival in some countries. By now it seems obvious that many ill effects follow a growth of income inequality. Economic development does not guarantee a better life for the world's poor in Third World nations. Where a nation starts the race in meeting basic needs for its population has much to do with the outcome. Being more affluent is by far the most important predictor of ability to meet basic human needs. Remember that high levels of income among the world's nations, however, stem from their global economic positions. The vulnerability of LDCs to multinational dominance drives the last nail in the coffin. The mix of world position and penetration is a major force in keeping basic human needs from being met. The more a country is penetrated by multinationals and the further it is from the core, the less able a country is to guarantee the survival of its population. LDCs with high multinational domination have much higher numbers of people whose basic human needs remain unmet.

In summary, neo-conservative, traditional axioms of free-market theory have been preached and promulgated by such core agencies as the IMF and the World Bank. Yet following these axioms has had almost no positive effect for the great majority of poor nations. On the basis of case histories and quantitative studies involving large numbers of countries, we can see that it is questionable whether countries in the periphery should seek their fortunes through economic development via foreign assistance. This is true, at any rate, for economic growth as it is thought of and practiced in today's minority of rich industrial nations. What has worked for core countries in the past may have no relevance for underdeveloped economies in today's world. LDCs are primarily dominated and penetrated by these power centers to begin with.

The agents of fortune launched by core nations—huge and omnipotent multinational corporations—occupy and dominate the hinterland much like the Roman Legions of a past era. Such economic dominance is more subtle and less visible than military occupation. Yet the effect of economic dominance is no less pervasive. Financially, it saddles the people of the Third World with ruinous debt that can never be paid back. The world system acts at the same time to impose hidden taxes to support the very agents of LDC repression—the multinationals themselves. The poor of host countries are forced to continue laboring for less than subsistence wages. This demand is imposed upon the poor by their own ruling elites to feed their own voracious demand for profit. Foreign firms must above all have the desires of their branch plants satisfied. The very economic growth hoped for by many of these nations is prevented when multinationals set up subsidiaries. Investment capital is bled off, preventing local businesses from starting up or surviving. Profits are sent back to the core. Rich nations help to maintain military dictatorships in LDCs in order to keep subjugated populations from revolting. The political and military machinations for protecting core multinational investments abroad have been vividly portrayed in great detail elsewhere.[97] The means have nearly always been terrorism, intimidation, repression, torture, and outright warfare by Third World governments against their own people. An unmistakable pattern follows multinational penetration. When foreign corporations enter a country, an increased likelihood of its citizens suffering political violence is the end result.[98]

As a reward for their third-class citizenship in the world economy and global workforce, the poor in LDCs get even less to eat—which can eventually end in outright starvation. Their babies die of malnutrition. All the while they are treated to the spectacle of their affluent ruling elites rubbing shoulders with visiting corporate and banking leaders from rich Western nations. These affluent elites unanimously praise the miracles brought by economic growth and the new global workforce. Yet decline has set in among LDCs as a result. Relative income inequality has grown at a rampaging pace in their societies. The result has been discontent, violence, brutal domination—and sometimes full-scale revolution.

There is an obvious moral dimension to these colorless statistics. Most Americans have no idea of the ill effects that the corporations of our nation inflict upon poor countries. It is hard to see the negatives behind the great availability, cheaper prices, and endless products in our consumer oriented, mall-to-mall society. In a "shop 'til you drop" mentality, little thought is given to where consumer goods have come from, who produced them, and at what price in human suffering such products have been bought from the Third World. Heedless of the wounds we give the rest of the world, Americans have been content to believe our government and our corporations have brought only good to poor countries.

A few who realize the harm done by the world economy agree that the supposed benefits to poor nations may not be all that tangible. Yet one might argue that even if the Third World loses, we gain. In a hard-nosed, practical manner, core nations may have been lucky to undergo early development and dominance. Yet will we now squander our economic inheritance gained by the pluck and hard work of our forebears? Although it is a pity that these countries are destined to never-ending poverty by impersonal market forces, is it our fault? Although multinational domination is hard on people in poor nations, does it not at least increase the economic well-being of those in the advanced core countries? Experts claim that the global economy benefits the average American. We no longer have to do the dirty work of manufacturing, which has led to work injuries, pollution of the environment, and depletion of resources. Since many other countries now do this job for us, we can concentrate on cleaner, more lucrative areas such as the information industry. Most Americans will be better off because of greater purchasing power. This is literally due to cheap foreign goods. Higher U.S. income is made possible by corporate profits returned to our country as stock dividends.

There seems to be no end to such rationalizations. In the next chapter, which looks at U.S. income inequality, we will see that most of these enrichment assertions are false. Rather than benefiting from the suffering prevalent in the Third World, most Americans have been on a downward slide of real income while a small minority of our population has been able to enjoy even more opulence. Like those in the Third World, the majority of U.S. citizens have caught the disease of failing economic health. Not only has real income slid, relative income inequality has also become worse. The cause of most of this decline has its locus in the rise of the global economy. Its effect has been to take money away from lower- and middle-income persons while thrusting even more into the hands of the very wealthy.

Chapter

5

APPLE PIE AND ECONOMIC PIE:
THE AMERICAN PATH TO
A SMALLER SLICE

At the end of World War II, American dominance of the global economy was clearly beyond dispute. Although the war had cost our nation many lives and casualties and had depleted local natural resources, the wartime economy had left the United States in hyperdrive. Our industrial capacity had been raised to a dizzying height of efficiency and volume. In steel production, for example, only the United States had undamaged plants to meet sharply renewed demand as the world went about rebuilding its bombed-out cities and factories. By 1950, American companies were making 45 percent of the world's steel.[1] The war had also firmly yanked the United States out of its prolonged depression of the 1930s. Fully one third of the nation's Gross National Product (GNP) was generated by federal spending during the war. The rapid speedup of technological breakthroughs was another unrecognized byproduct of the war. Over 82 percent of the major inventions and discoveries came from the United States in the decade of 1940–1950.[2] The effect was to leave our country in a strong position to maintain its economic supremacy for many decades to come.

All this was translated to boom times and good times for American citizens, at least for a quarter of a century. Between the end of the war and 1973, the average weekly earnings of forty-year-old men grew by nearly 3 percent per year. This was after the effect of inflation was subtracted. In the two terms of office held by Dwight Eisenhower (1953–1961), real family income increased by nearly one-third.[3] During the period from 1945 to 1970, the standard of living of the average American worker climbed into the stratosphere. This was true for all wages, whether they were paid by the hour, week, or year.[4] Relative income inequality also went down. The economic pie got larger. Magically, every worker's piece of it did also.

The number of workers with earnings below the poverty level went down during this same period. The New Deal of the Depression had introduced some welfare measures and Social Security. Yet it was especially during the 1950s and 1960s that the social safety net for workers was solidified and legitimized. Unemployment and health insurance became the norm, together with sick leave and liberal vacations. Nearly all of these costs were paid by employers, whose profits were setting records. It was a time of hope, of great confidence in the future. After living through the pain of financial depression and the horrors of war, America had finally seemed to achieve its salvation and just reward. Parents looked forward to the future. They were confident that their children would be rewarded even more than they in what seemed an endless cornucopia of consumer goods.

"MOURNING" IN AMERICA

Despite the signs of economic deterioration discussed in previous chapters, there are many apologists of unflinching optimism still with us. They continue to argue even today that it is once again "morning in America." The incredible growth in wealth and affluence that was the earmark of the decades after World War II remains undiminished in the eyes of a few. Most such claims, however, are politically motivated. For example, during the election of 1988 George Bush sought to ride into the presidency on the coattails of the Reagan legacy of economic "growth." Bush was fond of pointing out that during the last five years of the Reagan presidency, 17 million new jobs had been created in America. When the typical quality and pay of these new jobs was taken into account, however, such an assertion was open to fierce challenge. But, to be fair, there were significant goals attained by the White House during the 1980s. When Reagan assumed office, the annual inflation rate was well over 10 percent, conditions in financial markets were chaotic, and short-term interest rates were higher than at any time since the Civil War.[5] In an attempt to combat these trends, the Federal Reserve drove the prime rate up to an incredible 21 percent in the early 1980s. This acted to give the nation its sharpest recession since the Great Depression of the 1930s. Yet the result in the eyes of President Bush was to save thousands of dollars of mortgage interest costs for later home buyers since the rates went down soon after. During the 1981–1987 period, the share of a typical family income that went to pay the mortgate of an average-priced house slid from 40 percent to 31 percent.[6]

Yet when the slide of real income is taken into account, the alleged gains during the twelve year Reagan-Bush presidencies were illusory. Over the 1975–1985 period, median prices for first homes rose 125 percent while the average income of married couples aged twenty-five to twenty-nine who were renters rose only 80 percent. What this meant was that fewer couples in the 1980s could even qualify for mortgages. In 1975, three-fourths of married couples aged twenty-five to thirty-five met the criteria for an 80 percent mortgage. By 1985 the figure was below one half of their number.[7] Younger couples were simply priced out of the housing market during the last decade. Even with more favorable mortgage rates in the 1990s, the waning of real income for many young adults has put the basic American dream of owning a home beyond their means. According to the Census Bureau, the difficulty faced in buying a house may be growing worse.[8] Measured in constant, inflation-adjusted 1993 dollars, the median income of renter families dropped a whopping 19 percent in the last two decades, while the median value of owner-occupied homes rose 10 percent to $86,500 in 1993. Since renters also saw the cost of their rent go up 12 percent in the same period, it was nearly impossible to save for a down payment on a house. In essence, three negative trends militate against home ownership: higher cost of homes, lower family income, and higher rent payments. Even one of these factors undermines home ownership, but the three together have a deadly impact.

Political agendas are still feeding the claim that average real family income (after subtracting for the impact of inflation) went up during the Republican presidencies of the last decade. This assertion, however, is made on a shaky premise. Some economists believe the Consumer Price Index (CPI) overstates the effect of inflation on purchasing power since the CPI includes the cost of current housing and mortgage interest rates. These experts maintain that this cost affects only a few first-time home buyers and not the majority of people who already own their own homes. If this factor is deleted, it can be shown that all income groups— from the poorest fifth to the richest fifth—actually had increasing real income during the Reagan years. Some groups did better than others. The elderly over sixty-five years old increased their real income by nearly 19 percent, well-educated women by 20 percent, and the richest fifth of income groups by 13 percent.[9] Predictably, the poorest fifth income group saw the least rise in real income—a paltry 1.5 percent over eight years. This comes to about two-tenths of 1 percent per year, and only after the figures are altered in such a way as to exclude the basic costs of shelter that we all need.

This arithmetic is very dubious. It seems that some politicians are more than eager to start tinkering with such basic indexes as the CPI when it fails to give them the figures they want. When the measurements are left alone, a markedly different picture emerges. The true relative and absolute income of Americans will be examined shortly. For now, it needs to be pointed out that another way to manipulate statistics into a more rosy picture is simply not to comment on the embarrassing data. Few presidential advisors during the past decade brought up how well America was doing in reducing its trade deficit, its budget deficit, its consumer debt, or its corporate debt load.

The true impact of these unmentionables is just now beginning to be felt by most Americans in the 1990s. By the presidential election of 1992 it was obvious to most voters that they were economically worse off than in past decades. This undoubtedly helped Democrat Bill Clinton to be elected president. It also helped sweep in a Republican majority in both houses during the 1994 election, however, as frustrated voters remained impatient with the continuation of their economic plight.

A fuller analysis of the politics of income inequality will be addressed in chapter 7. For now, it is important to note that the severe recession which arose in the early 1990s also contributed greatly to disenchantment with our political leaders and with government in general. What is worse—from the perspective of common citizens—is that the so-called economic recovery left the pay of most workers either lower or virtually untouched. Profit rates for companies have never been greater. Since the severe recession of the early 1990s, unemployment has fallen dramatically, corporate profits have soared, inflation has remained low, prices have stabilized, and productivity gains have been continuous and steady. It has been especially true that productivity growth in the 1990s has been robust (measured in value of output per hour in the non-farm business sector, adjusted for inflation), which has led to record levels of corporate profits.[10] The average after-tax profit rate for nonfarm businesses was 7.5 percent in 1994, compared to an average of 3.8 percent in the 1952–1979 period.[11] This meant that corporate profit rates were the highest they had been in a quarter century. By the close of the second quarter in 1995, the return on equity for Standard & Poor's list of five hundred major blue-chip companies was running at an annual rate of 20 percent—the best ever for corporate America.[12] During 1994, while corporate profits were accelerating, the average wage of workers went down 3 percent after adjusting for inflation and deteriorating fringe benefits.[13] Indeed, over the 1989–1995 period and even during the recovery years after 1991, wages have been stagnant or declining for the vast majority of the work force—encompassing 80 percent of working men and 70 percent of working women, including male college graduates.[14] Given these glaring inequities, it is no wonder employees and voters are angry and unhappy.

The Decline of American Affluence

The silence of our political leaders regarding America's long term economic skid has been deafening. Yet the signs can no longer be ignored. Perhaps the drop of America from its top position as the most vital economy in the global marketplace was inevitable. Luck and good fortune had left our nation with no damage to its plants after World War II, but our allies and previous enemies rebuilt their industrial might. By using the latest and most technologically advanced machines, many industries in Japan and Germany became dangerously competitive with U.S. business from the day they went on line. Also, the entire postwar American strategy was to rebuild these economies to stimulate free trade and contain communism. Their success was part of the plan. Yet as these countries and a variety of other European and less-developed countries (LDCs) built up their industries, the weakness of older, more vulnerable U.S. plants was painfully obvious. Consider the world's largest firms. In 1956, forty-two of the top fifty were American. By 1980 this number had dropped to only twenty-three.[15] At the start of this decade, there were only five U.S. corporations among the world's ten largest multinational corporations.[16]

Other nations began to produce the same goods that the United States did. A greater variety of goods at more competitive prices was the result. In an era of free trade, this meant that Americans could buy from foreign sources goods that were cheaper (and of higher quality) than those made in the United States. Figure 5.1 vividly reflects the rise in imports to the United States as a proportion of the value of American-made goods. By 1992 Americans were buying $2 worth of imports for every $3 worth of goods made here. At the end of World War II, the ratio was less than $1 of imports for every $10 of American-made goods.[17] Translating such data into more immediate terms, this has meant great losses in jobs. Plant closings went up in the United States as foreign competition for America's purchasing dollar rose. Decline ensued in one industry after another. While imported steel was only 2 percent of the American market in the 1950s, it is 20 percent today. In 1960, America made three-fourths of all the world's cars. Today we account for only one-fourth of global automobile production. During the mid-1980s, the high-tech semiconductor industry lost $2 billion while twenty-five thousand employees were laid off in the computer industry as competition with foreign firms sharpened.[18]

Fully three-fourths of American goods must now face competition from foreign sources. We have not met this challenge with good grace. The very nature of our trade has shifted more toward the profile of a Third World country as we increasingly export food, raw materials, lumber, and unprocessed goods. A crucial sign of the relative advancement of

FIGURE 5.1   U.S. Imports (as a Percent of Manufacturing GDP)

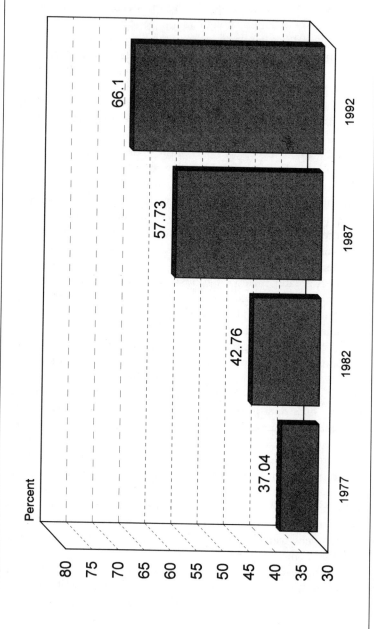

Source: Economic Report of the President, 1995, tables B12, B22.

a nation is its ability to produce its own capital goods, if not to export them. Capital goods are those products used to make other goods—machine tools, construction equipment, plant presses, robots, etc. Again, U.S. decline is unmistakable. In just a decade, imports of machine tools rose from less than 10 percent to almost 42 percent.[19] In a like manner, some of our most advanced industries—such as semiconductors, computer chips, consumer electronics, and the like—have found more fertile ground on foreign shores. While Japan and the European Economic Community (EEC) continue to erode our high-technology base of world sales, the battle for the middle ground of manufactured goods has already been lost.

Our abrupt skid as an economic powerhouse is very obvious when yearly trade deficits are looked at. Figure 5.2 dramatically illustrates the trend. From the end of World War II until about 1970, the United States continuously exported more than it imported in dollar terms. Trade deficits first appear in 1971. Yet in hindsight they were relatively benign, hovering at around $30 billion per year until 1982. Directly as a result of disastrous economic policies introduced in the 1980s, our deficit climbed to $170 billion by 1987. Only in 1988 did the trade deficit finally begin to sink to a smaller gap of $137 billion.[20] Much of this was due to severe deflation of the dollar, which had lost a third of its value over the year. Most experts at that time were very worried that the deficit did not shrink even more. A lower dollar value should have produced expanding exports (as American-made products became available at cheaper prices) and contracting imports (as goods from abroad became more expensive since dollars were worth less).

Incredibly, the loss in the value of the dollar continued at a furious pace in the mid-1990s, especially against the yen and the mark, yet our trade deficit again expanded to reach an all-time record. The hemorrhage was so severe that there was a speculative sell-off in dollars during the mid-1990s which could not be stopped despite repeated attempts by central bankers and our Treasury Department. Indeed, the dollar lost nearly one-fourth of its value against the yen during the first two years of the Clinton presidency.[21] The decade slide between the mid-1980s and mid-1990s was not much better. Ten years ago 250 yen bought a dollar, but today it takes fewer than 100.[22] In a nutshell, the stubbornness of the trade deficit's staying power—even against a declining dollar—is ample proof that America has lost its competitive edge. It also stands as a warning that other countries are less willing to finance our deficits by continuing to buy our Treasury bonds and underwrite our spiral of debt. Incredibly, officials from such Third World countries as Brazil and Argentina have now begun to question U.S. Treasury Secretary Robert Rubin about the soundness of America's economic stability.[23] Both

FIGURE 5.2   U.S. Trade Deficit (Billions of Dollars)

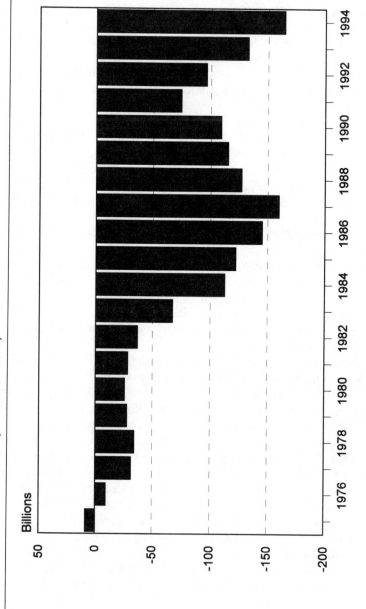

Source: Economic Report of the President, 1995, tables B105, B107

nations are astutely aware of signs pointing to instability and financial turmoil, having previously defaulted on their public debts as well. The major point: LDCs with poor credit ratings—nations we have helped prop up economically in the past—are now openly questioning whether the United States might be a shaky borrower. This development is especially ominous given the fact that $2.5 trillion of the $4 trillion in debt our nation has contracted since 1980 must be refinanced by the turn of the century.

The deficits had really began to do serious damage by 1980. Our country was bled of more than $1.5 trillion over the following fifteen years. While our imports continued to grow to all-time highs, our exports have tended to wither on the vine. The average yearly gush in red ink is $100 billion, which translates to $400 per person per year. Although this sum may seem paltry, nearly one-fourth of all nations in the world have a gross domestic product per person less than this amount. We must earn money, like any other nation, by selling our goods to the world. The buying spree can thus go on only so long before our pockets are empty. Many would argue that this has been the case for quite some time, and that we have been coasting on our past global dominance.

Much of the decline in exports and the explosion in the trade deficit can be traced to an unwillingness or inability on the part of American industry to invest in itself. While most industrial countries were forced to upgrade, rebuild, and modernize after 1945, our physical plants remained relatively unchanged. Excess money was used to build new factories in other countries. While this decision need not affect a corporation's profit sheet, it does reduce the competitive stance of our country. During the last decade, average annual investment was one-third less than in the previous three decades. Today, Japan spends nearly 60 percent more and Germany about 42 percent more (as a percentage of GDP) on nondefense research and development than does the United States.[24]

A National Science Foundation survey looked at the performance of America's two hundred leading industrial companies (which account for 90 percent of all corporate research and development spending).[25] After allowing for inflation, the data show that these firms by the end of the 1980s were no longer increasing R & D spending , although they had been raising such spending by nearly 6 percent each year in real dollars during the previous ten years. The study faults leveraged buyouts as the major culprit, but the result is the same whatever the cause—layoffs of scientists, engineers, and researchers at a time when the economy can least afford it. The long-term health of the economy is under attack from this brain drain due to games of jeopardy being played on Wall Street. Cutbacks of scientists in private industry continued into the 1990s. Over three thousand scientists were shed within two years by U.S. pharmaceutical firms while the chemical industry laid off sixteen thousand scientists in just one year.[26]

Another major avenue of investment is the willingness a society has to educate and train its own people. Even here we are falling behind. About 40 percent of German college students graduate with degrees in science and engineering, while fewer than 10 percent of American college students do.[27] At the same time, given our sharp turn toward paper entrepreneurialism, the proportion and number of business majors and MBAs has risen dramatically over the past decade. Although we now have more paper shufflers and dealmakers than we need, the oversupply seems to have no effect upon their pay or ability to win jobs. By contrast, the unemployment rate among scientists with Ph.D.s tripled in the 1990s, and is now the highest for all professions. Young scientists who do manage to acquire a job after ten years of arduous university training typically begin their careers at an annual salary of $18,000 to $20,000 per year. On average, this is one-half to one-third the starting salary that a newly minted Bachelor of Business degree recipient will command upon graduation—with half the college education.[28]

The increase in federal funds for scientific research had stopped by 1987, and budgets for both civilian and military research had flattened at the beginning of this decade. Because of dried-up funds, agencies that fund basic research are much less generous and more conservative in their orientation. The National Institutes of Health (NIH) and the National Science Foundation (NSF), for example, report that they awarded research grants to half of all applicants twenty years ago whereas today's award rate is barely above one in four.[29] A rhetorical question must accordingly follow. Given this erosion in both basic and applied scientific research, will the United States see a slowing of its technological base and a crumbling in its standard of living in the next century?

One telling sign of this erosion is the shift in technology transfer. American media, politicians, and corporate officials are quick to claim that the real action in the global marketplace is at the high-technology end of the trade spectrum. We are told that America never needs to worry because of our great lead in science. Yet even our prior dominance of technological innovation is in jeopardy. Nearly twice as many patents are granted to foreigners than to U.S. firms or individuals. Our greatest technological competitors have been out-producing us with innovations and inventions since the mid 1960s (Germany) and mid 1970s (Japan).[30]

A dramatic example of our technological tumble lies with high-definition television (HDTV). Unfortunately, while Japanese consumers are saving for their first sets, many Americans have not even heard of it. The new technology will give 1,125 lines of definition compared to the current standard of 525. The result will be clarity similar to that of films we see in theaters. Both Europe (spending $250 million on research alone) and Japan are feverishly working with well-developed projects and a number of consortia to bring this new technology on line in the 1990s. Although HDTV is to be the next generation of television, with an estimated $50 billion in sales potential, the United States has dragged its feet in efforts to develop it.[31] The stakes are colossal in a variety of other ways as well. The high-technology spinoffs from HDTV involve telecommunications, computer graphics, semiconductors, medical diagnostic equipment, and military operations. Since Japan had already spent $700 million in the 1980s to develop the needed hardware, it may have the leading edge with this future technlogy.[32]

With many countries making products only the United States and a few other countries manufactured a short time ago, the fight over relatively dwindling markets has become cutthroat. Given our failure to invest in modernization, our penchant to borrow and spend beyond our means, and our lack of support for product development through research, it should come as no surprise that American productivity has suffered a massive decline. Productivity is measured as the total value of output in the economy divided by the number of hours worked. In the 1947–1973 period American productivity rose 3.1 percent per year, with the increases slipping to just 1.1 percent in the following two decades.[33] Productivity actually dipped to an average −.8 percent for 1979 and 1980. Although it was to briefly spurt up again in 1983 because of a stupendous military spending spree, it quickly slid lower in the next few years and was in the red once more by the end of the last decade. Productivity also took a steep dive in the recession years of 1990 to 1991, which was to be expected and is typical of cyclical changes. Although there has been a sharp rebound in productivity since this time, such spurts are also typical of economic recoveries. It is simply too early to tell whether productivity growth may be slightly better in the 1990s than in the 1980s, given the data we now have in hand. It is safe to conclude, however, that any growth in productivity for the forseeable future will not come anywhere close to the financial dynamism prevalent before 1980.

The effect of the downturn in productivity is registered directly in lost take-home pay. A 1 percent decline in productivity typically yields a 2 percent growth in the proportion of low-paid workers.[34] Most important, however, is our past failure to compete with established industrial powers that pay similar wages. Comparing the United States to ten other industrial countries on productivity in the 1973–1979 period shows us to be only one country above the cellar. (The United Kingdom is in the unenviable last position.)[35] This dismal performance continued in the 1979–1985 period when the United States was literally last among eleven advanced industrial countries in the average annual change in output per hour. By the 1985–1992 period, and despite a very bad recession, the United States had managed to climb to fifth place in productivity increases, providing a much needed competitive edge that may be too little and too late.[36]

Much of the short-lived spurt in productivity in the 1980s was due mainly to a huge increase in defense expenditures. Rather than being a cause of productivity growth—even in the short run—military spending is akin to hauling water with a hole in the bucket. The United States has long been a heavy spender on military hardware, justifying these gargantuan levies as a necessary cost of the fight to contain communism. Despite the fall of the Iron Curtain and the collapse of most communist regimes around the world, our defense budget still remains at harmful levels, and is too high for the U.S. economy to continue to shoulder in the coming century.

Increasing the military budget was a particularly important goal of the Reagan presidency. With almost no opposition, his conservative coalition managed to drive up the budget for the Department of Defense to an unheard-of level for a nation at peace. Spending for defense more than doubled during the 1980s, growing at a rate three-and-a-half times the real growth rate of GNP. There were some reductions in real military spending introduced by President Bush and early in the Clinton Administration, but by the mid-1990s the cutbacks had been largely reversed. Moreover, the new budget proposal pushed by the Republican majority in Congress and approved by President Clinton actually allocates $10 billion more to military spending than the Pentagon has requested.

What might an appropriate and realistic military budget be? The Center for Defense Information, which is largely staffed by retired Pentagon brass who are now free to express their honest opinions, recommends $50 billion less than the projected $210 billion per year suggested by President Clinton. While this amount is only 3 percent of the $1.5 trillion federal budget, it is 2.5 times the level of federal spending on natural resources and the environment, slightly less than the spending on education and training, and more than the total welfare spending for all low income people.[37] Clinton's 1997 projected military budget is 1.6 times what Germany and 2.7 times what Japan spends on a per person basis, despite the fact that these nations no longer need to be protected by us:

> In past decades much of the U.S. military budget was designed to protect Germany, France, Japan, Korea, and other U.S. allies from supposedly aggressive communist nations. Today, the Soviet threat is gone, and our allies both in Europe and Asia, who have economies at least as strong as our own, are capable of defending themselves. . . . Despite the U.S.S.R.'s demise

as the only adversary whose nuclear forces came close to approximating ours, the Pentagon plans to keep operating 18 nuclear submarines, 500 land-based ICBM's (inter-continental ballistic missles), and more than 100 planes equipped with nuclear weapons. All told these forces can deliver 3,000 hydrogen bombs, each of which has many times the destructive power of the bombs that destroyed Japan's cities of Hiroshima and Nagasaki.[38]

Contrary to some claims, military spending acts to retard the economy rather than to speed up growth. There are major questions about the long-term benefits of military research and development. Massive defense spending may actually *harm* high technology by creating bottlenecks in production, encouraging inefficiency, and diverting human and capital resources away from pressing social problems.[39] It may be true that defense spending does increase employment somewhat, but it demands large amounts of capital to do so because of the skilled labor, expensive raw materials, and technical staff needed to build sophisticated weapons. Thus, although 21,000 jobs are created for every $1 billion spent on guided missiles, the same amount spent on education would yield 71,000 jobs.[40] One estimate shows that decreasing military spending by $35 billion and reallocating this money to programs for housing, transportation, and education to rebuild America's economy would create 250,000 new jobs.[41] We now also know, for example, that for each $100 spent on the military, there is $16.30 less spending on consumer durable goods (e.g., cars, appliances), $11 less on producer durable goods (e.g., factories, machines, business equipment), and $11.40 less on homes.[42]

The real damage is done to the economy at large. This is particularly true since the military drains off any future ability our country may have to function and compete in the global economy. The evidence that has been gathered on this topic is compelling and of one voice. The more a nation spends on its military, the lower its investment rate, productivity, and budget for civilian research and development (R&D) tend to be. We actually spend more in the United States on military appropriations than we do on private domestic investment. By assigning a greater priority to the military, more than one-third of our scientists, engineers, and R&D money is shunted away from civilian and commercial needs.[43] To put it bluntly, the United States has been outperformed in economic terms by just about all industrial countries because of our higher military spending. Figure 5.3 chronicles the dismal performance of the United States in washing its own economy down the Pentagon drain.[44]

Nearly all of the record military buildup in the last fifteen years was done without any increase in taxes. Indeed, a large-scale tax cut was instituted exactly at a time when military spending shot through the roof. The

FIGURE 5.3   Military Drain on Investment Capital

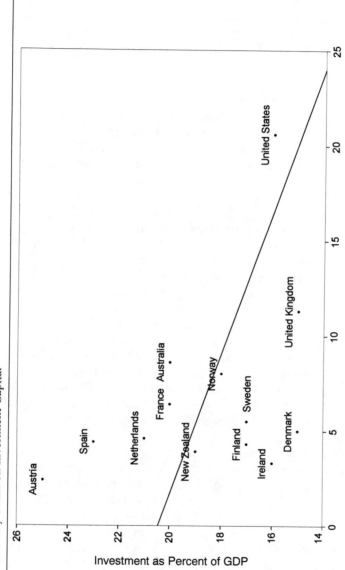

Source: Data from World Bank, 1994.

large holes that would have resulted in the ceiling plaster of our federal budget had to be patched up. America borrowed enormous amounts from abroad to stem the tide of red ink. Massive growth in the national debt was a direct result of giving too much to the military.

The data which show the extent of this run-up of debt has already been presented, but perhaps not enough has been said about its ultimate effect. America has been unable to earn money abroad by selling its own goods. Yet we insist upon buying more from other countries than we can afford. To attract more funds from Newly Industrialized Countries (NICs), Japan, and other nations awash with new trade surpluses, U.S. interest rates were raised. This made our country a haven for investment by foreign powers. Other countries began to spend the money our nation needed to invest in itself. In the early 1960s less than 3 percent of the direct foreign investment of other industrial countries was in the United States, but this had risen to one-fourth by 1978.[45] To pay for our yearly average $100-billion trade deficit, foreigners have had to buy $20 billion in U.S. real estate and businesses, $26 billion in U.S. Treasury notes, and $80 billion in corporate stocks and bonds.[46] Unfortunately, this means foreigners get the profits. Just as our multinationals were able to drain money from LDCs in the 1950s and 1960s, it now seems America is in danger of becoming the victim of its own overseas investment game plan.

How much of a run has there been on the store? The figures are alarming. By the start of this decade, the foreign claim on U.S. assets was starting to approach $2 trillion, which reflects more than a tripling since 1980. Today, nearly 5 million Americans work for foreign-owned companies doing business within our boundaries. Fully one in every twenty American workers employed in the United States now receives a paycheck from a foreign company, as does one in nine of our citizens employed in manufacturing. In 1980, for every $3 invested abroad by American companies there was slightly more than $1 invested in the United States by foreigners. While the value of U.S. direct investment abroad ($486.7 billion) is still slightly ahead of foreign direct investment in the United States ($419.5 billion), the gap has now all but disappeared. Put differently, the holdings of U.S. companies abroad in the 1980–1992 period increased by 125 percent, but the holdings of foreign companies in the United States during the same years went up over 500 percent. In short, our property and businesses have been getting bought up at a rate four times higher than the rate of acquisitions of U.S. firms abroad.[47]

There are minor benefits to the foreign buyout of U.S. assets. Some direct investment in the United States by foreign companies—like Toyota's new factory in Kentucky—may include the creation of jobs. Unfortunately, figures reveal that fully 84 percent of the money influx has gone to buy existing U.S. firms.[48] Typically, large staff reductions accompany most takeovers through mergers or buyouts. It is worrisome to witness the growth of foreign control over U.S. jobs and the loss of profits that would once have gone to American firms. What may be even more dangerous is that only 17 percent of foreign penetration is in a fixed, concrete form. Most foreign holdings of U.S. securities are highly liquid, such as U.S. bonds, portfolio stocks, cash bank accounts, and the like. The money can be gone overnight—in the click of a computer key. In a word, foreign interests may now have the power to destabilize the U.S. economy or make things much worse in a time of financial crisis

Is the crisis already here? The answer is given in the insolvency scandal of the savings-and-loan industry. Much of the precarious situation of this industry is due to deregulation, which allowed questionable loans to be made in a shaky pyramid of debt. Estimates vary as to how much the bailout will ultimately cost to guarantee deposits in many of these failing banks. Some projections are as high as $500 billion. The failures should not be that surprising in the face of the huge volatility in domestic and international economic markets. According to figures compiled by the Federal Deposit Insurance Corporation (FDIC), banks failed in the 1980s at a rate comparable to that of the Great Depression. Whereas an average of only ten banks per year failed in 1943–1981, an average of 137 banks per year failed in the 1981–1990 period. The amount of money involved in these failures over the past decade was just short of $250 billion.[49]

All of these economic indicators come together to give America a loud and clear message: Our nation's financial well-being is at stake. There is a common denominator in one trend after another—that of decline. Yet if we look only at macro or structural trends with global dimensions and national repercussions, it becomes easy to dismiss the very real pain that this decline caused for people. A percentage slide in productivity, a bank failure in Texas, or national deficits that approach numbers none of us can truly comprehend may prevent us from seeing how we are affected on a daily basis. Because of the workings of the global economy and the machinations of economic elites directing American multinational corporations, however, the financial security of most Americans has plummeted. The direct consequences have been increasing poverty, decline in real income for most people, and an explosion in relative income inequality.

THE GROWTH OF POVERTY

Given such a deterioration in postwar trends—together with more than a decade of conservative conomic policies protecting corporations, financiers, and stock owners at the expense of ordinary citizens and workers—it should come as no surprise that poverty has been increasing in the United States. In an era of unfettered corporate growth, mega-mergers, and speculative mania on the world's stock markets, not much is heard about how the other half lives. Unfortunately, the poor are still with us. Their ranks have swelled both relatively and in absolute numbers.

The U.S. Census Bureau gathers yearly statistics on the extent of poverty, a very important indicator of financial well-being. It classifies people as poor if their money income is below a certain threshold The level of poverty is revised each year to account for inflation. It is calculated from the Department of Agriculture's economy food plan, estimating what it costs to feed a typical family. At the time of this measurement's introduction in the early 1960s, about one-third of the budget of a poor family went for food. The cost of the market basket is accordingly multiplied by three to obtain the poverty cutoff point. These thresholds vary somewhat depending upon a family's characteristics. Obviously, a larger family has more mouths to feed and thus needs more money. The elderly need less because their food consumption normally goes down as they age. For those with income in 1993, the poverty threshold varied from a low of $6,930 (one person over sixty-five years old who is living alone) to $32,003 (a family of nine or more persons). For the average family of four persons, the guidelines assign them to poverty status if they received less than $14,904 in income a year. This means, of course, that a comparable family receiving $14,905 would *not* be poor.

To be sure, there are those who argue that our welfare benefits are too generous and that the way we measure poverty artificially inflates the number of the poor. Former President Ronald Reagan believed eligibility cutoffs similar to these were too high since they allowed social assistance programs to serve persons other than the "truly needy." Some critics correctly point out that the income of those we consider poor does not include the dollar value of transfer payments and in-kind benefits such as food stamps and Medicaid—nor the equalizing effect of federal income tax policies.[50] The alleged benefits to poor persons of changes in our tax structure will be discussed in a later section. A more detailed discussion of the problems and pitfalls in subtracting the value of in-kind benefits is also available elsewhere.[51] Among the most obvious weaknesses is the difficulty of valuing such benefits in monetary terms. While this approach may be straightforward for food stamps, recipients cannot eat medical benefits. Moreover, a researcher has to take great care in making adjustments to the data to avoid bias. Beeghley and Dwyer appropriately point out that, when measuring inequality and including the impact of transfer payments to the poor, researchers should also add the effect of income transfers that benefit the middle class and wealthy (e.g., tax policies such as mortgage interest deductions, exclusion of capital gains, employer pension and health insurance contributions, etc.).[52] When such middle-class benefits are included in analysis, the net effect cancels out any alleged redistributive benefits to those with low income.

To be sure, the way poverty is currently measured is far from perfect. The Census Bureau, which gathers the yearly poverty statistics in its March Current Population Survey, admits that several improvements could be made in the data.[53] Aside from including in-kind benefits such as food stamps when measuring income, taxes need to be subtracted to derive a better index of real disposable income. Spending for food no longer represents one third of a family budget according to the Current Expenditure Survey (CEX). Most families now require a lot more than three times their food budget to adequately function in today's society, since childcare, housing, transportation, and health costs are relatively more expensive. No adjustments are made in income for people in different regions of the country, despite very real variations in the cost of living from one region to the next (it is much cheaper to live in the South and/or in rural areas).

The various methodological and conceptual difficulties which have mounted in the thirty-some years since the measure's inception now demand remedy. Accordingly, and by special invitation of Congress, the National Research Council established a panel to address concerns about how poverty is measured. Their report is now complete and contains a variety of recommended changes to meet some of these concerns.[54] For example, the experts now agree with conservative critics that the dollar value of food stamps, subsidized housing, school lunch programs, and home-energy assistance should be counted as income. In 1992, 36.8 million persons were officially designated as poor under current measurement techniques. Counting these in-kind benefits would reduce the poverty count by 4.2 million individuals (11.4 percent). Yet, it is only fair to also subtract state, federal, and payroll taxes (such as Social Security) from income, which adds 1.2 million to the poverty register. Another 2.7 million are added when the cost of work-related expenses (transportation, uniforms, dues, child care, etc.) is taken into account. A whopping 5.3 million cross the line into poverty when out-of-pocket expenditures for medical and health care are added into the economic equation. In all, the analysis concludes that if all of the panel's recommended changes were adopted by Congress, anywhere from 9.1 to 11.4 million additional persons could be added to the poverty count.[55] Although many of these changes are long overdue, it is highly unlikely they will be instituted by politicians currently engaged in slashing welfare programs. The belief by the majority in Congress is that the nation is now strapped for cash, and can no longer shoulder the burden of social programs as it has in the past. Ultimately, it is politically impossible for elected officials to admit that poverty is an even worse problem than we previously realized, and then go about the business of tearing down the nation's social safety net.

Most experts *are* of the opinion that poverty—as currently measured—is vastly understated. The nearly unanimous feeling is that the way poverty is defined by the government is far too low to be realistic in identifying those who are greatly deprived.[56] A study by Andrew Winnick, for instance, states that the official definition of poverty vastly understates the extent of the problem.[57] Instead, he presents alternative measures which lead him to conclude that as many as one-third of American families are poor. Lee Rainwater, in an analysis of U.S. poverty rates from 1949 to 1989, introduces a number of refinements in its measurement—such as an equivalence scale to account for need based upon different family sizes and ages of household heads. He also bases his definition of poverty on what the American public thinks it ought to be, which has consistently been half of mean household income. Even when noncash benefits such as food stamps are counted as income, Rainwater concludes that there has been a tangible rise in poverty during the 1970s and 1980s. By 1989, his measure categorizes 19.1 percent of persons as in poverty, a rate that is half again as large as the official level of 12.8 percent.[58] Using different methodology, Rodgers and Rodgers not only support the conclusion that there has been a real rise of poverty in the past two decades, but that it has become chronic and less transitory in nature.[59]

The surface statistics about duration of poverty suggest otherwise, since 60 percent of all persons whose income fell below poverty had spells that lasted no longer than one year. The Census Bureau estimates that in the early 1990s, 22 percent of Americans who lived in poverty during a given year managed to escape it in the next (this means, of course, that three out of four poor persons remained in poverty during the next year).[60] At the other extreme, a fifth of all poor persons live in poverty for more than seven continuous years. And while it is true that most persons escape poverty after a couple years, they do not go very far. Typically, persons who are newly risen above the poverty line rarely have income more than twice the poverty threshold. Hence, to rise out of poverty is not to gain hard-won middle-class status. Rather, low-income people hover just above the line and remain vulnerable to dipping below it from time to time.[61] The degree of poverty risk is also substantial. Over a decade, fully one out of every four Americans will live in poverty at least one year.[62]

The official statistics on poverty are not all negative. In contrast to their impoverishment in the 1960s and 1970s, the elderly are no longer heavily represented among the poor. This is one of the few bright spots among the dismal figures documenting a nearly universal rise in poverty. Because Social Security was indexed in the late 1960s to automatically increase payments to offset inflation, the proportion of elderly living in poverty is today only one-third the rate that it was three decades ago. The poverty rate among those sixty-five years old and over is essentially equal to that of working-age persons in the eighteen to sixty-four year old bracket (12.2 percent and 12.4 percent, respectively). This is a vast improvement over the past, especially during the early 1960s when stories of the elderly subsisting on dogfood were widely circulated.

In the analysis of poverty, much depends on where we start. Tracking the poverty line from 1959—the year in which poverty measurement first began—shows a huge decrease in both the number and percentage of the poor, at least until 1973. The number of persons in poverty went down from 39.5 million to just under 23 million in this fifteen year period, and the rate was actually cut in half (from 22.4 percent to 11.1 percent).[63] These figures are conveniently ignored, however, by conservative analysts when they claim that the War on Poverty during the 1960s did not work. A mere glance at the record proves otherwise. The War on Poverty, starting with the Johnson administration, eventually raised 15 million persons out of poverty from 1959 to 1973, while the rate of poverty plummeted from 22.4 percent to 11.1 percent. Because of such new antipoverty programs as free school lunch and breakfast programs for poor children, the Food Stamp program, and so on, hunger in America was virtually eliminated during the 1960s and early 1970s.[64] In addition to the legion of newly-introduced, ambitious antipoverty programs, the huge sums spent on the war in Vietnam helped keep people working. The decline in poverty also took place at the tail end of American hegemony, before it had become evident that the U.S. economy was faltering. In essence, manufacturing jobs were still relatively available and high-paying. It is more than a coincidence that the upturn in poverty started in the same year that the real wages of American workers started their skid.

The gains made for those in poverty during the 1960s and early 1970s faded during the 1980s. Instead, transfer programs were hollowed out and income erosion set in. Figure 5.4 traces another unmistakable rise in poverty, peaking in 1983 just at the end of the very severe 1982 economic recession.[65] While both the number and percentage of poor persons declined a few years later, by the start of the 1990s poverty had returned with a vengeance. By 1993, the last year for which data are available, there were still nearly 40 million people living in poverty in the United States. In that year, the poor comprised more than one out of every seven people. In the past fifteen years, the number of poor people in the United States has increased by over 50 percent—translating to 13 million more people pushed into poverty since America began its great experiment of slashing social welfare programs. The proportion of all persons in poverty went from 11.7 percent in 1979 to 15.1 percent in 1993, reflecting a rate rise of nearly 30 percent as well. The rate at which

FIGURE 5.4 U.S. Poverty Rates by Race (Percent of All Persons in Poverty)

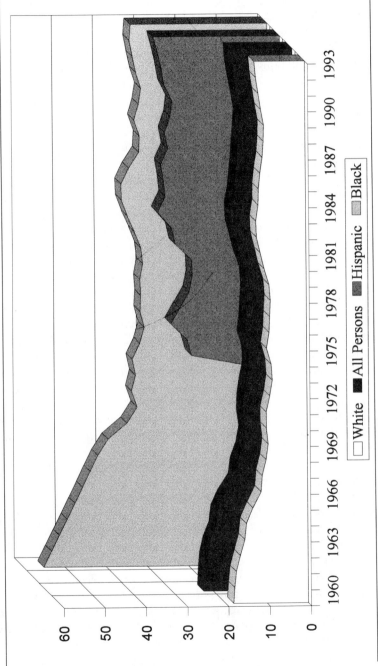

Source: U.S. Census Bureau, *Income and Poverty 1993*, CD-ROM.

poverty has increased in the past fifteen years is nearly double the rate at which our population has grown (16.3 percent). Today's poverty rate (excluding the recession-induced rate in 1983) is the worst within the last quarter of a century. We have to go back to 1965 to find a poverty rate higher than our current one. These figures are especially ominous in that the 1992 and 1993 poverty rates should have declined because of the economic recovery from the 1990–1991 recession. The fact that poverty is worse than ever despite economic growth signals hard times and hopelessness for literally millions of Americans.

The number of poor people today is greater than in any year since 1961. The 39 million plus Americans who suffer under poverty number well above Canada's entire population of 27.4 million. In fact, the large number of our nation's poor equals the entire population of Spain. Currently, only twenty-seven countries in the world have populations larger than the number of poor people in the United States. Equally menacing, the depth of poverty has increased dramatically in the past decade and a half. The Children's Defense Fund reports that life below the poverty line is even more harsh today than thirty years ago.[66] In 1959 the poverty level for a family of four was 57 percent of median family income. If this proportion had remained constant, the official child poverty rate at the start of the 1990s would have been nearly one in three children instead of one in five. Despite this increasing severity, one-third of all poor families do not get food stamps and three-fourths do not get any housing assistance. Estimates are that one in every eight American children under the age of twelve suffers from hunger—a total of 5.5 million children. Tragically, ten thousand children die from poverty-related causes in the United States each year.

Between 1979 and 1993 the proportion of the poorest families (those with income lower than half of the official poverty cutoff) grew by 63 percent. In 1979 only 3.8 percent of all persons were what could rightfully be deemed severely poor; by 1993 the percentage stood at 6.2. On the other side of the poverty cutoff, the near poor have also increased their presence. Fully one in four Americans has income only 50 percent above the poverty cutoff, an increase from about one in five in 1979. By 1993, the average amount needed to escape poverty for a family with at least one child was close to $6,500 (or a little more than $2,000 per family member). When the income deficit (poverty gap) is charted in constant 1990 dollars to control for the effects of inflation, it becomes crystal clear that the poor have been getting poorer. The mean deficit per person that poor persons needed to escape poverty has increased by about one-fourth since 1977 and is today greater than at any point in the measure's history (beginning in 1959).[67] It is important to note that the increase in the number, proportion, and severity of poverty is not accidental. These trends occurred at the same time as a major dismantling of the social safety net was taking place in the country. As a result, there was actually less help available to aid the indigent when they needed it the most.

The brunt of poverty actually falls on some people more than others, as can be seen in figure 5.4. We have long known that blacks suffer in poverty at much higher rates than do whites. In 1993, one of every three blacks was poor. This rate is nearly three times higher than for whites (12.2 percent). The black poverty rate also seems to be carved in stone, showing very little change since 1968.[68] Today, more than 30 percent of Hispanics are also poor, but their rate displays more volatility. In 1973, when data were first gathered on Hispanic poverty, their rate was 21.9 percent. In essence, the rate of Hispanic poverty is now 40 percent higher than it was two decades ago.

The poverty rate for whites also shows a worsening over the past two decades. Michael Harrington has provided evidence that the new element in this picture is the growth of poverty for white males who can no longer find jobs which pay enough to let them live in a decent manner, mainly because of deindustrialization.[69] Only 8.4 percent of whites were poor in 1973. During the 1980s the average yearly poverty rate for whites was 11 percent. During the first four years of the 1990s the rate has averaged 11.5 percent. Although the 1990–1991 recession has influenced the latest poverty figures for all groups, the severe recession of 1982 also influenced figures for the last decade. In short, poverty seems to be going up for just about everyone, regardless of race. Although the problem of poverty is worse among minorities, during the 1980s it broadened to include all segments of American society.

Despite the fact that a majority of American families may have been able to stay afloat in terms of real income within the past decade or two, nearly one out of eight families is now in poverty (12.3 percent). Prospects are especially grim for female-headed families. Over one-third of these families are poor, while over half of all poor families are headed by single females. The poverty rate for families with children has increased by over half— jumping from 14.2 percent in 1973 to 22 percent in 1993. In that year, nearly 16 million children were living below the poverty line, comprising over 40 percent of all poor persons.

In America today, one in every five children and nearly half of all black children are trapped under poverty. Despite higher poverty rates among minorities, however, it is important to repel cultural stereotypes and to stop blaming the victim.[70] At the start of this decade only 10 percent of poor children were black in a female-headed family on welfare in a central city.[71] Only one in forty-four, in such a family, had a mother who was a teenager. In fact, many of these are two-parent families, and both parents are working.[72] For example, half of poor family householders manage to work. Indeed, two of every three poor married-couple families have at least one worker—and in one-fourth of such families both spouses work.[73]

The figures become even more devastating when U.S. performance on helping its poor population is compared to that of other industrial countries. Despite the fact that the United States experienced more economic growth and less unemployment during the 1980s than most industrial nations, our poverty rates were twice as high as those of European countries.[74] Moreover, the United States had a much higher proportion than did other countries of poor households, with children, that remained poor for an extended period.[75] One in seven young households with children was locked into being poor for at least three years in the United States, compared to fewer than one in fifty for the former West Germany, France, Luxembourg, and the Netherlands. The depth of American poverty is more severe as well. We are first in comparison to seven other advanced nations in the proportion of children living in families with income 25 percent below the poverty line. While about one in ten U.S. children is this impoverished, comparable rates for other industrial countries hover at around 2 or 3 percent.[76] When government tax and transfer payments are calculated, only one in twenty single-parent families with children was lifted out of poverty in the United States, compared to a fifth in Canada, a third in the former West Germany, half in France, three-fourths in the United Kingdom, and 90 percent in the Netherlands.[77]

Lee Rainwater and Tim Smeeding have adopted most of the recommendations issued by the National Academy of Sciences panel on poverty measurement in their latest comparative research. In sum, they use disposable money income after adding the dollar value of in-kind benefits and subtracting taxes, while adjusting for household size and age of head. The poor are defined relatively as children under eighteen in families that receive less than 50 percent of a country's median personal income. This study yields the first glimpse of where child poverty is headed in the 1990s. Unfortunately, the trends appear to be a continuation of those begun in the 1980s. In a study of fourteen countries where at least two different measures were taken, the researchers found only two countries (Canada and Sweden) where there had been slight progress in reducing child poverty during the 1980s. Otherwise, there is either stability in the rates (Scandinavian and most European countries) or outright deterioration in children's economic well-being (the United States, the United Kingdom, and Israel). A major point of their study, however, is that nations can do a great deal to reduce child poverty even when market forces act to raise it. Figure 5.5 summarizes one of the main findings of their research, which essentially shows the impact government programs

FIGURE 5.5   Child Poverty Rates Before and After Government Programs

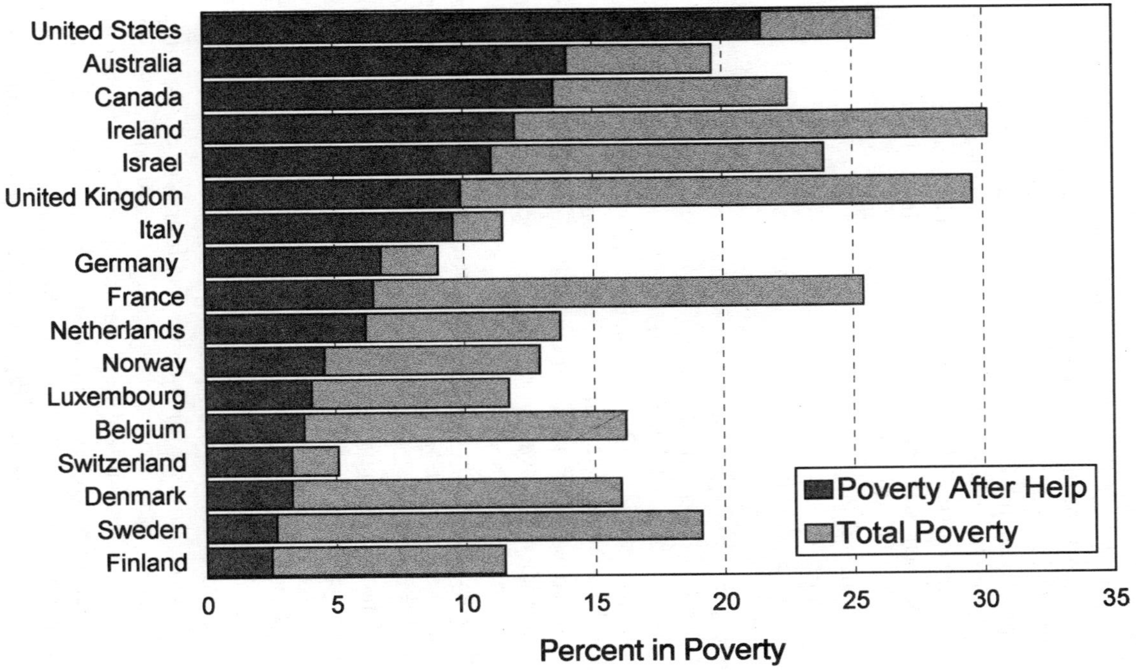

Source: Rainwater and Smeeding, 1995, table A2.

have in helping children escape poverty. It is quickly apparent that a couple of countries nose out the United States in its high rate of overall poverty (25.9 percent), such as Ireland (30.2 percent) and the United Kingdom (29.6 percent). France's rate is essentially equal to the U.S. rate at 25.4 percent. Part of the message from this graph is that child poverty is huge to begin with in a lot of wealthy, industrial countries. Although every country was able to reduce child poverty through a variety of government programs, the United States is not comparably effective in helping its poor children. After all is said and done, we lead every developed country in our post-transfer poverty rate—and by a large margin at that. All European countries (with the exception of Ireland) have been able to reduce their child poverty rates to below 10 percent while ours continues at an astronomical 21.5 percent. Seven of these countries have child poverty rates *under 5 percent*.

Figures such as these are often dismissed by apologists who claim that our poor are really well-off because the United States is such a wealthy country to begin with. This is certainly true in contrasting our poor children to those in Mexico and Bangladesh, but comparing our poor children to those of America's industrial peers yields a different conclusion. Measured by GDP per capita, many of these nations are now wealthier than the United States. Rainwater and Smeeding adjust their data for differences in Purchasing Power Parities (PPPs), which account for the availability and lower cost of goods in the United States. Even with this adjustment, the real spendable income of the typical poor child in fifteen out of seventeen other wealthy countries is actually higher than the income of an average poor child in America. Only in Israel and Ireland are poor kids worse off. In six of these countries—Switzerland, Sweden, Finland, Denmark, Belgium, and Norway— poor children have real standards of living at least 50 percent above that of poor children in the United States. In Germany, Luxembourg, Austria, and the Netherlands low-income children are at least 30 percent better off than those in the United States.[78]

What has produced this appalling tailspin for our nation's poor? According to one estimate, only about one-fourth of the rise in poverty rates for families with children is due to an increase in the percentage of single parent families. Moreover, much of that rise can also be explained by the numbers of newly divorced women with children—the majority of whom get off welfare after only two or three years. One study found that nearly one-third of the worsening poverty rates for children stems from increasing unemployment and lower earnings discussed above, while 42 percent is due to the declining effectiveness of government income programs in bringing families above the poverty line.[79] A congressional study has also sought to identify the major sources of this dramatic increase in poverty among children.[80] The study's researchers point out that the reason why nearly half of poor children are forced into poverty is because the work hours of one or more parents have been reduced. Only one-fifth of poor children have been driven into poverty due to the rise in numbers of households headed by females.

Partly to blame for the sharp rise in numbers of children in poverty is the systematic attack on social welfare programs that has been going on for the past fifteen years. One of the most widely publicized and infamous changes was the decision by officials in the Reagan administration to recategorize ketchup from a condiment to a vegetable in the federal school lunch program. This helped them pinch a few more pennies from the budget while depriving poor children of badly needed nutrition. There were ongoing slashes in the benefits paid for Aid to Families with Dependent Children (AFDC) as well as cuts in other basic welfare programs. Less dramatically, benefit levels have not been increased to offset inflation. A major cause of the rise of poverty among children has been the reduction in AFDC benefits paid to single mothers. After accounting for inflation, the average monthly AFDC benefit per family skidded from $676 in 1970 to $373 in 1993. For a three-person family, the median benefit has been essentially halved (–47.0 percent) over the past quarter century.[81] This drop in the effectiveness of AFDC support took place at a time when eligibility for AFDC tightened dramatically, so that only 60 percent of poor children received AFDC benefits prior to the end of the program in 1996 compared to 72–80 percent in prior decades.[82]

The amount of welfare dollars single-parent, female-headed families with children receive has never been generous in the United States. But the erosion of benefits is an added weight poor families must now bear. In 1973, nearly one in nine persons (11.8 percent) in such families was removed from poverty by cash public assistance (the great majority from AFDC). By 1990 this proportion had been virtually cut in half and stood at 5.9 percent.[83] Put differently, at the start of this decade only one of every seventeen persons was being removed from poverty. The variation from one state to the next is breathtaking as well, since each state is allowed to set its own need standards for poor families. For a family of three, maximum AFDC benefits ranged from $120 per month in Mississippi to $923 in Alaska. Not surprisingly, states where support is almost nonexistent fail miserably in removing families from poverty. In 1993, no state was able to do so on the basis of AFDC benefits alone. Figure 5.6 shows AFDC benefits in each state as a percentage of the 1993 poverty threshold for a one-parent family of three. Southern states are the worst offenders for nonsupport. The most generous states are in the northern Midwest, the urban Northeast, or on the West Coast.

FIGURE 5.6   AFDC Benefit for Family of Three as Percentage of Poverty: 1993

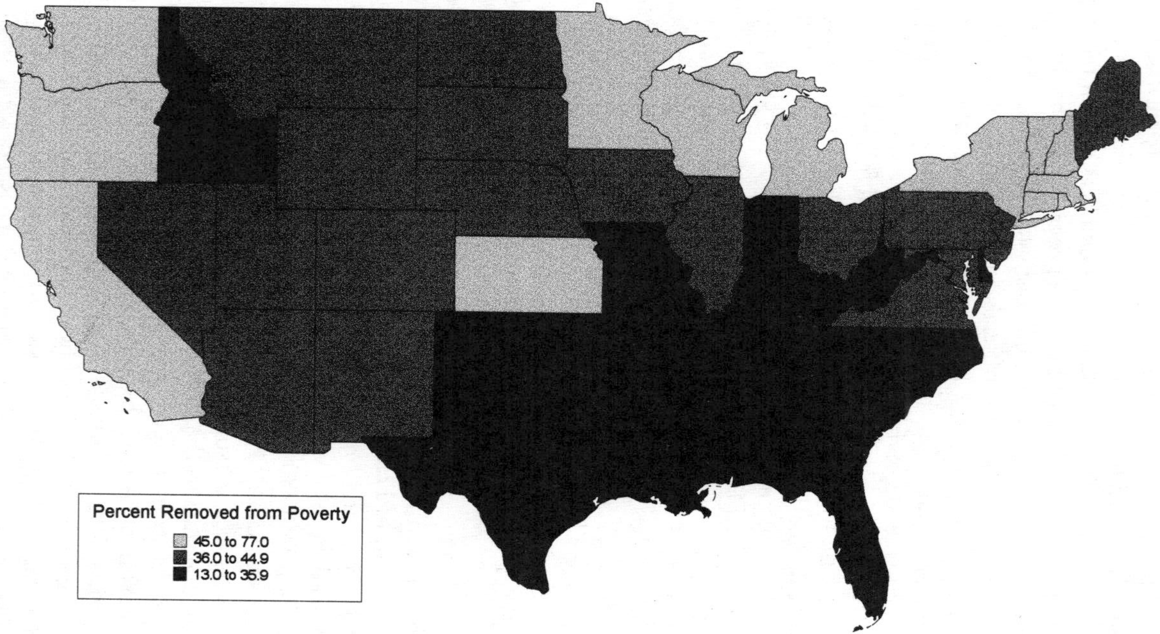

Percent Removed from Poverty

45.0 to 77.0
36.0 to 44.9
13.0 to 35.9

Source: Data from U.S. House of Representatives, 1994, table 10-11.

Much of the new poverty now emerging has hit the poorest of the poor. For example, in 1972 every single state paid some AFDC benefits to families with wages under 50 percent of the poverty line; by the end of the 1980s this was true of only thirty states.[84] As of 1994, no state paid enough in AFDC and food stamp benefits to raise a poor family of four above the poverty line. In fact, the maximum amounts for AFDC and food stamp assistance lifted such a family to only within 70 percent of the official poverty line of $14,763 in 1993.[85] The only area less generous than our welfare system is the so-called safety net set by the minimum wage, which remained unchanged at $4.25 per hour from 1990 to 1996. At this wage a full-time worker (forty hours per week for fifty weeks) would earn only $8,500 per year. A person would have to earn 26 percent more per year to lift a family of three out of poverty, or 42 percent more to remove a family of four from poverty. It is immediately apparent that working your way out of poverty in today's low-wage economy is impossible, although in the 1970s a minimum wage job earned workers enough money to do so.

The harshness of these changes becomes evident in a Congressional Budget Office study of how the nature of poverty has recently changed in the United States. There was an increase in the poverty rate of 41 percent over the past decade, just prior to the 1990–1991 recession. Principally as a result of the war on welfare, nearly 9 million more persons were pushed into poverty during the 1980s alone. The researchers conclude that almost half of this shift was due to ripping apart the social safety net, while one-fourth resulted from sheer population growth and another one-sixth came from a slide of job income.[86]

Since the social safety net has been shredded, it is not surprising that the number of homeless has been on the rise in the United States as well:

> In recent years, destitute men and women have become an increasingly common sight in parks, downtowns and suburbs. Their growing numbers, constant visibility and costly needs have pushed homelessness to the forefront of issues confronting local, state, and federal officials. "A lot of people thought the homeless thing was a short-term phenomenon," says Democratic Rep. Bruce F. Vento of Minnesota, the sponsor of a permanent housing plan introduced in the House in January. "What is becoming evident is it is becoming a permanent issue."[87]

Estimating the number of people within this category of dire need is difficult, and is frequently biased by political motives. An early estimate issued by officials under the Reagan administration, seeking to minimize the problem, came up with approximately 300,000 persons. By the end of the 1980s, most expert opinion put the number of homeless at between 560,000 and 680,000.[88] The Census Bureau came up with 400,000 as it valiantly tried to locate and count this population in the 1990 Census, but it admits that the numbers are shaky to begin with. Advocacy groups for the homeless say the number is between 700,000 and 3 million. A Columbia University study conducted by telephone survey (thus, biased toward a low estimate since the interviewees were not homeless) discovered that nearly 13.5 million Americans have been homeless for at least a few days during their lives—actually living in shelters or on the streets.[89] An additional 12.5 million have been able to stay off the street by moving in with others. The latest report by a task force organized under the Clinton administration judges the count to be 7 million, flatly stating that 600,000 persons are actually on the street on any given night.[90] Not included in this tally were millions of families who were on waiting lists for public housing, were seeking rent assistance, or had moved in with relatives or friends because they could no longer live on their own.

Approximately one-fourth of the homeless are families (mostly headed by women). Recalling some of the worst scenes in a Charles Dickens novel to describe modern-day America still cannot do justice to this calamity. In the worst-case scenario, there are 1.5 million children without a permanent roof over their heads in our country today. Research by psychiatrist Ellen Bassuck found that homeless preschoolers are much worse off than children in low-income families with permanent shelter.[91] She has documented that over half of these children have one or more developmental deficiencies, compared to 16 percent of children who have homes. As for the adults in this group, one in five admits to having been hospitalized for mental problems while half express or show signs of severe depression. The median incomes for homeless single people ($64 a month) and homeless families ($300) fall far below 50 percent of the poverty cutoff. It appears, from the data gathered so far, that their numbers have become swollen for a combination of reasons. In some cases homelessness may be primarily the result of deinstitutionalization of people incapable of caring for themselves (another example of the rapidly evaporating social safety net). In most cases, it is the sharp rise in poverty together with its deepening, frozen status that has cast so many people adrift. Another important contribution to homelessness has been the severe cutback of federally subsidized housing construction for low-income persons, which almost disappeared during the last decade.

Well over one-third of the homeless eat only one meal a day or less, while only a fourth get three meals a day. In 1991, more than 20 million Americans depended on soup kitchens or food banks for meals.[92] Poverty has taken its toll. The Food Research and Action Center in Washington, D.C., estimates that in any given month, nearly 5 million children will experience hunger. A separate analysis from the Center on Hunger, Poverty and Nutrition Policy at Tufts University believes that, in 1992, approximately 30 million Americans were hungry at any given time.[93]

Given all the facts discussed, there is little doubt that the economic status of the poor is declining. Despite the growing evidence of desperate poverty, frequent attacks on the poor have continued. For example, former President Ronald Reagan stated just before leaving office that the homeless are without shelter because that is their "choice." The poor, starving, and homeless are blamed for their lack of income, as if they are somehow defective. They are shamed for needing help. "Welfare queens" are said to retire in luxury by having children in order to live off the taxpayer. Irresponsible, promiscuous teenagers with raging hormones and not enough will power to "just say no" are accused of swelling welfare rosters when they get pregnant. Enough of a trend exists to perpetuate some of these stereotypes. For instance, there has been an undeniable rise in single-parent families due to the ever-growing number of divorces. All family households grew by 16 percent from 1970 to 1980, 11 percent from 1980 to 1990, and 3 percent from 1990 to 1993. For female-headed households, with no spouse present, comparable figures were 58, 25, and 10 percent. Put differently, one in nine children was living in a mother-only household in 1970, and this had mounted to nearly one in four a quarter century later.[94]

We also know that the creation of female-headed families has a great effect in casting these women and their children into poverty. Of all the poor families in the United States, one of every two is headed by a woman with no husband present. Yet there is little support for the stereotype that the mothers of these children are lazy, uncaring, and unwilling to work. In 1992, 54 percent of poor family householders worked, compared to 72 percent of all family householders.[95] The labor force participation rate of poor women is high, considering they have children and no other adult to help with child care. The majority of people who entered the AFDC program stayed in it for less than two years, although the potential for reentry at a later date remained elevated because of the severity of poverty.[96] There was no correlation in states between higher AFDC payments and higher rates of poverty among children. Nor was there any evidence that such welfare benefit levels encouraged out-of-wedlock births.[97] Studies consistently show that people do not go on welfare to make money and that they do not have babies promiscuously to receive higher benefits.[98] Nearly three-fourths of welfare families have only one or two children.[99] Although there is evidence that welfare benefits moderately increase divorce, the findings are not very consistent.[100] According to one estimate, welfare can account for at most only one-seventh of the growth in numbers of single mothers.[101]

At bottom, the attack on welfare programs has often been disguised racism built upon the false assumption that the real cause of poverty is the indiscriminate promiscuity of African-American women. While single black women with children form a relatively larger portion of poor families, African-Americans as a whole are more likely to be poor. What many analysts have failed to note is the dramatic rise of poverty and low income among white males in the United States. Females are still more likely than males to be poor, due to many of the factors discussed above. But for a time during the last decade the proportion of poor males grew at a rate nearly triple that of poor females.[102] Michael Harrington argues that much of the rise is due to the *new poverty* among young workers.[103] It may especially affect males, who can no longer find and keep high-paying industrial jobs that were once abundant. This assertion is in line with the previous deindustrialization argument which states that the erosion of good, well-paying jobs in favor of more part-time, low-wage service jobs has left a large hole in the typical American's pocketbook.

Many analysts, whether conservative or liberal, now seem to focus on the idea that poverty hinges on how well our economy does. A position is emerging that anti-poverty programs and the social safety net have less to do with economic well-being than does continued growth in the economy. Common sense would seem to support such an argument, since a thriving economy would tend to drive up wages and employment levels. Yet there is a need for caution before embracing the idea that we can "grow our way out of poverty (debt)." Such an argument is the same as the one used to justify development in the Third World. Economic growth did not work to benefit the masses in most LDCs. Nor is there any reason to believe that boom times will necessarily lead to less poverty in the United States. In a study of whether economic growth reduces poverty, it was found that increased real income was *not* associated with any decline in poverty.[104] Economic growth (such as technological improvements, increasing productivity, etc.) was important in the 1950s and 1960s. Yet its impact had dropped to almost nothing by the 1970s and 1980s. During this period, growing gaps in the income inequality distribution became a major force in rising poverty.[105] In prior decades, economic expansion—reflected by changes in unemployment and inflation—almost perfectly predicted a reduction in poverty. In the 1990s, the correlation has eroded to the extent that the poverty rate is nearly one-fourth greater than it should be, based upon economic expansion.[106]

There is much greater consensus on the need for full employment as an effective way to wean people from poverty. Cyclical business changes —with associated unemployment and employee layoffs—are a major cause of poverty. In a review of U.S. poverty research, Isabel Sawhill found that "slack labor markets and poverty tend to go hand in hand."[107]

If any factor emerges as the most potent in raising or lowering poverty, it is unemployment. This development bodes ill for the future, since U.S. unemployment rates have been rising for most of the past two decades. Our unemployment rate averaged 4.8 percent in the 1960s, 6.2 percent in the 1970s, and 7.2 percent in the 1980s.[108] Most experts acknowledge that the 8 million Americans who were counted as unemployed in the mid-1990s are really an underestimate, because of the way the government officially measures unemployment. You are not "unemployed" if you really want to work but have become too discouraged to continue looking for a job. Counting such individuals would add half a million to the jobless roster. Another 4.3 million who are working part-time because they cannot find full-time jobs should probably be added to the unemployment slate.[109]

It also needs to be noted that the unemployment rate is a monthly snapshot. The number of persons with at least one spell of joblessness during a year is usually two-and-one-half to three times as large as the official unemployment rate.[110] In any given year, therefore, it is possible that one out of every five American workers will be out of work at some point. Moreover, some decline may never be registered by unemployment statistics. A fall in the demand for labor may not result in outright layoffs so much as in more part-time work and/or reduced hours. The effect is the same. All have the result of whittling away at family income. More to the point, a downturn in the economy is not spread evenly over all income groups. The well-off are more insulated and may even benefit if their investments earn higher interest rates. The working heads of poor families, however, suffer relative income losses three times those of middle-income families.[111]

The point worth emphasizing in the debate about what factors are important in reducing poverty is that a growing mean or per capita income will not necessarily help the poor. When the economy is booming, the well-to-do benefit the most. On the downside, however, the poor are impoverished even more. Macroeconomic policies are important in helping the poor, therefore, but only to the extent that they encourage full employment.

In their rush to pay tribute to a laissez-faire doctrine, procapitalist conservatives have alleged that the war on poverty was a failure. For example, critics such as Charles Murray have accused social programs of creating welfare dependency, of encouraging the poor to remain so because of benefits, and of failure to eradicate poverty.[112] Some empirical evidence claims to show a U-shaped curve between welfare spending and the percent of people in poverty.[113] It has been asserted that benefits can actually rise to the point where they are attractive to people, who then choose to stay on welfare rather than work. A reanalysis of what is called the *Laffer Curve* indicates that this is a false relationship.[114] Instead, the tendency is in the direction we would expect. Poverty did decline in the 1960s as welfare spending rose. The proportion of poor people increased in the late 1970s and early 1980s when benefits leveled off and then went down. Part of the error in the early research was in using per capita income as a predictor of poverty. This is a gross average masking other factors such as recessions. Other miscalculations in the early research stemmed from looking only at federal (not state or local) spending, using spending per person of the entire U.S. population rather than of those who actually received aid, and ignoring in-kind benefits (e.g., food stamps).

Isabell Sawhill's thorough review of the literature yields no support for the idea that welfare *causes* poverty.[115] She notes that the safety net has worked admirably in some places (especially for the elderly, but also in medical assistance and compensatory education programs such as Head Start). In her summary of the research, the inescapable conclusion is that poverty would have been much worse were it not for the existence of income transfer programs. Despite all such governmental efforts, she sees continued poverty as stemming from past high unemployment and failure of average real income to grow over the past twenty years. Robert Moffitt has specifically examined the AFDC, Food Stamp, Medicaid, and public housing components of our welfare system. Although he found that AFDC and Food Stamp programs exerted a small influence on poor mothers in discouraging work, it is not nearly large enough to explain the high poverty rates among female heads of households. In fact, he reports that reducing benefits has little net effect on the labor supply, and he does not find any relationship between a viable degree of welfare support and the growth in numbers of female-headed families.[116]

It is proper to question the effectiveness of any government program. In the end, however, this should also include the various "wealth-fare" programs that benefit corporations and the very rich, and the exorbitant scale of Pentagon budgets. One can argue that by draining money needed for productive investment in our nation, such prolific spending serves to perpetuate poverty while increasing income inequality. Above all, the very welfare programs that conservatives are so fond of attacking, whether relatively effective or not, would be largely unneeded if more humane decisions were made in the private sector. The health of the U.S. economy—together with the jobs, income level, and poverty rates it produces—is dependent upon the behavior of huge corporations. Unfortunately, their decisions have led to moving plants and jobs abroad or into low-wage areas of our country. Their ultimate effect has been to reduce income as a whole and to drive up poverty rates.

The Decline in Wages

In an attempt to convince the American electorate that our country has been enjoying an unprecedented economic boom, conservative politicians often brag about the continuing increase in per capita income our nation experienced during the 1980–1992 era of Republican presidents. In essence, supporters argue that there was more money available in a theoretical sense for each man, woman, and child in the United States in the 1990s than ever before. The figures, often cited by supporters of Reaganomics as evidence that their economic policies have been working, are partially true. In 1992, per capita income stood at $15,033, over 13 percent higher in real dollars than in 1980. It should also be noted, however, that per capita income (in constant 1992 dollars, with inflation subtracted) peaked in 1989 at $15,904 and dramatically declined through 1992. Today, per capita income is no greater than it was in the mid-1980s.[117]

Aside from the recession-induced drop of the early 1990s, there are more reasons to be entirely skeptical about accepting these figures as "proof" that America has been enjoying economic growth and prosperity. Per capita income is calculated by dividing total money income by the entire population. Thus, per capita income is a measure that is similar to mean GNP per person, so often used to show how development policies have worked in LDCs. As seen in previous chapters, real GNP per person can increase in the Third World without necessarily benefiting the masses. The same can be said of the United States with regard to per capita income. The increase in per capita income is a single, gross average that hides a large-scale decline of real income for most Americans. While real average income has been largely stagnant or slipping since 1973 for most workers, some people at the top of the pay spectrum are definitely making more money. The real explanation lies in a larger dose of income inequality. What has been happening in the United States over the past few decades is a very fast rise of income for top businessmen, chief executive officers, and stockholders coupled with a serious dip for blue-collar, retail, and service workers. When persons at the richer end of the income distribution receive ever greater amounts of money, they will pull per-capita income up even while median income remains basically unchanged. This rise shown in higher per capita income has not filtered down to most Americans.

In fact, comparing *mean* income (the arithmetic average, where all income is added up and divided by total households) to *median* income (income at the exact midpoint, or fiftieth percentile, of the household income distribution) does show increasing income inequality. Using constant 1993 dollars, a household in the exact middle (the median) of the income distribution received $31,095 in 1980. This had risen a paltry 1.5 percent to $31,553 by 1992.[118] In the very same time period, mean income increased from $36,982 to $40,003—a gain of 8.2 percent. It is well known that the American income distribution is skewed to the right,[119] so it comes as no surprise that mean household income in 1992 was over one-fourth greater than median household income. What is more interesting is the fact that twelve years earlier the difference was only 18.9 percent. Under the Republican presidencies of Reagan and Bush, therefore, the difference in the proportion of median to mean income increased by nearly 42 percent. In essence, the rich are getting richer at a faster rate—especially the top 3 percent of all income recipients—while the growth of income for the majority of people is mostly nonexistent or quite modest.[120]

Folke Dovring believes this widening gap in median income divided by mean income is a measure that closely approximates the Gini ratio.[121] In practice, this usually produces a number between .900 (low inequality) and .700 (high inequality). The inescapable conclusion for Dovring is that household income inequality has been increasing and is projected to continue climbing in the future. He concludes that only one-fourth of the expansion in income inequality is due to changing demographic factors. In his view, the evidence is solid. The rich now get a larger share of personal income while the poor receive a lesser share than before. In real terms, the lower half of the American population now has less income than it had a decade earlier.

To begin to make sense of what has been happening to the income picture in the past two decades, it is important to distinguish wages, earnings, and income as defined in government statistics. Wages reflect the amount paid to production and nonmanagement workers in the private sector (excluding farm and government workers). Wages do not reflect the earnings of professionals, business managers, and others who draw salaries. Nonetheless, wages are paid to 80 percent of all those employed. Whether wages grow or decline will thus have a major impact on the economic well-being of most Americans. Earnings include wages, plus salaries and income from self-employment. Income figures cover an even wider gamut of sources—such as wages, salaries, government transfer payments, savings and bond interest, rental income, child support, alimony, and pensions. In short, income includes just about all the sources we are required to report on our income tax forms. By looking at all three data sources, we can see which groups in our economy are surging ahead and who is being left behind:

FIGURE 5.7   Weekly Industry Earnings (Constant 1982 Dollars)

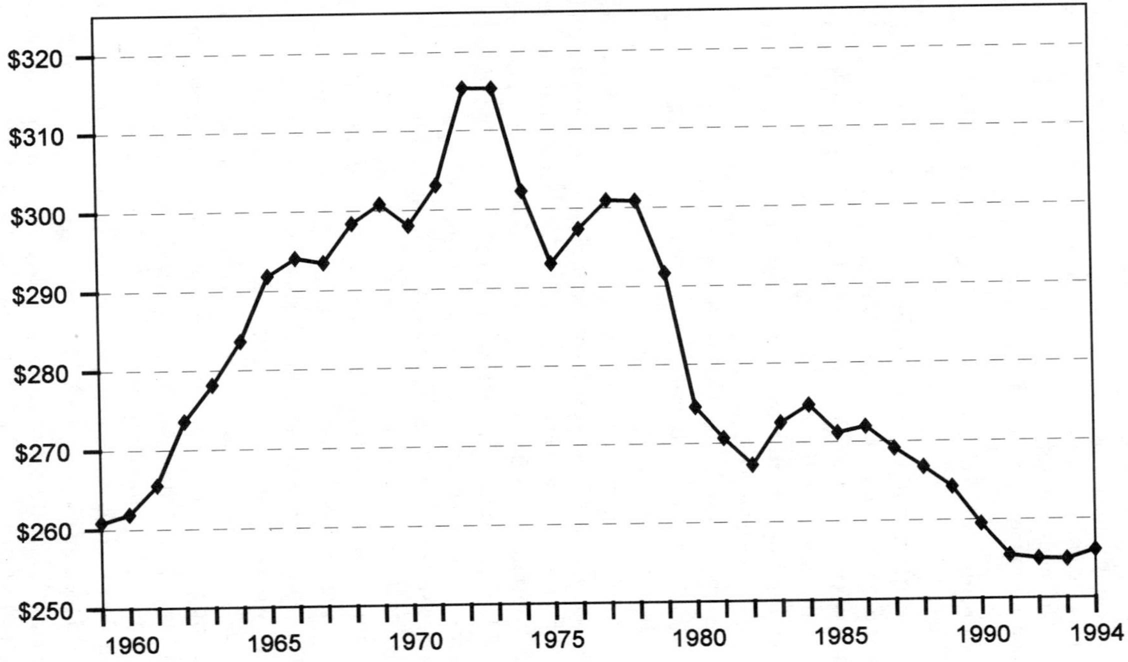

Source: Data from *Economic Report of the President, 1995*, table B45.

Since wages are falling and many government transfer payments (for example, welfare benefits) have not kept pace with inflation, much of the divergence between real wages and personal income must be a result of higher salaries for managers and professional workers, increased dividend and interest payments, and higher property income. Such forms of income are mainly concentrated among upper income individuals. The growth in various forms of non-wage income thus helps to explain why the distribution of income in the United States is growing more unequal.[122]

To begin with, since reaching its peak in 1972–1973 real average weekly earnings have fallen by nearly 19 percent through December of 1994.[123] A graphic illustration of this can be seen in figure 5.7, which traces the performance of average weekly earnings paid in the United States since the end of World War II. The Bureau of Labor Statistics (BLS), which collects and monitors such data, has converted the earnings into constant 1982 dollars to remove the effect of inflation and permit a fair comparison to earlier years. The facts speak for themselves. The average American worker is worse off today than at any time in the past third of a century. In terms of real earnings, today's typical worker actually earns less pay than workers did in 1960.

The picture gets worse. The decline in wages has been fairly widespread, although some sectors have managed to stay even or inch ahead slightly. Figure 5.8 tracks the performance of real average weekly earnings (in constant 1982 dollars) by private industry groups between 1970 and 1993.[124] Although the figures show general decline for all earnings recipients, some workers were able to eke out a few gains. Those jobs in the service sector—which are supposed to be the highlight of our emerging high-tech, information-age economy—are stagnant over the past twenty-three years, as is the average weekly pay of manufacturing jobs. Those in the mining industry have actually gained weekly earnings, but five other industrial categories have lost real income. Among the worst hit are retail workers. Already at a paltry $205 per week in 1970, their average weekly earnings skidded 30 percent to end at $143 in 1993. Considering the loss of manufacturing jobs and the explosion of retail jobs in the United States during this period, the implication for average weekly earnings for most American workers is forbidding.

In 1993 constant dollars, all males fifteen years old and over received a median income of $21,102 in that year—compared to $11,046 for females.[125] Full-time, year-round workers of both sexes were paid considerably more in 1993: $31,077 for men and $22,469 for women. These figures show continuing income inequality for women in our society. Full-time male workers still receive 38 percent more income than full-time female workers. While this fact is deplorable, there is a silver lining to the cloud. Twenty years ago, full-time women workers earned 57 percent of

FIGURE 5.8   Weekly Earnings by Industry (Constant 1982 Dollars)

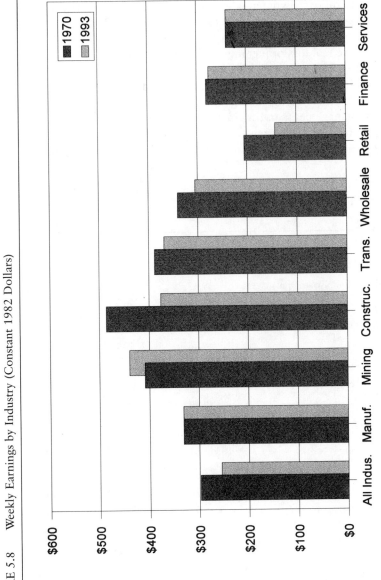

Source: U.S. Census Bureau, *1995 Statistical Abstract*, CD-ROM, table 661.

what full-time male workers earned. Today this gap has closed substantially and is at 72 percent. In fact, the average hourly earnings of female wage and salary workers is now 79 percent that of male workers.[126] Real income for women has shown a tireless climb. Median real income for year-round full-time women workers has grown larger and larger over the years, increasing by 13 percent between 1973 and 1993.

By contrast, income for male workers has been steadily getting smaller. The real loss of income was not limited to just one occupational spectrum—although we will see that manufacturing jobs were hardest hit. The slide in wages for males was evident across a wide range of jobs. Many of these, even including such professional employment areas as the law, business management, and engineering, have been widely regarded as erosion-proof. Table 5.1 reveals this notion to be false. In nearly every occupational group, whether white- or blue-collar, professional, or service

TABLE 5.1   Median Earnings in 1993 Constant Dollars by Occupation and Sex: 1982 and 1993

Occupation	1982	1993	Percent Change
Males			
All occupations	$23,236	$22,443	–3.4%
Executives, administrators, managers	39,479	40,335	2.2
Professional	37,461	40,505	8.1
Technical sales, clerical	25,475	24,406	–4.2
Service	9,861	10,795	9.5
Farming, forestry, fishing	5,565	8,416	51.2
Production, craft, repair	25,135	23,175	–7.8
Operators, fabricators, laborers	17,892	17,188	–3.9
Females			
All occupations	$11,617	$13,896	19.6
Executives, administrators, managers	22,056	25,282	14.6
Professional	20,702	25,865	24.9
Technical, sales, clerical	12,731	13,970	9.7
Service	5,175	6,684	29.2
Farming, forestry, fishing	2,715	3,106	14.4
Production, craft, repair	14,057	17,340	23.4
Operators, fabricators, laborers	11,311	11,182	–1.1

Source: U.S. Census Bureau, *Income and Poverty 1993*, CD-ROM, table P19.

category, white males have taken serious cuts in their pay or experienced virtual earnings stagnation between 1982 and 1993. This is especially evident when we compare the increased earnings of females in the same broad occupational groupings, which has been outstanding.

In 1973, median male income of full-time workers in constant 1993 dollars was $35,109. By 1993 this had shrunk to $31,077, or −11 percent. The decline has been mainly confined to white full-time male workers (−12 percent). African-American full-time workers did slide slightly, losing 3.2 percent after beginning at substantially lower salaries than white males. African-American full-time female workers gained nearly 19 percent in the two decades. Since the earnings of full-time white women workers were also increasing rapidly, however, the gap between African-American and white females actually widened slightly (in 1993 African-Americans earned 90 percent of white earnings, compared to 92 percent in 1979). According to the Census Bureau, the African-American/white ratio for median male earnings of full-time workers has improved only slightly, remaining at about three quarters of the white level over the past fourteen years. Nevertheless, the conclusion that the earnings of African Americans have not really suffered much in comparison to whites is highly controversial, and is challenged on a number of fronts.[127]

The loss in real dollars of over $4,000 for all full-time male workers may not have seemed very large over twenty years. It comes to a little over $200 dollars per year. Yet this same rate of dollar loss when applied to LDCs would have supported over one-quarter of a billion people in twelve of the world's poorest countries for each of the past twenty years. Perhaps there was no outcry because the change was so gradual and because inflation made it seem as if wages were actually going up. Yet the decline is very persistent and long term, spanning Democratic and Republican presidencies alike. It has continued to grow ever larger and been stubbornly resistant to any change. In the end it highlights the fact that the forces pushing American wages down may be difficult or impossible to remove.

No doubt there was some cushion to the fall. Many of these men are married to women whose take-home pay did go up. Not surprisingly, married-couple families with two adult workers earn substantially more than other types of families (especially female-headed single-parent families).[128] The increasing contribution of wives to family income has been considerable. Full-time female workers gained $2,606 in real median income during the last two decades ($130 per year), and more women are now working than ever before. Within fifteen short years, the number of married couples with children and two full-time workers in the paid labor force nearly doubled.[129] Today, 40 percent of married mothers with children six to seventeen years old work full-time, year round, while three out of four work either full- or part-time. Wives who work full-time now contribute a substantial 40 percent to familiy income.[130] Not only has the proportion of wives who work gone up, the number of hours they worked rose by one-third during the past decade as well. The bottom line: incomes for 60 percent of famlies would have declined without their contribution.[131]

Despite this glimmer of light, however, there has been a real loss of earnings for the majority of American workers in the past twenty years (after the disguising effects of inflation are removed). What has happened to the pay packet of the American worker relative to the global economy? We hear much of labor unrest around the world—a massive nationwide strike of workers in France, agitation in South Korea, unhappy workers in Mexico and China, street riots in many Latin American countries, and so on. Economic deterioration is clearly a reality. But how do we stack up in comparison to other nations? The fairest test would be to look at wages in the United States compared to wages in other industrial countries. The Bureau of Labor Statistics has conveniently indexed starting pay at 100 in 1982 U.S. dollars for seven industrial countries to make just such a comparison.[132] Figure 5.9 shows that wages increased in all countries. All levels should have gone up in this eleven-year period since inflation has not been removed. Yet the rate of climb is astounding. The hourly compensation for manufacturing workers doubled in all other countries except Canada and more than tripled in Japan. Even the United Kingdom, which registers dismal economic performance in most measures, easily outdistanced America.

There has been an even sharper erosion in the U.S. payscale compared to other countries just in the past few years. Much of this has been caused by a weakening of the dollar on global currency markets. In one year alone, the dollar lost a third of its value. In U. S. dollars, the 1992 average hourly pay for U.S. production workers at $16.17 was well behind most other industrial countries. While the United Kingdom ($14.60) was below us, Japan ($16.18) was dead even. All other European countries outperformed America in hourly wages, however, culminating in a rate approaching $26 per hour in Sweden and Germany. Fortunately, this is only part of the story. Because of comparatively lower taxes, and lower prices for readily available consumer goods, U.S. workers are still not in bad shape in terms of purchasing power parity. The average German wage is still nearly a fifth more than the average American wage, but this is better than the 60 percent difference before purchasing power adjustments. Figure 5.9B also shows the majority of our industrial peers close to or below our adjusted prevailing wage (Japan's hourly wage is actually cut by a third, shrinking to $10.78 per hour).

FIGURE 5.9   A. Index of Hourly Manufacturing Pay (U.S. 1982 Dollars)

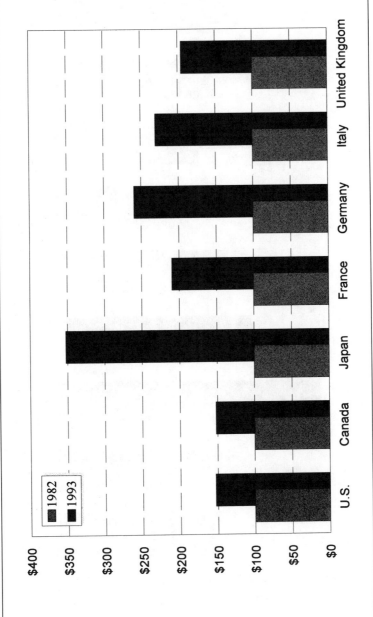

Source: Data from *Economic Report of the President, 1995,* table B111.

FIGURE 5.9   B. Hourly Pay of Production Workers (U.S. 1992 Dollars)

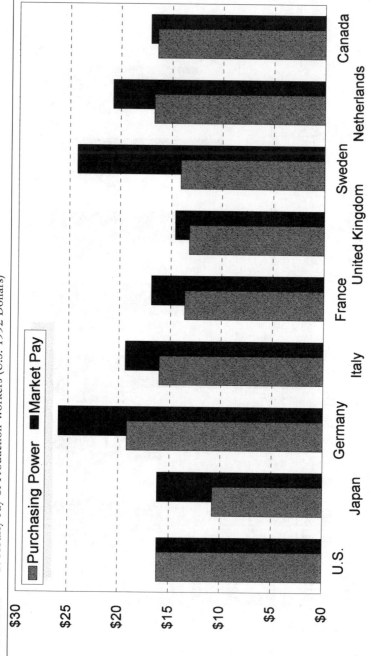

Source: Mishel and Bernstein, 1995, Figure 8C.

Yet there is no room for complacency about U.S. wages. The fact remains that we are losing ground at a rapid rate. The *Wall Street Journal* reports that during the mid-1980s our average hourly production-worker rate was twice as high as that of Britain or Japan, well ahead of those of Italy and France, and comfortably ahead of that of second place Germany. America had slipped to fifth place by the end of the 1980s, using unadjusted figures. While in 1985 Germany had a pay rate about three-fourths that of the United States, in 1988 the United States had a pay rate only two-thirds that of Germany.[133] Since that point, the U.S. hourly pay rate actually declined 0.1 percent per year (1989 to 1992) whereas our competitors increased their pay an average 2.3 percent annually.[134] Germany and Japan, in fact, gained over 3 percent per year.

Harrison and Bluestone report compelling figures from their analysis of Current Population Survey (CPS) data going back to 1963.[135] After removing the effects of business expansions and recessions, the researchers report that inequality in earnings increased by 18 percent for all workers in only a decade. Nearly two-thirds of this inequality could be laid at the feet of less pay per hour. Only one-third was due to the fact that some employees were working fewer hours (which may in itself have been less a result of personal choice than of lack of full-time jobs).

Many analysts have claimed that much of this growth in earnings inequality is caused by the huge influx of women (who have always been paid less) into the labor market. The rest can be blamed on the gigantic postwar baby boom invading the labor market all at once, which has also driven wages down because of oversupply. These explanations border on rationalization. The implication is that after such trends run their course, things will get back to normal and everything will again be fine, that once jobs are found for all the baby boomers and/or women who would have stayed home in an earlier epoch, pay rates will bounce back.

Such demographic factors can affect earnings to some degree and should not be completely dismissed. In a study of the 1947–1985 period, for example, Nan Maxwell concludes that demographic structures influence income inequality about as much as economic factors.[136] Frank Levy has estimated what happens to the real income of males as they age, comparing the results to estimates from prior epochs. In 1949, an average forty-year-old man increased his real income by 34 percent during the ten years it took to reach age fifty. During the 1960s a similar passage for a man going from age forty to fifty produced a 29 percent rise in real income. But a man who turned forty in 1973 actually earned 14 percent less in constant dollars by the time he turned fifty in 1983. Put differently, a young man who left his parents' home in 1973 now earns 25 percent less (rather than 15 percent more) than his father earned in 1973.[137] The relative supply of college educated persons has also fluctuated, causing wages to change.[138] Demographic trends—such as increases in women's labor force participation, and changes in birthrate, marriage ages, and family structure—have cushioned the impact of declining earnings for many Americans. Yet Levy in the end also sees a combination of cyclical and structural changes in the U. S. economy as partly responsible for changing earnings. Noting that the Gini ratio of household earnings has gone from .391 to .431 in the last two decades, he calculates that two fifths of the increased inequality is due to changed living arrangements while three fifths reflects economic factors.[139]

The truth is that pay has declined for just about every group (although it has hit younger people and white males harder). When looking at the impact of demographic variables alone, Harrison and Bluestone found:

> The baby boom and the growth in the number of female workers have had no *significant impact whatsoever* on the increase in inequality of wages. Indeed, men's and women's wages actually converged slightly in this period —owing more to declines in the average wage of males than to increases in the wages of women. The wages of white women and women of color are now almost indistinguishable. Put another way, all of the increase in inequality since 1975 must have occurred *within* age, race, and sex groups, not among them. Inequality is growing among whites as well as nonwhites, among the old as well as the young, and among women as well as men.[140]

Some groups have been much harder hit than others. All told, persons experiencing the greatest drop in wages include those with less than a college degree (three-fourths of the workforce), the bottom 80 percent of men, younger workers, and blue-collar workers.[141] Overall, between 1977 and 1994 the average hourly wage paid to American workers slid by over 13 percent. To hedge against this erosion, workers labored two more weeks per year than a decade earlier—helping to cushion the drop in real income. In 1993, the entry level hourly wage paid to a young male high school graduate was nearly one third less than in 1973 (–7.8 percent for male college graduates and –4.7 percent less for female college graduates). The rise in what has been called the *education premium* was much ballyhooed in the 1980s. This was a trend that showed college graduates earning an ever-greater proportion of money compared to non–college graduates. But this had ceased to exist by 1987. Although it still makes sense in terms of lifetime income to graduate from college, there is no longer any guaranteed payoff. A college graduate still makes more than a high school graduate, but real wages for this group have also been on a continuous slide since 1987—albeit at a lesser rate than for non–college graduates.

At this point, no one is safe from the ravages of our occupational marketplace. Even older, college-educated males well intrenched by seniority and experience, a previously sacrosanct group, are now getting the axe. Frank Levy believes the culprits for this new trend are the recession of the early 1990s, the restructuring layoffs by corporations, and a lackluster economy. Men aged forty-five to fifty-four years old with four years of college saw their median earnings dip by 10 percent in just three years—a decline far larger than that seen by younger men with college degrees.[142] The result is that people who are laid off typically never get new jobs that equal their earlier wages. For some, the culprit may be age discrimination. It is a truism that the older you are, the harder it is to find work. Studies vary, but the price of a permanent layoff is generally anywhere from a 20 to 40 percent reduction in pay for most workers. At the same time, workers who are spared the cuts become much more willing to accept no pay raise or even a pay cut. Earnings thus go down among far more workers than those who are actually let go.

The standard operating procedure for corporations in the 1990s has been to downsize in an effort to maximize profits. Previously, this meant firing blue-collar workers, but today the great majority of terminated employees are white-collar. Current thinking is that middle management (i.e., middle-aged males) is too bloated and needs serious cutting. Although white-collar salaried workers account for only 40 percent of all workers, they account for 62 percent of all job reductions now taking place.[143] The rate of downsizing among companies has been running about the same in the mid-1990s, during economic expansion, as it did during the 1990–1991 recession. The result is that even blue-chip American companies with stable profits have entered into the frenzy of job slashing. At the start of 1992 gigantic General Motors announced it was closing twenty-one North American plants as a means of eliminating 74,000 North American jobs. In that year, U.S. corporations were permanently reducing their workforces at a rate of 1,500 jobs per day.[144] The top three corporate job shedders since 1992 have been IBM (63,000), Sears (50,000), and AT&T (40,000).[145] In 1993 alone, corporate restructuring cost 600,000 mostly white-collar jobs. In only one month (January of 1994) another 108,000 job cuts were announced.[146] According to a 1996 poll conducted by the *New York Times,* nearly three-quarters of American adults say they have been touched to some degree by a layoff in the last fifteen years.[147] This includes one-fifth who have been personally laid off, another 14 percent who have had someone else in the household receive a pink slip, and an additonal 38 percent who know close friends, relatives, or neighbors who lost their jobs. The policy of employee sacrifices has worked for corporations, although the human toll is monstruous. After-tax profit rates for businesses in 1994 were the highest in a quarter century and were actually expanding in 1995. Unfortunately, this was achieved on the backs of hapless American workers rather than through investment and capital improvements.[148]

DEINDUSTRIALIZATION AND DECLINING WAGES

Much of the decline in wages and earnings cannot be attributed only to economic recession or a short-sighted temporary push for profits. The decay of wages and earnings is in large part caused by traumatic structural changes in American industries and corporations. So far, most of our discussion has been limited to looking at the wage performance among demographic categories or in large industrial groups—for example, comparing production workers from one country to those of another, or the manufacturing wage level in 1973 and at present. Although a very real decline is evident here, there has been another trend of massive proportions taking place in the United States. Up to now, *deindustrialization* has been alluded to only in passing. Yet this is one of the most significant economic factors to occur in the United States within the last few decades.

The decline in manufacturing jobs has been going on for quite some time now, along with a parallel increase in service employment. Service workers are found in a wide variety of occupations and industries, but are concentrated in the wholesale and retail trades.[149] By occupation, these include salesclerks, secretaries, bartenders, computer data processors, advertising agents, janitors, waiters, beauticians, and a host of other workers who are not directly involved in manufacturing a tangible product. In the past third of a century, the proportion of our workforce in goods-producing industries was halved (declining from 31 percent to 16.2 percent) while the proportion of employees in service-providing industries climbed from two out of three to four out of five workers.[150] The real crisis started to develop in the early 1980s when the expansion of jobs in goods production not only halted but began to get smaller.[151] Manufacturing jobs count for a lot. In 1993 the average weekly earnings of workers in manufacturing stood at $487, compared to $210 in retail trade—well over twice as much.[152] Not all service workers are poorly paid, but on the whole most of the growth in service jobs has been in such areas as fast food or janitorial services. While such low-paying jobs have been on the rise, the top-paying blue-collar jobs in heavy industry have been disappearing.

Lester Thurow believes that intense international competition has been one of the major causes of American wage decline. For example, this relationship can be brought into better focus when we realize it takes one million full-time U. S. employees to produce $42 billion worth of goods.

Thus, the trade deficit in an average year is akin to squeezing about 4 million workers out of manufacturing employment and into other jobs. Thurow presents evidence that manufacturing jobs are more highly paid *and* have a more equal distribution of pay than do service jobs. Given these facts, then, the decline in manufacturing *must* lead to more inequality in the distribution of earnings.

> In addition to paying higher wages, the exporting and import-competing industries generated a more equal distribution of earnings. . . . The meaning of these statistics is that when exports fall and imports rise to create a trade deficit, the distribution of earnings moves toward inequality. Jobs are lost in both exporting and import-competing industries and are replaced by jobs with lower, more unequal earnings in the rest of the economy. This factor is the principal reason for the observed decline in earnings of males. The industries that have been hit hardest by international competition—automobiles, steel and machine tools—are precisely the ones that have provided a large number of upper-middle-income male jobs.[153]

It should be pointed out that America is not the only advanced industrial country to undergo deindustrialization. In research involving seven wealthy nations (Australia, Canada, France, the Netherlands, Sweden, the United Kingdom, and the United States), Peter Gottschalk concludes that earnings inequality went up among male heads of households who were working full time.[154] In all countries, the mean earnings of young male workers (twenty-five to thirty years old) declined over time in relation to more experienced workers (forty to fifty-five years old). Moreover, over time all nations became more open to international trade, which accompanied the increased inequality. There was also a movement out of manufacturing and into service-sector employment in all countries. Gottschalk concludes that deindustrialization was affecting these countries in a similar manner by promoting earnings inequality—although technological change also contributed a lesser amount to the process.

There is little doubt that the erosion of wages is due mostly to the shift in the industrial mix within our country. Harrison and Bluestone estimate at least a fifth of the overall inequality in wages since 1978 has been caused by the disappearance of manufacturing jobs.[155] This has clearly been a major reason that real wages of men have been in a virtual free fall. The rise in numbers of low-income earners among men has not been due to the baby boom effect, education/job experience, or the unemployment rate.[156] In their earlier report to Congress, Bluestone and Harrison pointed out that while only one in four new jobs taken by white men in the 1970s was low wage, this proportion had risen to 97 percent by the first half of the 1980s.[157] In the latter period, the industrial Midwest lost a million middle- and high-wage jobs and gained only 900,000 new low-wage jobs.

Since the number of manufacturing jobs around the entire country actually went down, there were no new jobs to calculate whether any improvement had taken place in this sector.

In their analysis of CPS data, Harrison and Bluestone divided jobs into low-, middle-, and high-wage groups by first looking at the prevailing wages in 1973. This was America's best year for earnings, a level which has not been duplicated. Their research arbitrarily cast jobs with earnings at half the median wage as "low wage." "High-wage" jobs were defined as paying twice the 1973 median wage, since the earnings distribution is skewed so far toward high incomes. Revising these figures to constant 1986 dollars (to subtract the effect of inflation) allowed them to do a time-series comparison. Lastly, figures were converted to reflect year-round, full-time pay.[158]

The researchers then compared the types of new jobs created within three time periods, with highly instructive results. The 1963–1973 period could be described as our golden days when growth and increasing pay were the norm. In the 1973–1979 period there were several severe disruptions of the economy, including two OPEC oil embargoes, fierce inflation, and the costs of the Vietnam war coming home to roost. Nonetheless, momentum in the American economy still carried us along. By the 1979–1986 period, however, the rust had finally started to show on the body of the American economy. The authors also separately analyzed a number of categories (such as sex, age, race, location, type of industry, etc.) which have been popular competing explanations for our wage decline. None of them explained why the decline has taken place. Put differently, a slide is evident in all age, sex, race, region, and industrial groups.

There was a 10 percent decline in low-paying jobs between 1963 and 1973, while one-third of all new jobs created in 1979–1986 were low-paid. At the opposite extreme, the golden years saw nearly one in four (22.2 percent) new jobs created at a high pay scale, compared to only 14 percent in the latest period. For group after group the figures nearly always say the same thing. The proportion of low-paying new jobs is on the rise. The only exception is for the middle-aged group and in the New England/Northeast regions. Yet even here there is a large erosion of jobs at the top of the pay scale. The figures are also consistent with Harrison and Bluestone's major point. The changes in our economy are producing a polarized income picture. High-wage jobs have been increasing for most groups at the same time low-wage jobs are being spawned in record numbers (although not by nearly as much). The proportion of new jobs at the high end of the pay scale has doubled between the 1973–1979 and 1979–1986 periods (6.2 percent versus 13.9 percent). The proportions at the top end also doubled for men, women, those in the South, and services.

They have more than tripled for whites and for those in the Northeast and New England. In the end:

> The polarization of jobs is becoming increasingly universal, no matter what the color of workers' skin, their sex, their age, or for that matter, the industry within which they work. . . . Consider, for example, the position of men —historically the most privileged group in the workforce. . . . The proportion of net new employment that paid middle-level earnings to male workers—between $11,000 and $44,000 a year—has literally crashed, from nearly 78 percent between 1963 and 1973 to only 26 percent in the period ending in 1986. This is polarization with a vengeance. . . . Tragically, low-wage employment rose sharply among workers of color beginning around 1979 after more than a decade-and-a-half of improvement. Indeed, virtually all of the improvement experienced by black, Hispanic, and Asian workers between 1973 and 1979 disappeared in the 1980s. Younger workers were also hard hit after 1979. Almost three-fifths of the net new YRFT [year-round full-time] employment that went to workers under the age of thirty-five since then has paid less than $11,000 a year. . . . Not unexpectedly, the condition of the old industrial Midwest is the most extreme. It leads the nation in generating new low-wage jobs. What new employment has been created is *all* in the extremes of the distribution of wages—and fully 96 percent is in the bottom. There are no new jobs in the middle stratum at all! Deindustrialization, more prevalent in the Midwest than anywhere else, is plainly taking its toll. As well-paying manufacturing jobs disappear, new employment is almost entirely in the poorly paid jobs in the service sector, with a handful of new workers at the top. . . . Those workers traditionally most favored in the U. S. labor market—whites, men, and workers in the high-wage Midwest—are joining the low-wage segment in record numbers.[159]

The findings advanced by Bluestone and Harrison have been controversial and were challenged on a number of different fronts. Their earlier research was flawed methodologically by failure to look at full-time year-round workers only and by not employing a continuous set of years decycled to account for up and down swings in the economy. When Bluestone introduced these modifications using more current data, he revealed much the same U-turn pattern: middle-income jobs hollowed out while low-wage and high-wage jobs increased during the last decade.[160] Using multiple regression analysis, Bluestone further determines that popularly competing explanations for rising wage inequality are not viable. In particular, the business cycle has little effect on wage deterioration. Nor is there an effect from various baby-boom or baby-bust (contracting pools of younger workers) cohorts. Again, his strongest finding is that of deindustrialization. The reduced share in manufacturing employment has contributed heavily to stagnating wages.

Service sector employment increases inequality.[161] Using three separate measures of earnings inequality, Lorence and Nelson conclude that deindustrialization and the growth of service jobs is responsible for the bulk of rising inequality within 124 metropolitan statistical areas (MSAs) in the United States.[162] After controlling for business conditions, baby boom cohort effects, human capital, unionization, firm size, race, gender, and region, they find that expanding employment in service industries significantly increases earnings inequality at the lower, middle, and upper segments of metropolitan male workers. No impact upon earnings was discernible from a baby-boom effect, nor were there any business cycle influences over time. The effects of deindustrialization were particularly evident where they were theoretically expected, that is, among Rustbelt MSAs. The ill-effects of rising numbers of service jobs were cushioned somewhat by a less negative effect upon women's earnings as well as by some high-income jobs spawned for male service workers.

There is not much debate about the good jobs/bad jobs polarization. One exhaustive study classified six hundred different occupations from good to bad on the basis of seventeen measures of job quality (wage rates, fringe benefits, skill requirements, etc.). Although middle-class jobs declined while both good and bad jobs increased about equally during the last decade, the bad jobs actually were worse at the end than they had been ten years earlier. Yet optimists still see increased opportunity for workers with skills and education. They argue that workers with the appropriate technological expertise can reap huge benefits and higher salaries. While this is no doubt true to some extent, is the growth of "good " jobs enough to outweigh the explosion of "bad" jobs—while medium-income jobs disappear?

Nearly 9.5 million American workers were permanently displaced from their jobs in the 1980s. Well over half these workers held manufacturing jobs.[163] In the first half of the decade alone, almost one in eight workers lost jobs due to plant closings. Over 5 million businesses closed their doors during this period.[164] This increase in structural unemployment produced higher joblessness for men in general and for less-well-educated men in particular. The net absolute decline for semi-skilled blue collar machine operators and assemblers alone was 1.4 million jobs.[165] Meanwhile, nearly 80 percent of the net new job growth in the last decade (over 14 million new jobs) was located in the two lowest paying service sector industries—retail trade and services (business, health, personal).[166] In the end, both sexes who experienced job displacement in the 1990s were less likely to find full-time work at a comparable salary, but the impact was far worse on men than on women. Men were twice as likely to move from a full-time to a part-time job than the reverse. Even if displaced workers managed to win back full-time employment, it typically meant a substantial cut in pay (20 percent for men, 23 percent for women).[167] Trends such as these cannot help but lead to depressed wages.

Nearly all research supports the job polarization thesis. Using a more refined and exact analysis (where earnings of full-time year-round workers are divided into deciles [tenths] and tracked over twenty years) does show a declining middle with rapid expansion at both the low and high ends of the earnings spectrum.[168] The pattern for polarization is especially striking for white men during the past two decades, which becomes progressively more U-shaped over time. Strong growth in both high-wage and low-wage earners occurs while there are hefty declines among middle-income earners. Yet there is more growth in low-wage earners than in high-wage earners for white males. A less distinct but clearly evident polarization is also present for white women, but not for African-American men or women through the middle of the 1970s. Since that point, polarization has been evident for all four groups of workers.

Leann Tigges has looked at the absolute earnings decline in an attempt to isolate the major cause. Comparing data from two Census Bureau public use samples, she was able to trace what has happened to earnings over twenty years. Through multivariate regression analysis, the effects of sex, education, race, service versus manufacturing jobs, and employment in core versus periphery industrial sectors were compared. There was every reason to believe that demographic factors could have driven down earnings, since the labor force has grown less white, more female, and younger over time. Yet at the same time there has also been a massive shift away from manufacturing toward services, which could also cause a slide. The regressions effectively hold all of these variables constant simultaneously. Although there has been more earnings decline for younger males, deterioration was also apparent for older men.[169] The decline in earnings, then, cannot be laid at the feet of the huge influx of baby boomers into the labor force.

Tigges found that decline in the core versus the periphery of industry was also apparent. Although the association between working in service jobs and receiving low pay did not get worse over time, it did not get better either. It remained a very important factor in the slide toward lower earnings. She found no evidence that the increase in professional, technical, and managerial jobs over the twenty-year period acted to increase earnings. Much of the erosion in core industries was due to the increasingly negative influence of service work in this sector.[170]

Analysts are prone to argue that differences in findings are caused by different techniques and methods used in separate studies. Yet, the amount of agreement documenting a continued deterioration of earnings together with increasing inequality has reached a crescendo. A RAND study concludes that even using ten different measurements of earnings inequality, and examining total earnings versus wage and salary income combined, yields the same general conclusion.[171] This report describes a period of stable earnings inequality from 1967 to 1980 being replaced by growing relative earnings inequality after 1980. Inequality grew even for full-time year-round workers, as it did for all age and sex groups. The labor force entrance of younger workers was not primarily responsible for more inequality, nor was the influx of more women into the labor force.[172] Although the increase in service industry served to drive up earnings inequality, it would have grown much larger even without this shift away from manufacturing. Finally, the increase in earnings inequality was not reversed in the 1980s as the economy expanded.[173]

THE WORKING POOR

With the decline of wages and earnings, it will come as no surprise that many workers today do not make enough from their jobs to escape poverty. The phenomenon of the working poor emerged unmistakably during the 1980s, when the wage slide became so apparent. Tragically, it continues to haunt us well into this decade since none of the trends producing it have diminished. Definitions of the working poor vary somewhat, depending upon one's analytical perspective. The Bureau of Labor Statistics (BLS) defines the working poor as persons who worked (or were looking for work) during twenty-seven weeks of the year, and who lived in families with incomes below the official poverty line.[174] Over one in nine persons in the labor force during 1993 were living below the poverty line. Of these nearly 12 million workers, 70 percent (8.22 million workers) fit the category of working poor. The BLS study found low education to be the one characteristic most closely linked to a worker's poverty, although African Americans and women had higher probabilities of entering this cadre. Predictably, working women who were single parents also ran a high risk of being poor (at 20.5 percent, their poverty rate was four times that of working husbands and seven times that of working wives). When labor force problems encountered by this cohort in 1993 are examined, seven out of ten of the working poor experienced low earnings. Another four in ten experienced unemployment, while a smaller proportion underwent involuntary part-time employment.

A different approach by the Census Bureau is to look at full-time civilian workers aged sixteen and over who had yearly earnings below the poverty level for a four person family with two children. As of 1992, nearly one in six workers fell into the working poor category, that is, they earned less than $13,091 in that year.[175] The proportion of the working poor had risen by one third since 1979, when the comparable rate was at

12 percent. The proportion of working poor in the labor force increased for every demographic group with the exception of women workers with a bachelor's degree or higher. In sum, it did not matter what race, sex, age, or education level a worker had—all were more likely to be poor, despite working full-time, in the 1990s than they were at the end of the 1970s. Some groups, however, suffered more. The proportion of working poor rose by over 60 percent among men, and nearly doubled for workers eighteen to twenty-four and twenty-five to thirty-four years old. The proportion doubled for men with no high school diploma (going from 15.3 percent to 30.9 percent) and very nearly doubled for males with high school diplomas but no college education. Despite the sharp growth in numbers of working-poor males, however, women at all education levels were still more likely than men to fall into this category. African Americans and those of Hispanic origin are similarly disadvantaged.

Unlike many areas of social policy, there was a consensus here, developed across the political spectrum during the 1980s, that efforts should be made to help the working poor. Both conservatives and liberals agreed that those who worked full-time should not have to labor in poverty. As part of a growing effort to alleviate the poverty of this group, the Earned Income Tax Credit was expanded in 1986, 1990, and 1993. Legislation enacted under the Reagan and Bush administrations broadened Medicaid elgibility to include large numbers of children in working poor families, while a moderate rise in the minimum wage was passed in 1989.

But despite these efforts, the lot of the working poor is actually made worse by tax and transfer policies in the United States in comparison to our industrial peers. Inge O'Connor and Timothy Smeeding, using data from the Luxembourg Income Study, have traced what happens to the working poor in the United States, Canada, the United Kingdom, Sweden, and the Netherlands.[176] As with most studies using LIS data, the authors are able to adjust income using equivalence scales to measure need by family size, to focus on full-year, full-time earners of prime working age (twenty-two to fifty-five), and to measure disposable income after in-kind benefits (e.g., food stamps, housing subsidies) have been added and Federal and Social Security taxes have been subtracted. Poverty is again defined according to accepted international standards as less than half of the national adjusted median income for all households of all ages. To begin with, although full-time, year-round work is important in reducing poverty, it does not guarantee escaping it in any of the five nations studied. In the United States, the poverty rate for heads of households goes down from 14.1 percent to 4.6 percent among full-time worker heads. In three countries (Canada, the United Kingdom, and Sweden), the government package of taxes and transfers reduces poverty for the working poor. In both the Netherlands and the United States, government tax and transfer policies actually make things worse for working poor people. This is especially serious in the United States, however, since our poverty rates are nearly twice as large as those in the Netherlands. Depending upon household composition, this can be critical. The United States leads all countries with a 19.1 percent poverty rate for single-parent families whose head works full-time year round. (Second place Canada is at two-thirds of our rate, while Sweden's rate is actually less than 1 percent!)[177]

Poor parents are willing to work and to put in long, hard hours at that. A 1993 study by Shapiro and Parrott that looked at American workers with incomes below the poverty level found parents in poor families with children worked 1,475 hours—about three-fourths of what a full-time year-round employee normally works.[178] Comparing 1993 and 1977 data, the authors also conclude that 1993 workers were much more likely to be poor than ever before. The share of working families falling into poverty rose by nearly half in this time period. The rate had spurted to one in nine of all working families with children as the mid-1990s approached. Although part of the increase was due to a larger proportion of single-female-headed families, the poverty rate for married-couple working familes also increased by one-third. The authors see the culprit mostly as falling wages, especially for those at the bottom of the earnings spectrum. Yet other developments—less bound to market influence and more controllable through social policy—aggravated the growth in numbers of the working poor. In particular, large numbers of the working poor were denied partial AFDC benefits in the early 1980s, which has led to growing numbers of poor workers. Another major influence, however, has been the ongoing erosion of the minimum wage.

THE WAR WITH LABOR

Not surprisingly, because real wages and earnings have fallen for just about everyone, more employees are working longer hours to stave off poverty. This can have damaging results. Given the nonsupportive climate for working mothers and lack of family policy by the federal government, the stress level of families with children has increased dramatically. So has the neglect of our children, even in so-called middle-class married-couple families.[179] One of the main costs has been an unremitting time crunch on working parents. As of 1989, the average American worker put in 158 more hours on the job than the worker of twenty years ago—equivalent to working an extra month every year.[180] Paid time off to cover vacations, holidays, sick leaves, and so on fell roughly 15 percent in the last decade. Today, the average American worker gets a little more than three weeks off per year—compared to at least five weeks in most European countries. As a result, the average potential parental time with children declined by ten hours per week in white households and twelve hours in African-American households over the past twenty-five years.[181]

Harrison and Bluestone conclude that earnings inequality in the American labor force increased by a full 18 percent in little more than a decade.[182] This occurred even after the effects of inflation and downturns in the business cycle were removed. About one-third of this growth toward greater inequality was due to more part-time work. The other two-thirds stemmed from a growing polarization in wages and earnings, mostly because high-paying jobs were eliminated in the face of a more competitive global market. Looking only at the growth of inequality in earnings and ignoring the trend toward part-time work, the authors found that one-fifth of wage inequality was the result of the shift away from manufacturing jobs and into the service area. The rest they believe is due to *internal cutting of wages in occupations and industries.* Much of this was done with rollbacks, concessions, tiered contracts, elimination of high-wage jobs, paying less for newly created jobs, not replacing workers lost through attrition, forcing early retirements of top-paid staff, reducing or eliminating fringe benefits, refusing cost-of-living (COLA) raises, moving to a lower-wage section of the country, tinkering with the tax system to shift the burden away from the well-to-do and corporations, etc. Just about every conceivable means was used to force wages lower in order to keep profit margins up. Since earnings are the major portion of income for most of us, it will come as no surprise that mounting evidence points to a fullblown crisis for the great majority of the American workforce.

As of July, 1996, the value of the federal minimum wage—$4.25 per hour—was actually 26 percent below what it had been in the 1970s, once the effect of inflation is factored out. If the minimum wage had not been raised by Congress in the last half of 1996 it was projected to fall to a forty-year low—where it would have been equal to what it was worth in 1955. Even at the start of 1996, full-time year-round workers could not earn enough to escape from poverty. Indeed, a full-time minimum wage job would still leave a family of three $3,350 below the poverty line.[183] The cost of buying necessities for a family increased by almost 60 percent between 1981 and 1994, yet the minimum wage increased by only 12 percent during the same period. To stem this misery, Congressman Martin Sabo (D-Minnesota) introduced a bill to raise the minimum wage to $6.50 per hour in 1994, but failed to win its passage.[184] In February 1995, President Clinton proposed a far more modest raise of 90 cents, to $5.15 per hour, which would be implemented in two steps and which would not take full effect until 1997. Even this paltry bump was fiercely resisted by the Republican-controlled Congress. Election year politics eventually prevailed, however, as both the House and Senate finally passed the 90-

cent-per-hour pay raise that increased the minimum wage for 10 million Americans to $5.15 per hour.

Generally, the rationale for opposing such raises would be downsizing, layoffs, non-expansion of businesses or even failures. But new research now challenges this traditional view, having found that recent growth in state (eight states have lifted theirs above the federal level) and federal minimum wages did not result in job loss. Instead, jobs increased.[185] Polls reveal that 70 percent of voters, including a majority of Republican voters, support a higher minimum wage.[186] Raising the minimum wage would not only directly elevate the 10 million American workers currently at this level, but its ripple effect would raise another 15 million workers because employees slightly above this level would also be ratcheted up the pay scale. Nonetheless, several industries—including restaurants, retailers, food processors, the needle trades, and other non-durable goods manufacturers—remain adamantly opposed to increases of any kind because their profits are so dependent on a low-wage work-force.[187] Given this type of business opposition and political divisiveness in Congress, the working poor are not going to disappear, despite the newly raised minimum wage. It seems more likely that their ranks will continue to swell into the next century.

Stereotypes of minimum-wage workers as teenagers flipping hamburgers at McDonald's (three-fourths of such workers are adults) or as housewives earning pin money are still around but are not even remotely connected to reality. Instead, we have large numbers of desperate minimum-wage workers in the United States today whose income is absolutely necessary for the survival of their families. To begin with, landing a job at McDonald's may not even be that easy, depending upon who or where you are. One study which tracked workers in the fast food industry in central Harlem during 1992–1994 found competition for jobs in the fast-food industry was fierce.[188] Despite the fact that all jobs were at minimum wage, there were fouteen applicants for every successful hire. The researcher found that the oversupply of job seekers led to a "creeping credentialism" in the fast-food industry which would hire only high school graduates in their twenties. Incredibly, in this African-American neighborhood discrimination against hiring African Americans was still rampant. Recent immigrants were preferred by managers who believed these hirelings would regard $4.25 as a king's ransom in comparison to wages in their home countries. In short, it is not even necessarily true that anyone who wants a minimum wage job can get one.

This bleak reality is as authentic on Iowa farms as it is in the African-American ghettos of big cities. The *Washington Post* has documented how growing outsourcing in the automobile industry is being rapidly disseminated in the heartland, where family members are actually in dire need of such minimum-wage work.[189] In Guthrie Center, Iowa, "doing well" translates to a $15,000 annual income and means that you can keep the family farm. Farm wives have flocked to jobs which require light assembly and that typically pay as low as $5.50 per hour (replacing full-benefit $18 per hour jobs in Detroit). Although unenthusiastic about their own exploitation, these workers report that it pays for groceries. The typical $8,000 yearly income, for example, can be used to pay the $6,000 bill for a family's health insurance.

For millions of minimum- or low-wage workers, today's job market is a dead-end street. Rather than finding a springboard to higher-paying jobs, the working poor are now being sentenced to a lifetime of servitude at subsistence wages:

> Montie Lavoie can't quite remember all the bad jobs she has held. She washed dishes when she was 16, worked as a barmaid when she came of age, and then found a full-time job making windows—for $6 per hour. Claudia Burrow has a longer list: fry cook, nurse's aide, sales clerk, security guard, housemaid. . . . Sharon White went to work at 14 and spent 10 years as a warehouse shipper before the company went out of business. Then she went back to school to become a child care specialist. Then, needing more money, she got a job as a waitress for a caterer. Life at or near the minimum wage of $4.25 per hour has an unpleasant sameness everywhere in America. Lavoie, 36, and Burrow, 39, live on the high, empty western plains of Montana. White, 37, is from inner-city Baltimore. All three provide living proof of one of the nation's most hallowed truths: "You can always get a job, if you really want one." But what kind of a job? "For $6 an hour, you can't support a kid," Lavoie says. "For $6 an hour you can't even pay the rent." These women, and 13 million more low wage workers in America, have spent years—in some cases decades—trapped in labor's twilight zone. They have added a bitter corollary to the old saw: Yes, you can always find a job, but much of the time you can't earn a living.[190]

One of the primary causes for the continuing decline in wages and earnings, according to Bennett Harrison and Barry Bluestone, is the relentless war being waged on labor by American business.[191] The drop in the ability of the United States to compete in trade around the globe meant that profits went down for American corporations. As a response to the drop in profitability, American firms had a choice to make. A need existed to retool factories and invest in the most up-to-date technology money could buy. In this way, goods could be made more cheaply through a better, more efficient production process. According to free-market forces, firms would be able to sell more goods because their products could be made more cheaply and be of higher quality.

Unfortunately, another avenue was also available to protect and even enhance profitability. This strategy was to sidestep the need for large investment in new capital goods. The plan was to simply slash pay. Wages could be and were attacked on a broad front using a variety of means. By simply moving factories to the South or out of the country, firms pressured unions in the old northern industrial core to agree to wage rollbacks or two-tier wage contracts. Companies *can* save money if they break unions. There is no doubt that union workers earn more because of enhanced bargaining clout. Even in the mid-1990s their wages averaged $160 more per week—37 percent greater than non-union workers.[192] Moreover, although there is some debate, the majority of research indicates that unions have an equalizing effect upon earnings.[193] That is, the higher the percent of union membership in a given locale, the less inequality of earnings dispersion. In a sense, unions offer the best of both worlds—higher earnings for just about all workers, who are treated relatively equally.

Aside from outright decertification, two-tier contracts have done much to destroy the union advantage. Two-tier contracts simply protect the benefits of older workers. The union agrees with management that new workers who are hired will be forced to take much lower pay for doing the same job. The popularity of such contracts seemed to explode overnight during the 1980s. Almost one in ten labor contracts negotiated halfway through that decade involved a two-tier wage system—up from virtually zero in 1980. Not surprisingly, the two-tier system is more common in nonmanufacturing businesses where unions are weaker (17 percent), where government deregulation has taken place (35 percent in the airline industry), where profits in the firms are low, where intense competition exists from foreign imports, and where strong shifts in consumer demand take place (one-third of wholesale and retail trade now involves two-tier agreements).[194]

One of the results stemming from the active hostility employers have directed toward unions is a decline in their membership. As unions have been forced into wage concessions and two-tiered contracts by the threat of closed factories, they have become less effective. To the degree unions are less effective, fewer workers want to join them. Included in the arsenal of weapons companies have used to weaken or destroy unions has been out-sourcing orders abroad or through plants in the South, jobs that were previously done by a unionized work force. Campaigns to decertify unions have also been slick, well-orchestrated affairs that have met with some success. Even bankruptcy laws have been used to abrogate prior wage agreements reached with unions. When these strategies are considered together with growth in traditionally nonunionized service jobs and movement to right-to-work states, it is not surprising that membership in unions has declined. In 1975, union members made up almost 29 percent of all those employed, but this had reached a postwar low of 16 percent by 1994.[195] As union membership skidded, so did earnings. Indeed, the declining percentage of workers in unions was responsible for at least one-fifth of the rise in earnings inequality among male workers in the last decade.[196]

Another means of reducing labor costs in order to raise profits is to convert full-time work into part-time labor. In this way, various added costs such as health plans, pensions, and higher rates of pay can be avoided. The use of temporary worker agencies such as Kelly Services and Manpower, Inc, grew twice as fast as our nation's GNP during the last decade. In the early 1990s, Manpower was doubling its net income each year, and had revenues close to $2.8 billion.[197] Despite the fact that 40 percent of North American companies reported a worker shortage, many were refusing to expand their permanent labor force in favor of hiring "temps":

> Contingent labor—leased and temporary workers, involuntary part-timers, employees of subcontractors, and homeworkers—grew from 8 million in 1980 to 18 million by 1985. That number is nearly 17 percent of the total work force. If those whom the BLS [Bureau of Labor Statistics] considers to be "voluntary" part-time workers are added to the count, fully a quarter of the 1985 labor force could be considered contingent employees. . . . In slack periods, employers are less concerned with developing promotional ladders to keep their most prized employees and more interested in finding cheap and efficient ways of reducing the number of workers at the first sign of a downturn in sales. The use of contingent labor provides them with just such a mechanism. . . . The stagnation of real wages since the early 1970s also means that more and more families *need* whatever work their members can find, however "contingent." . . . Practically 100 percent of the net additional part-time jobs created in the United States since the late 1970s are held by people who would have preferred full-time jobs but could not find any. . . . That is, we should look to the behavior of employers and not employees to explain the growth in part-time jobs.[198]

The severity of the growing temporary-help phenomenon is in some dispute, but it is difficult to gain precise estimates since the government has only recently began to track what it now calls contingent employment. The Bureau of Labor Statistics (BLS) now plans periodic surveys of *contingent workers,* that is, those individuals who do not perceive themselves as having an explicit or implicit contract for ongoing employment. In its first report based upon a survey in 1995, it reports that the contingent workforce was fairly sizable—ranging upward to 6 million workers (about 5 percent of total employment).[199] Estimates vary, however, because analysts define contingent labor differently. The BLS offers three different definitions and presents data for all measurements. Their narrowest definition includes only wage and salary workers who have been in their jobs for less than a year and who expect their jobs to last an additional year or less (2.2 percent of all workers). The broadest definition—yielding the 5 percent figure—drops the time limitation and counts workers who regard their jobs as temporary and do not expect to continue in them. Yet even here, the liberal figure offered in the BLS report might reflect less of the problem than exists. There are an additional 6.7 percent self-employed independent contractors (many the victims of downsizing who have been unable to replace their jobs), 1.7 percent working "on-call," 1 percent working outright for temporary help agencies, and 0.5 percent who work for contract firms, for example, and who are out of a job once the contract expires. This totals nearly 15 percent of wage and salary workers.

Most revealing in the BLS study was the fact that no matter how contingent workers are defined, the great majority in all three groups wanted to have permanent rather than temporary jobs. This is the heart of the matter. Workers are increasingly being forced to take temporary jobs against their will because good-paying permanent ones no longer exist. Any benefits, such as health insurance or pension plans, are virtually nonexistent with these types of jobs. Temporary workers nationwide average $6.38 per hour, nearly one-fifth less than workers in service industry and one-fourth less than those in manufacturing who are permanent, full-time workers.[200] Capital wants a vulnerable and cheap labor force, but no longer only in clerical or service work. Temps now labor as computer analysts, lab technicians, nurses, engineers, architects, and so on. Almost half of all U.S. hospitals use temps daily. Hence, no area of employment has been shielded from the decline in earnings associated with the burgeoning temporary employment industry.

A major force encouraging workers to tolerate low wages has been unemployment. As mentioned previously, it has been on the increase in the past two decades and is arguably more severe and of longer duration today than at any time in the postwar period. In the war with labor, management has also won a battle by diminishing unemployment benefits. While layoffs increased, by 1990 fewer than four out of ten unemployed workers received benefits —which was a record low.[201] Coverage has simply deteriorated over the decades as eligibility requirements were toughened up. According to the Center for Budget and Policy Priorities, the unemployment system is weaker than at any time since World War II. Only eight states distribute benefits to 50 percent or more of their unemployed, while twenty-three states grant benefits to fewer than a third of their unemployed. Even for those who remain eligible, things can get tougher during economic downturns because of the weakened benefits. During 1991, at the depths of our last recession, an average of 300,000 workers exhausted their regular benefits each month. Never before in the history of the unemployment insurance program have so many workers lost benefits without qualifying for extended unemployment assistance.[202] Moreover, unemployment benefits that were originally tax-exempt became fully taxable during the 1980s. New federal laws now allow states to regulate their own benefits. During the 1980s, forty-four states responded by enacting new, harsher restrictions which made it difficult for unemployed workers to qualify for assistance:

> Some states raised the amount of income a worker had to earn to qualify for benefits. Other states laid down harsher laws disqualifying some unemployed from benefits altogether. Some did both. Many states also attempted to cut costs by reducing or freezing benefits. . . . Harsher penalties for quitting a job, being discharged for misconduct, or refusing an offer of "suitable work" while unemployed also limit those unemployed who can receive benefits. Many states now withhold aid from these workers for that entire spell of unemployment. In a nasty Catch-22, these "durational disqualifications" mean that the unemployed must first find a new job, work the required length of time, and earn the minimum amount to qualify for benefits again, which, once employed, they no longer need.[203]

Another weapon that big business has drawn from its arsenal to wage war against labor is tax policy favoring corporations over individuals. An increasingly unfair tax structure has contributed to the loss of income and growing inequality for families at nearly every income level, but especially among the poor and middle class. This has been accomplished through a variety of means which are more or less hidden from the American public. As we shall see, federal income taxes have not really increased to the degree that people believe, but the entire tax package bundle has shifted to burden the middle class even more in the past decade and a half. Before unraveling how this was done, however, it should be noted that the tax burden in the United States remains incredibly light in comparison to taxes of our industrial peers. The Organization for Economic Cooperation and Development (OECD) monitors tax revenue as a percent of a nation's Gross Domestic Product (GDP). From the last available data in 1991, the United States had a tax bite of 29.8 percent—third lowest among twenty-three industrial, higher-income countries (Turkey and Australia beat us out for the low end).[204] The average tax bite among these twenty-three nations is 39 percent—nearly one-third as much as the U.S. rate. There are, of course, the horror stories politicians and the media shout about when comparing our country to Europe. Sweden, Norway, the Netherlands, and Denmark all have rates hovering around half of their GDP. Perhaps this is a heavy tax burden, although polls in these nations show support for their taxation policies and an unwillingness among citizens to give up the social services that these revenues support. More interesting are nations far poorer than the United States that are willing to provide the necessary tax revenues to maintain the well-being of their citizens (Greece, Ireland, New Zealand, Portugal, Spain, etc.), whose rates hover in the 35 to 40 percent range. Not only is the tax burden on U.S. citizens comparatively light by international standards, the overall rate hardly changed during the last decade while it went up considerably in other OECD countries. If we were to raise our taxes enough to balance the federal budget, America would still be among the least taxed of OECD countries.[205]

Another caution needs to be issued: U.S. tax policies have not eroded the income of its citizens in a major way. Many economists (perhaps a majority) believe that talking about tax cuts clouds the real issue: declining pretax incomes. Tax cut proposals advanced by vote-currying politicans are almost never enough to meaningfully help middle-class or poor families. They carry a hidden cost of higher interest on mortgages and loans that eventually surfaces in the banking community, and they redirect attention and remedial action away from where it is really needed—on the loss of real income because of declining job earnings.[206] In the end, ceaseless talk of tax cuts is equivalent to putting a Bandaid on skin cancer. The remedy has no relationship to the disease.

Perhaps the debate should center around tax *increases*—at least for the very rich and for corporations. When all is said and done, the thrust of tax policy changes in the United States has certainly favored the rich and corporations at the expense of the poor and the middle class. In a nutshell, what has happened—especially since 1980—has been alterations of the federal income tax to make it less progressive, hefty increases in payroll taxes (Social Security), a lowering of corporate taxes, and an increasing reliance on state and local (more regressive) taxes to fund government operations and social programs. After federal taxes were paid, income was less for 90 percent of families in 1989 than in 1977. Twelve percent of the income loss can be attributed to changing federal tax policies that have hammered the middle class while lavishing huge tax benefits and cuts on the rich (the top 1 percent of income recipients).[207] Table 5.2 contains the elements of these changes, broken down by quintiles to show how the modifications actually affect families at different income levels.

Taken alone, the shift of federal income taxes away from progressivity (a state where higher income families pay proportionately more than lower-income families) does not seem large for any of the family income levels—although it is obvious that the poorest 80 percent of families have 0.1 or 0.2 percentage points less of their slice of the income pie because of the tax shifts favoring the rich. The big differences begin to emerge when actual amounts of money are calculated. The average 1995 adjusted family income for the poorest 20 percent is

TABLE 5.2   The Effect of Federal Income Taxes by Family Income Groups

	Change in Share from Lesser Tax Progressivity, 1977–1989	Percent Difference in Tax Payments, 1977–1995
Poorest fifth	−0.2%	−36.6%
Second fifth	−0.2	−0.6
Middle fifth	−0.1	+0.5
Fourth fifth	−0.1	+2.7
Richest fifth	+0.6	+2.2
Top 1 percent	+0.7	−7.9

Source: Mishel and Bernstein, 1995, tables 2.5, 2.8.

$8,391. These families now pay $285 less in taxes than in 1977, for a whopping 37 percent decline (mostly due to the expansion of the Earned Income Tax Credit in 1993, meant to help the working poor). At these families' level of income, $285 is a lot of money. The second quintile is $20 better off because of the changes than before, but the middle class (third and fourth quintiles) is definitely paying more. While the richest fifth is also paying more, the top 1 percent of families actually pay about 8 percent less than they did at the start of the 1980s.

Most of us would not quibble about policies that help the working poor. As was seen, families at this level need a boost due to the failure of the earnings marketplace. But it should be kept in mind that it really does not cost that much to help lower income families through tax strategies. In 1986, in a year of major tax law revision, individuals and families with incomes below $10,000 paid less than $5 billion in income taxes, which amounted to only 1 percent of all income taxes collected in that year.[208] Put in this light, it is fair to ask why our country's leaders have not done even more to help the poor.

It is especially for the tax changes favoring the richest 1 percent of our families that universal condemnation is due. While most middle-income Americans are paying more of the overall tax burden, the most affluent among us actually received major tax breaks from changes in the laws introduced throughout the 1980s. The drop in the tax rate for this privileged class translated to an average 1992 tax cut of $83,457—30 percent less than what they would have paid in 1977. The total cost of this tax cut for the richest 1 percent was $84 billion per year. According to tax expert Robert S. McIntyre, the tax cuts for the richest 1 percent can explain the entire increase in the size of the federal budget deficit.[209]

This point is a major key to the budgetary mess the United States finds itself in today. Although there was overspending during the last decade—especially on a questionable military buildup—the real problem stems from insufficient tax revenues because the well-off are now paying much less than they used to. Common citizens must either make up the difference or go without government services they really want. Aside from greater funding for the military, costs of the S & L bailout for privileged bankers who gambled away our money, and the ever-growing interest on the federal debt, spending for the rest of the national government actually declined in the past decade. The federal outlay had dropped from 9.9 percent of America's GNP in 1980 to 7.4 percent in 1990. This $130 billion decline means that the typical family got $1,260 less from the government each year in federal programs and services, at the very same time its taxes climbed.[210]

Much of the tax increase for middle- and working-class Americans has been hidden, coming in the form of exploding payroll (Social Security) taxes. In little over a decade and a half, the Social Security tax rate was raised nine times (an increase of 31 percent)—going from 5.85 percent in 1977 to 7.65 percent in 1994.[211] As a consequence, Social Security taxes take in far more than they pay out. By 1990 there was an annual surplus in Social Security collections of $69 billion per year, or 1.3 percent of the nation's GNP that was projected to rise to 2 percent by the mid-1990s.[212] Since this tax is capped ($60,600 in 1994), most of us pay it every week, while those earning more than the cap pay no tax on their additional income. For example, a person earning $600,000 per year would pay Social Security for only five weeks a year. Contrary to what a lot of people believe, our Social Security tax is not being banked and/or earmarked under some fund with our name on it that we can claim once we reach sixty-five. Nor is much of it being used to support our currently retired population. Instead, it is a hidden tax being spent as fast as it is collected, serving to mask the full magnitude of our nation's deficit:

> A chunk of your Social Security tax is being spent, every day, to pay for the cost of running the U.S. government. By decade's end, a half trillion of Social Security tax dollars will have gone to pay for everything from military hardware to congressional junkets . . . to the interest on the national debt. . . . In simplest terms, the people in Washington who write the tax laws and spend the money have substituted a portion of the Social Security tax for the federal income tax—without ever explaining what they were doing. What it means is this: If you are a working couple with two children, earning $40,000 in combined salaries, you are paying 6.2 percent of your income, in part, to fund government operations. And if you are a corporate executive like, say, Daniel P. Tully . . . chairman of the board of Merrill Lynch & Company, Inc. . . . who earned $5.2 million in 1992, you are paying seven-hundreths of one percent of your income, in part, to fund government operations. For more than two decades, a collection of Congresses and Presidents of both parties have transferred the overall tax burden in the United States from the people at the top of the economic ladder to those in

the middle and, in many cases, at the bottom . . . by dramatically increasing Social Security tax rates and then using the money for ordinary government programs.[213]

The growth in tax avoidance of our major corporations is another reason the nation's tax burden has been shoved unto the backs of common people. Whichever way it is measured, corporations have increasingly avoided their tax obligations. Between 1979 and 1993, the taxed profits of corporations as a percent of GDP fell from 7.4 percent to 4.6 percent; the 1993 effective tax rate—25.8 percent of actual profits—is about half of what corporate taxes were in 1947 and about two-thirds of their level in 1980.[214] The lessening of corporate liability has been planned, including the deduction of interest payments on their debt, accelerated depreciation, and a variety of other means providing special treatment and enhanced protection. Since they now pay less, we pay more. In 1954, corporations paid 75 cents in taxes for every dollar paid by individuals and families—an amount that fell to 20 cents on the dollar in 1994. The significance of corporate deadbeats cannot be overestimated. If they were paying taxes in the 1990s at the same rate they had been in the 1950s, nearly two-thirds of the federal deficit would disappear overnight.[215] The tax avoidance schemes were purchased by well-paid corporate lobbyists who have persuaded Congress over the years to grant major exceptions and to exclude certain industries (many times single companies) from a variety of tax obligations. Such corporate welfare has reached epidemic proportions, and will be revisited in the last chapter.

Lastly, with reference to the tax structure, since the federal government is doing less because it is actually unable to collect as much tax from rich corporations or wealthy individuals, state and local taxes have picked up much of the slack. The problem is that their approach to taxation is even more regressive than the federal government's. All taxpayers tend to pay equally, instead of the rich paying at a higher rate. Much of the reason is that states rely upon a mixture of tax sources, and are only partially funded by income tax. Sales and property taxes are especially regressive. When state personal income tax, state corporate income tax, property tax, sales tax, and excise tax are all combined, an inverse relationship between income and tax liability emerges. The poorest fifth of four-person families pays 13.8 percent of its income for these taxes. Thereafter, the state and local tax bite lessens swiftly as we climb the income totem pole. The second "fifth" pays 10.7 percent, the third "fifth" owes 9.5 percent, the fourth "fifth" must yield 8.4 percent, the next richest 15 percent gives up 7.7 percent, the next wealthiest 4 percent part with 6.9 percent of their income, and the richest 1 percent has to share only 6.0 percent of its income.[216]

THE DECLINE OF INCOME

The United States has largely failed to hold its own in the global economy. Our goods are overpriced and of shoddy quality. They come from antique plants and are made by a workforce using outdated technology. Because of these factors together with record U.S. interest rates and an overvalued dollar during most of the 1980s, our ability to export to the rest of the world has abruptly dropped. Severe trade deficits have been the inevitable result. They have literally turned the United States completely around. We have been humbled in the eyes of the world, going from the largest creditor nation to the largest debtor nation in five short years. The mounting trade gaps have left our nation in hock to other advanced industrial countries such as Japan, Germany, and even Great Britain. These countries have come to own more and more factories, bonds, treasury notes, stock portfolios, and real estate in the United States. In 1985, U.S. assets abroad with direct investment at market value totalled $1,288 billion compared to $1,160 billion of foreign assets in our country. Although roughly even, the United States still had an edge on investment abroad. By 1993 these figures were $2,647 billion and $3,155 billion, respectively. Today, there is about one-fifth more foreign investment in the United States than American investment in other lands.[217] Much of the profit from these endeavors, of course, is repatriated elsewhere and does not stay in the United States to percolate into our parched economy. The money that does stay may be reinvested, but again this serves ultimately to transfer funds out of our country in the form of dividends, profits, and interest to foreign shores.

As our competitors have invested in new factories with the latest and most efficient production techniques, we have slashed our R & D funds to the bone. Instead, an expanding pot of money has been handed to the military for esoteric projects with no commercial value. Not surprisingly, the productivity of the American workforce has also taken a nose-dive. Without the watering that any growing, healthy economy needs, the fruits of profit have withered on the vine. To meet the decline in profit and the drop in real money return, all actors in this drama of decline borrowed more money. National debt exploded, as did corporate and personal debt.

With the nation awash in red ink, conservative politicians and business gurus responded to the crisis by attacking "welfare freeloaders" and "overpaid union drones" rather than by dealing with the root cause of the crisis. The immediate impact was a drastic cut in the real income benefits of welfare and anti-poverty programs. This in turn led to a growth in numbers of the poor and homeless. Less recognized was the long-term loss of real earnings, previously described, that took place in America.

This slide has reduced the real wages of just about everyone in the labor force, regardless of race, sex, age, industry, or region of the country. As we saw, some have been more hard hit than others in the earnings melt-down, such as white males and workers in our core industrial regions. Manufacturing jobs literally disappeared. Unions were forced to accept humiliating losses, which led to a subsequent drop in membership as well. The pay of manufacturing workers has stopped rising and now lags far behind the wage scale in other industrial countries. It is true there has been job growth in the United States, but a large portion of this has been in service work with significantly lower pay. Quite simply, fewer good-paying jobs have been created in the 1980s and 1990s than in the 1960s and 1970s.

It should then come as no surprise that the real income of most Americans has gone down since its peak in 1973. As documented, there has been a long-term decline for males since that date, with only a slight gain for women. A few things need to be said before the data is analyzed in more detail. How we perceive the figures can depend very much upon the time frame we use. It is possible to look at the same data and have one analyst label it a "crisis" while another would see some "significant improvement." For example, the claim could be made that median family income (as measured in inflation-adjusted 1993 dollars) went up by nearly 6 percent between 1980 and 1990, from $36,912 to $39,086.[218] This statement is undeniably true. Yet the false prosperity of this decade was based upon mortgaging our future through heavy borrowing and by spending money we did not have. By the 1990s, these particular chickens had come home to roost. Within the next four years, even after recovery from the 1990–1991 recession, median income declined, reaching $36,959 in 1993. In short, there was virtually no gain over the past four-teen years. A more comprehensive yardstick would measure median income from the end of World War II to the latest date of data availability (1993). Figure 5.10A traces the incredible rise of real income through the golden 1950s and 1960s, and the leveling off of the standard of living which took place two decades ago. By the early 1970s, the climb had petered out. Real income, after taking out the effects of the up-and-down wave action, is approximately the same today as it was in 1973. There was some improvement in the latter part of the 1980s, but because these gains had been built upon a house of cards, they were all but wiped out in the first half of the 1990s.

Whether very significant income growth has occurred over the past twenty years also depends upon the choice of deflators which are used to subtract the effects of inflation (which enables yearly comparison in con-stant dollars). The Bureau of Labor Statistics (BLS) used the official Consumer Price Index (CPI-U) to measure changes in the cost of living

FIGURE 5.10   A. Median Income of Families (Constant 1993 Dollars)

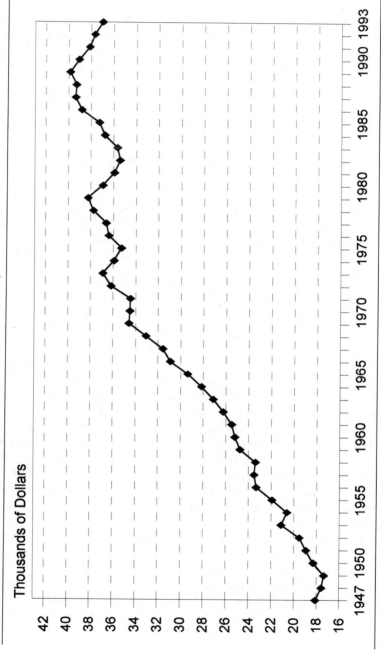

Source: U.S. Census Bureau, *Income and Poverty, 1993,* CD-ROM, table F3.

based upon the price of a market basket of goods and services for the urban population. In 1983 the BLS began using a revised method (CPI-U-X1) which attempted to correct for the distorting effects caused by the sharp rise in mortgage interest rates in that period. A comparison of both methods shows that the change actually results in a sharper trend toward growth in median family income between 1967 and 1990 than would have appeared if the old methodology had been kept intact.[219] Unfortunately, there is no longer any choice. Official statistics now reflect a more generous estimate of median income than in the past.

Some analysts doubt the wisdom of discontinuing mortgage rates in the market basket of goods (as a means of subtracting the effect of inflation) in favor of equivalent rent values. To begin with, mortgage rates are a good generalized reflection of the ravages of inflation, which reduces all incomes in a society. Mortgage rates are pegged to the prime rate which is closely dictated by the monetary policy of the Federal Reserve system. While nominal interest rates have fallen since the early 1980s, real rates remain high by historical standards. The run-up has helped aggravate existing income inequalities so that the wealthiest 10 percent of households increased their income by 50 percent through higher interest rates while the lower 80 to 90 percent of income earners have experienced declines in their real incomes.[220] Thus, it seems dangerous to minimize the effect of interest rates by downplaying the cost of mortgages. Mortgage rates ultimately remain an important factor for anyone in the real estate industry, but especially for first-time home buyers who have been having increasing difficulty in qualifying for such loans.[221] By adopting the new methodology, a younger generation is in effect being told that their shelter problems are of little concern.

The graph in figure 5.10B shows several important trends. There is a long-term gap in median incomes between non-Hispanic white families and African-American and Hispanic families in the United States. It seems as if these deprivations are cast in concrete. Virtually no improvement appears for minority families. No hopeful sign points to some closing of the income gap that has existed over the past twenty years. For that matter, real family income is getting worse for Hispanic families, and it has been virtually checked for African-American families. On average, white families have gained over $3,000 in constant dollars since 1972 (8 percent), but Hispanic families lost approximately the same dollar amount (–11.4 percent). African-American families saw nearly $800 (–3.6 percent) vanish during this period. Not only have minorities lost greater absolute dollar amounts, but their percentage declines have been much heavier as well since they start from a much lower base than do white families.

On the whole, there is some reason to suspect that a less sharp drop in real income would have happened were it not for the recession at the

FIGURE 5.10   B. Median Family Income by Race (Constant 1993 Dollars)

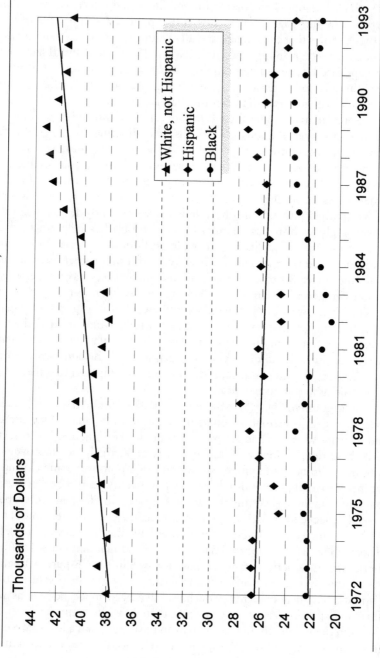

Source: U.S. Census Bureau, *Income and Poverty, 1993,* CD-ROM, table F3.

start of the 1990s. But the effect of this short-run downturn can also be overemphasized. Beginning in 1989, mean income in constant 1993 dollars declined steadily for six straight years among the bottom three-fifths of American families, while it declined in five of six years for the next highest fifth. Only the richest fifth of our population did not experience a consistent decline—although income bounced around quite a lot for this quintile. Again, adopting a longer time frame clarifies the picture considerably. Figure 5.11 shows an unmistakable shift toward greater income for the richest fifth, with declines in the proportion of total income going to each of the other four quintiles. Over the last quarter century, the richest fifth gained nearly 16 percent in its proportionate share of total family income in the United States. The next lower fifth experienced a wash, losing only 1.2 percent of their income slice. But the middle fifth, next lower fifth, and poorest fifth lost 11.3, 20.2, and 26.8 percent respectively in the proportion of income they received. Again observing families over the last quarter century, figure 5.12 shows a yawning chasm between the poorest fifth and highest fifth with respect to absolute income gains. The richest fifth displays an unstoppable climb of income (ending at a mean of $107,471 per family), whereas the poorest fifth is barely able to maintain a mean income of about $10,000 in constant 1993 dollars.

Figure 5.13 illustrates the decline in both absolute and relative terms of real family income between 1973 and 1992, in constant 1989 dollars.[222] The data is from the famous *Green Book* prepared yearly by the Committee on Ways and Means, U.S. House of Representatives. Several improvements are employed to measure income in a better way than the standard Census Bureau procedure. While both sources always present data in constant dollars (adjusted to take out the effect of inflation), adjustments are made by analysts for the Committee on Ways and Means to take into consideration the concept of need based upon family size. Income is also changed to add the cash value of food stamp and housing subsidies, while federal income and Social Security payroll taxes are subtracted (no adjustments are made for state and local taxes). When such adjustments are made, it is still obvious that a large segment of America's 68.5 million families have suffered. The poorest fifth of families has gone through a shocking experience. On the face of it, the drop in real income of $805 does not appear very large, but this is about a 13.3 percent decline in income for a segment of our people that is impoverished to begin with (figure 5.13A). At the other end of the scale, the richest fifth of American families saw a gain of about 11 percent in real income during the past two decades. The average family income for this category climbed from $66,364 (in 1989 dollars) to $73,487—a real income gain of over $7,000.

FIGURE 5.11 Share of Income Received by Each Fifth of Families, 1969–1993

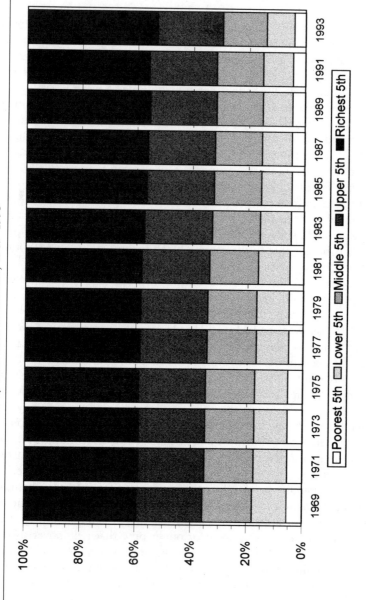

Source: U.S. Census Bureau, *Income and Poverty, 1993*, CD-ROM, table F2.

FIGURE 5.12   Inequality by Family Income (in Constant 1993 Dollars)

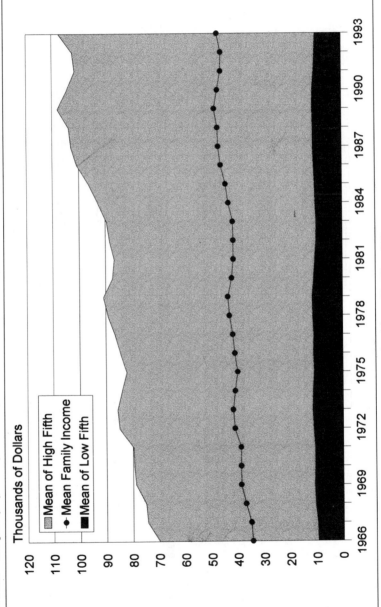

Source: U.S. Census Bureau, *Income and Poverty, 1993,* CD-ROM, calculated from tables F2a and F3.

FIGURE 5.13   A. Percent Change in Average Family Cash Income, 1973–1992

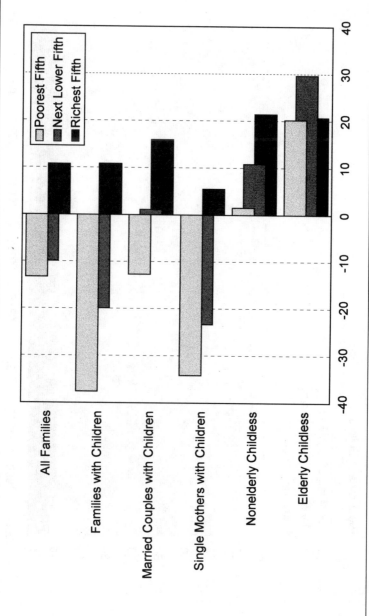

Source: U.S. House of Representatives, Committee on *Ways and Means*, 1994, calculated from table H23.

FIGURE 5.13 B. Change in Income of Families with Children, 1973–1992 (in Constant 1989 Dollars)

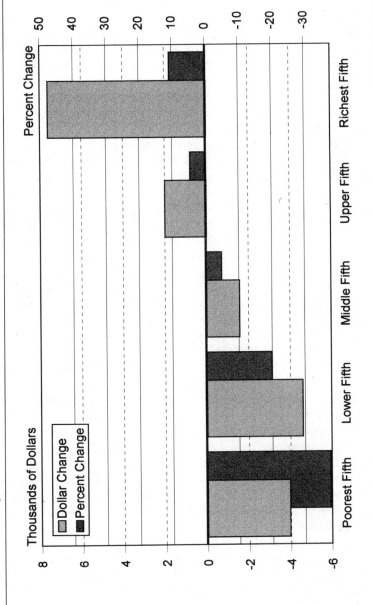

Source: U.S. House of Representatives, Committee on Ways and Means, 1994, calculated from table H23.

The figures also suggest that the hardest hit with relative and absolute income loss have been families with children (especially those headed by single mothers), as opposed to childless families—for which no decline emerged within the past twenty years. The lowest fifth of families with children experienced nearly a $4,000 loss over two decades (left vertical axis of fig. 5.13B)— amounting to about 38 percent of their income (right vertical axis of fig. 5.13B). The next lower fifth actually lost slightly more money ($4,615) although its proportionate loss was less severe (–20 percent) since this quintile started at a higher average income in 1973 than the poorest fifth. The middle fifth also lost 5 percent of its income while both the upper fifth (5 percent income rise) and the richest fifth (11 percent) gained absolutely and relatively.

Although the poorest fifth of married families with children did forfeit 12.7 percent of its money over the two decades, the remaining 80 percent of married families with children managed to stay ahead of the bill collector. This, however, has more to do with working wives, moonlighting husbands, and laboring teenagers than with any rise in husbands' incomes. The reason that the number of earners in families has been going up is not hard to fathom. In an attempt to stay abreast of loss in real income, wives in husband-wife households have entered the labor force in record numbers. It will come as no surprise that if more persons work in a family—including children who are also going to school—real median income will rise. For example, in 1993 median family income went up by 81 percent, going from a one-earner family ($26,193) to a two-earner family ($47,424). Income more than doubles between one-earner and three-earner families ($57,745).

Not taking account of the increase in working wives essentially hides another pothole in the American economy. There is nothing inherently wrong with more family members choosing to work—if it is a choice. Many women have greatly benefited from their increased labor force participation. It may have made more egalitarian relationships with men a reality, as women won their own financial freedom. Other benefits of work are harder to measure but do exist, such as pride in craftsmanship, intrinsic satisfaction, job-related friendships, service to humanity, and the like. Working wives have also removed the traditional burden which held men completely responsible for the financial well-being of the family. In all, there are many positives connected to wives who work—advantages to themselves, to their families and husbands, and to society in general. Given the decline in real income, however, especially for males and for young adults, there is some doubt that the major reasons that dual-earner families have increased are "voluntary."

One simulation shows that family income inequality among married-couple families increased by 8.6 percent over a sixteen-year period, but that it would have increased 50 percent more if married women had not worked.[223] In a recent, exhaustive review of previous studies, Judith Treas points to a major consensus in the research. Working wives have been an equalizing force on the income of husband-wife families. This has been especially important because of the cutbacks in social welfare benefits and the drastic fall in the earnings of younger males. In short, "Besides blunting overall inequality, women's greater workforce involvement has also worked to smooth out income disparities between age groups, between generations, and even over the life cycle."[224]

The most recent data released by the Congressional Budget Office (CBO) measures income differently than the Census Bureau does. All estimates measure income in constant dollars that subtract the effect of inflation. The CBO adjusts income on the basis of need as measured by family size, and measures after-tax income as well. Finally, the Congressional Budget Office regards a single person living alone as a "family," like a married couple or a single-parent unit. Thus, the CBO data corresponds more fully to what the Census Bureau would refer to as a "household"— all persons who occupy a housing unit, whether related or not. Probably the most significant advantage of the CBO data over that of the Census Bureau or the House Ways and Means Committee is that it analyzes income shares going to the top 1 percent of families. This is especially significant because it reveals a startling fact: the wealthiest 1 percent of our population (2.5 million people) now has nearly as much after-tax income as the bottom 40 percent of the population (approximately 100 million people).[225] In 1996, the average projected after-tax income for this privileged group was $438,000—comprising 12 percent of all family income in the United States. According to this study, the richest fifth of families receives as much after-tax income as the bottom 80 percent of families— fully half of all national after-tax family income. In the fifteen years between 1977 and 1992, real income nearly doubled for the richest 1 percent of families, went up by 28 percent for the top fifth, stagnated for the middle fifth, and declined by 17 percent for the poorest fifth.

It is no wonder, then, that a minority of persons in our society have been supporters of conservative, supply-side economic policies that favor hefty tax cuts for the rich. Those who are in upper-income brackets are obviously better off now than during the 1970s. Yet the shadow looming over this rosy scenario is the equally obvious loss in real income among the great majority of families. There has been an undeniable bipolarization taking place. Had the American economic pie been growing since 1973, the fact that the poorest of the nation's families who received less in absolute income would not have mattered so much. On the average, however, the trend has been to lower real income since 1973 for 80 percent of families in our country. Disastrously, this decrease has taken its toll on those least able to afford it—where the dollar loss has been the greatest. On the other hand, the major exceptions to this slide have been the richest families, who have actually seen their incomes go up. The implications for the future are ominous if these trends continue without letup. Given time, it is possible for the unrelenting deprivation of the Brazil model to emerge in the United States with all of its malevolent force.

JUMPING GINIS: THE SURGE IN RELATIVE INCOME INEQUALITY

With the growing gaps in real income among and within such a wide variety of groups (e.g., age, gender, industry, race), we can safely predict that relative income inequality has been on the rise in America as well. If anything, the statistics are even more unanimous on this point. The Gini score for the distribution of family income declined from .376 in 1947 to .348 in 1967—a drop in inequality of 7.5 percent in twenty years. Although this was not an incredibly large amount, at least the trend was toward more equality at the very moment that real income was increasing for nearly everyone. Since 1968, however, relative income inequality has again gone up rather steadily, rising 23 percent to a post–World War II record Gini score of .429 in 1993. What this means is that inequality is greater today than at any time in the past forty-six years. While it increased by less than 5 percent between 1968 and 1979, between 1980 and 1993 it rose an additional 17.5 percent. Figure 5.14A illustrates this climb in income inequality for families by race since 1968. The surge of inequality affected both African-Americans and whites equally, but there was more inequality among African-Americans to begin with. In 1966 (the first year of comparative data that is available) the African-American Gini ratio stood at .375 compared to .340 for whites, a difference of 10 percent. By 1993, the difference was nearly 16 percent (.482 and .416, respectively). Thus, although relative income inequality has been climbing for both white and African-American families, it is bounding upward at a much faster pace for African Americans.

Figure 5.14B also shows a startling reversal in the percent of income going to the richest 5 percent as opposed to the poorest 40 percent of families. By 1982, for the first time in twenty-five years, the share of national income going to the wealthiest 5 percent of families was actually greater than that of the bottom 40 percent. The gap kept widening rapidly after this point through 1993—the last year of data. Put differently, if income were divided in a totally equal manner, the top 5 percent was getting nearly four times more than it normally would, while the bottom 40 percent was getting almost three times less than would be expected.

FIGURE 5.14   A. Family Income Gini Scores: 1968–1993

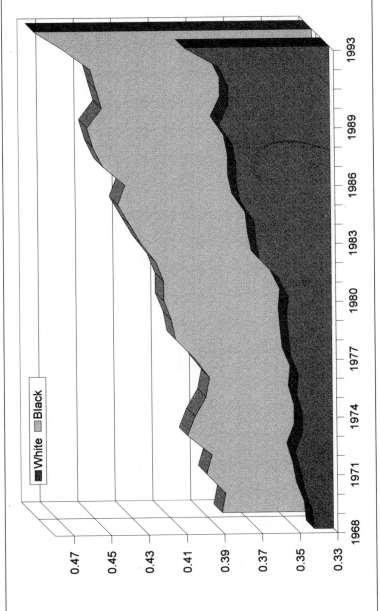

Source: U.S. Census Bureau, *Income and Poverty 1993,* CD-ROM, table F2.

FIGURE 5.14   B. Shares of Family Income: 1968–1993 (Poorest 40% and Richest 5%)

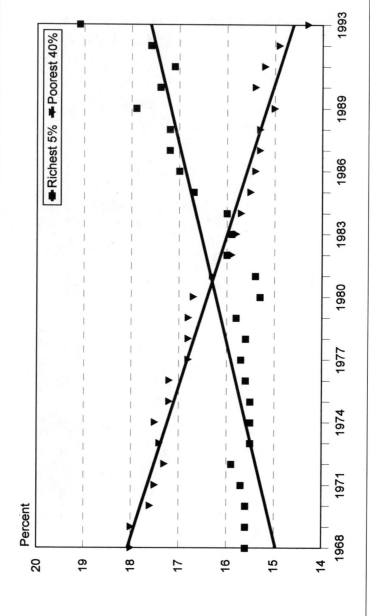

Source: U.S. Census Bureau, *Income and Poverty 1993*, CD-ROM, calculated from data in table F2.

When we look at quintiles, the full extent of the change seems to be understated because more of the instability is at the far ends of the income distribution (bottom and top 5 percent) than within the much broader 20 percentile range. Nonetheless, even here a growing amount of income dispersion at the extremes is also evident. To get a clear picture, figure 5.15A compares the size of the slice that each of the quintiles received from the family income pie of 1993. Extreme disparities are evident, showing the poorest fifth receiving five times less than it should under a fully equitable distribution system while the next lower fifth received half of its proportionate share. The middle and next higher fifth received slices that were roughly proportionate to their numbers, but gluttony abounds for the richest fifth—which receives almost half of all family income in the United States.

Figure 5.15B shows how the size of these shares has changed in the last twenty years. As with absolute income, the bottom 80 percent of American families also lost a proportionate share of their income. Thus, in addition to losing real dollars over two decades, each of the four lower quintiles saw a drop in the percent of total family income that accrued to them. The upper fifth (sixtieth to eightieth percentile) lost only 2.9 percent of its share, compared to -10.3 percent for the middle fifth, -16.8 percent for the lower fifth, and -24.5 percent for the poorest fifth. In stark contrast to this trend stand the richest fifth of American families, which saw the size of their slice increase by nearly 15 percent over two decades. Among this well-off cohort is the top five percent of American families, which performed twice as well as the richest fifth. The richest 5 percent of American families saw their proportionate share of the income pie rise by nearly one-third of its original size over the past twenty years. When income is adjusted to reflect family size, the bite of federal taxes and Social Security, and also the help of food stamps, subsidized housing, free school lunches, and so on, families with children have taken an even harder hit. The poorest fifth saw its pie slice shrink 36 percent in 1973–1992, compared to a 20 percent reduction for the next highest fifth (not shown).[226]

THE DECLINING MIDDLE CLASS

Since income erosion is evident just about wherever one looks, concern mounted during the last decade that our country's middle class was shrinking and might eventually disappear. The consequences, it was feared, would lead to political instability, crime, social upheaval, and the ushering-in of a new societal Dark Age. Blackburn and Bloom examined the question of a disappearing middle class head-on. By looking at the changing size of actual income classes, their study traced income inequality over

FIGURE 5.15   A. Share of Family Income by Quintile, 1993

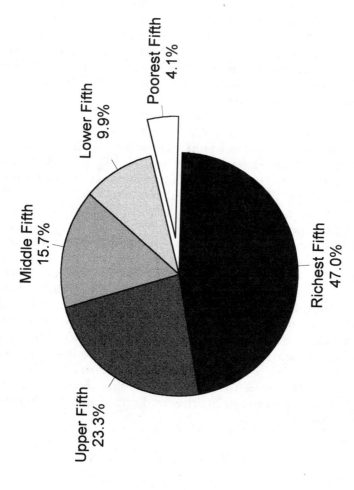

Poorest Fifth
4.1%

Lower Fifth
9.9%

Middle Fifth
15.7%

Upper Fifth
23.3%

Richest Fifth
47.0%

Source: U.S. Census Bureau, *Income and Poverty 1993*, CD-ROM, table F2.

FIGURE 5.15   B. Percent Change in Share Received by Each Quintile, 1973–1993

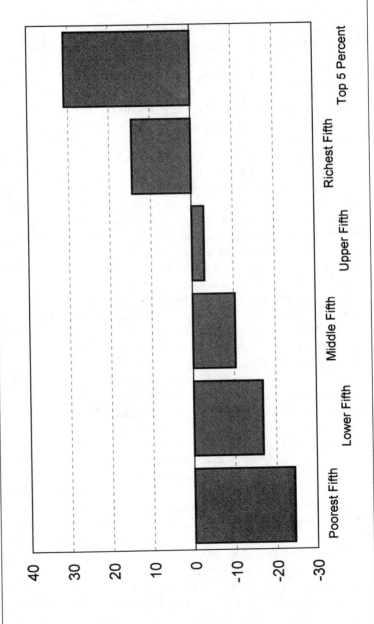

Source: U.S. Census Bureau, *Income and Poverty 1993*, CD-ROM, calculated from data in table F2.

nearly a two-decade period (1967 to 1984). Rather than using quintiles, these researchers chose a different way to categorize income classes, a method more sensitive to change. They define lower-class as having income less than or equal to 60 percent of median income, lower-middle-class as having income above 60 percent of the median up to the actual median, middle-class as having income above the median up to 160 percent of the median, upper-middle class as above 160 percent but less than or equal to 225 percent of the median, and upper class as having income above 225 percent of median income. Thus, their definition does not use fixed points but is relative—being pegged with median income. By these criteria, the upper class has gone up by half of its original proportion over the eighteen years of this study, taking 8.3 percent of all family income in 1967 but 13 percent in 1984.[227] On the whole, Blackburn and Bloom see the increasing proportion of women who work as being of some importance in reducing inequality for husband-wife families, but having almost no general impact for all families. The researchers also conclude that there has been a drop of nearly 16 percent in the number of middle-class families.[228] Even combining the three middle-class groups (upper-middle, middle, and lower-middle) fails to hide the decline. In 1969, these three groups made up 71 percent of all families—but were only 65 percent of all families in 1983.

Another study by Katherine Bradbury defines a family income of $20,000 to $49,999 as middle class—which comprises nearly half of all families.[229] Thus, her approach is to use fixed points (in 1994 dollars, her definition of the middle class would include people who roughly fall into the $30,000 to $72,500 range, after inflation is factored out). Comparing family incomes over eleven years shows an increase of 13 percent in low-income families. A corresponding decrease of about 10 percent occurred in the middle class while nearly 5 percent was added to the proportion of the upper class. A very detailed, exhaustive study by Lynn A. Karoly using more recent data for the 1963–1988 period basically agrees with Bradbury's finding of middle class decline. One of Karoly's caveats, using a variety of methodological techniques and refinements, is that the middle class really shrank by only 5 percentage points rather than by 10 because Bradbury used a less appropriate price deflator.[230] By also using the different price deflator to correct for inflation, Bradbury's conclusion that the lower class grew faster than the upper class is also reversed.

There is considerable disagreement about what constitutes the "middle class," but surprisingly sharp consensus that, however defined, this bedrock of American democracy was still being pulverized well into the 1990s. Although Kevin Phillips uses a variety of indices and measurements in his book on the declining middle class, he never identifies precise cutting points. In his view, however, the American middle class is approximately 45 percent of the population. It comprises half of the fortieth to fifty-ninth income percentiles and includes virtually everyone in the sixtieth to niney-eighth percentiles below the upper 1 percent. Viewed in this way, the American middle class is 5 to 10 percentage points smaller than it was a decade or two ago.[231] A more sociological approach focuses on both income and associated lifestyles, but adds little to clarification—other than to identify deindustrialization as the *raison d'etre* for middle-class decline.[232]

Some of the most recent data, together with a noteworthy encapsulation and discussion of alternative methods, are offered by Duncan, Smeeding, and Rodgers.[233] Their research adopts an approach similar to that of Blackburn and Bloom's earlier research by employing a relative definition, but introduces a number of refinements. Although they use a snapshot approach by looking at Census Bureau Current Population Surveys as they have changed between 1967 and 1989, they also introduce panel data (the 1967–1989 Panel Study of Income Dynamics, or PSID) to see what happened to actual families through a long time period. The lower range of the middle-income group was set at the twentieth percentile and the upper boundary at the ninetieth percentile. In 1989 dollars this was $15,000 to $50,000. Both sets of data show a steady decline in the fraction of middle-income households from nearly 60 percent in the late 1960s to about 51 percent in the late 1980s. Breaking the time period into the 1970s and 1980s and tracking both up and down mobility also yields no surprises. During the 1970s adults had roughly equal chances of moving up from the middle class as well as down from it. During the 1980s, a middle-class adult's probability of moving down (85 in 1,000) was larger than the chance of moving up (75 in 1,000). The researchers conclude that the probability of falling from middle income status—falling from grace—increased significantly during the last decade and was mainly attributable to declines in the earnings of males. Lastly, neither economic cycles nor demographic factors explained very much of this new downturn. Another part of their analysis (data unpublished) showed these trends persisting into the 1990s.

THE ROLE OF DEMOGRAPHIC CHANGES

By now it is plain that however we may choose to measure income inequality, whether in relative terms or absolute dollars, it has gone up by a large amount in the past two decades. Yet some experts believe this deterioration of real income is mostly due to changes in demographic factors. Frank Levy, for example, in his book *Dollars and Dreams,* does a good job of stressing the impact that a number of such changes have upon income inequality.[234] While family income has gone down in real terms, Levy points out that family size has also been reduced over the years (especially due to lower fertility). Today there are more persons living alone than at any time since the end of World War II. More women, particularly wives, are working than ever before. The shift toward female-headed families has also had a major impact on family income, since these units are much more prone to poverty and hardship. They are also heavy recipients of government transfer programs and more likely to have no earners, which affects the family income picture. Levy points out that the proportion of families with no one working nearly tripled in little over a quarter century (increasing from 5.4 percent to 15.1 percent). To account for these changes, he adjusts the income figures away from the Census definition to add the value of transfer benefits (such as food stamps, Medicare, employer health insurance, etc.) while subtracting taxes paid. In his view, the tax structure takes a bigger bite out of higher income families so that inequality is artificially increased by not looking at post-tax income. Once all of these adjustment factors are built into the data, it can be shown that family income was more equally shared in 1984 than in 1949. In essence, the poorest fifth actually received a slightly higher proportion of income while the richest fifth of families ended up with slightly less income.[235]

There are a number of reasons to challenge any conclusion that income inequality is actually less today than previously. To begin with, Levy compares almost all of his current (1984) income data to 1949. This was at a time when sharp inequality existed, before the long-term decline to 1973—at which point inequality started increasing again. In his most recent publication, Levy is still maintaining that the period of the 1950s and 1960s has provided social scientists with a "bad yardstick." In his view, the sharp run-up of real income during these two decades was an "aberrational period" in U.S. life.[236] Even if one does adopt a half-century time frame exclusively, however, it is still hard to be sanguine about the harsh income trends so manifest in the last two decades of this century— or to ignore the real pain and the negative consequences that they entail. I would argue that the relevant comparison should run from 1973 to the present. Otherwise we ignore the very large increase in real income and decline in relative inequality during the 1950s and 1960s. In a word, the U-shaped history of postwar inequality is made invisible.

More to the point, however, are the reasons behind the need for demographic adjustments. It is certainly true that changes in family structure have cushioned the loss of income since 1973. Yet this speaks volumes about the skid in real income, not about how much better off we are because more wives work now than ever before. No one disputes the beneficial impact of wives who work. Over the past twenty years, wives' earnings have become increasingly important in raising mean family income and in reducing inequality. Just in the past decade, wives' earnings were responsible for two-thirds of the increased income among couples.[237] In a review of nine studies that address the impact of working wives, it was found that all agree on one basic point: working wives equalize income (at least among whites). The strongest finding is that income inequality would have increased at an even greater rate if wives had not entered the labor force in such record numbers:

> There is a growing sentiment that the wife's paycheck is what enables many families to maintain a toehold on the middle rungs of the income ladder. . . . To sum up, the economic prospects of young men deteriorated at a time when young women commanded higher wages than ever. Labor force participation by married women undoubtedly permitted young families to achieve middleclass consumption standards that would have been impossible on one paycheck alone. The trade-off was that families lost some of their inflexibility to field additional earners, should the main breadwinner become unemployed.[238]

There is also strong evidence that income inequality would have been worse were it not for increasing welfare and transfer payments during the 1960s and early 1970s—at least for families headed by women and unrelated persons of either sex.[239] Rather than seeing such trends as benign, however, one can view them as proof that the decline of income has touched just about every one of us. It is equally valid to see many of these demographic shifts as being caused by the scaling-down of real income—thus masking the true savagery of the decline—as to interpret them as trends that hide more equality. The decline in fertility and average family size is a good example. This has allowed families to partially escape the full brunt of declining real income. Yet the fall in fertility is likely a result of the slide in real income, rather than a cause of the rise in real income. This is a classic example of the chicken-or-the-egg dilemma. My guess would be that there is less choice behind declining family size and later marriages as the financial well-being of families has waned. People react to adversity and bad fortune. If the cost of living rises and/or real income goes down, they will act in ways to protect themselves. One way would be literally to cut back on the mouths to feed or perhaps not to get married at all. In this sense, it is questionable to suggest that smaller families result in higher relative or per capita income when they may in actuality be caused by a drop of real income.

The same may be said of the increasing role welfare payments have played in the distribution of family income. Some authorities argue that in-kind benefits such as food stamps hide what would be a larger income among the poor. Yet, is it not more appropriate to ask why there has been such a sharp need for these benefits in the first place? The number collecting such benefits may expand because the ability of low-income families to survive has become more difficult. In a stagnant or declining economy, many husbands among the poor simply cannot find work. Or, if they do, it may pay less than a living wage. One means of coping might then be separation, divorce, or a ruse to produce the eligibility (female-headed household) needed to collect many welfare benefits.

There is an unstated premise underlying the argument that welfare benefits to the poor actually increase their real income. The implication is that the system is generous and compassionate in taking care of these unfortunates. In fact, it is so generous that we should perhaps cease to think of the poverty-stricken as "really that poor." By the same token, no one asks why the system has failed to provide the means for poor families to support themselves, why decent paying jobs are becoming more scarce, why recessions are more frequent and severe, and why earnings have been cut so severely in industry after industry from one occupation to another.

Lastly, these adjustments have been made, we are told, to more fairly estimate the true income of families. Why is it that in-kind benefits such as employer health insurance or the dollar value of food-stamps have been factored in as adjustments, but not capital gains? While the former benefit lower- and middle-income families, adding capital gains to family income would skew the distribution incredibly. Such large amounts of money accrue almost exclusively to rich families and without a doubt would show income inequality to be much worse than current government statistics show. Our government now even uses a deflator adjustment that removes the cost of home mortgages, reasoning that the record rates in the 1980s drove the consumer price inflators up above normal.[240] The experts believed that since this affected only a few first-time home buyers, mortgage rates should not be used. We can only imagine what a baby-bust family looking for their first house would think of this reasoning.

Researcher Katherine Bradbury offers a clear description of the appropriate role played by demographics in this slide of real income.[241] Her study controls for most of the factors that could possibly confuse the reasons why such a loss developed in the first place. Dividing family income into low-, middle-, and high-income brackets, she reveals an increase of 13 percent at the low end, a decrease of about 10 percent in the middle class, and an added 5 percent in the upper class over an eleven year period. But critics have cautioned that such a slide is worse than it appears because of changing demographics. There *has* been a decrease in the proportion of nuclear families. Real income *has* steadily increased for unrelated individuals. Singles *do* make up a larger portion of our population with income. When Bradbury added this group to families and recalculated the proportions, however, the significance of her findings did not change. The middle class still declined by 11 percent, while the lower class grew by 12 percent. The upper class does shrink, however, by a small 2 percent. Smaller family size, the baby boom, population shifts to the

South, more working wives, and greater numbers of female-headed households all affect real family income as well. The strength of their impact was calculated by Bradbury. In a word, the effect of these demographic variables in causing the income decline was very small. Cuts of income resulted in all these groups.

Surprisingly, the effect of the baby boom was the opposite of what many would expect. The baby boom actually drove up family income slightly, mainly because this cohort is now entering middle age (forty to fifty years old, a time when income peaks). Very young families under age twenty-five, where income is lowest, have actually gone down in proportion by over one-third. Thus, although real income went down for every family whose head was under sixty-five years old, the fact that so many boomers were surging all at once into middle age and peak income actually neutralized the effect of age.

When taken alone, these changes do have some impact upon a declining middle class and loss of real family income. Many of these demographic variables exhibit strong effects. Looking at family income, for example, shows that nearly one-third of the impact of demographic variables is due to the increase in female-headed families while one-sixth is caused by smaller families. Yet 42 percent of all of the demographic effect—which resulted in *higher* family income—can be attributed to working wives. The bottom line is that these variables tend to cancel one another out. Of all the loss that occurred in family income, only 10 percent can be explained by shifts in demographic characteristics. Fully nine-tenths of the decline in real family income cannot be explained by any of these demographic factors. Bradbury surmises that much of the decline may be due to wounds suffered in our industrial heartland as manufacturing jobs vanished, but she does not examine this factor directly.

Lynn Karoly has also traced the role of demographics as a cause of increasing income inequality over two decades using more updated data (1967–1987).[242] Her results are substantially the same as Bradbury's. Although there has been a significant growth of younger families (baby boomers), single households, and female-headed families, on balance they have little impact on increasing inequality. All age and headship groups experienced a rise in income inequality. While the growing youthfulness of our population in particular had nothing to do with increasing inequality, changes in family composition did have some effect. The trend away from married-couple families and toward single-parent families or unrelated heads could explain one-third of increasing inequality over the past two decades.

To sum up, the great majority of research in the 1980s and 1990s has continuously pointed to one fact that is impossible to deny. Real income has dwindled for the great majority of Americans at the very time that relative inequality has skyrocketed. Using a variety of techniques and definitions, most studies agree that poverty is once again haunting the land as the proportion of poor families grows relentlessly larger. There is also basic agreement that the rich have grown even more wealthy, at least since the early 1970s. There is less agreement about the reasons for the slide in the financial well-being of most Americans. Debate has centered on shifts in basic demographic characteristics, such as more working wives, the youth and size of the baby boom, more female-headed families, regional shifts, and the like. While the importance of these factors taken one by one is undeniable, when considered together they tend to neutralize one another. The root causes of the decline are not easy to see, and argument continues to swirl among various viewpoints.

Although the general effect of demographic variables as a whole is weak, there is still no direct proof that large scale economic changes are behind the decline. There is plenty of evidence regarding the association, however, between declining income, growing relative inequality, and the gutting of our industrial infrastructure. America has literally undergone a transformation in its basic mode of production. Manufacturing is rapidly disappearing while service work is on the rise. Whole regions such as the Midwest have suffered long-term decline as a result of this shift. Yet the economy as a whole has also suffered. We have witnessed mounting personal, corporate, and governmental debt. A gaping trade deficit refuses to go away. Losses in productivity preceded a slash in wages. Cutbacks in R & D and a decline in new plant investment have robbed future generations of income. All the while, mergers, buyouts, and outright purchase of U.S. businesses and property by foreign powers continues to erode our nation's capacity to generate income for its citizens. Scandalous defense budgets dumped into an endless vacuum have sucked away the very lifeblood of America's economy, rather than providing the security that was promised.

There is a great deal of face validity to the argument that the decline of America in the global economy has been at the root of income loss and growing inequality. In the next chapter, we will examine some of these economic and demographic factors as they affect income locally within the United States. There is a rich history of research into the variation in income inequality between regions, states, and smaller localities that is worth exploring. By looking at such forces on a more local level, we may be able to get a better view of the interaction of shifts in global and national forces upon income inequality. We will also see clear evidence of a very wide spectrum of inequality in the United States. It seems an axiom of life that as some fail, others will prosper. Yet because of such sharp geographic variation, at least part of the general impact of America's slip toward financial mediocrity may have remained hidden.

Chapter

6

SUNSET IN THE SUNBELT:
HOW GEOGRAPHIC VARIATION
HIDES INCOME INEQUALITY

Many sections and communities in the United States suffer from low income. Other areas are cursed with high relative inequality. Although all Americans are aware that income inequality exists in our country, we persist in viewing our society as middle-class. While much of our population is well-off or at least financially comfortable, pockets of poverty remain effectively hidden from the mainstream. Many of them are in the rural South. Surprisingly, however, even such areas of affluence as New York, Atlanta, Dallas, Miami, Los Angeles, and San Francisco are host to virulent inequality. By 1990, virtually every state in America had a household income Gini ratio greater than India's or Ethiopia's. The income inequality that we tend to see as a problem of the Third World is just as much a part of our culture as of less developed countries (LDCs)—at least on a relative basis. Put in a slightly different context, Americans are subjected to higher household income inequality rates than China, Pakistan, India, Ethiopia, Indonesia, South Korea, and Ghana. Over 80 percent of the world's population (in countries where relative income inequality data exist)—numbering 3.1 billion people—lives under less relative inequality than do people in the United States. In the end, income inequality is not just a Third World problem.

In the past few decades Americans have also experienced a tremendous caldron of economic shifts between and within a variety of geographic boundaries. Such changes can be seen as partially hiding the average decline of absolute income since 1973. The growth of relative inequality is also easily masked because some sections of the nation have become obviously more affluent. Nonetheless, by looking at smaller units, such as regions, states, and counties in the United States, we can see uneven economic development. An attempt has been made in this chapter to piece together the causes which increase or decrease both absolute and relative income inequality for various regions within the United States.

Any analysis of regional differences must bow in the direction of historical forces and geographic specializations. Because of early settlement, the northeast seaboard rose to early dominance in this country. Emerging at the same time was a southern economy geared toward cotton growers with large plantations worked by slaves. The West soon became the breadbasket of the nation, while manufacturing took root in the Northeast. Industry in this region was first thrust into world competition by needs arising during the Civil War.

The dramatic rise in the economic fortune of southern states is one of the most significant trends in this century. The South has long been an underdeveloped region in comparison to other parts of the United States. It was particularly devastated by the Civil War. Because the South was treated as a country occupied by an unfriendly, hostile foe, it was impossible for this region to develop in the last century. A combination of internal factors, which are discussed below, also stopped any major growth in the South's industrial base. But this region's successful climb out of an economic swamp first became evident in the 1930s. New Deal programs and the spread of the defense industry were major prods to development and expansion. Retirement of relatively wealthy northerners to states such as Florida helped to disperse income. The huge out-migration of poor southern African Americans to northern industrial cities drained some poverty from the South as well. The South has been this century's Horatio Alger story of regional economics. Since the end of World War II, and especially beginning in the 1970s, the region has experienced an explosion of growth.

WHISTLING DIXIE TO AULD LANG SYNE: THE SOUTH GROWS RICHER AS THE NORTH DECLINES

One of the rallying cries heard long after the Civil War in the South was that it would one day rise again. It would reemerge from the calamity and defeat of the war to challenge Yankee supremacy. Few believed this would be in terms of renewed military power, but there were many embittered souls in the South who longed for a day of reckoning. They hoped a time of retribution would arrive some day in the future. Many southerners have celebrated the decline of the North with joyful glee. They see the consequences as doubly sweet. Income in the North has stagnated or gone down as manufacturing jobs have virtually disappeared. At the same time, there has been a large-scale upsurge of manufacturing, business relocation, and industrial investment in the South. One consequence has been more jobs, many of which would have stayed in the North thirty years ago. As a result of the greater availability of jobs—often at higher pay than was thought possible by locals in their recent past—income has gone up in the South.

The movement toward greater equality of income between our country's geographic regions can be easily seen in table 6.1. When we look at personal per capita income as a percent of the average for the country as a whole, two related tendencies leap out. The dominance of the northeast as an economic core has dwindled in the past one hundred fifty years while the resurgence of the South in this century has been obvious. While in 1840 per capita income in the Northeast was 135 percent of the nation's average, this gradually lessened to 105 percent by 1980 (although it bounded up again to 116 percent by 1993). The effect of the Civil War on the South is easily seen in the data, but after the turn of the century this region started its long but steady climb out of America's economic basement. In 1900 the South had a per capita income only half that of the United States as a whole, but this gradually increased to 92 percent by the time of the last Current Population Survey (CPS) in 1994. Lastly, the effect of the Gold Rush, railroad expansion, and land giveaways through the Homestead Act can be seen in the West, which saw a dramatic loss of per-capita income set in before the turn of the century. Much of this decline in per-capita income may also have been due to a huge influx of poor people seeking their fortunes. Many chose to follow the popular adage of a century ago: "Go west young man, go west!"

TABLE 6.1　　Personal Per Capita Income in Regions as a Percent of the U.S. Average: 1840–1993

Region	1840	1880	1920	1960	1980	1993
United States	100%	100%	100%	100%	100%	100%
Northeast[a]	135	141	132	114	105	116
North Central[b]	68	98	100	101	100	98
South[c]	76	51	62	80	86	92
West[d]	—	190	122	105	103	101

Source: Agnew, 1987, p. 92; U.S. Census Bureau, *Statistical Abstract 1994*, CD-ROM, table 699.

a. Northeast Region: Maine, Vermont, New Hampshire, Massachusetts, Rhode Island, Connecticut, New York, Pennsylvania, New Jersey.

b. North Central Region: Wisconsin, Illinois, Michigan, Indiana, Ohio, Iowa, Kansas, Minnesota, Missouri, Nebraska, North Dakota, South Dakota.

c. South Region: Delaware, Maryland, District of Columbia, Virginia, West Virginia, Kentucky, Tennessee, Arkansas, Louisiana, Mississippi, Alabama, Georgia, Florida, South Carolina, North Carolina, Texas, Oklahoma.

d. West Region: Idaho, Montana, Utah, Colorado, Wyoming, Washington, Oregon, California, Nevada, Arizona, New Mexico, Alaska, Hawaii.

Despite this gradual growth, just after World War II the South was still far below the nation's average income level. These sixteen states and the District of Columbia have long been poor kin to the rest of the country. Figure 6.1 portrays the United States in a map of median household income earned in 1989. The South still has relatively more states in the lowest income quartile (47 percent) than do the other three regions (23 percent in the West, 17 percent in the North Central, and none in the Northeast).

The latest median household income figures by region, which cover 1994, show that the South is the only region continuing to gain income (2.9 percent in real income gain over the previous year).[1] Median household income now stands at $30,021 in this region. Southern household income is still less than income in the Northeast ($34,926), the North Central ($32,505), and the West ($34,452). Yet the South is now within 93 percent of the median household income for the whole country ($32,264), which is the closest it has been to absolute parity in its entire history. If the strength of this trend continues, the South will finally pull even with the rest of the nation by the dawn of the next century.[2]

A more complete picture of increases in absolute income can be seen in table 6.2. The data in this table is in constant 1989 dollars, so comparisons can be made to earlier dates with the effect of inflation removed. Median household income does remain lower in the South, but the growth during the 1970s for this area was much greater than for other regions. The seventeen states of the South had an average increase in median household income of nearly 14 percent in the 1969 to 1979 period. This pace was much higher than the Northeast (an actual decline of –0.4 percent), and comfortably ahead of the Midwest (5.4 percent) and the West (6.2 percent). Trends in the 1980s, however, do not bode so well for the South. This region placed third out of the four areas in percent change in median household income, growing by a much slower rate during the 1980s (4.5 percent). The South is also home to the nation's four lowest income states—Mississippi, West Virginia, Arkansas, and Louisiana.

The effects of deindustrialization and the farm crisis are particularly prevalent in the Midwest, where household income actually declined by –1.4 percent during the decade. In fact, three of every four states in the Midwest show a loss in median household income over the past decade—a rate well below that of the third of southern states experiencing income decline.

The overall rate of increase for the West (8 percent) essentially reflects the phenomenal growth in California (17 percent). The large population in this state is actually equivalent to all of Canada's, and thus pulls up the median for the entire region. The sparsely populated states of

FIGURE 6.1   Median Household Income, 1989

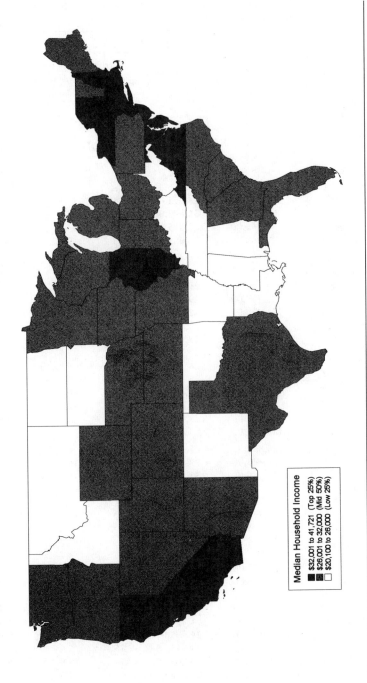

Median Household Income

■ $32,001 to 41,721 (Top 25%)
▨ $26,001 to 32,000 (Mid 50%)
☐ $20,100 to 26,000 (Low 25%)

Source: Data from U.S. Census Bureau, *America's Income—Changes Between the Censuses,* 1993.

TABLE 6.2   Median Household Income and Percent Change by States
and Regions: 1969–1989

| | Median Houshold Income | | | | |
Region and States	1989	1979	Rank	1979–1989 Percent Change	1969–1979 Percent Change
United States	$30,056	$28,220		6.5	5.7
Northeast	33,826.00	29,006.00	1	16.6	−0.4
Maine	27,854.00	23,151.00	27	20.3	0.6
Vermont	29,792.00	24,783.00	20	20.2	−1.1
New Hampshire	36,329.00	28,508.00	7	27.4	4.7
Massachusetts	36,952.00	29,450.00	6	25.5	−2.2
Rhode Island	32,181.00	26,973.00	13	19.3	−0.5
Connecticut	41,721.00	33,643.00	1	24.0	−1.7
New York	32,965.00	27,895.00	11	18.2	−4.4
Pennsylvania	29,069.00	28,285.00	23	2.8	5.1
New Jersey	40,927.00	33,178.00	3	23.4	1.7
North Central	29,334.00	29,748.00	3	−1.4	5.4
Ohio	28,706.00	29,750.00	26	−3.5	1.9
Indiana	28,797.00	29,462.00	25	−2.3	4.9
Michigan	31,020.00	32,212.00	15	−3.7	2.4
Illinois	32,252.00	32,376.00	12	−0.4	6.0
Wisconsin	29,442.00	29,626.00	22	−0.7	4.6
Minnesota	30,909.00	29,762.00	17	3.9	8.0
Iowa	26,229.00	28,150.00	37	−6.8	13.5
Missouri	26,362.00	26,109.00	35	1.0	8.1
Kansas	27,291.00	27,417.00	30	−0.5	15.0
Nebraska	26,016.00	26,685.00	38	−2.5	14.2
South Dakota	22,503.00	22,045.00	47	2.1	8.6
North Dakota	23,213.00	25,626.00	44	−9.4	17.9
South	26,832.00	25,668.00	4	4.5	13.6
Delaware	34,875.00	29,904.00	9	16.6	2.1
Maryland	39,386.00	33,984.00	4	15.9	6.9
District of Columbia	30,727.00	27,164.00	18	13.1	4.2
Virginia	33,328.00	29,282.00	10	13.8	12.2
West Virginia	20,795.00	24,405.00	50	−14.8	19.5
Kentucky	22,534.00	23,401.00	46	−3.7	13.7
Tennessee	24,807.00	23,697.00	40	4.7	13.6
Arkansas	21,147.00	20,467.00	49	3.3	21.4
Louisiana	21,949.00	25,516.00	48	−14.0	24.0
Mississippi	20,136.00	20,269.00	51	−0.7	23.4
Alabama	23,597.00	22,905.00	42	3.0	13.4
Georgia	29,021.00	25,190.00	24	15.2	9.0
Florida	27,483.00	24,591.00	29	11.8	9.0
South Carolina	26,256.00	24,651.00	36	6.5	14.6
North Carolina	26,647.00	24,265.00	34	9.8	9.8
Oklahoma	23,577.00	24,716.00	43	−4.6	19.1

TABLE 6.2   *(Continued)*

Region and States	Median Household Income				
	1989	1979	Rank	1979–1989 Percent Change	1969–1979 Percent Change
Texas	27,016.00	27,997.00	33	–3.5	18.0
West	32,270.00	29,879.00	2	8.0	6.2
Montana	22,988.00	25,839.00	45	–11	10.4
Idaho	25,257.00	25,613.00	39	–1.4	8.8
Wyoming	27,096.00	33,503.00	32	–19.1	32.5
Colorado	30,140.00	30,256.00	19	–0.4	14.1
Utah	29,470.00	29,611.00	21	–0.5	10.9
Nevada	31,011.00	30,516.00	16	1.6	2.0
New Mexico	24,087.00	24,566.00	41	–1.9	10
Arizona	27,540.00	27,562.00	28	–0.1	6.8
Washington	31,183.00	30,777.00	14	1.3	7.2
Oregon	27,250.00	28,118.00	31	–3.1	7.7
California	35,798.00	30,569.00	8	17.1	4.4
Hawaii	38,829.00	34,306.00	5	13.2	2.1
Alaska	41,408.00	42,586.00	2	–2.8	14.5

Source: U.S. Census Bureau, *America's Income—Changes Between the Censuses,* 1993.

Montana and Wyoming did very poorly as energy prices for coal and oil crashed. In the end, over two-thirds of the states in this region experienced a decline in median household income during the 1980s.

It is only in the Northeast (a nearly 17 percent increase) that we see astounding income growth. All nine states in this region saw real increases in median family income, and in all but one state the growth was in double digits. For an allegedly declining core, the Northeast displays amazing vigor in money growth. This can also be seen in the expansion of the business service industry containing well-paid professional jobs which remain with core multinational headquarters in the major cities of this area. Increased defense spending and the import boom of the 1980s also benefited cities and states on both coasts.[3]

In summary, during the last decade, nearly half of all states saw median household income go down in real constant dollars after inflation is removed. Moreover, five times as many states experienced declines in incomes in the 1980s as in the 1970s. The severity of this decline was also greater in the 1980s. None of the few states which saw a loss of income in the 1970s experienced a decline greater than 5 percent. During the 1980s, however, two states had declines between 5 and 10 percent while four states showed a loss higher than 10 percent. How you viewed the 1980s, then, would depend on where you lived. Many citizens of southern, western, and midwestern states would tend to see doom and gloom, while northeasterners would understandably paint a rosier scenario.

Although the total growth of income in the South over the past two decades continues to be ahead of the rest of the country in percentage terms, this is only one side of the proverbial coin. It is obvious that southern states start from a much lower income base, so that the same amount of dollar increase in this region would yield a higher percentage increase than in the North. This is exactly what happens in the data. Between 1979 and 1989 households gained more in real absolute income in the West ($2,391) and the Northeast ($4,820) in comparison to the South ($1,164). As with LDCs in comparison to core industrial countries, because such absolute dollar differences are so great to begin with, closing the income gap in real dollars is harder to do than the percentage increase figures indicate.

Nonetheless, measuring increases in real median income (after inflation is subtracted) over the last two decades shows healthy growth in many southern states. Of the twenty-one states posting gains in real median household income in both of the last two decades, six are in the South. Maryland, Virginia, Georgia, Florida, North Carolina, and South Carolina all have growth rates equal to or greater than the national average. Change for most of the South is in the right direction. Especially before 1980, southern growth was spectacular:

> For example, gross regional product nearly doubled between 1960 and 1975, while industrial output more than doubled. Between 1970 and 1976 fast-growing Texas added more nonfarm jobs than Michigan, Illinois, Ohio, and Massachusetts combined. In 1976 the industrial output of Texas alone exceeded that of Australia. Between 1970 and 1977 per capita income increased at a national average of 71 percent, while Arkansans and Mississippians saw their personal incomes rise by 86 and 83 percent, respectively. These states had traditionally been the South's slowest growing, and any significant increase would appear dramatic when considered on a percentage basis. Still, even in more economically advanced Florida, per capita incomes grew by 69 percent during this period as compared to only 58 percent improvement in New York. . . . By the 1970s the region was attracting approximately half of the nation's total annual foreign industrial investments.[4]

Southern banks were soon bulging at their seams with a new torrent of cash. Asset growth in the largest five southern banks grew six times faster than the five largest United States banks during the last decade.[5] By 1983, the South had overtaken the North in terms of commercial bank deposits—having the largest amount among the four regions. In both absolute and relative terms, therefore, the South's deficit of investment capital has mostly disappeared. Although southern banks continue to be conservative, preferring to allow loans for regional expansion to originate in northern banks, southern banks are definitely in a position to encourage their own investment should the need arise.

In order to understand the internal dynamics of the economic changes happening across the South today, we need to know a little about the area's history. This region was once host to a plantation economy based upon slavery and gross income inequality. The interests of a small planter aristocracy were constantly at odds with those of both poor whites and African Americans held in bondage. Matters did not improve after the Civil War. Although slaves were now legally "free," they were kept in a position of subservience and dire poverty through Jim Crow laws, denial of the ballot through poll taxes or literacy tests, violence, lynchings and any other means that worked.[6] Poor whites, although not much better off economically, at least were less likely to be the victims of brutal physical intimidation and attack. Yet they were also held back through dependency on agriculture in an area of poor soil. A sharecropping system had emerged in the South that operated on rules similar to those of serfdom in medieval Europe. Few were lucky enough to own land directly. Even fewer could farm much of the land, which was marginal. Too much concentration upon only a small number of cash crops, such as tobacco or cotton, also took a predictable toll. As is common in market economies, prices tended to fluctuate wildly. At best, this led to insecurity. More often than not, however, financial ruin was unavoidable. It was hard to pay creditors when crops were largely worthless. Frequent bankruptcies ended in the loss of farms. The land more often than not reverted to a few large landowners, some of whom were remnants of the plantation aristocracy.

Thus, the South was a largely agrarian society with only sparse manufacturing well into the twentieth century. What little industry did flourish in the region was connected with textile and clothing manufacture, which was also associated with low wages and unstable prices. In a word, the southern economy was similar to that of an LDC on the international scene. The South functioned as a periphery region for a more wealthy core located in the North. The Yankee aristocracy was made up of big business tycoons who had made their wealth in heavy industries such as steel, railroading, shipping, petroleum refining, and the like. It was this circle and its northern economic interests that placed the South in a "most favored colony" status for a century after the Civil War. Although almost all industrial development, economic growth, and financial advancement took place in the North, there were still some trickle-down benefits for the lagging, economically dependent South.[7]

The South had to be content with the leftovers from the feast of northern economic expansion. It could do so because there were always a few industries and factories which were nearly obsolete, operating on the fringe of profit. While such companies could not survive in the high-wage mecca of the north, they would do rather nicely south of the Mason-Dixon line. Low production costs, especially bottom-of-the- barrel wages, kept some firms from going under. This dismal fact of life also guaranteed a slow but steady growth of industry in the South, as weaker core firms were spun off to this region. At the same time, northern interests were also served by keeping expenses down while southern spinoffs operated as feeders to their more lucrative, profitable companies in the core.

It was especially after the Civil War that the manufacturing belt around the Great Lakes was formed, linked to the newly emergent railroad empires of the Northeast. This formed an American economic core which remained dominant at least through the 1950s. Regional growth rates and geographic specializations were set by the core, which demanded massive raw materials from the hinterland. These resources were made over by industry and exported, both abroad and back to the periphery, as finished goods. John Agnew gives a concise account of the shifting fortunes of America's regions.

> The North constituted a core, but the periphery was not united. Rather, it consisted of a West into which vast investments were placed . . . and a South which was drained of its resources rather than developed through investment. After the Civil War, the South sank into a tributary condition as the most backward section of the national economy. . . . The South had lost a war and paid the penalty by becoming dependent on the North and by missing out on the western bonanza. National income during the period from the Civil War until the 1890s was systematically redistributed from the periphery to the core. In particular, the South helped pay for northern industrialization either through higher prices for domestic goods or through duties on imported products. The export of southern cotton underwrote northern economic expansion rather like collective agriculture was later to underwrite Soviet industrial expansion. Economic subordination of the South [was needed] for American overseas expansion.[8]

Regional and historical conditions thus play a large role even today in the amount of income we end up taking home in our paychecks. Original settlement patterns, ethnic groupings, racial barriers, regional specializations, natural resources, the fruits of war, outright domination by economic giants, and like factors all enter into the formula which can fix a region's income level. Yet there is still room for local forces to work as well. At least within the South, a large degree of influence has been exerted by native elites which has slowed the region's economic development. There has long been a conservative bias operating in southern banks, for example, which acts to favor a powerful landowning class.

Interest rates were kept high and industry was at times actively discouraged from entering. The fear was that wages would be raised if many firms were to compete for labor.[9]

The companies that were allowed to come in had to pass an anti-union, low-wage litmus test given by local economic elites. This gate-keeping function persists even today. Major local resistance is often aroused when a unionized plant with good paying jobs wants to set up shop. The Phillip Morris Tobacco Company was frozen out of Greenville, South Carolina, because its new plant and its twenty-five hundred workers would be unionized. Mazda Motor Company was actively discouraged from building its new plant in the Greenville/Spartanburg area for the same reason, even though three thousand direct jobs and another three thousand support jobs would have been created in the new auto factory:

> Some business leaders feared that it would upset the existing wage scale and force the multitude of textile and apparel manufacturers in the area to raise their wages. After Mazda ultimately chose the Michigan location, in part because of the highly skilled (and unionized) workforce in that state, the Spartanburg County Development Association cheered the decision in its monthly newsletter by noting: "It is our considered view that the Mazda plant would have had a long-term chilling effect on Spartanburg's orderly industrial growth. An auto plant, employing over 3,000 card-carrying, hymn-singing members of the UAW [United Auto Workers] would, in our opinion, bring to an abrupt halt future desirable industrial prospects."[10]

It *is* with the question of wages that the major clue to these large regional income inequality differences can be found. The South has long prided itself on being a safe place for businesses, where wages are low, taxes are almost nonexistent, and the institutional climate is hostile to unions.[11] The norm of low wages grew out of the slavery-ridden plantation system which in turn gave way to share-cropping after the Civil War. Low pay was built upon racism and chronic poverty, so that workers were desperate for whatever little income could be gained from an economically barren system. This deprivation acted to attract the very industries that were seeking low-wage workers. Thus, the system was to perpetuate itself in a vicious circle. Some northern core industries first started relocating in the South to take advantage of sub-marginal wages by the late 1800s. New England textile firms went to the rural areas of North Carolina, South Carolina, and Georgia at the turn of the century in search of cheap labor.

This pattern was to repeat itself endlessly throughout the twentieth century, peaking during the 1970s and 1980s. Southern politicians and community spokesmen have actively and aggressively sought to attract industries to their home towns. As they have competed with one another and especially with northern states, southerners have repeatedly sold business firms on their superior location by stressing a favorable business climate. The package making up a favorable business climate always contains the same items: cheaper wages, no unions, low taxes, abundant raw materials, fewer energy costs, growing consumer markets, tax holidays, a free or subsidized infrastructure (roads, schools, utility hookups, etc.). Yet it is especially the low wages that are continually mentioned:

> The salesmen have included southern governors, state development boards, local chambers of commerce, and assorted industrial front men. All have sung the virtues of southern industrial development to United States and foreign investors. In a world guided by business maxims, where profit is the guiding principle to social and economic life, southern boosters found that their product, the South, sold best when it was packaged as a place where labor was abundant and cheap, where unions were not welcome, and where local communities were willing to offer a large variety of financial inducements to attract new industry. A measure of the success (whether real or imagined) of this public relations program has been the highly publicized population explosion in the Sunbelt states. The decade of the 1970s was truly one of romanticizing the wonders of southern growth and development.[12]

The wisdom of courting factory transplants, especially from foreign lands, needs to be questioned. By the mid-1990s, eleven Japanese auto factories were estimated to be producing three million cars per year—30 percent of all cars and trucks sold in the United States. The hidden costs, especially to the South which has been most aggressive in wooing Japanese investment, are very high. By the time Toyota finally agreed to build an assembly plant in Georgetown, Kentucky, that state had contributed $300 million in lower taxes and other incentives. This amount will pay the plant's entire wage bill for two or three years. And again, the lure is reduced employee benefit costs in the assembly plants plus lower wages in supplier plants (averaging just 60 percent of the costs of American car manufacturers). For every foreign automobile company assembly plant in the United States, there is an army of third-tier suppliers made up of low-paid, part-time, and contingent workers.[13]

FLIGHT FROM THE SOCIAL WAGE: "THE SOUTH SHALL RISE AGAIN"

Some argue that American corporations have also entered the transplant frenzy by simply folding up their tents from northern, frostbelt states during the past few decades. The goal of these companies was to avoid high pay scales and cantankerous unions. Rather than submit to what some executives see as inflated wages, business policy has instead been geared toward relocating in the South and West. Many states in these regions have right-to-work laws which make it difficult to unionize and to strike.

There is much truth in this view. For example, critics claim that in addition to fleeing high northern wages, business has been actively avoiding the social wage that has typically gone with their pay package. Corporations and individual workers have been paying taxes for community amenities, social services, education, welfare benefits, and the like all along. These forms of the social wage are much more evident in northern states than in the South. For example, it is clear that unions are much more dominant in northern states. In the South at the start of the 1990s, only 14.5 percent of workers in manufacturing were unionized, compared to 19.5 percent in nonsouthern states. There are larger welfare payments outside the South ($2,351 per person in 1990) than inside that region ($1,564). While $798 per person was spent on health services in southern states in 1990, the comparable amount was $879 in nonsouthern states. In this year, the average weekly unemployment benefit pay was $144 in the South but $161 outside the region, while average AFDC payments were $237 and $407, respectively. Average per pupil expenditures on education from kindergarten through twelfth grade is $1,243 in the South but $1,401 in other regions. All such expenditures plus other social services translate to a higher cost of doing business. In the eyes of corporate accountants geared to the bottom line, this is a major sin. Rather than pay for such a social infrastructure, firms have opted to move abroad or at least to the South or West for what they deem a "better business climate." In poorer regions of the country, companies can pay lower wages that may still be higher than the local scale.

It has definitely been in the sunbelt—whether in the deep South or in the Southwest—that more jobs have been created. Jobs have disappeared quickly from the Northeast and industrial Midwest at the same time (also deepening poverty, inequality, and residential segregation between African Americans and whites).[14] Looking just at one ratio—jobs lost divided by jobs created—shows the sunbelt to have been a big winner in competition for earnings.[15] Lorence and Nelson also conclusively demonstrate the greater disappearance of manufacturing jobs together with a climb in low-paid service jobs in rustbelt metropolitan areas in comparison to other regions.[16] But the map in figure 6.2A dramatizes just how desperately the South could still use higher earnings. In 1980, the average hourly pay of southern production workers in manufacturing industries was 89 percent of their northern counterparts; by 1993 this had crept up only to 93 percent. Even in 1993 this region continued to have a greater proportion of states in the bottom quartile of earnings than did other areas.

FIGURE 6.2 A. Average Hourly Earnings, 1993

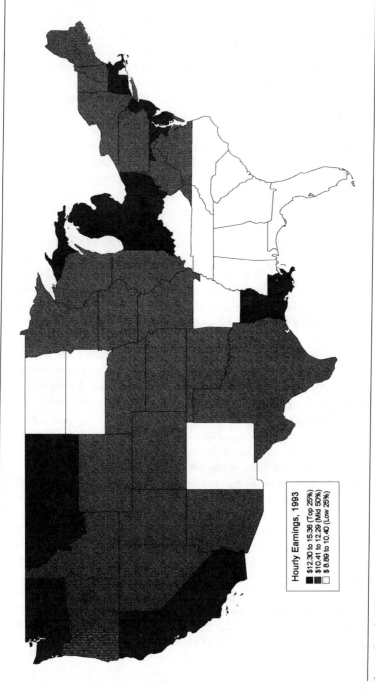

Hourly Earnings, 1993

- $12.30 to 15.36 (Top 25%)
- $10.41 to 12.29 (Mid 50%)
- $ 8.89 to 10.40 (Low 25%)

Source: Data from U.S. Census Bureau, *Statistical Abstract 1994*, CD-ROM, table 1238.

On the other hand, the change in earnings between 1980 and 1993 heavily underlines a related trend. The East North Central Census region (made up of Michigan, Wisconsin, Illinois, Indiana, and Ohio—our industrial heartland) dropped considerably in average earnings. Between 1980 and 1993, the average earnings of manufacturing workers in these five states increased by 56.4 percent (unadjusted for inflation), compared to 60.1 percent in all of the North and 64.2 percent among southern states. Indeed, southern states have done quite well in posting wage gains relative to the North over the past thirteen years. Nearly two of three southern states are in the top half of the pay gain distribution. In a ranking of all states, North Carolina is seventh with an 82.5 percent gain, South Carolina tenth (75.3 percent), Georgia eleventh (74.7 percent), Virginia twelfth (74.4 percent), Tennessee thirteenth (69.9 percent), Delaware fourteenth (69.3 perent), Maryland fifteenth (68.6 percent), and Mississippi sixteenth (68.4 percent). Although southern manufacturing jobs posted healthy wage gains, there was a hidden cost to the victory. The increase was made possible largely by cutting back on jobs or moving them out of the North. Within just six years during the last decade, work in manufacturing had fallen 8.2 percent around the country. The loss came to 1.8 million jobs.[17] Such extensive deindustrialization has also had a depressing effect on wage rate gains in the industrial heartland for those who were able to keep their jobs. Among the five core states for Great Lakes heavy industry, Ohio (twentieth, at 63.9 percent) and Michigan (twenty-fifth, at 61.3 percent) were barely in the top half of the wage-gain distribution, while the other three states were in the bottom half. Indiana ranked thirty-fifth (55.1 percent), Wisconsin fortieth (51.6 percent), and Illinois forty-third (50.1 percent).

While the larger percentage increases in hourly manufacturing earnings over the past decade in the South are undeniable, other ways of looking at the figures may take some of the gloom away for the industrial heartland. Despite decades of deindustrialization and vanishing jobs, conditions remain better with respect to manufacturing earnings in the North. Northern states still had higher average hourly manufacturing earnings in 1993 ($11.90) than did the South ($10.94), while their absolute dollar raises have been higher in the past thirteen years as well ($4.37 and $4.24, respectively). Much of this is no doubt due to the fact that northern wages started at much higher levels. Thus, even with smaller percentage raises, many of the increases tend to be larger than in the South.

There is little room for cheer in most of the data. The decline in manufacturing has been most prevalent in the northern states and is directly related to the failure of American corporations to invest in themselves and to compete in the world market. The manufacturing sector of

FIGURE 6.2   B. Percent Increase in Average Hourly Earnings, 1980–1993

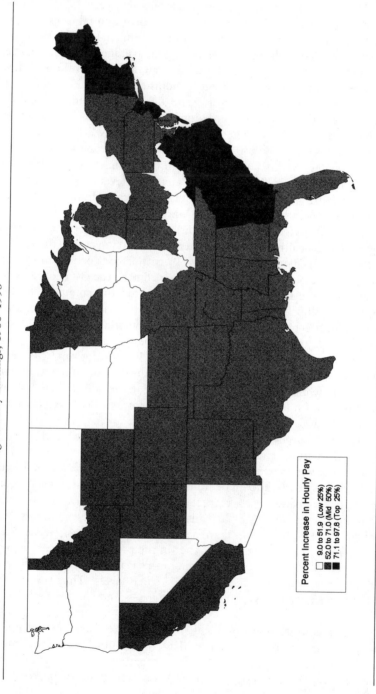

Percent Increase in Hourly Pay

☐ 9.0 to 51.9 (Low 25%)
▨ 52.0 to 71.0 (Mid 50%)
■ 71.1 to 97.8 (Top 25%)

Source: Data from U.S. Census Bureau, *Statistical Abstract 1994,* CD-ROM, table 1238.

the United States has been very concentrated in only nine states, seven of which are in what is now called—by way of summing up this decline—the Rust Belt (New York, New Jersey, Pennsylvania, Ohio, Michigan, Illinois, and Indiana). At the same time businesses have been fleeing the North, many have left our nation altogether. When runaway shops did stop and rebuild on this side of the Rio Grande, it was nearly always because of the cheaper wage structure of the South. Many experts believe that the decentralization of manufacturing production led to the economic revitalization of the South. Yet there was less financial benefit in wages than would have occurred if these industries had stayed in the North. One in three manufacturing jobs in the South was below the national wage average for all production workers.[18] A quick glance at the map in figure 6.2A tells the whole story. Most of the production work in the South in 1993 was still being done at wages below the prevailing national norm. Despite the fact that many of these new jobs are in recently built factories within high-technology industries, the pay level is quite low. The appeal of this regional periphery made up of the South and Southwest stems from the same concerns that led U.S. corporations abroad—greater profit. The low wages, hostility toward unions, minimal sums spent on social services, and cheaper energy costs translate into greater dollars for corporations leaving their old northern homes in the industrial core. Were it not for the attractiveness of the South, the national erosion of wages and real earnings would have been even worse as plants relocated to Mexico, Thailand, or wherever a compliant workforce could be bought at the cheapest rate.

THE BLOOM IS OFF THE MAGNOLIA BLOSSOM!

Because of this stress on low wages, then, the South has ended up attracting the very industries and businesses that are unwilling or unable to pay decent wages. While there has been sharp job growth, the rise of income has not been as rapid. For one thing, although job growth has been higher than in the North, manufacturing jobs actually increased at a slower rate than population growth in the South.[19] More job growth actually occurred in low-end service work located in major cities, just as it did in the rest of the country. The South has also been most attractive to industries at the end of their product development stage, when all innovation, creativity, and technological breakthroughs have long since been wrung out of the manufacturing process. In such non-growth or declining types of industries, long and stable production runs are needed with minimal costs. At this point, cheap labor becomes a major factor in keeping profits alive. Branch plants of this kind that lead some critics to view the South as an economic periphery, working to serve the interests of core conglomerate corporations located in the North. The same charge has been leveled at such high wage nations as Japan, who have actively sought the cheaper pay scale here than at home.[20]

The prevailing footprint of plant relocations and startups in the South proves this point. William Falk and Thomas Lyson have looked at changes in all southern counties between 1970 and 1980, concluding that 80 percent of all growth in manufacturing jobs took place in nonmetropolitan areas where the pay scale was at its absolute lowest.[21] Most of these jobs were from branch plants of northern corporations. Although urban and metropolitan counties have gained many more jobs, they tend to be in service and high technology industries which pay more than jobs in the non-unionized rural hinterland among poor African Americans— where corporate control is at its absolute maximum. The huge gap of earnings inequality between metropolitan and nonmetropolitan areas has continued in the 1990s. An updated analysis by Lyson and Tolbert which examines earnings inequality between residents of metropolitan and non-metropolitan areas for the entire country shows a continuation of these trends through 1991.[22] Using three different earnings inequality measurements, they find a continuous increase in inequality over the past quarter century. Although increases occurred in both metropolitan and non-metropolitan areas, they were greater outside metropolitan areas.

In the rural South, two-thirds of all new jobs are manual in nature.[23] This area has been described as a stagnant backwater of the American economy. Few of these jobs are what we would normally describe as good—most are part of the secondary labor market (janitors, food service workers, factory assemblers, etc.). Most manufacturing has been limited to low-end production. Thirty-five percent of the manufacturing work-force in southern rural counties, for example, is in textile and clothing firms. Such industries have a stunted opportunity structure for workers. The great majority of employees are at the unskilled, blue-collar level. Their pay reflects this reality. Even in urban areas, Falk and Lyson conclude that many of the new jobs are in poor-wage, low-skill service work.

Yet, because poverty and deprivation have existed for so long in the South, especially in rural areas and particularly among African Americans, the labor force has been even more docile and undemanding. Poor southerners have flocked to apply for jobs that pay much more than what they were used to, but much less than comparable workers were making in unionized, northern factories. In a word, the South's poor were grateful for these jobs:

> Even in the Sunbelt era southern workers continued to respond to the neopaternalism of antiunion employers such as Nissan Motors. Although

one United Auto Workers leader told Newsweek that Nissan's desire to hire "hard working country people" really amounted to an attempt to employ "peasants," the company's guarantees of lifetime employment, its efforts to involve employees in management decisions, and its recreational and fitness programs drew 100,000 applications for 2,600 jobs at its Smyrna, Tennessee, plant. Union officials were perplexed at the attitude of one Nissan worker who insisted: "If the company treats the workers well . . . there's nothing a union can offer." Such assertions were particularly confounding in light of the fact that this worker could have earned a starting wage that was $2.50 per hour higher if he had been employed in a unionized automobile plant in the North.[24]

What this means, in the end, is progress with a small "p." There has been undeniable improvement. The South is better off in terms of higher income, economic growth, and in closing what seemed an eternal income gap with other regions in the United States But critics point out that the gains have been nothing in comparison to what they could have been if the South had not been so hostile to organizing workers. It is riddled with so much dire poverty that the belief that *any job is a good job* is still unchallenged. Because of what amounts to an institutional addiction to low wages, the South has remained one of the nation's major economic problem areas.[25] Over a twenty-year period during the 1960s and 1970s, Mississippi, Kentucky, Arkansas, the Carolinas, and Georgia all saw declining real per-capita income. In 1993, only five of the seventeen states (29 percent) in this region had a per-capita income above the national average.[26] James Cobb concludes from his analysis of the southern economy that without doubt this region's nonunion climate has helped to depress wages.[27] Industrial development not only failed to generate large-scale prosperity, but it has also left a large number of southerners trapped in poverty. Figure 6.3 illustrates the swollen poverty rates of southern states in comparison to other regions. Over 39 percent of the nation's poor live in the South, while this region has the highest percent of its population in poverty (17.1 percent)—more than the West (15.6 percent), the Northeast (13.3 percent), or the Midwest (13.4 percent).[28]

Table 6.3 lists both the worst and the best dozen states with regard to personal poverty in 1993. All but one of the twelve worst states are southern, while low-poverty states tend to be scattered throughout the United States. Even more dramatically, the highest poverty states of Mississippi and Louisiana, together with the District of Columbia, have poverty rates that include fully one fourth of their populations. These rates are three times higher than those in such low-poverty states as Hawaii, Connecticut, and Alaska (8–9 percent).

FIGURE 6.3   Percent of Persons Living in Poverty, 1993

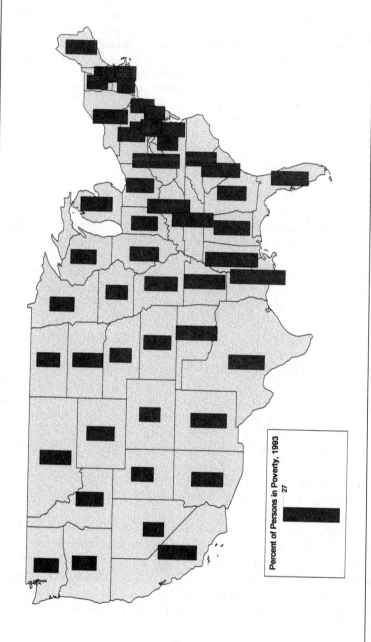

Percent of Persons in Poverty, 1993

27

Source: Data from U.S. Census Bureau, *Income, Poverty, and Valuation of Noncash Benefits,* 1993, table D.

TABLE 6.3   Poverty Rates of Persons by State, 1993

Lowest States by Personal Poverty		Highest States by Personal Poverty	
Hawaii	8.0	Texas	17.4
Connecticut	8.5	Florida	17.8
Alaska	9.1	California	18.2
Maryland	9.7	South Carolina	18.7
Virginia	9.7	Tennessee	19.6
Nevada	9.8	Oklahoma	19.9
Colorado	9.9	Arkansas	20.0
New Hampshire	9.9	Kentucky	20.4
Vermont	10.0	West Virginia	22.2
Delaware	10.2	Mississippi	24.7
Iowa	10.3	District of Columbia	24.7
Nebraska	10.3	Louisiana	26.4

Source: U.S. Census Bureau, *Current Population Reports,* Series P60-188, table D.

Conditions are better in the cities. Analysis reveals urban and metropolitan areas have picked up the great bulk of good, high-paying jobs in the past decades. In fact, southern cities such as Atlanta, Miami, Houston and Dallas are often referred to as "Jewels in the Crown"— locations where growth and opportunity beckon to all. These giant metropolitan areas acted as magnets to attract high-technology industry and expanding business during the past two decades. The growth rate of southern metropolitan areas has been twice that of metro population increase in the nation as a whole.[29] Yet there was a downside to such glitter and hoopla as well. Heavy in-migration from poor rural areas, especially among impoverished African Americans, drove up income disparities in these cities while overloading the limited social services that did exist. Atlanta soon became known as the homicide capital of the Deep South, but was challenged for this title by Houston—with a murder rate two-thirds higher than New York City's.[30] John Agnew has offered a vignette of Houston that is anything but flattering: no zoning regulations, a huge poor population, America's largest concentration of illegal migrant workers from Mexico (excluding California), depletion of the city's water table, heavy traffic congestion, high air pollution, abysmal education, and high unemployment (20 percent) in an area that is supposed to have abundant jobs.[31] Miami has become the drug-smuggling capital of the United States, offering a Swiss-type banking haven for capital flight and laundered money coming from Latin America. All the while, racial tensions and conflict are higher here than in any other American city. With the fall in the price of oil, Dallas and Houston were among the five worst real estate markets in the country by the end of the 1980s.[32] Millions of square feet of newly built office buildings went unoccupied in their downtowns for a prolonged time as did entire developments of subdivisions, townhomes, and condos.

In a word, the sheen is off the golden egg of southern industrialization. A variety of developments in the past decade combined to challenge the wisdom of unrestrained economic growth. To begin with, the southern economy remains heavily dependent upon the investment of others—whether wealthy multinationals from Japan or northern core firms which may decide to keep going further south—into Mexico or other LDCs. Commentators on the southern growth cycle now agree this region is no longer insulated from the problems of northern wage earners. Despite lower pay, many firms are relocating out of the South into the Third World. Southern textile industries, for example, still suffer from an exposure to intense foreign competition. Although some domestic protection exists, it has not been enough to stem the tide of business out-migration.

The precarious nature of southern development is evident in its almost complete reliance on northern bank loans and investment funding.[33] The hand that giveth can take away. Gary Green notes that these investors have increasingly chosen to flee our country in search of even cheaper wages, resulting in deindustrialization in the South as well.[34] More than 750,000 jobs were lost in the textile industry alone in just the past decade. The businesses that do stay must automate and reduce their reliance on paid workers to remain competitive with imports, which again serves to slash the work force. Lastly, external financing drains profits away from the South back to the North or to Japan. These funds, if resulting from native investment, could be used for even more regional expansion. Although there has been dramatic growth of investment in the South and a sharp rise in the amount of southern commercial bank deposits, Green reports that the rate of commercial loans has not even kept pace with population growth in this region.

A number of other conditions leave the South more exposed to economic collapse than other regions. The area is overloaded with military installations and bases. By the 1970s the defense industry was the largest single employer in many southern states. In that decade, southern states held nine of the top twenty-one spots in dependency upon employment related to defense spending.[35] When the Pentagon budget was finally trimmed somewhat in the early 1990s, the South (along with California) bled the most. Military spending is only one of six pillars holding up the roof of the southern economy, although it is an important one.[36] Overreliance on low wages may ultimately backfire as multinationals pull out in search of greener pastures in LDCs. Real estate values have already taken a dive as in-migration to the South slowed (mostly due to an economic spurt in the Northeast). Energy as an economic pillar crumbled during the 1980s as the price of oil scraped bottom. Although this helped the rest of the nation, it aggravated conditions in petroleum producing states. In one year, for example, Texas registered a $6 billion budget deficit—the largest of any state in U.S. history—while Louisiana and Oklahoma were losing $50 million in tax revenues for every one-dollar drop in the price of oil.[37] The farm crisis spread out of the Midwest in the 1980s to enfold southern agriculture as well. Income from agriculture has been smothered by a long-term slide in world food prices, although it was driven up in the short term by severe drought. Lastly, many believe the confidence put in high-technology industrial growth as a panacea for economic ills is grossly misplaced. Critics challenge whether there have been any meaningful financial benefits for those southerners who assemble electronic parts. They continue to be paid at marginal levels.

Definite clouds have loomed over the horizon in the Sunbelt. Even during the national expansionary period of the 1980s, manufacturing output, construction, and employment declined in Georgia while vital services and the trade sector remained flat. In North Carolina, hours worked per week by industrial employees dipped while Louisiana encountered consecutive years of negative economic growth. The gross regional product for the entire Southeastern Division sank by 42 percent in only one year. In this twelve-state southern region, all leading economic indicators were flat toward the end of the last decade, with almost no growth in the region. At that time, the South was the slowest growing region in the country in retail sales.[38]

THE GRAND CANYON OF INCOME INEQUALITY:
GAPS OF INCOME WITHIN REGIONS AND BETWEEN COUNTIES

One of the major results of market systems that everyone agrees upon—whether Marxist, capitalist, economic nationalist or World System advocate—is that the process of development is uneven. In our discussion of the Third World as it ties in with wealthy industrial countries, it was obvious that major differences exist among poor countries. In this sense, the model of core/semiperiphery/periphery seemed to fit best with what has actually been taking place in the world economy. The dependency of LDCs on core investment could be seen in their penetration by multinational corporations. As a result of this development, income inequality rose as elites were enriched while labor was forced to work at subsistence wages to keep production costs low.

The parallels with the South are obvious. Applying the core/periphery model to this region seems fitting. Gary Green, for example, presents detailed evidence that most finance capital for industrialization in the South came from northern banks in the core.[39] William Falk and Thomas Lyson prove beyond doubt that southern development seems to have literally bypassed rural areas, especially where large numbers of African Americans live.[40] Falk and Lyson admit that cities in the South have reaped greater benefits from the influx of northern or foreign firms, but even in metropolitan areas greater income inequality has been an unfailing byproduct of development (especially for racial minorities).

Quite some time ago Nobel-winning economist Gunnar Myrdal argued that the process of development is by its very nature unstable.[41] It spawns conflicting forces as investment gushes into underdeveloped regions. *Spread effects* are forces which act to make a poorer region better off. Examples include the introduction of advanced production technology, creation of a small cadre of skilled workers, greater profits for local business elites, a larger market for regional products, and so on. *Backwash effects* make the poor region worse off. Because of superior technology and financing in the rich core, for example, the periphery has no chance to build its own industrial base to compete with the core's giant corporations. The periphery can become locked into a subordinate status so that locals are forced to work for core branch plants at much lower wages than in the core. Profits are drained away from the periphery, also quashing any chance of independent, local development in the region. The rich region can thus become even more wealthy while the poor region suffers further economic setbacks.

What makes Myrdal's theory different from modern world systems theory is that it at least holds forth the possibility that a backward region may possibly win as it goes through development. It is possible that this could be taking place in the South as the real income of the region increases. The drama has yet to play itself out. What is more certain, however, is that backwash effects do appear in any area that goes through modernization. Uneven regional development can rob the income of some groups, which then sink further into the mire of poverty. This has been especially true of the South. In a review of industrial migration to this region, David Smith found that city slums and rural poverty remained largely untouched—again most markedly among African Americans.[42]

Research indicates that the rural South, when judged on a variety of measures of social well-being, is the most ill-equipped region of the country.[43] It is the highest of any region in:

1. The number of banks that quit financing farm loans
2. The percentage of farmers who went out of business or declared bankruptcy
3. The number of farmers with delinquent Farmers Home Administration loans
4. Dependency on off-farm employment to make ends meet
5. Counties with chronic low income (92 percent of all U.S. counties in this group are in the South)
6. Poverty levels (21 percent of rural southerners live below the poverty line—a proportion recently growing larger)

7. Low levels of educational attainment (nearly all of the nation's 489 counties with the lowest proportion of high school graduates are in the South)
8. Rates of functional illiteracy (difficulty in basic reading and writing comprehension)

It is becoming more apparent that the development strategy built upon low wages in the South has backfired. Low-wage manufacturing jobs were supposed to yield to high technology once industries were firmly established. Instead, multinational branch plants are being relocated to LDCs. Over half of the 1.5 million jobs lost in the South during a five-year period last decade were in manufacturing.[44] The counties with the fastest rates of employment growth are also those with the lowest percentages of workers in manufacturing. This is especially serious for the rural South, whose employees are just as dependent on manufacturing as workers in the industrial heartland of the Midwest.[45] The rural South is completely helpless to compete for topnotch industry that does remain in the United States, since it demands an educated work force with highly developed technical skills.

On the basis of these economic forces alone, then, we could expect huge regional differences in income to emerge and persist in the United States. To the extent that the South remains a periphery to a northern core, income will be low and show all the gross inequities that are so obvious in the Third World. The social and cultural conditions that feed into this inequality, however, are not directly economic but are very important all the same. As a result of the slave plantation heritage in the South, there is more conservativism in the region, and this may have led to anti-union hostility. There is certainly more racism. In the rural South, two of every five African-American men are without jobs, cannot find full-time jobs, or cannot earn enough money to raise themselves above the poverty line.[46] There is continuing, strong employment discrimination directed against African Americans in the rural South, and this works to impoverish them even further. This is not surprising given the evidence of much higher levels of racial prejudice and discrimination than in the North.[47] And in the South, the higher the proportion of African Americans in a locale, the stronger the discrimination. Even today there are large numbers of rural African Americans concentrated in many southern counties. We can thus expect major income differences between areas in this region—if only due to race. Indeed, one study of the 354 counties in the four states of the deep South (South Carolina, Georgia, Alabama, and Mississippi) found that racial composition caused from one-sixth to one-third of total income inequality.[48] Not only did higher proportions of African-American residents drive up inequality, this force was worse in counties experiencing rapid industrialization than in those stagnating or undergoing a decline in manufacturing.

Because of pronounced cultural differences, region itself has an important impact on the level of income. Stuart Holland in his book on the interplay between regions and economics argues at great length that income differences between areas are due to more than just the normal functioning of market forces.[49] Among other important variables that he highlights are slavery, federal protectionism, and regional specialization of products difficult to duplicate elsewhere (such as sugar, cotton, and tobacco). Support for this view was found in a study of 180,000 men whose army enlistments had expired.[50] In comparison to human capital variables, such as education and work experience, about one-sixth of earnings differences are due to the region where the worker was employed. A study by Falk and Rankin of counties known as the "Black Belt" in the South reveals sharp differences in householder earnings caused by regional location in comparison to other human capital variables. These counties run from Virginia and Maryland down the coast through the Carolinas and Florida, and then across Georgia, Alabama, Mississippi, Louisiana, and Texas. They have historically had large African-American populations and also tend to be poor, rural, and agricultural. Southern African Americans earn only 81 percent of what their nonsouthern counterparts earn, while African Americans in Black Belt counties earn 82 percent of what all southern African Americans earn (nonmetropolitan southern African Americans are at the bottom of this hierarchy).[51] In the end, over three-fourths of the earnings differences among African Americans are due to regional residence, while only a fifth are caused by individual characteristics (age, gender, education, etc.). African-American householders outside of the South earned $2,563 more than southern African-American householders, while non-Black Belt householders in the south earned $1,960 more than did African Americans living in Black Belt counties.

The importance that region has upon income differences begins to become more apparent when we look at how U.S. counties fare today. All 3,141 counties were ranked by median household income received in 1989 from 1990 census of population data. Although more recent per capita income figures are available, they do not directly reflect the spread of economic well-being as well as the household income figures do. Nor are they based upon the extensive coverage and depth for small area analysis that is possible in the decennial census figures. Lastly, per capita data is analagous to a gross or crude average that could hide true deprivation in families because of the presence of a few super-rich persons—especially at the smaller county level.

Table 6.4 lists the three hundred poorest counties in the United States, which is about 10 percent of all counties. A brief glimpse at this list reveals that fully three-fourths of these counties are in the South. Most are also rural. A glance at table 6.5 shows that with a few major exceptions (Alaska), a great number of the three hundred richest counties are suburban counties of major metropolitan areas, nearly all of which are in the North. All U.S. counties were arbitrarily divided into the highest quartile (25 percent) for rich counties, the middle 50 percent to comprise middle-income counties, and the lowest quartile of family income to designate poor counties. In 1989, the poorest fourth of counties had an average median family income of less than $17,131 while the richest quartile showed an average median household income of $32,839—nearly twice as much.

The profile of these counties is quite instructive. For instance, the poorest quartile of counties again shows a heavy regional bias toward the South. This section of the country has 1,425 counties, or 45 percent of all U.S. counties. Yet the South has over two-thirds of the 785 poorest counties. In short, it has over 50 percent more poor counties than would be expected on the basis of the normal proportional representation in each region. In all, 37.1 percent of all counties in the South fall within the lowest quartile (poorest fourth) of U.S. counties based upon median household income. This proportion is twice as large as the next poorest region of the country, which is the Midwest, with 18.8 percent of its counties in the bottom quartile. Incredibly, fewer than 1 percent of counties in the Northeast fall into the cellar position.

Table 6.6 shows the breakdown of these counties by the four regions. The percentages of poor, middle, and rich counties are listed beside the percent change from what should be expected on the basis of normal regional representation. All regions outside the South are heavily under-represented on poor counties. At the other extreme, among the fourth of wealthiest counties, the South has a –46 percent gap while the northeastern and western regions are over-represented by such counties. It is very clear that the Northeast and the West are especially fortunate, having a relative absence of poor counties and an abundance of counties in the top quartile of family income.

All the counties have been mapped for each state by median household income (figure 6.4). To contrast areas, household income was divided into a bottom quartile, top quartile, and middle 50 percent. The map key is geared to the U.S. distribution as a whole. It is thus easy to see how the counties in each state compare with the national norm. By visualizing the figures in this way, it is apparent that most states have at least a few rich counties, regardless of what section of the country they are in. Yet the highest-income counties continue to appear most often in the Northeast, Upper Midwest, and West while the low-income counties concentrate in southern states and in portions of the Southwest (Texas and New Mexico).

TABLE 6.4   Median Household Income of Poorest 300 U.S. Counties, 1989

Rank	County	Income
1	Owsley, KY	$8,595
2	East Carroll, LA	$9,791
3	Holmes, MS	$9,809
4	Kalawao, HI	$10,000
5	Starr, TX	$10,182
6	Jefferson, MS	$10,267
7	McCreary, KY	$10,598
8	Tunica, MS	$10,965
9	Wolfe, KY	$11,000
10	Shannon, SD	$11,105
11	Clinton, KY	$11,348
13	Hancock, TN	$11,822
13	Zavala, TX	$11,822
14	Jackson, KY	$11,885
15	Wilkinson, MS	$11,910
16	Tensas, LA	$11,931
17	Lee, AR	$11,949
18	Greene, AL	$11,990
19	Magoffin, KY	$12,160
20	Dimmit, TX	$12,222
21	Maverick, TX	$12,262
22	Breathitt, KY	$12,383
23	Wilcox, AL	$12,437
24	Choctaw, OK	$12,451
25	Lee, KY	$12,461
26	Wayne, KY	$12,560
27	Chicot, AR	$12,680
28	Humphreys, MS	$12,696
29	Knox, KY	$12,697
30	Clay, KY	$12,732
31	Madison, LA	$12,792
32	Sumter, AL	$12,811
33	Clay, WV	$12,855
34	Claiborne, MS	$12,876
35	Cumberland, KY	$12,989
36	Mora, NM	$12,993
37	Issaquena, MS	$13,005
38	Presidio, TX	$13,016
39	Costilla, CO	$13,057
40	Phillips, AR	$13,071
41	Bell, KY	$13,078
42	McDowell, WV	$13,141
43	Searcy, AR	$13,221
44	Morgan, KY	$13,229
45	Sharkey, MS	$13,304
46	Todd, SD	$13,327
47	Knott, KY	$13,329

TABLE 6.4   *(Continued)*

Rank	County	Income
48	Guadalupe, NM	$13,350
49	Webster, WV	$13,371
50	Avoyelles, LA	$13,451
51	Brooks, TX	$13,509
52	Tallahatchie, MS	$13,593
53	Duval, TX	$13,602
54	Pushmataha, OK	$13,613
55	Monroe, AR	$13,633
56	Leslie, KY	$13,692
57	Oregon, MO	$13,705
58	Clay, GA	$13,709
59	Quitman, MS	$13,730
60	Ripley, MO	$13,740
61	Perry, AL	$13,769
62	Coahoma, MS	$13,780
63	Evangeline, LA	$13,797
64	Wayne, MO	$13,815
65	Keweenaw, MI	$13,821
66	Lafayette, AR	$13,849
67	Harmon, OK	$13,880
68	Elliott, KY	$13,890
69	Atoka, OK	$13,898
70	Pemiscot, MO	$13,911
71	Fentress, TN	$13,924
72	Randolph, GA	$13,972
73	Hall, TX	$13,987
74	Bolivar, MS	$14,020
75	Woodruff, AR	$14,024
76	Frio, TX	$14,059
77	Apache, AZ	$14,100
78	Menominee, WI	$14,122
79	Ziebach, SD	$14,129
80	Walthall, MS	$14,135
81	Coal, OK	$14,177
82	Conejos, CO	$14,188
83	Noxubee, MS	$14,205
84	Yazoo, MS	$14,234
85	Menard, TX	$14,271
86	Kemper, MS	$14,315
87	Corson, SD	$14,324
88	Franklin, MS	$14,341
89	Sunflower, MS	$14,431
90	San Saba, TX	$14,462
91	Dickens, TX	$14,484
92	Calhoun, WV	$14,496
93	Hale, AL	$14,508

(Continued on next page)

TABLE 6.4   (*Continued*)

Rank	County	Income
95	Mellette, SD	$14,539
95	Gilmer, WV	$14,539
96	Lake, MI	$14,562
97	Buffalo, SD	$14,566
98	Worth, MO	$14,568
99	Willacy, TX	$14,590
100	Dewey, SD	$14,599
101	Lee, VA	$14,618
102	Edwards, TX	$14,639
103	Menifee, KY	$14,650
104	Lincoln, WV	$14,659
105	St. Landry, LA	$14,670
106	Taliaferro, GA	$14,700
107	Jim Hogg, TX	$14,704
108	Huerfano, CO	$14,730
109	Bullock, AL	$14,745
110	Harlan, KY	$14,774
111	Alexander, IL	$14,786
112	Metcalfe, KY	$14,815
113	Red River, LA	$14,831
114	Sioux, ND	$14,838
115	Shannon, MO	$14,910
116	West Carroll, LA	$14,924
117	Zapata, TX	$14,926
118	Fulton, AR	$14,950
119	Catahoula, LA	$14,956
121	Rockcastle, KY	$14,967
121	Johnson, TN	$14,967
122	Whitley, KY	$14,979
124	Casey, KY	$14,993
124	Pickett, TN	$14,993
125	Allendale, SC	$15,013
126	St. Francis, AR	$15,029
127	Letcher, KY	$15,112
129	Edmonson, KY	$15,134
129	San Augustine, TX	$15,134
130	Newton, AR	$15,139
131	Martin, KY	$15,142
132	Pike, MS	$15,149
133	Franklin, LA	$15,159
134	Rolette, ND	$15,163
135	Hughes, OK	$15,168
136	Monroe, KY	$15,214
137	Leflore, MS	$15,219
138	Wheeler, OR	$15,224
139	Johnston, OK	$15,264
140	Lawrence, KY	$15,273

TABLE 6.4   (*Continued*)

Rank	County	Income
141	Marion, TX	$15,288
142	Richland, LA	$15,298
143	Lawrence, AR	$15,337
144	McPherson, SD	$15,345
145	Carter, MO	$15,357
146	Mason, TX	$15,366
147	Roane, WV	$15,375
149	Dixie, FL	$15,380
149	Attala, MS	$15,380
150	Dunklin, MO	$15,388
151	Montgomery, MS	$15,396
152	Hudspeth, TX	$15,401
153	Collingsworth, TX	$15,421
154	Jefferson Davis, MS	$15,442
155	Norton, VA	$15,460
156	St. Helena, LA	$15,475
157	Hardin, IL	$15,498
158	Coleman, TX	$15,519
159	Putnam, MO	$15,549
160	Jefferson, OK	$15,553
161	Cottle, TX	$15,583
162	Lowndes, AL	$15,584
163	Haskell, OK	$15,592
164	Stewart, GA	$15,606
165	Barbour, WV	$15,607
166	Sierra, NM	$15,612
167	La Salle, TX	$15,615
168	Pulaski, IL	$15,625
169	Calhoun, GA	$15,640
170	Macon, AL	$15,642
171	Stone, AR	$15,655
172	Floyd, KY	$15,661
173	Amite, MS	$15,669
174	Hart, KY	$15,671
175	Luna, NM	$15,684
176	De Baca, NM	$15,686
177	Desha, AR	$15,719
178	Okfuskee, OK	$15,738
179	Kinney, TX	$15,750
180	Wright, MO	$15,770
181	Eastland, TX	$15,774
182	Lewis, KY	$15,775
183	Natchitoches, LA	$15,778
184	Johnson, KY	$15,782
185	Benton, MS	$15,794
186	Adair, KY	$15,809
187	Sullivan, MO	$15,826

(*Continued on next page*)

TABLE 6.4   (*Continued*)

Rank	County	Income
188	Allen, LA	$15,838
189	Saguache, CO	$15,853
190	Scott, TN	$15,858
191	Yalobusha, MS	$15,885
192	Rowan, KY	$15,922
193	Bath, KY	$15,940
194	Concho, TX	$15,942
195	Scotland, MO	$15,944
196	Quitman, GA	$15,972
197	Leake, MS	$15,975
198	Conecuh, AL	$15,992
199	Hickory, MO	$16,010
200	Acadia, LA	$16,022
201	Bienville, LA	$16,043
202	Butler, AL	$16,054
203	Estill, KY	$16,056
204	Mingo, WV	$16,066
205	Swain, NC	$16,068
206	Caldwell, LA	$16,069
207	Claiborne, LA	$16,073
208	Marion, MS	$16,084
209	Fulton, KY	$16,087
210	Crowley, CO	$16,088
211	Childress, TX	$16,091
212	Wayne, MS	$16,095
213	Jasper, MS	$16,130
214	Karnes, TX	$16,155
215	Mississippi, MO	$16,159
216	Douglas, MO	$16,187
217	Perry, KY	$16,202
218	Taylor, GA	$16,210
219	Red River, TX	$16,217
220	Clay, AR	$16,219
221	Perry, MS	$16,230
222	Washington, LA	$16,246
223	Butler, MO	$16,285
224	Las Animas, CO	$16,286
225	Marshall, OK	$16,292
226	Dickenson, VA	$16,292
227	Iron, MI	$16,307
228	De Soto, LA	$16,315
229	Adams, OH	$16,318
230	Kiowa, OK	$16,322
231	Dooly, GA	$16,326
232	Boyd, NE	$16,329
233	Wilcox, GA	$16,333

TABLE 6.4   (*Continued*)

Rank	County	Income
234	Braxton, WV	$16,359
235	Tyrrell, NC	$16,363
236	McCurtain, OK	$16,413
237	Ozark, MO	$16,417
238	Early, GA	$16,421
239	Grundy, TN	$16,425
240	Campbell, TN	$16,450
241	Summers, WV	$16,457
242	Carter, MT	$16,458
243	Crenshaw, AL	$16,460
244	Scott, AR	$16,470
245	Dallas, AL	$16,493
246	Aurora, SD	$16,497
247	Kenedy, TX	$16,500
248	Montgomery, AR	$16,503
249	Charles Mix, SD	$16,541
250	McCulloch, TX	$16,544
251	La Paz, AZ	$16,555
252	Culberson, TX	$16,559
253	Howell, MO	$16,564
254	Telfair, GA	$16,573
255	Copiah, MS	$16,583
256	Wheeler, GA	$16,585
257	Dent, MO	$16,594
258	Bryan, OK	$16,610
259	Mercer, MO	$16,629
260	Carroll, MS	$16,639
261	Jackson, AR	$16,641
262	Newton, TX	$16,656
263	Musselshell, MT	$16,661
264	Dallas, MO	$16,673
265	Prairie, MT	$16,694
266	Hidalgo, TX	$16,703
267	Randolph, AR	$16,719
268	Schuyler, MO	$16,729
269	Donley, TX	$16,747
270	Graham, NC	$16,754
271	Texas, MO	$16,757
272	Fayette, WV	$16,774
273	Motley, TX	$16,780
274	Russell, KY	$16,788
275	Sabine, LA	$16,790
276	Lake, TN	$16,804
277	Cocke, TN	$16,818
278	Powell, KY	$16,828
280	Cibola, NM	$16,848
280	Gregory, SD	$16,848

(*Continued on next page*)

TABLE 6.4   (*Continued*)

Rank	County	Income
281	Tangipahoa, LA	$16,849
282	Poinsett, AR	$16,858
283	Bennett, SD	$16,864
284	Adair, OK	$16,886
285	Emmons, ND	$16,892
286	Forest, WI	$16,907
287	Izard, AR	$16,910
288	Benson, ND	$16,917
289	Mahnomen, MN	$16,924
290	Benton, MO	$16,925
291	Warren, NC	$16,937
292	Cedar, MO	$16,939
293	Wheatland, MT	$16,946
294	Wirt, WV	$16,951
295	Trinity, TX	$16,963
296	Taos, NM	$16,966
298	Jenkins, GA	$16,967
298	Winn, LA	$16,967
299	Hempstead, AR	$16,986
300	Seminole, OK	$17,007
	Mean of Poorest 300	$15,012

Source: U.S. Census Bureau, *USA Counties 1994*, CD-ROM.

The average median family income of all U.S. counties in 1989 stood at $23,979. Among counties in each region, the Northeast led the pack with an average of $30,690 compared to the West's $25,991 and the North Central's average of $24,073. The difference from the southern average of $22,261 (which is well below the norm) is dramatic. In essence, going by counties alone, the average median household income of northeastern counties is nearly 38 percent above those of southern counties. Although select areas in the South may be doing better (especially the Virginia and Maryland suburban counties around the District of Columbia and the Flordia coastal counties), the large number of poor counties in this region pull down the average by quite a bit. An analysis of variance on the data also shows that the differences among regions are much greater than differences in median household income between the counties within each region. In other words, each region is fairly homogenous. Counties in the South are fairly equal in being low-income. Counties in the other regions are fairly similar in being higher-income. The big difference continues to be found between a uniformly low-income South and the other three more affluent regions of the country. This finding continues to dramatize the deep sectional differences of

TABLE 6.5   Median Household Income of Richest 300 Counties, 1989

Rank	County	Income
1	Fairfax, VA	$59,284
2	Morris, NJ	$56,273
3	Somerset, NJ	$55,519
4	Los Alamos, NM	$54,801
5	Hunterdon, NJ	$54,628
6	Howard, MD	$54,348
7	Nassau, NY	$54,283
8	Montgomery, MD	$54,089
9	Putnam, NY	$53,634
10	Rockland, NY	$52,731
11	Loudoun, VA	$52,064
12	Douglas, CO	$51,718
13	Bristol Bay, AK	$51,112
14	Falls Church, VA	$51,011
15	Fairfax, VA	$50,913
16	North Slope, AK	$50,473
17	Fayette, GA	$50,167
18	Fairfield, CT	$49,891
19	Prince William, VA	$49,370
20	Bergen, NJ	$49,249
21	Suffolk, NY	$49,128
22	DuPage, IL	$48,876
23	Sussex, NJ	$48,823
24	Marin, CA	$48,544
25	Westchester, NY	$48,405
26	Santa Clara, CA	$48,115
27	Juneau, AK	$47,924
28	Calvert, MD	$47,608
29	Valdez-Cordova, AK	$47,500
30	Manassas, VA	$46,674
31	San Mateo, CA	$46,437
32	Charles, MD	$46,415
33	Norfolk, MA	$46,215
34	Lake, IL	$46,047
35	Collin, TX	$46,020
36	Orange, CA	$45,922
37	Monmouth, NJ	$45,912
38	Hamilton, IN	$45,748
39	Chester, PA	$45,642
40	Middlesex, NJ	$45,623
41	Ventura, CA	$45,612
42	Livingston, MI	$45,439
43	Fauquier, VA	$45,222
44	Ketchikan Gateway, AK	$45,172
45	Anne Arundel, MD	$45,147
46	Contra Costa, CA	$45,087
47	Tolland, CT	$45,019

(Continued on next page)

TABLE 6.5   (*Continued*)

Rank	County	Income
48	Kodiak Island, AK	$44,815
49	Stafford, VA	$44,661
50	Arlington, VA	$44,600
51	Waukesha, WI	$44,565
52	Washington, MN	$44,122
53	Anchorage, AK	$43,946
54	Richmond, NY	$43,861
55	Middlesex, MA	$43,847
56	Montgomery, PA	$43,720
57	Williamson, TN	$43,615
58	Chesterfield, VA	$43,604
59	Gwinnett, GA	$43,518
60	McHenry, IL	$43,471
61	Oakland, MI	$43,407
62	Bucks, PA	$43,347
63	Sitka, AK	$43,337
64	Poquoson, VA	$43,236
65	Middlesex, CT	$43,212
66	Prince George's, MD	$43,127
67	Kendall, IL	$42,834
68	Fort Bend, TX	$42,809
69	Johnson, KS	$42,741
70	Ozaukee, WI	$42,695
71	Litchfield, CT	$42,565
72	Rockwall, TX	$42,417
73	Kenai Peninsula, AK	$42,403
74	Aleutians East, AK	$42,384
75	Carroll, MD	$42,378
76	Burlington, NJ	$42,373
77	Dutchess, NY	$42,250
78	Dakota, MN	$42,218
79	Wrangell-Petersburg, AK	$42,020
80	Rockingham, NH	$41,881
81	Union, NJ	$41,791
82	Harford, MD	$41,680
83	Alexandria, VA	$41,472
84	Frederick, MD	$41,382
85	Spotsylvania, VA	$41,342
86	Cobb, GA	$41,297
87	Mercer, NJ	$41,227
88	Will, IL	$41,195
89	Geauga, OH	$41,113
90	Plymouth, MA	$40,905
91	Scott, MN	$40,798
92	Matanuska-Susitna, AK	$40,745
93	Hanover, VA	$40,683
94	Hartford, CT	$40,609

TABLE 6.5   (*Continued*)

Rank	County	Income
95	Honolulu, HI	$40,581
96	Hillsborough, NH	$40,404
97	York, VA	$40,363
98	Nantucket, MA	$40,331
99	St. Charles, MO	$40,307
100	Columbia, GA	$40,122
101	Kane, IL	$40,080
102	Anoka, MN	$40,076
103	Pitkin, CO	$39,991
104	Warren, NJ	$39,929
105	Hendricks, IN	$39,892
106	James City, VA	$39,785
107	Prince of Wales-Outer Ketchikan, AK	$39,495
108	Rockdale, GA	$39,389
109	Gloucester, NJ	$39,387
110	Orange, NY	$39,198
111	Queen Anne's, MD	$39,190
112	Carver, MN	$39,188
113	Solano, CA	$39,113
114	Jefferson, CO	$39,084
115	Manassas Park, VA	$39,076
116	Cherokee, GA	$39,052
117	Macomb, MI	$38,931
118	Baltimore, MD	$38,837
119	Maui, HI	$38,771
120	New Castle, DE	$38,617
121	Skagway-Yakutat-Angoon, AK	$38,583
122	New Haven, CT	$38,471
123	Washington, WI	$38,431
124	Oldham, KY	$38,416
125	New Kent, VA	$38,403
126	Platte, MO	$38,173
127	St. Louis, MO	$38,127
128	Medina, OH	$38,083
129	Essex, MA	$37,913
130	Delaware, OH	$37,896
131	Placer, CA	$37,601
132	Passaic, NJ	$37,596
133	Henry, GA	$37,550
134	Alameda, CA	$37,544
135	Bristol, RI	$37,539
136	New London, CT	$37,488
137	Fairbanks North Star, AK	$37,468
138	Kauai, HI	$37,425
139	Powhatan, VA	$37,394
140	Delaware, PA	$37,337
141	Hancock, IN	$37,333

(*Continued on next page*)

TABLE 6.5   (*Continued*)

Rank	County	Income
142	Arapahoe, CO	$37,234
143	St. Mary's, MD	$37,158
144	Porter, IN	$37,142
145	Douglas, GA	$37,138
146	Santa Cruz, CA	$37,112
147	Campbell, WY	$37,055
148	Washington, RI	$36,948
149	Eagle, CO	$36,931
150	Denton, TX	$36,914
152	Albemarle, VA	$36,886
152	Roanoke, VA	$36,886
153	Chittenden, VT	$36,877
154	Shelby, AL	$36,852
155	Snohomish, WA	$36,847
156	Napa, CA	$36,773
157	Summit, UT	$36,756
158	Warren, OH	$36,728
159	St. Croix, WI	$36,716
160	Forsyth, GA	$36,642
161	Saratoga, NY	$36,635
162	Ottawa, MI	$36,507
163	San Benito, CA	$36,473
164	Washtenaw, MI	$36,307
165	Sonoma, CA	$36,299
166	Elbert, CO	$36,273
167	Virginia Beach, VA	$36,271
168	Goochland, VA	$36,239
169	Wake, NC	$36,222
170	Sweetwater, WY	$36,210
171	Camden, NJ	$36,190
172	Clinton, MI	$36,180
173	King, WA	$36,179
174	Kent, RI	$36,070
175	Haines, AK	$36,048
176	Cecil, MD	$36,019
177	Lapeer, MI	$35,874
178	Newport, RI	$35,829
179	Merrimack, NH	$35,801
180	Olmsted, MN	$35,789
181	Worcester, MA	$35,774
182	Chesapeake, VA	$35,737
183	Eaton, MI	$35,734
184	Grundy, IL	$35,728
185	De Kalb, GA	$35,721
186	Santa Barbara, CA	$35,677
187	Hennepin, MN	$35,659
188	Seminole, FL	$35,637

TABLE 6.5   (*Continued*)

Rank	County	Income
189	Lake, OH	$35,605
190	Henrico, VA	$35,604
191	Sherburne, MN	$35,585
192	Sarpy, NE	$35,575
193	King George, VA	$35,556
194	Washington, OR	$35,554
195	Monroe, MI	$35,462
196	Clackamas, OR	$35,419
197	Monroe, NY	$35,337
198	Boulder, CO	$35,322
199	Summit, CO	$35,229
200	Douglas, NV	$35,209
201	Aleutians West, AK	$35,187
202	Greene, OH	$35,116
203	Davis, UT	$35,108
204	Boone, IL	$35,103
205	Monroe, IL	$35,086
206	El Dorado, CA	$35,058
207	Johnson, IN	$35,035
208	San Diego, CA	$35,022
209	Los Angeles, CA	$34,965
210	Clay, FL	$34,860
211	Prince George, VA	$34,825
212	Boone, IN	$34,652
213	Clarke, VA	$34,636
214	Oconee, GA	$34,566
215	Essex, NJ	$34,518
216	Cumberland, PA	$34,493
217	Boone, KY	$34,485
218	Colonial Heights, VA	$34,472
219	Brazoria, TX	$34,418
220	Woodford, IL	$34,375
221	Clay, MO	$34,370
222	Queens, NY	$34,186
223	Hampshire, MA	$34,154
224	Warrick, IN	$34,069
225	Calumet, WI	$34,050
226	Ulster, NY	$34,033
227	Collier, FL	$34,001
228	Lander, NV	$33,988
229	Midland, MI	$33,948
230	Canadian, OK	$33,855
231	Windham, CT	$33,851
232	Mecklenburg, NC	$33,830
233	Outagamie, WI	$33,770
234	Atlantic, NJ	$33,716
235	Elko, NV	$33,715

(Continued on next page)

TABLE 6.5   (*Continued*)

Rank	County	Income
236	Williamson, TX	$33,695
237	King William, VA	$33,676
238	Culpeper, VA	$33,523
239	Monterey, CA	$33,520
240	Clayton, GA	$33,472
241	Box Elder, UT	$33,468
242	Wright, MN	$33,456
243	San Bernardino, CA	$33,443
244	San Francisco, CA	$33,414
245	Albany, NY	$33,358
246	Northwest Arctic, AK	$33,313
247	Morgan, UT	$33,274
248	Humboldt, NV	$33,269
249	Uinta, WY	$33,259
250	Lancaster, PA	$33,255
251	Union, OH	$33,244
252	Salem, NJ	$33,155
253	Clear Creek, CO	$33,149
254	Ontario, NY	$33,133
255	Ocean, NJ	$33,110
256	Paulding, GA	$33,085
257	Riverside, CA	$33,081
258	Botetourt, VA	$33,079
259	Madison, AL	$33,048
260	Lexington, SC	$32,914
261	Northampton, PA	$32,890
262	Woodford, KY	$32,858
263	Wilson, TN	$32,852
264	Strafford, NH	$32,812
265	Frederick, VA	$32,806
266	Morgan, IN	$32,762
267	Racine, WI	$32,751
268	Dane, WI	$32,703
269	Cook, IL	$32,673
270	York, PA	$32,605
271	Benton, WA	$32,593
272	Palm Beach, FL	$32,524
273	Leavenworth, KS	$32,500
274	Putnam, OH	$32,492
275	Wayne, NY	$32,469
277	Clermont, OH	$32,465
277	Monroe, PA	$32,465
278	Storey, NV	$32,457
279	Lehigh, PA	$32,455
280	Warren, IA	$32,452
281	Butler, OH	$32,440
282	York, ME	$32,432

TABLE 6.5   (*Continued*)

Rank	County	Income
283	Rappahannock, VA	$32,377
284	Kent, MI	$32,358
285	Tarrant, TX	$32,335
286	Sacramento, CA	$32,297
287	Cumberland, ME	$32,286
288	Jefferson, MO	$32,281
289	New York, NY	$32,262
290	Montgomery, TX	$32,254
291	Teller, CO	$32,209
292	Nevada, CA	$32,200
293	Linn, IA	$32,137
294	Park, CO	$32,102
295	Berks, PA	$32,048
297	Ramsey, MN	$32,043
297	Kitsap, WA	$32,043
298	Dukes, MA	$31,994
299	Mercer, ND	$31,969
300	Rensselaer, NY	$31,958
	Mean of Richest 300	$38,702

Source: U.S. Census Bureau, *USA Counties, 1994,* CD-ROM.

TABLE 6.6   Household Income of Counties by Region, 1989

	Region				
Income Category	Northeast	North Central	South	West	Total Number
Poorest 25 percent	0.9	18.8	37.1	12.8	782
Percent different from expected	−2,600	−33.2	32.6	−94.7	
Middle 50 percent	37.8	56.6	45.8	53.8	1,571
Percent different from expected	−32.3	11.6	−9.1	7.1	
Richest 25 percent	61.3	24.6	17.1	33.3	785
Percent different from expected	59.2	−1.4	−45.9	25.0	
Total counties in region	217	1,055	1,425	444	3,141

FIGURE 6.4   Median Household Income by County, 1989

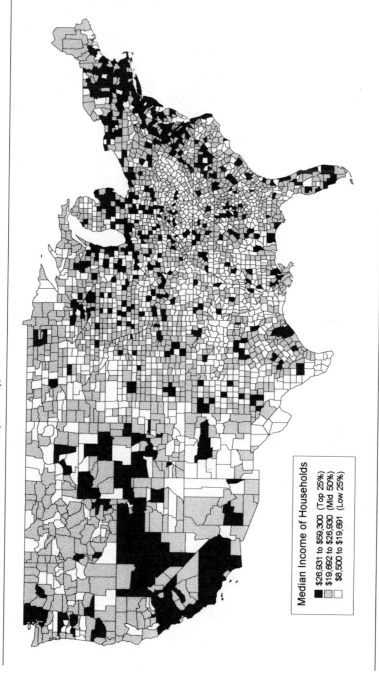

Median Income of Households

■ $26,931 to $59,300 (Top 25%)
▨ $19,692 to $26,930 (Mid 50%)
□ $8,500 to $19,691 (Low 25%)

Source: Data from U.S. Census Bureau, *USA Counties 1994*, CD-ROM.

TABLE 6.7   Mean Values of Selected Characteristics by County Income
Groups, 1990

Characteristic	County Income Groups		
	Poor	Middle	Rich
Percent African-American	13.3	7.3	6.4
Percent high school graduates	59.9	70.3	77.9
Percent college graduates	9.7	12.6	19.1
Percent of households headed by a female	15.3	12.5	12.8
Percent unemployed	8.7	6.8	6.3
Percent of total payroll in manufacturing	24.5	28.5	28.7
Percent rural/farm	8.3	7.3	2.9

Source: Data from U.S. Census Bureau, *USA Counties, 1994,* CD-ROM.

income in the United States. Southern incomes remain lower than incomes elsewhere in the nation. Perhaps progress has been made by a few states or a few counties in the South. Yet the majority of the counties in this region are saddled with low to middle incomes in comparison to counties outside the region.

Poor counties also tend to be nonmetropolitan. Almost 98 percent of the poorest fourth of counties are located ouside any metropolitan area, while nearly three-fourths of all U.S. counties are nonmetropolitan.[52] On the other hand, rich counties tend to be situated in a metropolitan complex—although usually on the suburban county fringe of a lower income central city county.[53] Two out of three counties in the top quartile of household income are metropolitan—a proportion half as large as would be expected if rich counties were distributed in proportion to overall metro/nonmetro percentages. Table 6.7 contains the mean averages for a number of different variables which help us identify what other forces might be operating to increase or decrease family income in these counties. In a study of earnings for both African-American and white males between the ages of thirty-five and fifty-four, it was found that some of these factors had an important impact on take-home pay.[54] Human capital variables such as education can have a positive effect on income, but regional differences together with race were found to have a major influence upon earnings. This was especially pronounced between metropolitan and other counties, and in areas where African Americans were heavily concentrated.

Huge differences exist between rich and poor counties on degree of urbanization, which again highlights the association between ruralness and low income.[55] The rural farm rate is three times higher in poor counties (bottom quartile on median household income) than in rich counties (top quartile). The same is more or less true with the percent of residents who are African-American. Poor counties have twice the proportion of African Americans as do rich counties. There is an evident schism in unemployment between poor and rich counties as well, with rich counties having an unemployment rate 28 percent lower than poor counties. The role of human capital as measured by education can be seen in high school completion rates. But it is especially with college graduates that huge differences emerge. Rich counties have twice the proportion of those twenty-five years old and over who have completed four years of college in comparison to poor counties. Although there are no dramatic gaps in the percent of the labor force payroll going to those employed in manufacturing, rich counties do have a slight edge in manufacturing dominance. Again, however, existing differences may be masked by regional effects. While manufacturing jobs in the North may be associated with high income counties, evidence already cited suggests that the reverse case actually holds true in the rural South. Thus, the effect can appear to be canceled out if region is not held constant.

Other factors associated with the South come together to make differences even sharper. The rural dimension in combination with high proportions of African Americans has a lethal effect on economic well-being in counties. In a study of 778 nonmetropolitan southern counties, it was found that employment in manufacturing was a significant negative predictor of median family income—as were percent African American, percent rural, unemployment, and distance from a metropolitan area.[56] One study has found that nearly two-thirds of counties where there are large numbers of farms owned by African Americans are in the Persistent Low Income (PLI) category.[57] Counties in this category are frozen in poverty. They have been consistently ranked in the bottom quintile of all nonmetropolitan U.S. counties in the past thirty years on the basis of per capita income. Of the 130 counties in the study cited, there was no instance when median African-American income was higher than median white income. In most of these remote, poor, poverty plagued counties, whites' income was generally twice that of African Americans. The average size of farms owned by African Americans in this area is anywhere from one-third to one-half the size of all southern farms. Poverty is much greater among the African-American rural farm population while education is at a dismally low level in comparison to that available to rural whites in the region. This concentration of poverty among southern counties can be clearly seen in figure 6.5, which shows a swath running through the Deep south and into a few southwestern states.

FIGURE 6.5 Percent of Persons in Poverty by County, 1989

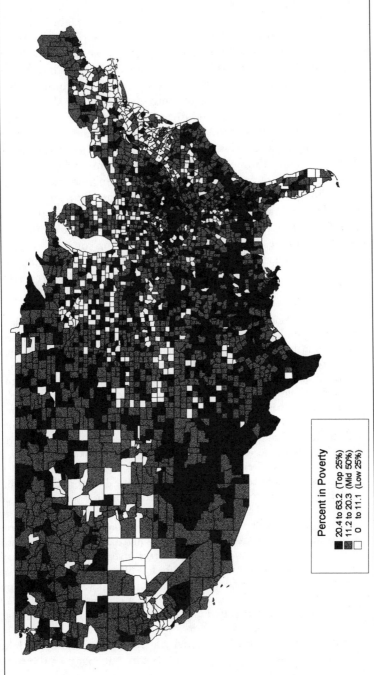

Percent in Poverty

■ 20.4 to 63.2 (Top 25%)
▨ 11.2 to 20.3 (Mid 50%)
□ 0 to 11.1 (Low 25%)

Source: Data from U.S. Census Bureau, *USA Counties, 1994,* CD-ROM.

The only positive finding in these studies is that even in such rural southern counties, education is sharply related to higher median family income. This at least raises the idea that improvement might be possible, but the lack of education in this area can only be described as appalling. Louis Swanson, in a review of the painful human cost to the low income of the rural South, speaks movingly of the inability this area has shown in its futile effort to eradicate ignorance and illiteracy. The result of such a chronic lack of education translates into dead-end job skills. In this periphery hinterland, even if an employment structure offered job opportunities, rural southerners would be unable to take advantage of openings that involved higher level skills. Since the jobs are not there to begin with, the question is moot. Rural southerners' poor incomes, which according to the evidence are driven even lower by the presence of manufacturing plants in failing industries such as textiles, translate into poverty. Over one-third of all nonmetropolitan African-American families in this region have incomes below the poverty line (56 percent of African-American, female-headed families). Swanson calls this "suffering" on a massive scale, which is hard to disagree with. Grinding, constant poverty devastates self-esteem. It gives rise to fear and despair, depression and worry, family stress and violence. Yet it has even deadlier effects:

> Once again America is backsliding with regard to nutritional status. Numerous studies . . . have consistently found that poverty is directly linked to malnutrition as well as a host of other health risks. . . . The rural poor are over 65 percent more likely to consume diets which are inadequate for multiple essential nutrients. The rural poor were more than twice as likely as were the United States nonpoor to experience severe levels of dietary inadequacy. Of the various regions, the Southern poor were consistently in the worst categories. . . . The South was found to be the region in which the rural poor had the highest prevalence of biochemical deficiencies. . . . The rural poor were almost three times as likely to be growth-stunted as were children from nonpoor families. The South was considered to be the most affected region. Consistent with dietary and biochemical findings, the Southern rural poor exhibited the highest prevalence of low height-for-age. . . . The worst is saved for last—infant mortality and low birth weight. . . . The findings suggest [they are] considerably higher in rural poor counties than the rest of the nation. While the national infant mortality has declined and then levelled off in recent years, the rate has actually increased in rural poor counties. Data for low birth weights points to a similar disparity. . . . A characteristic of Southern rural poverty is hunger, malnutrition, and higher than average levels of infant mortality. This is one dimension of the human suffering that is part of the rural South in crisis.[58]

Confusion may set in when so many factors are associated with low or high income levels. But it is possible to sort out many of the forces which act to raise or lower median family income in U.S. counties by using multiple regression analysis. It was believed that some variables in the South would have more impact there than outside the region. In particular, given stronger racism in the South plus less income payoff for manufacturing jobs in this region, race was expected to have a greater negative effect upon income here than outside the South. The association between manufacturing jobs and higher income was also expected to be weaker, nonexistent, or perhaps even reversed. Thus, separate regression runs were made for all U.S. counties, nonsouthern counties, and southern counties. Percent of the population that is African-American, when used alone, is weak as a predictive variable. It does not reflect the greater likelihood of discrimination that results when a large proportion of African Americans is concentrated in a given area. To tap this dimension, counties were divided into low and high by race on the basis of the proportion of their population that was African American. Nationally, 12.3 percent of our population is African American. Counties with an African-American population less than or equal to this percent were designated as low on race, while other counties were ranked as high. Table A1 in the Appendix lists the results of the three regression runs.

How does each variable work to raise or lower the economic well-being of families in U.S. counties? What are their impacts—taken alone or as a group—on median family income? To begin with, seven predictive variables were used in the equations based upon our discussion and the results of previous research. For U.S. counties, *whether a county was part of a metropolitan area or not* was expected to heavily influence median family income. The *percent of a county's population living on rural farms* is less a geographic factor, however, than a labor force indicator. In addition to results of a variety of southern studies, Frank Levy identifies the shift away from the farm (notorious for low income) to manufacturing jobs in northern cities as one of the most important contributors to rising income within our century.[59] In this movement, low paid jobs in agriculture disappeared through mechanization, but gains tended to be positive. Manufacturing and service jobs in urban areas actually paid more in real earnings, even after the higher cost of city living was subtracted. This effect was measured by *percent of total payroll going to those employed in manufacturing.* Education, as measured by the *percent completing high school,* has faithfully been associated with higher income. The crucial effect of *race,* already discussed, was expected to depress family income— as was the *percent of female-headed households* (associated with low family income and poverty in the last chapter). Lastly, prior research has found that *percent unemployed* leads to lower real income

Of all of the variables used in the analysis, the impact of African-American dominance was the only one that tipped the scales in the opposite direction to what was expected. Those counties that were ranked with higher proportions of African Americans than the nation at large (more than 12.3 percent of their population) actually showed higher median household income. Among all seven variables, however, this was nearly the *least* predictive in its impact upon household income, having only about one seventh of the clout that the most predictive variable (percent of high school graduates) displayed. The results of African-American dominance also reflect a shift in results obtained in 1980 on median family income, which did show the expected inverse relationship (again weak). It may be that the effect of race upon income—holding constant many other relevant variables—is actually weakening or disappearing, at least when the units of analysis are counties. For indivual level data, discussed earlier in chapter 5, the results are dramatically different.

All other variables were also statistically significant, but in the expected direction. Another factor showing a diluted contribution to median household income, although in the expected direction, was the percent unemployed in the civilian labor force. Counties with higher unemployment rates tend to have lower median household incomes. The weakness of this variable as it affects income, however, can be seen especially in non-southern counties, where it is not statistically significant. Among all U.S. counties, the unemployment rate has only one-ninth the impact of the most important predictive variable.

By a wide margin, the strongest predictor of household income was the percent who had completed high school. Education carried nine times the weight that unemployment did in determining household income. The next most important variable was whether a county was part of a metropolitan area. Higher household incomes are strongly associated with metropolitan residence, being over six times as important as unemployment and carrying about 70 percent of the weight of education in determining household income. Among nonsouthern counties, the importance of this variable is virtually equal to that of education. At the opposite extreme are counties dominated by high percentages of people living in rural areas and on farms. Counties such as these tend to have low household income, although the importance of this variable is very small among southern counties in comparison to nonsouthern counties.

The effect of female-headed households was also in the expected direction and statistically significant. Counties with higher proportions of family households headed by women tend to have lower household income. The effect of this variable is almost twice as strong in the South, which should not be surprising given the lower levels of AFDC support in this region. In the nation as a whole, it is the fourth most important predictor, but in the South it is the third. Last, the dominance of manufacturing is positively related to higher household income as well, being roughly twice as important as unemployment in its impact on income. Counties with higher proportions of their total labor force payroll going to those employed in manufacturing also tend to have higher median household incomes.

Whatever the varying importance of these forces, however, all of them taken together can account for almost two-thirds of the entire variation in median household income in U.S. counties. Their importance in fixing a family's income is beyond dispute. The finding that manufacturing continues to play a positive factor in raising income provides fresh ammunition for critics who charge that deindustrialization is working against the welfare of most Americans. It also adds credence to the fact that development in the United States has been uneven—and that severe income deprivation continues to exist for southern households in comparison to those outside that region.

The comparison of southern with nonsouthern counties also conveys a few minor differences between the two geographic areas. Surprisingly, percent unemployed has no predictive significance among nonsouthern counties although it is important within southern counties. The negative impact of female-headed families is nearly half again as large in southern counties as in nonsouthern counties. High percentages living on rural farms has much less of a depressive weight (about one-fourth) on household income in the South as it does among counties in other regions. The most noteworthy conclusion in looking at these income determinants, however, is that they tend to act pretty much the same no matter which county you live in—regardless of region. Education is vastly important no matter where you are. Opportunity also tends to beckon in the form of the bright lights of big cities. In conclusion, the uniformities in what contributes to higher household income are more impressive than geographic differences—although some variation does exist between regions and counties.

RELATIVE INCOME INEQUALITY: WHERE THE GRASS IS GREENER

David Smith wrote an entertaining and informative book describing inequality in a variety of important social variables.[60] Sharp differences among geographic areas tend to exist along many lines of inequality other than just income. Yet a very surprising but persistent trend emerged in his data comparing American states. He found that relatively rich states—as measured by per-capita income—do not necessarily have an edge with respect to indicators of social pathology (violent crimes, rates of venereal disease, narcotics addiction, illiteracy, etc.). In essence, these are common problems that surface in big cities. As we have seen, areas where metropolitan populations dominate tend to have much higher household incomes. One point Smith makes is especially worth repeating: higher absolute income will not necessarily translate into a better life. The extremes of affluence may bring their own problems of social pathology—such as divorce, alcoholism, higher levels of stress-induced illnesses, and the like. Although poor states are nearly always high on social pathologies, rich states are not necessarily low on these indicators.

Again, much will depend upon how income is shared. With reference to the Third World, it was apparent that a high Gross National Product (GNP) per capita did not necessarily translate into meeting basic human needs. This is also true of U.S. states and counties. It is possible for areas with high levels of absolute income to display a large number of social blights such as crime. A variety of vices tend to flourish wherever income is poorly and unfairly shared. This is especially so when the contrast between the extremely wealthy and the desperately poor is blatant to begin with.

It is therefore crucial to look at relative income inequality in comparing regions in the United States. It is possible that states or counties that are modest or even poor with respect to absolute family income will fare much better than richer areas when relative income inequality is examined. Also, can we say with impunity that a rich area is well-off if high income is not shared very equally among its people? Popular films such as *Down and Out in Beverly Hills* have vividly contrasted the very rich with the destitute and homeless. The reality of such a caricature can be found in any major U.S. city. Bag ladies, panhandlers, vagrants, drifters, grate-dwellers, and the like collide daily with the wealthy in cities such as New York. Contact may be only momentary as the affluent glance at the poor while being whisked away in their limousines, but the glimpses have staying power for both groups. These images add credence to one particular viewpoint: What good is a high level of income for a community if it is enjoyed by only a few persons?

The development argument sold as a panacea for LDC economic woes has been used to rationalize regional variation in the United States as well. The idea has been popular that less developed states will pass through an inverted U-shaped experience as income growth sets in. As applied here at home, the usual scenario predicts that a rural, low-income, farm state will have a fairly equal distribution of income. As such areas develop and grow—usually via industry relocation with its associated high-technology and service jobs—real income goes up but relative income gets more skewed and uneven. Finally, as development reaches its ultimate stage, high incomes will be shared more equally among the population. Once past the threshhold stage of development, then, everyone benefits from higher income.

On the surface, there is much support for this view. The South has been going through a modernization and development sequence since the end of World War II. As real income has gone up for this region, relative income inequality has dropped. Table 6.8 compares the four separate regions of the country by state on household income Gini ratios, calculated using the same techniques as in earlier works.[61] There has been a definite coming together of relative income inequality among the four regions. In 1949, the South had the highest degree of inequality with a family income Gini ratio of .4488, which was nearly 22 percent greater than the most equal region, the northeast. By 1979, the gap had closed to about 10 percent.[62] Although the South still displayed greater inequality, it was not so far removed from the Northeast. More important, the South saw a 15 percent decline in its relative income inequality take place in these thirty years while the Gini ratio in the Northeast declined only 6 percent. Clearly, the sharing of family income within the South is becoming more equal at the same time that absolute income has been going up.

What has happened in U.S. states since 1980? First, the good news is that income inequality in the South has continued to shrink in relationship to the other three regions. Its household income Gini ratio was only 7 percent greater in 1990 than that of the West, which had the most equality of any region (the differences among the three nonsouthern regions, however, are almost imperceptible). The remaining trends do not bode well for a bright and glorious future. Alaska was virtually the only state that actually saw a reduction of its relative household income inequality, which fell by only one-third of 1 percent. All other states and the District of Columbia experienced a surge of inequality during the 1980s. By contrast, only about half of all nonsouthern states experienced an increase of relative family income inequality in the 1970s while only three of the seventeen states in the South did so. The run-up of inequality during the 1980s should not be surprising given the trends discussed in previous chapters. Yet the magnitude of the change is chilling. The average state growth of inequality was nearly 6 percent during the decade, and every region averaged over a 5-percent surge in its Gini score (table 6.8). The states with especially high increases of 8 percent or more were scattered throughout the nation: Wyoming (10.02), West Virginia (10.0), Pennsylvania (9.94), Ohio (9.55), New York (8.94), Illinois (8.79), Michigan (8.62), Connecticut (8.42), and Arizona (8.41). For the most part, states in the South actually saw slightly lower increases in household-income inequality than did states in the other three regions.

With the move toward growing income inequality taking place in the nation as a whole, the trend toward relative equality among states and counties in the South may not continue at such a rapid pace. Perhaps new dimensions of inequality will arise to replace the old patterns. Although the Northeast was the most egalitarian with respect to family income during the 1970s, inequality among households increased the most in this

TABLE 6.8   Gini Scores of Household Income by State and Region:
              1979–1989

State	1990 Gini	Rank	1980 Gini	Percent Decade Change
Mean for all U.S. states	.4274		.4039	5.83
Northeast	**.4202**		**.3945**	6.47
New Hampshire	.38457	1	.37175	3.45
Vermont	.39355	3	.38312	2.72
Maine	.40621	8	.37896	7.19
Rhode Island	.41724	19	.39931	4.49
Massachusetts	.42478	27	.40036	6.10
New Jersey	.42578	28	.39803	6.97
Connecticut	.43245	32	.39887	8.42
Pennsylvania	.43337	33	.39417	9.94
New York	.46404	48	.42596	8.94
North Central	**.4187**		**.3951**	5.96
Wisconsin	.40011	5	.38434	4.10
Iowa	.40832	11	.38927	4.89
Indiana	.40863	12	.38103	7.24
North Dakota	.41285	14	.39343	4.94
Nebraska	.41363	15	.39736	4.09
Minnesota	.41365	16	.39466	4.81
South Dakota	.41938	20	.40496	3.56
Kansas	.42348	25	.40422	4.77
Ohio	.42459	26	.38759	9.55
Michigan	.42633	30	.39248	8.62
Missouri	.43594	35	.41054	6.19
Illinois	.43752	37	.40219	8.79
South	**.4453**		**.4221**	5.47
Delaware	.40280	6	.39955	.81
Maryland	.40675	10	.39010	4.27
Virginia	.42209	23	.40325	4.67
South Carolina	.42593	29	.40726	4.58
North Carolina	.42793	31	.40569	5.48
Georgia	.44358	39	.42494	4.39
Oklahoma	.44500	40	.42677	4.27
Florida	.44793	41	.42877	4.47
Tennessee	.44817	42	.42402	5.70
Arkansas	.44820	43	.43169	3.82
West Virginia	.44854	44	.40777	10.00
Alabama	.45449	45	.42974	5.76
Kentucky	.45518	46	.42197	7.87
Texas	.45543	47	.42390	7.44
Mississippi	.47033	49	.44174	6.47
Louisiana	.47426	50	.44466	6.66
District of Columbia	.49370	51	.46452	6.28

TABLE 6.8   (*Continued*)

State	1990 Gini	Rank	1980 Gini	Percent Decade Change
West	.4171		.3945	5.75
Utah	.39323	2	.37262	5.53
Alaska	.39400	4	.39522	−.31
Wyoming	.40616	7	.36917	10.02
Hawaii	.40673	9	.39522	2.91
Washington	.41072	13	.39278	4.57
Nevada	.41541	17	.39651	4.77
Idaho	.41649	18	.38944	6.94
Oregon	.42065	21	.39651	6.09
Colorado	.42066	22	.39682	6.01
Montana	.42269	24	.39177	7.89
Arizona	.43522	34	.40145	8.41
California	.43743	36	.41651	5.02
New Mexico	.44258	38	.41388	6.93

Source: Data from U.S. Census Bureau, Population Division, Income Statistics Branch.

region between 1979 and 1989. Among all states, just short of 28 percent had increases of 7 percent or more in their household income Gini scores during the 1980s, but 44 percent of northeastern states fell into this category. The South saw only 18 percent of its states experience this degree of inequality growth, while it led all other regions in the proportion of its states with the *least* growth of inequality. During the 1980s, nearly half of all southern states saw their Gini ratios climb by less than 5 percent—a rate surpassed only by the Midwest (58 percent). Only about one-third of states in the Northeast and West fell into this low inequality growth category. In the South, then, the growth of median household income in relation to the rest of the nation has been accompanied by a less rapid upsurge in relative inequality—which seems a double blessing.

But before the champagne is poured, we must be made aware of the distance yet to be travelled for the South to emerge as a full and equal partner with other parts of the country. Figure 6.6 maps the states by their household income Gini scores. A fairly solid tier of dark grey, indicating high inequality, sweeps across the South. With only a few exceptions, this region continues to contain nearly all of the states in the top 25 percent that are most unequal in terms of household income.

Counties were also mapped according to an Income Disparity Index which combined the two most extreme elements of the income distribution.[63] A Z-score was calculated for the percent of families in each U.S. county living under poverty in 1989. What a Z-score does is locate a county in comparison to the mean of all counties, expressed in units of the standard deviation of the variable being looked at. For the most part,

FIGURE 6.6   Household Income Gini Scores, 1989

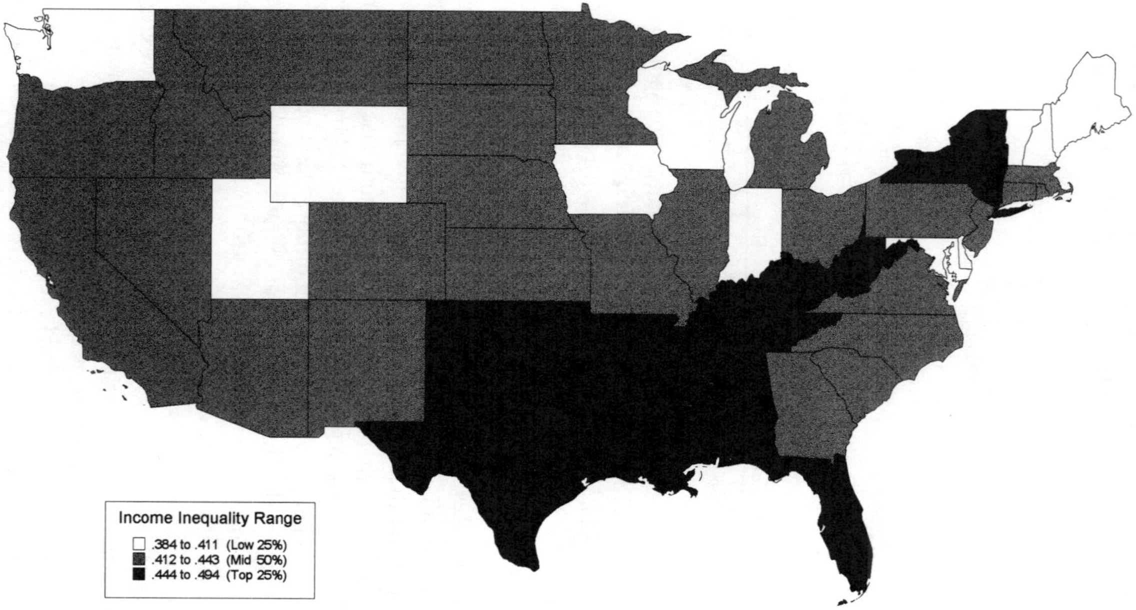

Income Inequality Range

.384 to .411 (Low 25%)
.412 to .443 (Mid 50%)
.444 to .494 (Top 25%)

Source: From data in table 6.8.

a Z-score varies from plus to minus 3.0. A z-score of plus one (+1.0) for a particular county on the percent of families below poverty would mean that county's poverty percentage is greater than 84 percent of all the other counties. A Z-score of +0.5 translates to a county having a poverty rate greater than 69 percent of all other counties. A Z-score of plus two (+2.0) means the county is higher than nearly 98 percent of all other counties on poverty. It is obvious that at least with the percentage of families under poverty, a large negative Z-score would be most desirable. A –2.0 Z-score, for example, means that 98 per cent of all other counties have larger percentages in poverty, while a –1.0 Z-score means 84 percent of counties have higher poverty (a Z-score of 0.0 means that a county is exactly equal to the average of all counties on the percentage of poverty).

To make this discussion a little less hypothetical, it is instructive to compare actual geographic areas. For example, the District of Columbia has a Z-score of +.04 on the percent of families living below poverty, which means that 51.6 percent of all counties had lower proportions of families under poverty. Prince George's County in Maryland is essentially a suburban area which adjoins the District and is a part of the metropolitan area comprising our nation's capital. It has a Z-score on this variable of -1.326, meaning that 90.8 percent of all other counties have higher levels of family poverty. The pattern of central city with high poverty and adjoining suburban counties with low poverty is fairly typical throughout the United States.

A second major part of the Income Disparity Index reflected the income distribution at the other end of the spectrum, among the wealthy. It was possible to calculate the percent of families with income at or above $75,000 in 1989, a category which can easily be designated as "rich." While variation on this variable was substantial, with some counties having no families with income at this level while one highly blessed county had over 40 percent of its families with income this high, the average was fairly modest. Among all U.S. counties, only 5.7 percent of all families had income of $75,000 or more in 1989. As for family poverty, a Z-score was also calculated for this variable to better estimate the degree to which a given county departed from the norm. Again, sticking with the District of Columbia example, the District had a Z-score of rich families of +2.815. This means that the percent of rich families in the District is higher than 99.75 percent of all other counties in the nation—i.e., there are a lot of rich families living in the District of Columbia. Neighboring Prince George's County in Maryland is also quite well-off, with a Z-score of +2.93 (99.83 percent of all counties have a lower percentage of rich families).

The nice thing about Z-scores is that they are standardized. At bottom, they measure the exact degree to which a given county departs from the average on a given variable. Z-scores have also been retranslated from whatever their original units were (such as dollars, percents, numbers, etc.) into what are now expressed as standard deviations. Standard deviations are standard units which are equivalent to one another. In this manner, it is possible to compare apples and oranges. Z-scores can be mathematically manipulated. In particular, they can be added and divided. For the Income Disparity Index, both the Z-score for percent of families in poverty and the Z-score for percent of rich families were added, and then divided by two. Thus, a county with a high positive Z-score on family poverty and a high positive Z-score on family wealth would have a distorted income distribution, with high proportions of both rich and poor families—together with an anemic middle class. At the other extreme, a high negative Z-score on these two variables would indicate a county with little poverty and a low proportion of rich families. This county would thus have a large, healthy, middle-class population made up of families with average income. Because we are used to thinking of scores that are positive (+) as "good" while scores that are negative (–) as "bad," one last adjustment was made to the index to make it easier to understand. All scores were multiplied by –1.0. In effect, the signs are reversed, so that scores that are large and positive now become desirable while scores that are large and negative reflect extreme income disparity.

Counties were divided on the basis of their Income Disparity Index score into a low fourth (the most unequal), a middle group of 50 percent (more moderate inequality reflecting average levels of poverty and wealth), and a top fourth (counties with little poverty, few rich families, and high proportions of middle income families). Their distribution is graphically illustrated in figure 6.7, which shows that counties which are the most egalitarian on a relative basis are chiefly concentrated in the Upper Midwest and in a few states of the Great Plains. There seems to be a dearth of egalitarian counties in the South, the Southeast, the Southwest, and especially California.

Table 6.9 shows the distribution of relative income inequality groups as measured by the Income Disparity Index by region. If high inequality counties were distributed equally, each region would have 25 percent of its counties in this category. As with low absolute income, however, the South is also burdened with more inequality. Nearly one-third of its counties have a fairly lopsided income distribution, while over one-fourth of northeastern and western counties do (only 8 percent of midwestern counties are highly unequal). Put another way, half again as many counties in the South are highly unequal as would be expected if this region conformed to national patterns. The Midwest would have to triple its number of unequal counties to reach what would be expected on a

FIGURE 6.7   Income Disparity Index by County, 1989

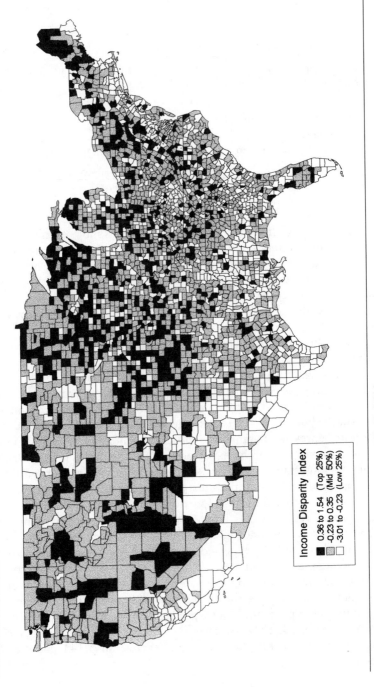

Income Disparity Index

■ 0.36 to 1.54 (Top 25%)
▨ -0.23 to 0.35 (Mid 50%)
☐ -3.01 to -0.23 (Low 25%)

Source: Data from U.S. Census Bureau, *USA Counties,* 1994, CD-ROM.

TABLE 6.9   Income Disparity Index of Counties by Region, 1989

Income Disparity Category	Region				Total Number
	Northeast	North Central	South	West	
Lowest 25% (most unequal)	28.1	7.8	36.8	26.4	785
Percent different from expected	11.1	−221.6	32.2	5.1	
Middle 50%	42.9	42.2	55.2	55.4	1,571
Percent different from expected	−16.7	−18.6	9.4	9.7	
Highest 25% (most equal)	29.0	50.0	7.9	18.2	785
Percent different from expected	14.0	50.0	−414.7	−37.0	
Total counties in region	217	1,055	1,425	444	3,141

Source: Data from U.S. Census Bureau, USA Counties, 1994, CD-ROM.

national basis. At the other end, where income is more evenly shared in low inequality counties, the South only has 8 percent of its counties in this category while all other regions have proportions of egalitarian counties two to six times this average. Put differently, the South has only a third of the highly equal counties that it should have on the basis of its numbers, while the Midwest has twice the number of egalitarian counties as would normally be expected. Without doubt, the heartland is home to the middle class while the South continues to be haunted by huge gaps between rich and poor.

Once again, we can gain a sharper view of what high and low relative inequality areas are like by looking at the top and bottom three hundred counties ranked by the family Income Disparity Index. Table 6.10 starts at the bottom with the most unequal counties. While the South and West are represented among the most unequal counties roughly in proportion to what would be expected on the basis of their numbers, two other regions stand out. The Northeast has twice the proportion of unequal counties among the most unequal three hundred as would be expected on the basis of its proportion of all counties (13.3 percent and 6.9 percent, respectively). The Midwest has a proportional representation among the most unequal three hundred (10.3 percent) that is only a third of what would be expected on the basis of normal county distribution among regions (33.6 percent). In a word, severely unequal counties tend to be located disproportionately in the Northeast. That region is also half again more heavily metropolitan (39 percent) in comparison to all counties (26 percent). Not surprisingly, then, the three hundred most unequal counties are also twice as likely to have smaller percentages living on farms in rural places (3.6 percent) as all 3,141 counties (7.3 percent).

TABLE 6.10   Family Income Disparity Index Scores of the 300 Most Unequal
U.S. Counties, 1989

Rank	County	Disparity Index
1	Falls Church, VA	−3.00459
2	Fairfax, VA	−2.89082
3	Westchester, NY	−2.67888
4	Montgomery, MD	−2.65882
5	Morris, NJ	−2.64491
6	Starr, TX	−2.62683
7	Shannon, SD	−2.59930
8	Fairfield, CT	−2.53088
9	Marin, CA	−2.51605
10	New York, NY	−2.48257
11	Tunica, MS	−2.46526
12	Rockland, NY	−2.46361
13	Nassau, NY	−2.44398
14	Somerset, NJ	−2.42116
15	Valdez-Cordova, AK	−2.40461
16	East Carroll, LA	−2.39477
17	Hunterdon, NJ	−2.39343
18	North Slope, AK	−2.38608
19	Howard, MD	−2.25596
20	Arlington, VA	−2.21255
21	Bergen, NJ	−2.20361
22	Los Alamos, NM	−2.10217
23	Holmes, MS	−2.08523
24	Ziebach, SD	−2.02887
25	Owsley, KY	−1.98228
26	Maverick, TX	−1.93215
27	Putnam, NY	−1.92419
28	Todd, SD	−1.87115
29	Zavala, TX	−1.85762
30	Santa Clara, CA	−1.84884
31	Dimmit, TX	−1.84785
32	San Mateo, CA	−1.78629
33	Bristol Bay, AK	−1.77021
34	Lake, IL	−1.74654
35	Alexandria, VA	−1.73502
36	Norfolk, MA	−1.72425
37	Fulton, GA	−1.71644
38	Fairfax, VA	−1.70081
39	Issaquena, MS	−1.69740
40	Juneau, AK	−1.69582
41	Apache, AZ	−1.67984
42	Orange, CA	−1.67228
43	Monmouth, NJ	−1.67192
44	Tensas, LA	−1.66581
45	Pitkin, CO	−1.66176
46	Dillingham, AK	−1.66005
47	Wilcox, AL	−1.65554

(Continued on next page)

TABLE 6.10 (*Continued*)

Rank	County	Disparity Index
48	Jefferson, MS	−1.63813
49	Anchorage, AK	−1.63429
50	Collin, TX	−1.62927
51	Contra Costa, CA	−1.62805
52	Middlesex, MA	−1.59583
53	Menominee, WI	−1.59273
54	Essex, NJ	−1.58849
55	Humphreys, MS	−1.57411
56	Hudspeth, TX	−1.56966
57	Dewey, SD	−1.56105
58	McKinley, NM	−1.55282
59	Chester, PA	−1.53102
60	Oakland, MI	−1.51754
61	Presidio, TX	−1.50359
62	Suffolk, NY	−1.50222
63	Buffalo, SD	−1.48913
64	Lee, AR	−1.47760
65	Hidalgo, TX	−1.47522
66	Sharkey, MS	−1.47458
67	DuPage, IL	−1.47032
68	Greene, AL	−1.46650
69	Wolfe, KY	−1.45137
70	Coahoma, MS	−1.45122
71	Bolivar, MS	−1.44497
72	Union, NJ	−1.43713
73	District of Columbia	−1.42792
74	Richmond, NY	−1.42792
75	Sioux, ND	−1.41891
76	Willacy, TX	−1.41891
77	Mercer, NJ	−1.39494
78	Perry, AL	−1.39109
79	Loudoun, VA	−1.37677
80	McCreary, KY	−1.37059
81	Montgomery, PA	−1.36444
82	Magoffin, KY	−1.36323
83	Rolette, ND	−1.35956
84	Madison, LA	−1.35761
85	Williamson, TN	−1.35327
86	Knox, KY	−1.34139
87	Douglas, CO	−1.33144
88	Webb, TX	−1.33109
89	La Salle, TX	−1.32462
90	Frio, TX	−1.30382
91	Orleans, LA	−1.29940
92	Cameron, TX	−1.29782
93	Bethel, AK	−1.29683
94	Sunflower, MS	−1.29353
95	Kenai Peninsula, AK	−1.29126
96	Northwest Arctic, AK	−1.29044

TABLE 6.10 (*Continued*)

Rank	County	Disparity Index
97	Zapata, TX	−1.28440
98	Bennett, SD	−1.28126
99	Aleutians East, AK	−1.28096
100	Ventura, CA	−1.28068
101	Martin, KY	−1.25984
102	Wilkinson, MS	−1.25050
103	Breathitt, KY	−1.24645
104	Hamilton, IN	−1.24272
105	Middlesex, NJ	−1.23652
106	Claiborne, MS	−1.22670
107	Los Angeles, CA	−1.21895
108	Leflore, MS	−1.21623
109	Sumter, AL	−1.21571
110	San Francisco, CA	−1.21438
111	Kodiak Island, AK	−1.20689
112	Dallas, AL	−1.19899
113	Yazoo, MS	−1.19087
114	Guadalupe, NM	−1.18677
115	Allendale, SC	−1.18351
116	Corson, SD	−1.17491
117	Phillips, AR	−1.15872
118	Jim Hogg, TX	−1.15736
119	Noxubee, MS	−1.15670
120	Clay, KY	−1.15580
121	Nome, AK	−1.15579
122	Tallahatchie, MS	−1.15009
123	Essex, MA	−1.14685
124	Mellette, SD	−1.14578
125	Knott, KY	−1.14213
126	Alameda, CA	−1.13418
127	West Feliciana, LA	−1.13247
128	Hartford, CT	−1.12634
129	Washtenaw, MI	−1.12232
130	Bronx, NY	−1.12034
131	Duval, TX	−1.10649
132	Quitman, MS	−1.10286
133	St. Landry, LA	−1.10258
134	Lake and Peninsula, AK	−1.08733
135	Lee, KY	−1.08450
136	Glasscock, TX	−1.08288
137	Bell, KY	−1.07964
138	Karnes, TX	−1.07037
139	McPherson, NE	−1.06819
140	Morgan, KY	−1.06792
141	San Juan, UT	−1.06689
142	Chicot, AR	−1.06123
143	Fayette, GA	−1.06119
144	Jackson, KY	−1.06018

(*Continued on next page*)

TABLE 6.10 (*Continued*)

Rank	County	Disparity Index
145	Edwards, TX	−1.05902
146	Ketchikan Gateway, AK	−1.05372
147	Avoyelles, LA	−1.05084
148	Yukon-Koyukuk, AK	−1.04428
149	McDowell, WV	−1.04154
150	Rockwall, TX	−1.03908
151	Catahoula, LA	−1.03600
152	Lowndes, AL	−1.03393
153	Brooks, TX	−1.03166
154	Sussex, NJ	−1.03132
155	Mora, NM	−1.02910
156	Washington, MS	−1.02787
157	Hancock, TN	−1.02700
158	Johnson, KS	−1.02342
159	Lawrence, KY	−1.01470
160	Kings, NY	−1.00549
161	Bullock, AL	−1.00226
162	Elliott, KY	−.99890
163	Passaic, NJ	−.99791
164	Clay, WV	−.99570
165	Honolulu, HI	−.99529
166	Clinton, KY	−.99206
167	Anne Arundel, MD	−.98813
168	Cobb, GA	−.98607
169	Navajo, AZ	−.98383
170	Jackson, SD	−.97610
171	Wade Hampton, AK	−.97601
172	Evangeline, LA	−.97407
173	Wrangell-Petersburg, AK	−.97346
174	Geauga, OH	−.97034
175	Val Verde, TX	−.96840
176	James City, VA	−.96769
177	Leslie, KY	−.95744
178	Fort Bend, TX	−.95418
179	Wayne, KY	−.95193
180	Borden, TX	−.94936
181	Calvert, MD	−.94885
182	Clay, GA	−.94633
183	Tolland, CT	−.94061
184	Santa Cruz, CA	−.93978
185	Brewster, TX	−.93860
186	Natchitoches, LA	−.93859
187	Matanuska-Susitna, AK	−.93150
188	Lynn, TX	−.93082
189	Glacier, MT	−.90695
190	Hale, AL	−.90529
191	Mingo, WV	−.90254
192	Desha, AR	−.90204

TABLE 6.10 *(Continued)*

Rank	County	Disparity Index
193	Williamsburg, VA	−.90110
194	Harris, TX	−.90003
195	Fauquier, VA	−.89891
196	Franklin, LA	−.89290
197	Hayes, NE	−.89215
198	Madison, MS	−.88722
199	Middlesex, CT	−.88676
200	St. Francis, AR	−.88239
201	Pike, MS	−.88156
202	Quitman, GA	−.87958
203	Cibola, NM	−.87478
204	Randolph, GA	−.87451
205	Calhoun, GA	−.86937
206	Big Horn, MT	−.86705
207	Prince William, VA	−.85668
208	Santa Barbara, CA	−.84542
209	Dutchess, NY	−.84494
210	Culberson, TX	−.84273
211	New Haven, CT	−.84227
212	Plymouth, MA	−.84003
213	Bucks, PA	−.83985
214	Suffolk, MA	−.83210
215	Cook, IL	−.82700
216	Ozaukee, WI	−.82256
217	Macon, AL	−.82102
218	Monroe, AR	−.81947
219	Talbot, MD	−.81943
220	Carter, MT	−.81478
221	San Saba, TX	−.81460
222	Wayne, MI	−.81347
223	Perry, KY	−.80654
224	Pemiscot, MO	−.80650
225	Goochland, VA	−.80270
226	Prince George's, MD	−.80254
227	Alexander, IL	−.80147
228	Harlan, KY	−.79913
229	Manassas, VA	−.79725
230	Hennepin, MN	−.79015
231	Charles, MD	−.78926
232	Adams, MS	−.78558
233	Collier, FL	−.78446
234	Whitley, KY	−.77994
235	Jim Wells, TX	−.77900
236	Dawson, TX	−.77501
237	Red River, LA	−.77039
238	Walthall, MS	−.76869
239	Dooly, GA	−.76213
240	Orange, NC	−.76189

(Continued on next page)

TABLE 6.10   (*Continued*)

Rank	County	Disparity Index
241	Childress, TX	−.75870
242	Lafayette, AR	−.75512
243	Acadia, LA	−.75344
244	Livingston, MI	−.75041
245	Richmond, VA	−.74872
246	Lincoln, WV	−.74861
247	Saguache, CO	−.74810
248	Burlington, NJ	−.74597
249	Conejos, CO	−.74403
250	Harmon, OK	−.74376
251	Summit, UT	−.74134
252	Fulton, KY	−.74067
253	Dallas, TX	−.74050
254	Delaware, PA	−.73968
255	Morehouse, LA	−.73744
256	Pointe Coupee, LA	−.73375
257	Sitka, AK	−.73368
258	Webster, WV	−.73316
259	Stewart, GA	−.73132
260	Menard, TX	−.72892
261	Arapahoe, CO	−.72873
262	Norton, VA	−.72846
263	Delaware, OH	−.72734
264	Richland, LA	−.72715
265	Letcher, KY	−.72646
266	Macon, GA	−.72368
267	Woodruff, AR	−.72200
268	Litchfield, CT	−.72194
269	Petroleum, MT	−.72043
270	Midland, MI	−.71908
271	Camden, NJ	−.71904
272	Uvalde, TX	−.71651
273	Dade, FL	−.71573
274	De Kalb, GA	−.71549
275	East Baton Rouge, LA	−.71499
276	Pecos, TX	−.71007
277	San Diego, CA	−.70113
278	Lancaster, VA	−.70099
279	Shelby, AL	−.70017
280	Menifee, KY	−.70008
281	Marion, GA	−.69874
282	Midland, TX	−.69869
283	Iberville, LA	−.69666
284	Floyd, KY	−.69444
285	Queens, NY	−.68685
286	Monroe, NY	−.68468
287	Palm Beach, FL	−.68400
288	St. Louis, MO	−.68358

TABLE 6.10   (*Continued*)

Rank	County	Disparity Index
289	Boulder, CO	−.68036
290	Choctaw, AL	−.67799
291	King, WA	−.67767
292	Taylor, GA	−.67706
293	Deaf Smith, TX	−.67647
294	San Augustine, TX	−.67544
295	Burke, GA	−.67426
296	Collingsworth, TX	−.67343
297	Greene, GA	−.67070
298	Fairbanks North Star, AK	−.67029
299	St. Helena, LA	−.67008
300	Johnson, KY	−.66786

Source: Data from U.S. Census Bureau, *USA Counties, 1994,* CD-ROM.

At the other end of the spectrum, looking only at the three hundred most egalitarian counties in the nation, the opposite trends prevail. These counties—which tend to have lower proportions of families either in poverty or with very high incomes—are three times less likely to be metropolitan than the three hundred most unequal counties. The three hundred most equal counties also have three times the proportion of their population living in rural areas on farms than the bottom three hundred. Where are they? It seems the Midwest is destined to be solidly middle class while the South is doomed to be feudal. Almost three of every four counties in this top three hundred category are in the North Central region. This proportion is seven times the size of its representation among the bottom three hundred, and twice as large as would be expected on the basis of chance. The South, which should have nearly 45 percent of the top three hundred most equal counties if all things were equal, has less than 11 percent (a four-to-one ratio). The other two regions have roughly the same presence among the three hundred egalitarian counties we would expect based upon the overall distribution of counties among the four regions.

A great number of the three hundred counties with high inequality are also in major metropolitan areas. Indeed, just glancing at the fifty most unequal counties shows them to be either huge central city counties (New York) or suburban counties of major metropolian areas such as New York City, the District of Columbia, San Fransisco, Los Angeles, and so on. New York County has the tenth highest family Income Disparity Index among all of the nation's counties. Unfortunately, nearly 1.5 million people lived in this income-skewed locale in 1990. Although it is home to many wealthy people, we also know it plays host to a teeming

TABLE 6.11   Family Income Disparity Index Scores of the 300 Most Equal
              U.S. Counties, 1989

Rank	County	Disparity Index
2	Yellowstone National Park, MT	1.53963
2	Loving, TX	1.53963
2	Kalawao, HI	1.53963
4	Hooker, NE	1.00958
5	Juab, UT	.92922
6	San Juan, CO	.88741
7	Bland, VA	.82261
8	Keith, NE	.81721
9	Hodgeman, KS	.81527
10	Ottawa, KS	.79993
11	Eddy, ND	.79247
12	Union, IN	.78792
13	Arthur, NE	.78588
14	Osage, MO	.77412
15	Chattahoochee, GA	.76651
16	Jackson, CO	.76364
17	Whitley, IN	.76261
18	Clark, KS	.75905
19	Polk, NE	.75635
20	Van Wert, OH	.75568
21	Fountain, IN	.74836
22	Garfield, WA	.74827
23	Lake, MN	.74812
24	Jay, IN	.74711
25	York, NE	.74573
26	Sedgwick, CO	.74515
27	Juniata, PA	.74492
28	Hancock, IA	.74063
29	Franklin, ID	.73880
30	Steuben, IN	.73800
31	Divide, ND	.73346
32	Huntington, IN	.73302
33	Essex, VT	.73031
34	Wabaunsee, KS	.72884
35	Dodge, NE	.72869
36	Wasington, IA	.72341
37	Cuming, NE	.71580
38	Forest, PA	.71549
39	Ohio, IN	.71525
40	Gosper, NE	.71455
41	Jefferson, NE	.71407
42	Wyandot, OH	.70825
43	Hamilton, IA	.70344
44	Fallon, MT	.70212
45	Seward, NE	.70048
46	White, IN	.69644
47	Clark, ID	.69494
48	Cloud, KS	.69402

TABLE 6.11 *(Continued)*

Rank	County	Disparity Index
49	Elk, PA	.68644
50	Hamilton, NY	.68644
51	Wheeler, NE	.68466
52	Cook, MN	.68432
53	Perry, PA	.68009
54	Brown, MN	.67969
55	Oneida, ID	.67922
56	Noble, IN	.67721
57	Saline, NE	.67702
58	Sweet Grass, MT	.67641
59	Daggett, UT	.67567
60	Beadle, SD	.67415
61	Wichita, KS	.67164
62	Gasconade, MO	.66889
63	Rush, KS	.66649
64	Coffey, KS	.66612
65	McPherson, KS	.66382
66	Madison, NE	.66335
67	Colfax, NE	.66220
68	Columbia, PA	.66148
69	Dodge, WI	.66086
70	Scott, KS	.65854
71	Blackford, IN	.65584
72	Morrow, OH	.65570
73	Warren, IA	.65291
74	Marion, KS	.65283
75	McCook, SD	.65184
76	Roseau, MN	.65060
77	Cass, NE	.64939
78	Henry, OH	.64883
79	Howard, NE	.64679
80	Alleghany, VA	.64652
81	Nance, NE	.64629
82	Lawrence, IN	.64626
83	Price, WI	.64480
84	Republic, KS	.64470
85	Haskell, KS	.64320
86	Waupaca, WI	.64123
87	Schuyler, NY	.64092
88	Williams, OH	.64065
89	Paulding, OH	.63717
90	Bond, IL	.63713
91	Otoe, NE	.63456
92	Coryell, TX	.63368
93	Rawlins, KS	.63352
94	Sargent, ND	.63239
95	Boundary, ID	.63191

(Continued on next page)

TABLE 6.11 (*Continued*)

Rank	County	Disparity Index
96	Mitchell, IA	.63171
97	Henderson, IL	.63081
98	Grant, SD	.62959
99	Swift, MN	.62850
100	Henry, VA	.62780
101	Sioux, IA	.62687
102	Weston, WY	.62662
103	Mercer, OH	.62659
104	Ness, KS	.62580
105	Wayne, NE	.62542
106	Moody, SD	.62475
107	Harrison, IN	.62370
108	Osage, KS	.62192
109	Iowa, IA	.62185
110	Millard, UT	.62167
111	Lake, CO	.62085
112	Perry, IN	.62049
113	Dundy, NE	.62017
114	Barber, KS	.61986
115	Boone, IA	.61925
116	Fulton, OH	.61552
117	Wyoming, NY	.61538
118	King, TX	.61533
119	Wasatch, UT	.61293
120	Columbia, WI	.61161
121	Elmore, ID	.60990
122	Harvey, KS	.60961
123	Anderson, KS	.60883
124	Fulton, IN	.60852
125	Johnson, NE	.60746
126	Pottawatomie, KS	.60617
127	Mitchell, KS	.60616
128	Pierce, NE	.60577
129	Saunders, NE	.60436
130	Emery, UT	.60229
131	Fayette, IN	.60227
132	Uinta, WY	.60166
133	Griggs, ND	.59818
134	Winnebago, IA	.59728
135	Jasper, IA	.59672
136	Perry, MO	.59669
137	Gunnison, CO	.59523
138	Steele, ND	.59494
139	Hardy, WV	.59459
140	Caribou, ID	.59444
141	Lake of the Woods, MN	.59307
142	Pembina, ND	.59301
143	Deuel, NE	.59233

TABLE 6.11   (*Continued*)

Rank	County	Disparity Index
144	Phillips, CO	.59214
145	Schuylkill, PA	.59198
146	Pasco, FL	.59187
147	Green Lake, WI	.59186
148	Charlevoix, MI	.59182
149	Ransom, ND	.59167
150	Dixon, NE	.59078
151	Hardin, IA	.58993
152	Douglas, IL	.58943
153	Randolph, IN	.58839
154	Wabash, IN	.58796
155	Garden, NE	.58781
156	Trempealeau, WI	.58753
157	Red Lake, MN	.58594
158	Meade, KY	.58557
159	Northumberland, PA	.58431
160	Adams, PA	.58413
161	Alexander, NC	.58333
162	Turner, SD	.58148
163	Hamilton, NE	.58102
164	Marshall, IN	.58075
165	Grundy, IA	.58015
166	Iron, WI	.57910
167	Putnam, OH	.57907
168	Dodge, MN	.57863
169	Freeborn, MN	.57862
170	Moniteau, MO	.57842
171	Meade, KS	.57820
172	Grayson, VA	.57752
173	Woodson, KS	.57723
174	Smith, KS	.57719
175	Randolph, NC	.57632
176	Putnam, IN	.57632
177	Pierce, ND	.57562
178	Nelson, ND	.57511
179	Jefferson, MO	.57450
180	Alger, MI	.57443
181	Beaver, OK	.57323
182	Spencer, IN	.57263
183	Washington, IL	.57247
184	Rich, UT	.57111
185	Franklin, MO	.57100
186	Sauk, WI	.57043
187	Fillmore, NE	.56945
188	Presque Isle, MI	.56914
189	Hernando, FL	.56755
190	Marshall, IL	.56724
191	Montgomery, IN	.56682

(*Continued on next page*)

TABLE 6.11   (*Continued*)

Rank	County	Disparity Index
192	Harper, OK	.56630
193	Scott, IL	.56577
194	Fremont, IA	.56536
195	Jersey, IL	.56485
196	Adams, NE	.56443
197	Dickinson, KS	.56419
198	Manitowoc, WI	.56339
199	Glades, FL	.56188
200	Valley, NE	.56164
201	Carroll, IN	.56152
202	Cheyenne, NE	.56137
203	Watonwan, MN	.56054
204	Wright, IA	.56010
205	Sumner, KS	.55984
206	Wells, IN	.55873
207	Henry, IA	.55818
208	Traill, ND	.55713
209	Manassas Park, VA	.55650
210	Iroquois, IL	.55641
211	McLeod, MN	.55547
212	Harney, OR	.55490
213	Saline, AR	.55445
214	Dawson, NE	.55391
215	Amherst, VA	.55381
216	Chattooga, GA	.55370
217	Meade, SD	.55336
218	De Kalb, IN	.55242
219	Richardson, NE	.55185
220	Lee, IL	.55165
221	Jo Daviess, IL	.55070
222	Shelby, IA	.54974
223	Sullivan, IN	.54949
224	Harper, KS	.54936
225	Fulton, PA	.54889
226	Sheboygan, WI	.54779
227	Warren, PA	.54657
228	Clay, NE	.54548
229	Butler, IA	.54503
230	Lucas, IA	.54454
231	McDowell, NC	.54447
232	Carroll, OH	.54400
233	Marquette, WI	.54391
234	Lincoln, SD	.54226
235	Florence, WI	.54181
236	Covington, VA	.54136
237	Mercer, ND	.54115
238	Kearney, NE	.54084
239	Prince George, VA	.54062

TABLE 6.11 (*Continued*)

Rank	County	Disparity Index
240	Cameron, PA	.54030
241	Norton, KS	.53965
242	Mills, IA	.53909
243	Gibson, IN	.53855
244	Anderson, KY	.53805
245	Beaver, UT	.53799
246	Chickasaw, IA	.53795
247	Lincoln, WI	.53752
248	Garfield, UT	.53712
249	Warren, IN	.53662
250	Dakota, NE	.53638
251	Cass, IA	.53592
252	Howard, MO	.53556
253	Stanly, NC	.53537
254	Waseca, MN	.53499
255	Rutherford, NC	.53468
256	Jefferson, KS	.53450
257	Box Elder, UT	.53419
258	Thayer, NE	.53417
259	Cumberland, IL	.53356
260	Union, SD	.53303
261	Bullitt, KY	.53160
262	Marshall, KS	.53158
263	Marinette, WI	.53104
264	Skamania, WA	.53092
265	Wahkiakum, WA	.53035
266	Bremer, IA	.53012
267	Houston, MN	.53004
268	Oneida, WI	.52979
269	Sullivan, PA	.52823
270	Kingfisher, OK	.52820
271	Washington, KS	.52804
272	Clay, IA	.52739
273	Red Willow, NE	.52719
274	Daniels, MT	.52670
275	Rowan, NC	.52575
276	Benzie, MI	.52372
277	Herkimer, NY	.52351
278	Platte, NE	.52257
279	Renville, ND	.52203
280	Prairie, MT	.52161
281	Armstrong, PA	.52147
282	Cooper, MO	.52133
283	Perkins, NE	.52090
284	O'Brien, IA	.52054
285	Ray, MO	.52044
286	Shawano, WI	.52044
287	Benton, MN	.51844

(*Continued on next page*)

TABLE 6.11   *(Continued)*

Rank	County	Disparity Index
288	Auglaize, OH	.51760
289	Carroll, VA	.51683
290	Darke, OH	.51682
291	Clearwater, ID	.51680
292	Cheyenne, KS	.51646
293	Augusta, VA	.51596
294	Crawford, MI	.51580
295	Cerro Gordo, IA	.51539
296	Crawford, IL	.51537
297	Martin, IN	.51533
298	Crook, OR	.51503
299	Washington, NY	.51475
300	Fond du Lac, WI	.51457

Source: Data from U.S. Census Bureau, *USA Counties, 1994,* CD-ROM.

population of derelict, homeless, and deprived persons. All told, there are nearly 83 million people in the most unequal three hundred counties. Thus, although the bottom three hundred counties are less than one-tenth of all counties, one-third of the nation's entire population in 1990 is crowded into them. In a word, although this group of top inequality counties seems small at only 10 percent of all counties, an average of 276,000 people live within each of their boundaries. One out of every three persons is exposed to hard-core relative deprivation on a daily basis in the United States. Not all of the data is dismal, however, since nearly 6.5 million persons live in the three hundred most egalitarian counties as well. But because these egalitarian counties also tend to be rural and in the Midwest, only 2.6 percent of our nations' inhabitants are lucky enough to reside in them. Each county in this group had an average population of 21,369 in 1990. This was considerably below the national average of 79,182 for all U.S. counties and far below the average population of the three hundred most unequal counties (one thirteenth of their size).

Many of the same variables associated with high absolute income are connected to relative income inequality as well. A profusion of studies exists that explores the existence of an inverted U-shaped curve between development and inequality. Looking at U.S. states, for example, we have known for quite some time that relative income inequality goes up with percentage of nonwhite population, the dominance of farming in a region, percent living in urban areas, greater participation of women in the labor force, and residence in the South. On the other hand, income inequality falls in states where employment and educational levels are high, where large proportions of workers are employed in manufacturing, and where income is high.[64] There is copious evidence that even within metropolitan areas, higher income is related to lower relative income inequality.[65] We also are now aware that the larger the size of a metropolitan area, the greater will be its degree of relative income inequality.[66]

In most research, the staying power of the Kuznets curve has been phenomenal. The correlation between development and reduced income inequality has also been found in U.S. congressional districts and in a large sample of three hundred counties in twelve eastern states.[67] Table 6.12 lists all such variables associated with relative income inequality by county groups. All of these indicators have been highly predictive of income inequality for U.S. states.

The association of these indicators at the county level with relative income inequality continues to be strong. For example, education as measured by the percent completing high school is associated with lower family income inequality, as is percent of a county's total payroll paid to those employed in manufacturing jobs. Among the factors acting to increase relative family income inequality are the presence of African Americans, percentage of households headed by females, and the unemployment rate.

All of these factors were regressed with family Income Disparity Index scores for U.S. counties to find out what weight each would have in predicting relative income inequality. Two dummy variables were added to account for the impact of African-American dominance (0=below average proportion of African Americans in the population, 1=above average) and being part of a metropolitan area (1=SMA, 0=non/SMA). To fully test whether the development factor was also operating at the

TABLE 6.12   Mean Values of Selected Characteristics by County Income Disparity Groups, 1990

Characteristic	Income Disparity Groups		
	Low (Unequal)	Middle 50%	High (Equal)
Percent African-American	17.7	7.5	1.7
Percent high school graduates	65.7	69.6	73.7
Percent college graduates	16.0	13.1	12.0
Percent of households headed by a female	17.8	12.8	9.8
Percent unemployed	8.1	7.2	6.2
Percent of total payroll in manufacturing	23.4	28.9	29.1
Percent rural/farm	3.8	6.1	9.8

Source: Data from U.S. Census Bureau, *USA Counties, 1994,* CD-ROM.

county level, both median family income and median family income squared for each county were included as independent predictive variables. In this manner, we can tell whether a Kuznets curve actually does operate for U.S. counties. According to prior findings at larger geographic levels (U.S. states and congressional districts), there was a good possibility that development might lead to lower relative income inequality in the long run. The inverted U-shaped curve proposed by Kuznets would predict low inequality among low-income counties, followed by rising relative income inequality as incomes in counties went up. Finally, the process would supposedly end in counties rapidly becoming more equal as they reach the highest level of affluence. As in prior research, development is measured in the regression formulas as median income and median income squared. Table A2 in the Appendix contains the results of this analysis.

To begin with, the percent of a county's total payroll going to those holding jobs in manufacturing proves to be a hardy influence on equalizing family income. For all U.S. counties, and for all counties both within and outside the South, this indicator is strong, positive, and in the expected direction. The more manufacturing dominates the payroll of a county, the more equal the distribution of family well-being as measured by the Income Disparity Index. This major finding once again says a great deal for those concerned with the negative ramifications of deindustrialization on the American pocketbook. Not only do industrial jobs promote higher absolute income, they reduce relative income inequality. The beneficial impact of this factor is hard to overemphasize. Based upon these findings, we may fear for the survival of the once thriving middle-class American lifestyle. As these jobs disappear, real income goes down and relative income inequality soars.

An even stronger predictor of family income equality in counties is a low percentage of families that are headed by females. As this percentage climbs, relative family income inequality becomes worse. Female single parenthood has long been associated with poverty and/or low income, so this finding is not surprising. It is twice as important in its effect on promoting inequality, however, as the heavy concentration in manufacturing mentioned above in promoting equality. While macro forces such as deindustrialization remain important, the vitality of micro units such as the family have a major impact on relative income inequality. Other scholars have commented at great length on the changing nuclear family as it has drifted away from the traditional husband/wife structure, but time does not permit an exploration of this theme. Suggested remedies proliferate, however, from both liberal and conservative analysts, who are mostly at odds with one another: make divorce more difficult, enforce better child support collection, increase welfare support for single-parent families, decrease welfare support, wage war against the trend of births to unmarried mothers, make birth control more easily available to the young and abortion services more affordable to the poor, emphasize religious and pro-family values more emphatically, teach better sex education in the schools, etc. One fact that nearly all sides agree upon, however, is that the typical economic well-being of mothers and their children in single parent families is abysmally low. This condition, left unchecked, is a threat to the very existence of a strong American middle class.

The effect of most of the other variables in predicting family income disparity fluctuates among regions. The percent of a county's population with at least a high school education is related to greater equality, as expected. The degree to which a given county is metropolitan also has a statistically significant (although weak) impact in promoting greater income inequality. African-American dominance also has a weak but significant effect. Counties with a higher proportion of African Americans than the national norm are also more likely to be equal. This may reflect counties that are uniformly higher in poverty, where everone is in the same sinking boat. Or it may reflect the presence of unionized African Americans in northern counties where they are heavily represented in manufacturing industries. The dynamics remain to be explored. Last, before we proceed to the major finding, unemployment or the percentage living on farms had no statistically significant effect on promoting or hindering income equality at the national level. These variables are inversely related to equality, however, in counties outside of the South.

Among all U.S. counties, and within counties both in the South and outside it, the inverted-U curve is dominant but in the opposite direction as predicted by development theory. The Kuznets relationship would predict that a poor county is likely to have a fairly high degree of inequality. Theoretically, as we enter the group of counties with a medium level of family income, income inequality should go up. Finally, as we ascend into the realm of wealthier counties, the Kuznet hypothesis predicts that income inequality will begin to decline once again. By plotting the mean score of the Income Disparity Index against the average median family income for each decile grouping of counties, we can see a distinct inverted-U curve emerge in the graph of figure 6.8. But it is in the *opposite* direction from the prediction of the Kuznets hypothesis. The Y-axis, which measures the Income Disparity Index, portrays higher positive scores as indicative of counties with more equal, middle-class, family income distributions. These counties would tend to be low on family poverty *and* low on percentages of relatively rich families. It is evident from the graph that poor counties, in the lowest deciles, have negative Income Disparity Index scores, as do wealthy counties at the other end of the decile range. It is counties with medium family incomes that have the most equality.[68]

FIGURE 6.8   Income Disparity Index for U.S. Counties by Family Income, 1989

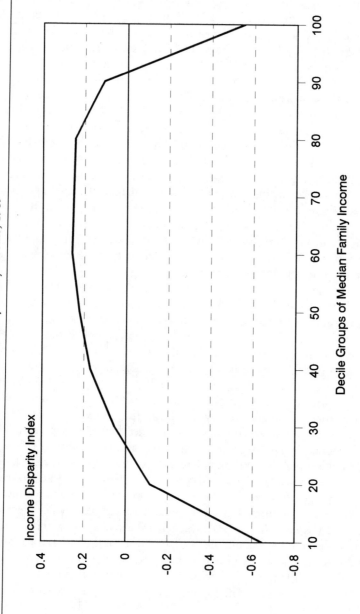

Source: Data from U.S. Census Bureau, *USA Counties, 1994,* CD-ROM.

The few variables examined in this analysis end up predicting well over three-fourths of the variation in the Income Disparity Index for all U.S. counties. This is highly critical in that, while there are unidentified explanatory variables that may have an effect upon inequality, they are likely to be few in number and of little weight. In short, we are dealing with the most important predictive variables, ones that have shown a major influence on equality in the past. Yet, at least among the 3,141 U.S. counties, median family income acts in a way exactly opposite to that predicted by the Kuznets perspective. There is a development effect—but in the reverse direction. Poor and rich counties are most unequal, while middle-income counties are most equal. There appears to be a major difference between development forces in small areas—such as counties—and in larger areas, such as countries.

In the end, the differential impact of development variables upon and between various sizes of geographic areas needs to be explored in later research. The point that must be emphasized here is that when development in U.S. counties is looked at, it does not work for the majority. At the state level, where there is an association between high income and low relative inequality, the large gaps between counties are hidden. There are frequently a few monolithic counties that can sway figures for an entire state, which may hide pockets of poverty in outlying areas that do not share the obvious wealth of metropolitan areas. There is indeed a statistically significant (but small) negative association between higher family income and lower income disparity for counties (–.094). Yet as with our illustration of storks and fertility, we have found that when other factors are looked at, the impact of rising income does not produce less relative inequality. In fact, beyond middle-income counties the reverse happens.

For the large majority of U.S. counties, it is also true that even the spurious association of higher income/lower inequality is not very evident. The nonlinearity of the relationship can be easily seen. For example, the mean Income Disparity Index of family income is –.303 for the poorest 25 percent of counties and .216 for the 50 percent of counties categorized as middle-income. For the wealthiest fourth of counties based upon median family income, the average Income Disparity Index again plummets to –.129. In other words, improvement in relative inequality may not occur despite rising family income. The exact opposite (growing inequality) is especially likely to occur in affluent counties that are higher in median family income than three-fourths of the less well-off counties. This finding is in line with a study by Orley Amos which also finds regional income inequality increasing among counties once development has been completed, that is, relative inequality continues to soar as income increases even more among the very richest counties. In other words, as in the findings reported here, inequality follows a pattern of increase-decrease-increase that is the exact opposite of the Kuznets curve.[69]

At bottom, huge numbers of people in the United States suffer from an income inequality ratio that resembles that of the Third World. By 1990, for example, virtually every state in America had a household income Gini ratio greater than India's or Ethiopia's. We are not used to thinking of ourselves as being in the same league as Asia. Yet with respect to relative income inequality this is certainly true for large numbers of our population. The good news, of course, is that America is still a wealthy nation. Relative income inequality here does not have nearly the life-and-death repercussions that it most certainly has in the Third World. Since we start at such a high level of income, most of our poor are much better off than the poor in many other countries of the world. Also, there is cause for optimism in the ongoing growth of real income and lessened inequality in the South. The time for this region of the country to become an economic equal to the rest of the country is long past due. No one can fault the South for its effort and its partial success as it has tried to engineer its own economic growth. It has even been argued by many critics that were it not for the low wage structure, conservative climate, and hostility toward unions in the South, even more American manufacturing jobs would have been lost to foreign lands.

Yet the South remains vulnerable because of its success. The attraction of this region has been its low wages and anti-unionism. In this era of the global factory, it is getting beaten at its own game of offering the most for the least by desperately poor Third World nations. Thus, the South can no longer compete for industry and is in great danger of losing the marginal factories it does have. Deindustrialization is an even greater threat to the South than it is in the North. We have already seen that a peripheral industrial belt exists in some of the poorest counties in the South, made up of industries at the very end of their product cycle. And without needed internal resources—such as a well-educated, skilled labor force—it can never hope to win high-technology firms like those which flourish in the north. In other words, in the smorgasbord of inducements for factory relocation it can offer only a Kmart line to many corporations that must shop at Saks. It is also true that although progress and growth in the region are obvious, they have left large segments of the underprivileged wholly untouched. The rural population, and especially African Americans, have been forced to continue to do without.

In the final analysis, the low-wage southern strategy may fail. In a way, it is the same as buying a cheap, low-quality car instead of a more expensive one of higher quality. At the start, both may seem to work equally well. Yet over the long run, it may cost more to buy a cheaper product than a better one that is initially more expensive. Continued durability and long-range reliable performance will in the end yield more in an investment, even if the starting cost is greater. Many decisions by American business firms support this view:

> The belief that investors will seek out pockets of cheap labour in poor regions is not quite consistent with the actual practice of plant location in contemporary America. The South has certainly attracted new industry seeking to economize on labour costs. But the region has not gained much in modern industries such as electronics—high wage, capital-intensive activities—that could make important contributions towards improving the South's income levels, despite the fact that parts of the South appear at first sight to be highly favourable for electronics manufacturing. . . . This is partly because what matters the most to a modern industrial corporation may be linkages with other related firms, suppliers of components and sources of scientific or technological development which tend to point to a northern or west-coast location. It is also because leading U.S. companies are going multi-*national* rather than multi-regional, locating more investment and employment abroad (e.g., in Europe) than in Appalachia or the deep South. The United States is part of a world-wide economic system.[70]

A common trait has repeatedly been shown to be linked to the trends toward a loss of income and an upswing of inequality. This condition operates along with the importance of regional differences, and is crucial to both North and South. The presence of manufacturing jobs remains a crucial part of our economic well-being. This is true whether we are speaking of the United States as a whole or of forces operating within each of its regions. The findings consistently emphasize the very great impact that manufacturing has in raising income levels within counties. Median family income levels are highest in locales where manufacturing jobs are abundant. Lower relative income inequality is strongly predicted by manufacturing employment as well. Where industrial employment exists, income tends to be higher and more equally shared.

Within the context of the world economy, the persistence of this finding is shattering. The prognosis for the United States must remain grim if our deindustrialization continues. For every job that is lost in manufacturing to foreign shores or fails to develop because of investment in new plants abroad, our income shrinks. What little remains is divided in an even more unfair manner than before. We are told that having other countries do the "dirty work" of actually producing things in factories is beneficial to us. It reduces pollution and the drain of resources in our own country, while removing boring, strenuous, and/or dangerous jobs. Yet this is more a bald assertion rather than a proven, absolute fact. All evidence indicates that manufacturing jobs are very desirable. They pay more than the service jobs that are replacing them. Manufacturing employment promotes relative income equality as well. Industrialization is closely connected with a strong middle class and the absence of extreme income inequality. With its disappearance, we may expect many of the social ills so typical of impoverished Third World countries to develop. At that point, it will truly be a time of "mourning in America."

FOOTLOOSE FACTORIES, MEXICO, AND NAFTA

How far south is south? The "better business climate" in the southern United States is not enough for the many American businesses that have continued their low-wage flight further south into Mexico. Encouraging deindustrialization in the South as well as in the frostbelt has been the recent passage of legislation to include Mexico in the North American Free Trade Association (NAFTA). The *maquiladora* region of Mexico on its northern border has always been a popular stopping off point for U.S. factories searching for cheap labor and nonexistent governmental controls. But certain restrictions on import tariffs have acted as a brake on wholesale relocation of factories. Recently, this barrier ceased to exist as Congress and President Clinton negotiated and approved an agreement with Mexico to reduce such trade tariffs to zero over the next few years. The last impediment to relocation for American corporations in search of cheap labor has been effectively removed.

Who benefits? Huge multinational corporations are able to garner even larger profits through a variety of means under NAFTA and the General Agreement on Tariffs and Trade (GATT) mentioned in earlier chapters. To begin with, such agreements and trade treaties offer U.S. firms more access to foreign markets. Even if Mexico is a much poorer country than the United States, it still offers 91 million potential consumers. China has over one billion people while India is rapidly approaching this mark. To gain access to foreign markets, the United States must reciprocate by opening its borders to world trade. A provision of NAFTA not frequently talked about requires companies in its member nations (Mexico, the United States, and Canada) that want duty-free treatment to make products that have specific percentages of their content made in North America.[71] Thus, U.S. automakers who shift their plants to Mexico for the lower wages can still gain protection against Japanese and other Asian producers. Of course, corporations are also able to avoid the expense of environmental protection, higher taxes, and higher wages, which are standard costs of doing business in the United States. The *maquiladora* region in Mexico is now said to have the highest levels of toxic exposure ever seen on earth.

The most comprehensive and detailed critique of free trade is offered by Ravi Batra, who in his book *The Pooring of America* recounts numerous objections to America's open-door polices. To begin with, he offers empirical proof that the free trade policies in place (especially since 1973) exactly coincided with the downturn of wages in our country, despite a continuing growth in productivity:

> Anyone who blames America's declining real wages on declining productivity growth, does not understand economics. Ford Motors today produces roughly the same number of cars and trucks as in 1975 with only about half the number of employees. This means that worker productivity there has practically doubled in eighteen years; yet Ford workers have lower real earnings (adjusted for inflation) than they did in 1975. Can we blame the drop in their living standard on their lower productivity? Of course not.[72]

In Batra's view, the real culprit is free trade. His historical examination of the United States and comparative analysis of our industrial peers shows that nations have prospered economically with a modicum of protection offered by reasonable tariffs. Throughout our history until about 1970, the United States was a comparatively closed economy. Not until 1973 did America really enter into a relatively unrestricted free-trade status, when the value of our trade (the value of imports added to the value of exports) to GNP ratio surged beyond 13 percent. When this ratio is plotted against the average annual percent growth rate in GNP for advanced industrial countries, the relationship becomes starkly apparent (figure 6.9). It is obvious that as nations become ever more open to free trade, their economic growth deteriorates.

This relationship holds true for developing countries such as Mexico and for the successful Newly Industrialized Countries (NICs) of East Asia. While America, Germany, France, and Italy all initially developed and prospered behind protectionist barriers, so did South Korea and Japan.[73] From 1940 to 1980, Mexico's ruling elite protected the country's vulnerable economy by limiting imports, and was rewarded with an average annual increase in output of 6 percent. Since then, and especially after NAFTA was signed in 1994, Mexico's standard of living has gone into a virtual free fall. In December, 1994, Mexico was forced to devalue its peso. As a result, Mexico's economic output dropped by over 10 percent in only a three-month period during 1995 while the decline in its Gross Domestic Product was the largest since it began recording this measure. Austerity measures introduced to stabilize its economy resulted in a quadrupling of interest rates, a massive surge in unemployment, government spending cuts, higher energy costs, and tax increases. All these forces, of course, deepened its recessionary tailspin.[74] Interest rates

FIGURE 6.9   Dominance of Trade Produces Lower Growth

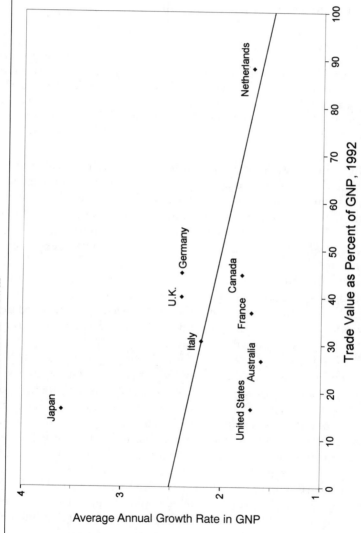

Source: Data from World Bank, 1994, tables 1, 13.

passed 100 percent for small- and medium-sized businesses while operating costs doubled. The results were devastating for average Mexican workers:

> "Everything's mortgaged now, even the cooking pots," said Nereo Valencia, 23, who runs a family farm and a tortilla business with his six brothers. They had to shut down several tortilla machines because they couldn't afford spare parts. Their herd is down to 25 cattle from its normal 150, he said.[75]

Following the official startup of NAFTA on January 1, 1994, the trade deficit mushroomed overnight as Mexico bought now cheaper U.S. goods at a faster pace than they exported products to our country. Exports from Mexico did grow by 17 percent, but the value of imports jumped to 26 percent—leaving Mexico with a trade deficit of $7 billion.[76] The Salinas presidency was also rocked by a number of high profile political assasinations and a full-scale rebellion in the southern state of Chiapas in direct protest against the activation of NAFTA. As the value of the peso tumbled, Salinas authorized the Treasury Ministry to buy up the surplus to stabilize its price with money it did not have. As the saying goes, "when the tire hit the pavement," outright panic ensued. Foreign investors realized Mexico was on the verge of bankruptcy and was not the stable, violence-free democracy they had been led to believe it was. The sell-off of Mexican stocks, bonds, and notes was phenomenal.[77] So serious was the run on the bank, it threatened to bring down financial institutions in the United States and other wealthy nations that had bought heavily into the "Mexican Miracle." U.S. officials scrambled to put together a $53 billion loan policy to bail out the Mexican economy while saving our own at the same time. The plan—complete with U.S. loan guarantees—has been attacked as a bailout for the pro-NAFTA crowd whose trade treaty is seen as the direct cause of the debacle.[78]

Under NAFTA, Mexico currently offers runaway firms a level of wages and benefits one-tenth the level of those in Canada and the United States.[79] Indeed, since the peso devaluation on December 20, 1994, there has been an added drop of 40 percent in labor costs in Mexico.[80] The new bargain-basement wage rates have acted as an even further stimulus to U.S. manufacturers seeking a haven of low pay. This has been one of the major objections voiced by NAFTA critics in the United States. A lot of our current jobs, especially at the low end of the wage spectrum, will simply migrate south of the border.

It is too early to get precise figures on how NAFTA has affected employment trends, but there are some early estimates. Patrick Buchanan, an anti-free-trade hawk and contender for the Republican presidential nomination in 1996, believes NAFTA has so far cost 340,000 American jobs and a trade deficit of $17 billion in 1995.[81] Such claims may be more inflated political rhetoric than argument with a sound, factual basis. The estimates of many experts, however, are in this same ballpark. The range seems to run from a net job loss of 300,000 to half a million over a ten-year period. Although this impact may appear minimal to a U.S. labor force with 126 million workers, the loss is hard-hitting in certain industries and specific regions.[82]

Preliminary indications show that the job loss direction is negative, supporting the predictions of NAFTA critics. As part of winning approval for NAFTA, the U.S. Department of Labor was required to offer a program to provide worker training assistance for companies that must lay off employees as a result of competition from Mexican and Canadian imports. In the first six months of 1994, 167 companies with 23,734 workers asked for such assistance (many affected companies do not seek help). Most are small manufacturing plants making low-tech parts, or are in the apparel industries.[83] For the first six months of 1995, the comparable job loss due to the trade pact had risen to 38,148 people.[84] Such numbers *are* on track for a group of studies that predicted Mexico's entrance into NAFTA would culminate in the loss of half a million jobs over the next decade.[85] Although this is less than 0.5 percent of the U.S. workforce, it puts downward pressure on the wages of remaining jobs. According to one estimate from a UCLA professor, the job drain translates to a loss of $1,000 a year per person for 70 percent of the U.S. workforce. Moreover, a *Wall Street Journal* poll of CEOs found 40 percent planning to move some production to Mexico in the near future, while one-fourth admitted they intended to use such a threat as a bargaining chip to cut wages here in the United States.

Aside from ploys such as these, have U.S. firms really benefited? One lobbying group went back to sixty-six firms to interview company officials who had publicly and explicitly voiced high expectations for the trade pact. Nearly 90 percent of the officials said they had not made significant progress in fulfilling their job and export projections in the first twenty months after Mexico entered NAFTA.[86] The trade deficit that Mexico initially experienced with the United States rapidly disappeared after its peso devaluation, which was similar to imposing a 50 percent tariff. U.S. exports to Mexico now cost twice as much, which puts an enormous dent in the demand for our consumer goods. Much of the value of our remaining exports to Mexico is actually created by American-owned companies in the *maquiladora* region, and their products are destined for later export to the United States after assembly. Using the predictions of NAFTA supporters—that there would be twenty thousand new jobs created for every $1 billion increase in U.S. trade with Mexico—this means that so far nearly 140,000 Americans have lost their jobs because of the trade pact.[87]

In the end, neither Mexican nor U.S. workers seem to have benefited from the brave new world of free trade, although one suspects that profits will continue to soar for major multinationals. Yet we do not have to be helpless victims. America need not be doomed to sink to a second-rate status. Our income level may not have to fall so severely. We do not necessarily have to suffer the consequences of an emerging world economy. There are a number of strategies that we can pursue as a country, or even as individuals acting alone, to prevent further income erosion and the spread of inequality. The remaining chapter will consider ways to overcome what too many of us may see as dark forces beyond our control.

Chapter

7

DEALING FROM THE TOP OF THE DECK: SOME ALTERNATIVES AND STRATEGIES FOR NEEDED CHANGE

It is plain that huge income inequality persists in the world and in the United States. At the global level we have seen the gap of real and relative income between nations increase since the end of World War II. This has occurred despite decades of development, international programs promoting growth, massive private investment, and countless plant start-ups by core multinational corporations in the Third World. While LDCs were promised that the influx of firms from industrial countries would help their economies and enrich their populations, their situations actually got worse. Basic human needs were sacrificed in these lands. All the while, scarce resources were squandered to attract foreign industries with give-away tax programs, a costly buildup of infrastructure, and questionable export expansion policies. Authoritarian Third World governments have forced a continuation of low wages to keep foreign funds coming in. Promised benefits to workers failed to develop. Instead, severe economic decline, depression, suffering, and starvation have worsened during the 1990s in most countries of the periphery. To pay off the mountain of debt which was first urged upon LDCs by core banks, austerity is now demanded by these very same lending institutions. In poor countries of the Third World, this means that already severely deprived people must make do with even less. Income inequality has become worse, the rich richer and the poor poorer.

None of this great transfer of wealth between the Third World and wealthy industrial countries has been of benefit to the workers and average citizens of the United States. Rather, within our country real income has declined since 1973, and relative income inequality is now greater than at any time within the past forty years. As the social safety net has been shredded in America, the ranks of the poor have swelled. Homelessness is rampant. A once healthy middle class has become increasingly squeezed as more and more families have descended into low income. Baby boom children now in the work force have suffered the most dramatic deterioration, with much lower real wages. As a result, they are increasingly locked out of the American Dream—a home of their own. At the same time, the ranks of millionaires and overpaid CEOs of our nation's major corporations have swelled. While the firms they head are increasingly unable to compete in the world marketplace, they have continued to rain undeserved pay increases and opulent stock options on themselves. In the end, a small minority of Americans has flourished and thrived because of conservative economic policies and the emergence of the world economy. These are the very wealthy. They are the elite upper crust of our nation's economic giants, who continue to garner huge profits from bloated, nonproductive defense spending. They are the stockholders who have reaped huge dividends from a policy of plant relocation in the periphery—whether in the American South or in the Third World. As America's factories have been robbed of any hope of raising productivity because of shutdowns and refusal to retool, high-paying manufacturing jobs have virtually disappeared. The United States has become the victim of deindustrialization.

Only rarely has there been any public recognition given to these harsh facts of life. When the problem is discussed, often there is a tendency to blame the victim. Americans are accused of being addicted to imports, of being spendthrift and careless about going into debt, of being lazy and unproductive, of expecting unrealistically high salaries, of not saving enough, of wanting too much. Although there may be some kernel of truth in a few of these assertions, the culpability of U.S. multinational businesses and large-scale political policies are conveniently forgotten. There is almost never any connection made by analysts between the emerging world economy, the flight of U.S. investment capital to LDCs, and the slide of income at home. In the never-ending chase for higher profits, wages are relentlessly squeezed. This onslaught led plants to relocate where a docile work force was willing to work for next to nothing. Although the most obvious results can be seen in Third World countries, there is mounting evidence that wages have been steadily pushed down in the United States. More and more women, poor rural farmers, African Americans, illegal immigrants, and other easily exploited groups are being swept into our domestic labor force. The jobs are often in low-level service work at subsistence pay. Through various antilabor strategies, employers have successfully attacked higher wages by employing part-time workers, relocating to the South, ignoring Green Card restrictions, and the like. There has emerged in America a "peripheralization" of our own labor force.[1] We are poorer because of it.

Not all change has been for the worse. The silver lining in the storm clouds gathering on the horizon has been a consistent increase in the real income of women. Although the median family income of African Americans has dropped in the past ten years, a sizeable upper-middle class of well-educated minorities now exists in our nation. While the plight of the underclass in urban America is now more urgent, the division between the haves and have-nots in our society is today less purely racial than it was in the past. If anything, white males have taken the brunt of the loss in real income, as has the industrial heartland of the Midwest. The South has made steady progress in its uphill climb in real income, while its relative inequality has shrunk in comparison with the rest of the nation. The bedrock of the South's economy is still tenuous because the low-end industry migrating to the region offers low wages. Yet the bottom line has still produced economic growth in the short run.

Such shifts have worked to obscure the general, long-term, overall ebbing of income in the United States. Although certain segments and a few regions of our society have financially benefited in the past decade or two, the trend on the whole has been downward for the average American. Moreover, because of the changes in the global economy toward greater efficiency, higher productivity, cutthroat competitiveness, and the flight of our own investment capital abroad, the future looks bleak. There does not seem to be any meaningful activity, planning, or policy-making on the domestic front that assures us that the income crisis is being addressed. Instead, the theme appears to be "business as usual." Nonproductive mergers and leveraged buyouts create mega-corporations with enormous debt. Such organizations are not suitably positioned to compete effectively in the world market. We have robbed Peter to pay Paul in order to finance this runaway expansion. Funds which normally would have gone into R&D and product development have been pointlessly sacrificed to the god of growth. With its over-reliance on military technology, our nation has less and less to offer consumers in the world marketplace. What we do sell tends to be shoddy and overpriced because U.S. policies rely on heavy international borrowing, cutbacks in retooling, high interest rates, and trade deficits. As a result, America's share of world markets in one industry after another has dropped like a rock.

The implications of income deterioration are ominous. There is a temptation to whistle in the dark, loudly proclaiming that America can never be like Brazil. Yet our national economy seems capable of spawning only low-paying jobs in service industries. America continues to undergo explosive growth in violence, drug abuse, crime, illiteracy, and lowered educational attainment. The ghettos now bristle with the latest in sophisticated automatic weaponry, which can wreak havoc if unleashed. A replay of the urban riots of the 1960s would seem like a Sunday picnic compared to what could develop in the future if the powder keg explodes. Nutrition is deteriorating rapidly in parts of the country, while more people are being excluded from minimal health care. The poor, derelicts, bag ladies, assorted petty criminals, and numerous deinstitutionalized mentally handicapped persons wander our streets without shelter or help. Can we really persist in fooling ourselves that all is well in America? Are we really better off than less developed countries which—although perhaps poorer on a per capita basis—at least provide minimal care for their needy?

There is room for anger here as well. We can only hope the awesome growth of the "me-first" attitude and the politics of greed will finally abate. This is unlikely to occur, however, all by itself. Stereotyped notions about the poor need to be challenged. The idea that persons are poor because they are lazy or incompetent or unfit needs to be questioned. Prejudicial notions—for example, that the homeless are out on the streets because they want to be—deserve to be held up to ridicule. Countless rationalizations are offered to explain away the rise of such revolting gaps in living standards. There is a growing need to salve guilt as the poor are deprived of even more while the rich accumulate greater wealth than they can reasonably spend. While in Rio, for example, my wife and I were told that the slum dwellers in the *favellas* were not so badly off. After all, many of them had televisions in their shacks! How long will it be before we are told by apologists for the rich that the lack of food among the poor is good for them because they won't suffer from weight-related health problems? How bad must our society become before we are willing to help the less fortunate?

There is a temptation to yield to despair after looking at countless statistics, all of which seem to be pointing downward. Global developments and shifts in world markets appear far beyond our control. There is a tendency to believe we must remain the victims of such all-powerful forces. Since these omnipotent influences are beyond our control, why even try to change things? The major actors on the world and national economic stages are, indeed, large and monolithic. Core governments, major corporations, bank consortiums, OPEC, the IMF, and the World Bank, to name a few, have all occupied the spotlight at one time or another. If lone individuals seem to make a difference, they are usually of the stature of the U.S. president, the Federal Reserve chairman, the CEO of General Motors, and the like. We mere mortals seem to be completely out of the picture.

PERCEPTIONS OF INEQUALITY

Confounding the difficulties of introducing needed change is the failure to realize that it is necessary. How can this be, in the light of the severe inequities which developed over the past fifteen years? Indifference and apathy to the growing gulf of inequality and the injuries it can spawn may stem from lack of awareness. In every society, there is an ideological basis for viewing the structure of inequality—why and how resources are divided among the people. The ideological basis is always status quo oriented, explaining that the differences that exist serve a purpose and ultimately are good for society as a whole. But American society is atypical of advanced capitalist countries in its naive faith that it is somehow classless. Americans are not aware of social class, or they tend to reject the very existence of social class. They fail to see our system of stratification that has deeply rooted disadvantages based upon family connections and workers who are exloited by an economic system of production that chips away at their pay:

> They have little awareness . . . of a comprehensive social system that determines the distribution of opportunities and other social benefits. They are not aware, in other words, of the highly organized system of social power connecting the economy, the professions, education, family, health, consumption, interaction in primary and secondary groups, politics, government, and law.[2]

One of sociology's crowning achievements as a discipline has been the recognition and development of social class as a concept. It is one of the few areas of academic specialization where this has occurred. Yet even some sociologists now claim that the idea of social class is irrelevant.[3] While the existence of social class has been ably defended,[4] the very fact that its explanatory power has been called into question hard on the heels of an explosion in inequality over the past two decades is significant. The myopia surrounding structured inequality extends into American academics as well as into the general citizenry. The whole area of social class analysis has been burdened with over-complexity, theoretical fine points, and intellectual hairsplitting—leaving it weakened as an analytic tool. Sociological studies which employ only social class perceptions are too narrow to encompass the broader gamut of beliefs and attitudes concerning inequality.[5]

A sole reliance upon social class as a measurement—especially in the Marxian sense, which implies a well-developed class-consciousness—may lead some sociologists to conclude that inequality is dead or dying in the United States. This is far from the case. To be sure, there are historical and cultural reasons why it is difficult for Americans to recognize the embedded nature of our social and economic inequality. Our nation was founded by people fleeing religious and class oppression. The long history of the United States, with its open frontier, promoted a social levelling that emphasized individualism and self-reliance. These attitudes and beliefs gradually became part and parcel of our country's basic value system.

James Kluegel and Eliot Smith, in their book *Beliefs about Inequality,* offer the most complete and well-developed picture of how people in America tend to view their stratification system. The basic foundation of this belief structure was derived from extensive empirical research by the authors of a representative sample of American adults (2,212 respondents). In their view, a "dominant stratification ideology" justifies economic inequality because of a persistent, universally held belief that there is plenty of opportunity to advance through hard work.[6] From this premise, two deductions follow: (1) individuals are personally responsible for their own economic fate; (2) where people end up depends upon their efforts, so any resulting distribution of income is fair and just. This perspective is conservative in that it stresses *individualistic forces* not *structural forces* (such as monetary or trade policy, deindustrialization, growing deficits, and the like) as primarily responsible for where one stands in terms of income distribution. In this way, the dominant ideology of America opposes structural solutions to inequality (welfare, unemployment insurance, affirmative action, reindustrialization, etc.) because there is little recognition of macro-level forces that affect our economic well-being. Most Americans, therefore, do not even look beyond the individual for reasons why a person is poor or rich. If you are poverty-ridden, it must be because you are lazy or have not adequately trained yourself for the labor market. Conversely, while there is some recognition that the wealthy start with more advantages, the rich are mainly seen as deserving because they have earned their position through hard work, risk, and sheer talent.

Beliefs about inequality are more complex, however, than just a disarming faith in the rags-to-riches myth. Reality does intrude. On the basis of day-to-day living a multitude of examples pop up to expose unfairness to middle-, working-, and lower-class Americans. For many decades, powerful unions achieved huge gains for workers; the civil rights and feminist movements built an awareness of institutionalized gender and race inequities; static or declining incomes led to declining faith in the system; the spread of higher education promoted a healthy skepticism and critical thinking about ideological beliefs that once went unquestioned, and so on. For Kluegel and Smith, however, these challenges are more like potholes in the road than guidelines for serious questioning of the status quo. Many forces keep us in ignorance. Our economic and political leaders, in order to retain their privileged positions, promote beliefs that justify existing inequality by their control of political, business, educational, religious, and cultural institutions. Actual occupational mobility, until the last fifteen years, was far more upward than downward. There is a very human tendency to want to attribute your own success in life to personal ability and effort. Many are also uncomfortable with the notion that people may be mere pawns of powerful, outside forces:

> Finally, there may be motivational reasons for refusing to acknowledge inconsistencies. In everyday life people have many other concerns besides those that stem from economic injustice. They may enjoy their work, must care for their families, often worry about threats to their safety from hostile foreign powers or from criminals. To entertain the thought that one is the victim of an unjust social order or that the social order victimizes others (with oneself in some degree responsible for the injustice) is profoundly distressing. Anger, frustration, shame, and other emotions may lead one to seek social change, but they may also lead to avoiding the issues.[7]

In the end, the dominant stratification ideology prevails. But it does exist alongside challenging beliefs that see injustice as stemming from structural causes or from basic inequality at the start. We do get fleeting glimpses of an uneven playing field. The tendency is to suppress these ideas, which are dissonant with the prevailing system—to compartmentalize them as a means of reducing feelings of ambivalence. Challenges to the economic status quo are possible, then, under certain circumstances, such as occurred in the Great Depression of the 1930s. There could be political ramifications if unemployment gets too high, if real wages skid too low, if too many plants are relocated to Mexico, if downsizing and layoffs get out-of-hand, if too many benefits and stock options are lavished upon CEOs, and so on. In short, there is some hope that our ideological blinders could get ripped off if conditions get bad enough.

It is instructive to contrast the perceptions of inequality among our citizens with those of other industrial nations. We now have detailed data from a great number of nations on how inequality is viewed. One collaborative effort, the International Social Justice Project (ISJP), involved over twenty researchers from twelve different countries. Using the same questionnaire and similar methodological procedures, representative samples were taken in Great Britain, Germany, Japan, Holland, the United States, Hungary, Bulgaria, Czechoslovakia, Slovenia, Poland, Russia, and Estonia. All surveys were completed in the spring and fall of 1991, yielding recent, comparative international data on issues of economic justice. In looking at the United States, Great Britain, and Germany together, it was discovered that three common factors were used by people in all three countries to evaluate the fairness of how income is distributed. First, other advanced capitalist countries express beliefs that some inequality is justified—for example, those who work harder deserve more. A second factor encompasses equal outcomes (governments should place a ceiling on high incomes; the fairest way to distribute wealth would be to give everyone equal shares, etc.). A third factor that people consider in fair allocations is distribution based upon need (especially housing and health care). Thus, ordinary people in a variety of advanced industrial countries use similar multiple criteria in making judgments about how economic resources should be distributed.[8] Because of our historic economic individualism, there is significantly lower tolerance for government-guaranteed minimum support of low-income people in the United States than in Germany, Japan, Great Britain, or the Netherlands. Of the five nations, we are also lowest in supporting a limitation on income at the top of the distribution.[9]

There were dramatic similarities among all thirteen nations in the ISJP study. In looking at the reasons people offer as the causes of wealth and especially of poverty, there is a mixture of individualist and social explanations—similar to what Kluegel and Smith discovered in the United States. In all countries, then, there is support for *split consciousness,* the idea that people maintain divided selves by compartmentalizing beliefs that are at odds with one another. In the end, they do not see one set of causal factors as exclusive of or an alternative to the other group of explanatory variables.[10] Among all thirteen countries in the ISJP study, Kluegel and Mateju found strong evidence of "divided selves"—a simultaneous adherence to egalitarian and inegalitarian beliefs about economic justice. Some variation did exist among these countries, however. Citizens of the United States, West Germany, and Britain emphasized a stronger meritocratic orientation. One example is that inequality is necessary to reward those who work harder. Less dearly held values with a welfare orientation were also part of the belief system in all countries, but especially in the Netherlands and Japan (for example, size of family should influence pay).[11]

Evidence from another collaborative research effort by seventeen nations, the International Social Survey Programme (ISSP), adds validity to the ISJP findings that people regularly maintain opposing dual perceptions of how income should be distributed. The International Social Survey Programme is made up of voluntary study teams in member nations, each of which undertakes to run a short, annual self-completion survey containing a common set of questions asked of a representative, nationwide sample of adults. The specific topics change from year to year, but attempts are made to replicate each module every five years or so. Among the topics under investigation have been social networks and support systems (1986), family and changing gender roles (1988), work orientations (1989), religion (1991), role of government (1985, 1990), and social inequality (1987, 1992). The first social inequality module, done in 1987 in ten countries, examined opinions and attitudes toward inequality in the categories of "rich and poor" as well as "privileged and underpriviledged." The replication (with the addition of a few more variables) was completed in 1992 among seventeen nations.

Being the most recent and detailed, the 1992 module contains the data we are most interested in. In the United States, the replication was made by means of reinterviewing the 1991 General Social Survey (GSS) sample in 1992. Final sample size was 1,273 persons. The GSS is essentially an almost annual personal interview survey of U.S. households conducted by the National Opinion Research Center (NORC) at the University of Chicago. This source of secondary data is highly regarded among scholars, is of top-grade quality, and is continuously funded by the National Science Foundation. The first GSS survey in the United States took place in 1972. Since that time, more than 25,000 respondents have answered approximately 1,500 questions.[12] Several dozen questions asked during 1987 and in 1992 tapped attitudes toward the rich and the poor, equality and inequality, and beliefs about income redistribution and economic justice.

It is impossible to fully describe the huge gamut of questions posed in the 1992 ISSP replication carried out in the GSS in the United States. As have previous scholars, however, I selected a variety of representative questions to elicit inegalitarian (meritocratic) beliefs along with questions eliciting belief in redistribution and egalitarianism. Four items measured support for inequality (meritocracy). Respondents were first asked to rate how important they thought hard work was for getting ahead in life. They used a five-point Likert-type scale: 1. Essential; 2. Very important; 3. Fairly important; 4. Not very important; 5. Not important at all. For the second item, they were asked to choose among five responses— 1. Strongly agree; 2. Agree; 3. Neither agree nor disagree; 4. Disagree; 5. Strongly disagree—to the statement: "The way things are in [Respondent's country] people like me and my family have a good chance of improving our standard of living." The same five-point response spread was used to reply to a third statment: "Large differences in income are necessary for [Respondent's country] prosperity." Finally, for the fourth item respondents were asked to respond to the statement: "Generally, how would you describe taxes in [Respondent's country] today? First, for those with high incomes, are taxes: 1. Much too high; 2. Too high; 3. About right; 4. Too low; 5. Much too low?" Obviously, answers with low scores on all four of these questions would reflect beliefs and values that essentially support inequality. The thrust of this orientation is that some people should receive more in the scheme of things, because the system as it exists tends to reward effort, talent, and hard work. To mess with the status quo by taxing the rich too heavily may destroy the very inequality that created the nation's overall prosperity.

The results to these questions are graphed in figure 7.1 for the ten capitalist countries of the ISSP study. In nearly all cases, the United States leads among wealthy industrial nations in support for a capitalist status quo. In this graph, the lower bars represent more intense agreement with the four items. For instance, the United States has the lowest mean response (1.76) to the item professing belief that hard work will help a person get ahead in life. In other words, Americans tend to agree with this idea more strongly than do people of the other nine wealthy industrial countries of the West. Our nation is third (below only Australia and Austria) in the faith expressed in the chance to improve our standard of living. With regard to statistical significance, Sweden is ahead of the United States in support for the notion that inequality is necessary for prosperity, while we tie with Australia and Italy (but are again ahead of six other nations) in believing that prosperity flows from large differences in income. Lastly, the United States is second from the top (below former West Germany) in supporting the idea that taxes are too high for the rich—which shows a tendency to support the idea that the wealthy should be allowed to keep more of their income. Among its peers, then, the United States shows some of the strongest support for inequality of all the wealthier, capitalist countries.

Support for income equality and redistribution has also been measured in a series of four questions that I have chosen to highlight from the 1992 ISSP inequality study. Respondents were asked to use the 5-point Agree/Disagree scale to evaluate the following statement: "Inequality continues to exist because it benefits the rich and the powerful." A second statement of the same genre read: "Differences in income in [Respondent's country] are too large." Whether a respondent perceived income differences as too large or not, the next statement focused on the government's role in adjusting such income gaps: "It is the responsibility of the government to reduce the differences in income between people with high incomes and those with low incomes." Lastly, a common strategy to reduce income differences between the wealthy and poor is to tax the latter very lightly. Respondents were asked whether they felt taxes were too high for those with low incomes. Responses to all four items are presented in figure 7.2.

There is much less support for egalitarian beliefs and income redistribution in the United States than elsewhere. We are second from bottom (barely nosed out by Sweden, where there is room for legitimate complaint) in the proportion who believe that inequality continues to exist because it benefits the rich and powerful. Our nation is about in the middle in regarding income differences as too large, which is interesting given our higher Gini ratios. Regardless of income gaps, Americans are sharply opposed to any effort on the part of government to reduce

FIGURE 7.1　Support for Inequality via Meritocracy

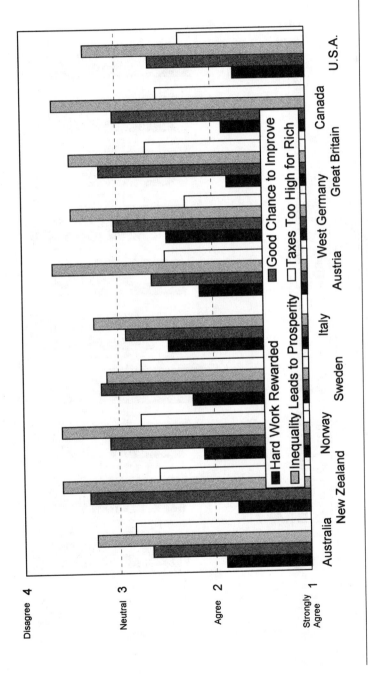

Source: Data from *International Social Survey Programme Data and Documentation 1985–92,* CD-ROM.

FIGURE 7.2   Support for Equality and Redistribution

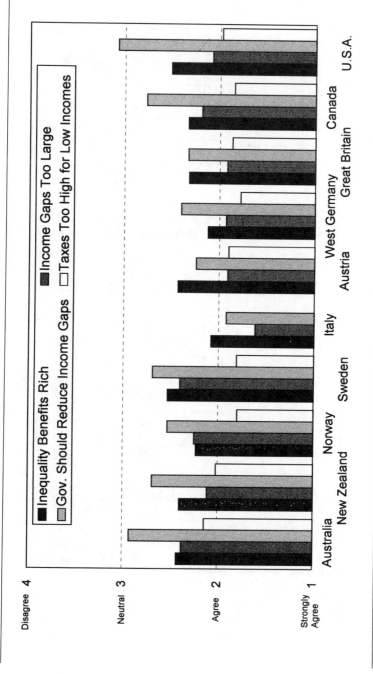

Source: Data from *International Social Survey Programme Data and Documentation 1985–92*, CD-ROM.

income differences. We lead all countries in our tendency to disagree with this idea, and ours is the only country where a large proportion of the population does not agree with that item. Lastly, despite the sorry state of those under poverty, Americans are not very likely to believe that taxes are too high for those with low incomes. Among all ten nations, we are the third *least* likely to see taxes as being too high for the poor. In the end, there is not as much support for income redistribution in the United States as in other wealthy industrial countries—despite our greater inequality and higher rates of poverty.

This is bad news for any hope of a political initiative to tackle through government programs the problems caused by inequality. Populist leaders who seek to build a momentum for needed change in this area are likely to encounter a stone wall of resistance. There is simply a great deal of denial in the American populace that growing income inequality could be a problem for the nation, for our families, and for our individual well-being.

There is some limited consolation in the data. What may exist is the potential for resistance to growing inequality—although the indications are sparse because of a lack of longitudinal data. Kluegel and Smith report that, at least until the start of the 1980s, a variety of studies indicated long-term stability in American attitudes toward inequality. The authors found deeply rooted optimism surrounding the belief that there is personal opportunity for economic advancement in the nation. (Large majorities still believe that hard work will win advancement, but there has been some erosion of the support for this view over the last three decades). Moreover, the American view of the causes of poverty was remarkably stable from the mid-1960s into the 1980s. That is, individual factors were used by most Americans to explain why people are poor (lack of effort, poor skills, bad attitudes, etc.).[13] Has the reluctance to see more structural causes of inequality finally dissipated after another fifteen years of burgeoning income disparity and an explosion in poverty rates?

A few limited indications can be seen in changes which took place between 1987 and 1992 in the repeated items used in the ISSP module on inequality. Although a longer time period—such as a decade—would have been preferable, a five-year period can at least tell us which direction trends are running and give a rough estimate of their magnitude. The same eight items were used as in figures 7.1 and 7.2 to measure the dual sets of inequality beliefs. Table 7.1 shows the change in the four items that support inequality as a result of merit considerations. There is virtually no movement between the 1987 and 1992 responses on the belief that hard work is necessary for advancement. Well over a third of respondents in both samples said it was "essential" while another half said it was "very important." There has also been very little change in the proportions who

TABLE 7.1   Change in Support for Inequality via Meritocracy, 1987–1992

Item	Category	1987	1992
Hard work rewarded:	Essential	37.4	37.7
	Very important	52	50.4
	Total	89.4	88.1
Good chance to improve living standard:	Strongly agree	19.2	10.7
	Agree	53	44
	Total	72.2	54.7
Inequality leads to prosperity:	Strongly agree	5.6	4.2
	Agree	26.9	22.2
	Total	32.5	26.4
Taxes for high-income people:	Much too high	6.3	7.5
	Too high	12	10.2
	Total	18.3	17.7

Source: Data from Davis and Smith, 1993.

feel that taxes are too high for wealthy people; about 18 percent in both years answered "much too high" or "too high." But some movement is evident in the other two indicators. There has been a moderate erosion in the proportion who believe that large income differences are necessary for America's prosperity. The proportion agreeing with this idea slipped from one-third to a little over one-fourth in only five years. Although a loss of one percentage point per year may seem minor, it is considerable in the context of changes in public opinion, which tend to be frozen over long periods of time on a wide variety of issues. Page and Shapiro found that well over half (58 percent) of the 1,128 repeated policy questions used in the General Social Survey showed no significant opinion changes over time. When using data from the General Social Survey (where the ISSP items have been measured), they point out that opinion changes of 6 percent or more do tend to be statistically significant.[14] In this context then, there has been significant decline in the proportion who believe that inequality leads to prosperity. There is absolutely no room for doubt with the fourth item, which measures the proportion of the American public who believe there is a good chance to improve their living standard in the United States. The percent who agreed with this view plummeted from 72 percent to less than 55 percent in five short years. Changes of this magnitude in the realm of public opinion are nearly unheard of. It dramatizes the deep loss of faith that Americans have suffered in their economic way of life. While a bare majority still continue to cling to the belief of great opportunity, based upon the depth and severity of this slide, massive disillusionment is inescapable. It is likely that, by the end of this century, a majority of Americans will no longer believe there is much economic opportunity left in our land.

When we consider support for equality and redistribution, there has been even more change, reflecting a growing dissatisfaction with how income is shared in the United States. Only one item did not meet the 6 percentage point cutoff in concluding a significant change, and it missed by only a fraction of a point. Nonetheless, huge numbers in both 1987 (70 percent) and 1992 (75 percent) believed taxes for low-income people were either "too high" or "much too high." The other three measurements did see significant change, reflecting increased sentiment for a more equitable dispersion of income. A majority of Americans (58 percent) now agree with the statement that inequality continues to exist because it benefits the rich and powerful. Over three-fourths believed income differences in the United States were too high in 1992, compared to "only" 58 percent in 1987). Government intervention to reduce income inequality has never been popular in the United States, yet now nearly 40 percent of Americans agree with this sentiment compared to 29 percent in 1987. Some caution on government involvement in redistribution, however, is in order.[15]

Considering that our nation is the epitome of laissez-faire economic policy, rugged individualism, and a Social Darwinist belief in the survival of the fittest, there seems to have been an earthshaking shift over five short years. Thus, from the viewpoint of the eight attitude measurements used by the ISSP project, there is a large potential for change. People see that income inequality has increased, that their opportunities have narrowed, and that some are more unfairly rewarded than others in a biased

TABLE 7.2   Change in Support for Equality and Redistribution, 1987–1992

Item	Category	1987	1992
Inequality benefits the rich:	Strongly agree	13.8	17.8
	Agree	35.6	40.5
	Total	49.4	58.3
Income differences are too large:	Strongly agree	14.9	27.7
	Agree	43.1	49.4
	Total	58	77.1
Government should reduce income gaps:	Strongly agree	6.8	9.5
	Agree	21.9	28.8
	Total	28.7	38.3
Taxes for low-income people:	Much too high	29.4	30.7
	Too high	40.3	43.9
	Total	69.7	74.6

Source: Data from Davis and Smith, 1993.

contest for scarce resources. The powder keg of change may be in place, awaiting a spark.

Thus, with the decline in standard of living for most of us and a sharpening tilt of the economic playing field that has given an even more unfair advantage to the wealthy, we can ask whether large-scale political discontent might emerge. This is the central thesis in two books by Kevin Phillips, *The Politics of Rich and Poor* (1990) and *Boiling Point* (1993).[16] Phillips does a good job of outlining the massive rise of inequality in our country and the political backlash it will inevitably produce. Extremism, discontent, instability, and rhetorical stridency may be spawned from the frustration that a middle-class electorate feels as it sees its way of life rapidly disappearing. Discontent and financial uneasiness worked to great advantage in the 1992 presidential election of Bill Clinton. Once he started focusing on issues of economic insecurity and identifying with the pain of a threatened middle class, his campaign thrust ahead to victory over incumbent George Bush.

Whereas the sharp recession of 1991 helped defeat George Bush, a prolonged financial recovery and relative economic stability helped Bill Clinton retain the presidency in 1996. In this election, Democrat Clinton was able to resonate with the fears of voters by portraying himself as a stalwart defender of the middle class. Republican candidates were castigated by political strategists as marauders of Social Security and Medicare, programs dear to the hearts of working Americans. These fears also served to weaken the Republican majority in the House of Representatives.

Without doubt, apprehension over maintaining an adequate middle-class lifestyle also figured largely in the shattering off-year election of 1994. It can be argued that continued voter unhappiness over income deterioration fed a large-scale defection from the political status quo. After all the ballots were tallied, Democrats lost ten seats in the Senate and fifty-four seats in the House of Representatives. Capping the debacle was the loss of eleven state governorships, so that Republicans held half again as many state governerships as Democrats (thirty to nineteen). When the smoke cleared, Republicans had a controlling majority in both houses of Congress. In the House of Representatives, it was the first time they held a majority since 1954. To add insult to injury, not one Republican senator, governor, or House member was defeated for reelection.

There was a lot of anger feeding this election, as there is likely to be in future political contests. In 1994, because Democrats were in control of both houses and the presidency, the backlash against business as usual was deservedly directed against them since they were so identified with lack of progress. Exit polls showed much of the massive switch in allegiance was caused by voters who saw President Clinton and the Democratic Party as too identified with big government and higher taxes.[17]

Public opinion surveys continually show cynicism and dispair reaching new depths with regard to our leaders, politicans, and governmental institutions. Just prior to the 1994 election, a *Washington Post*/ABC poll found that four out of five people agreed that "those we elect to Congress in Washington lost touch with the people pretty quickly," while two out of three believed "public officials don't care what people like me think."[18] An underlying theme has been loss of confidence in government. While during the 1950s three-quarters of the public said they trusted the government "to do what is right" all or most of the time, by 1994 this ratio had dropped to one in four.[19] Such jumps in alienation and profound loss of confidence in government did not bode well for incumbent politicians going into the 1994 election. By this time, three out of every four Americans were saying they wanted to see major changes in the way federal government works.[20] For the most part, according to a number of polls conducted just after the November 8, 1994, election, the message was disarmingly simple: less is more. In a nutshell, voters wanted less welfare, less taxes, and less government.[21]

How angry were they? Much of the discontent was centered among white men, who voted nearly two to one in favor of Republican candidates. This is the very group that has suffered a massive drop in income in the past two decades, that has encountered job insecurity and layoffs, and whose lack of advancement is perceived to be caused by favoritism toward women and minorities via governmental pressures and programs. But other groups fed the anger as well:

> Said Andrew Kohut of the Times-Mirror Center on the People and the Press: "No matter what you were angry about, you voted for the Republicans. If you were angry about your standard of living declining, you voted Republican. If you were angry at Bill Clinton, you voted Republican. If you were angry at government in general, you voted Republican."[22]

Many analysts believe, however, that the volatility of the voters' mood may not only continue but get much worse if no visible progress is made with the country's slide in standard of living. Republicans could be in for a massive abandonment by voters who perceive that no significant changes are being made. There are a number of reasons why a Republican controlled Congress may not be able to deliver for impatient voters. Above all, the deterioration of wages for most workers (and especially the angry, white males making up the new GOP power base) may simply be beyond government control, even if politicians were inclined to get involved.[23] Kevin Phillips speculated that the new volatility and impatience of American voters would allow only twelve to eighteen months of grace before going for the throat of newly elected politicians. Moreover, he argued much of the Republican Party's success in 1994 was due to a precarious combination of low turnout and unprecedented disillusionment with the two-party system—a weak framework for a new politics. Since Bill Clinton received *his* comeuppance and repudiation in 1994, the electoral spotlight was clearly on the Republican-controlled Congress in 1996.[24] Disillusionment with the Republican Congress was enough to lose them ten seats in the House of Representatives as a result of the 1996 election, although they were able to retain a slim majority while picking up two more Senate seats. Polls also showed a larger than ever gender gap as women tended to abandon GOP candidates in favor of Democratic contenders.[25]

This brings us squarely to the interpretation of the 1994 election results. Republicans claimed that their victory was a resounding afirmation of their "Contract with America" and a mandate for change. But polling in the last weeks of the campaign found overwhelming majorities of Americans had never heard of the contract, were unfamiliar with its major provisions, or were uninfluenced by its content. Even a month after the Republican sweep, only 27 percent of the public had heard or read anything about the Contract with America.[26] Rather, six out of ten persons in a *Time*-CNN postelection survey said the 1994 election was a repudiation of Democrats, while only one in six said it was an affirmation of the Republican agenda.[27]

> What the numbers suggest is that dissatisfied Republicans voted in droves, dissatisfied Democrats stayed home, and dissatisfied Perot voters shifted Republican. The admixture was roughly one part ideology, two parts disappointment. . . . But the ideological right needs to be wary too. A right-wing promised land is not what the voters are dreaming of, and this vote is not an irreversible mandate. Democrats promised two years ago they would make government work better. The people were not convinced and supported Republicans who promised that less government would work better. If the Republicans can't prove that—and quickly—watch out, because this electorate does not reward abstractions.[28]

The crucial importance of the disenchanted and alienated should be emphasized. Although they stay away from the polling booth because they feel it has few real choices to offer, their inaction figures prominently in electoral outcomes. Of the people who are eligible to vote, only a fraction register and an even smaller percentage actually vote. In 1994, fewer than half of those eligible to vote actually did so—comprising well over 100 million people. A much lower turnout of minorities and the poor also undoubtedly served to benefit GOP candidates.[29] When accounting for those who did not vote in 1994, Republican House candidates were endorsed by only 17 percent of elgible voters.[30] Nonvoters disagreed with

the GOP agenda and saw government as functioning much more effectively. They were also less keen on welfare bashing and were more than
twice as likely to agree that "poor people have hard lives because government benefits don't go far enough" than Republican voters (49 percent to
25 percent).

The public was not enamored of Republicans or their performance
after they took control of Congress in 1994. After watching the Congress
during its first one hundred days, just about as many voters disapproved
of Republican policies (43 percent) as approved of them (44 percent),
while the level of discontent with the way the nation was dealing with its
problems remained the same.[31] An Associated Press poll found that by
August of 1995, most Americans doubted that the House of
Representatives had produced much real change, while nearly three out
of four did not trust the Republican-dominated congressional plan to
reform Medicare, the most salient issue of the session.[32] Only one year
after the so-called "tidal wave" of the Republican Party takeover, a host of
public opinion surveys were showing a self-destructive turn in the GOP's
agenda. Sixty percent of the public felt Republican budget cuts had gone
too far, while three out of four were against cuts in Medicare. Voters
opted more for Democratic candidates (45 percent) than Republican contenders (41 percent). While a majority of Americans were dissatisfied
with the course of politics over the previous year, only 7 percent blamed
President Clinton, while over a third blamed Congress. A number of
other items also indicated that the electorate was standing well to the left
of the current Republican mainstream—and was unhappy about the
direction the GOP was taking the nation.[33] One year after the 1994
sweep, Americans were actually more inclined to trust President Clinton
than the GOP to do a better job with the main problems facing the
nation.[34]

> No opposition Congress and House leadership (at least since the invention
> of polling) has lost so much credibility so quickly. . . . According to a new
> poll . . . disdain for Washington, national politics and the current party sys-
> tem has hit record highs this year, foreshadowing yet another round of elec-
> toral volatility and insurrection in 1996 . . . in just eight months of new
> party control, Gingrich is not only the first speaker to be rated in polls as
> too extreme. He has also . . . managed to garner negative ratings at a pace
> heretofore matched only by mass murderers. Now, the GOP Congress, hav-
> ing misread November as a mandate rather than another negative landslide,
> is in danger of becoming the newest symbol of inadequacy. . . . Most Wash-
> ington insiders just didn't get what the 1994 revulsion was: genuine voter
> anger, a three-decade buildup of disenchantment. Such disillusionment isn't
> going to be overcome with cosmetic reforms, smaller-government speeches
> written by unsafe-meat lobbyists, term-limit gimmicks and Republican

influence-peddlers replacing Democratic influence-peddlers at the trough. The next reform wave has to wash a lot deeper.[35]

The public's disappointment in the performance of the Republican congressional majority should not be read as a ringing endorsement of President Clinton or the Democratic Party. If anything, the attitude seemed to be "a pox on both your houses." Given their druthers, a very large segment of Americans would like to see a third party emerge to effectively challenge both Republicans and Democrats—whom they see as offering no meaningful alternatives or solutions to our national problems. There is a lot of evidence indicating a desire for a third party. Between 1966 and 1994, both Democratic and Republican registration declined steadily while voters registering as independents rose from 3.9 percent to 12.4 percent.[36] When people were asked about their support of a third party candidate, 26 percent said they would like to see an independent elected president, compared to 35 percent favoring an unnamed Republican candidate and 32 percent favoring Clinton.[37] In short, interest and support is very high. Would people really flock to a third candidate if given the opportunity? Ross Perot's strong showing in the 1992 race would seem to suggest this. One poll found that 60 percent of respondents did not see that a vote for a third party candidate was wasted, while a mammoth 80 percent agreed that voting third party sends an important message that the system needs to be changed.[38] For many who support this trend, it is a viable means to just say no. While Perot's support dropped from one in five voters in 1992 to 8 percent in the 1996 election, this trend should not be interpreted as satisfaction with the political status quo. Perot was not allowed to participate in the nationally televised debates in 1996 as he had been in 1992. Moreover, many supporters of a third-party movement were not interested in Perot's candidacy. But even with his slide in support, Perot garnered enough votes, together with other third-party candidates such as Ralph Nader, to prevent Bill Clinton from being reelected with a clear majority in 1996. And nonparticipation of eligible voters rose to a near record high in 1996. Less than half of those eligible to vote did so. The 49 percent participation rate was the lowest since 1924, and the second lowest since 1824.[39]

Although we are now in an era when the electorate's patience with politicians is worn thin, the abrupt drop in support for Republican initiatives in the 1994–96 Congress was awe-inspiring. To begin with, voters did not support the Contract with America, as party leaders assumed, because most voters never knew about it or were only vaguely aware it existed. Two of the eleven easiest items were rapidly passed and signed into law because nearly everyone from both parties agreed on them: (1) subject the operation of Congress to laws it has passed for the nation; (2)

stop mandating programs for states without funding them. The line-item veto has also passed, again because no serious opposition existed. Two items which were intensely controversial eventually went down in flames: an amendment to the Constitution that would require a balanced budget and a limit to the number of terms House and Senate members could serve.

Some provisions of the Contract were popular with voters, as surveys suggested. Nearly everyone wants to see the United States achieve a balanced budget in the abstract, although people are less sure about such a serious step as amending the Consitution to reach this goal. It is also true that a lot of oxen get gored by the required budgetary cutbacks, so that when citizens see the damage, their support is often transformed into opposition.[40] Welfare is everyone's favorite whipping boy, and the public as well as politicians support reform efforts and cutbacks. But despite the low standing of welfare assistance, people do have ambivalent attitudes and are willing to see increased spending to help the poor—especially children in poverty.[41] Indeed, in polls taken after the election of 1994 it was found that the majority of Americans want to maintain or increase spending on poverty programs overall.[42] This made reform efforts much more difficult to attain, since some of the cutbacks were seen by the public as punitive and mean-spirited. While House Speaker Newt Gingrich was advocating orphanages for babies born to unwed mothers, wholesale cuts were passed in AFDC, the school lunch program, the WIC program to ensure good nutrition for poor pregnant women and children, the jobs training program, the Earned Income Tax Credit for the working poor, and so on. AFDC was discontinued as a program of support for poor families at the end of 1996. While control of family support systems was spun off to states, new federal laws mandated a five-year lifetime limitation on public assistance funded by the federal government.

The Contract was also perilously out of alignment with many other voter priorities. Both President Clinton and Congress wanted to increase defense funding, which was rapidly accomplished (more later)—although there was little public support for this move. A major effort was launched to cut spending on crime while it is at the very top of twenty items in spending priority among survey respondents.[43] Limitation of lawsuits against corporations for such things as product liability was never high on any voter's wish list. The Contract's regulatory reforms were seen as attempts to restrict the power of the federal government to regulate health, safety, and the environment—perceived as a giant step backward by most Americans. Lastly, although many of us would love to see our taxes cut, few believe this is possible if we balance the federal budget. Some analysts see this as a naked bribe offered to the electorate, that is, we would be more willing to vote for Republicans by appealing directly to our self-interest at the expense of the nation's solvency.[44] While a $500 tax cut per child seems generous on the surface, when all other give-backs are considered, more than half the dollar benefits of the GOP plan would go to families making more than $100,000 per year (less than 5 percent would go to families making less than $30,000 per year).[45] As a matter of fact, the public is more willing to accept tax *increases* to balance the budget rather than to accede to cuts in Social Security, Medicare, or education funding.[46]

Reams of analyses and commentaries have been written about the ensuing budget mess and the associated attempts to hammer out agreement on welfare reform, Medicare changes, tax policies, and the like. By the approach of the 1996 presidential election, there had been a breakdown of any meaningful change because neither party could agree with the other. Republicans and Democrats were simply unable to find any common ground to tackle the nation's important problems. In 1995, for example, the federal government was forced to shut down twice because there was no agreement on the budget. As a means of pressure, Republicans simply refused to allocate temporary funds to keep the government in operation. Impasse, confusion, and disarray dominated the political landscape, while the economic plight of many Americans grew worse. There is some hope that political compromise and a modicum of civility will return to national politics as a result of the 1996 presidential elections. The major actors of confrontation remain, however; President Clinton was reelected while the Republican party retained control of Congress. Nonetheless, the GOP margin of power in the House of Representatives was eroded, and right-wing extremism became more muted in the face of criticism by the electorate. At least for poverty programs, there is little left to argue about. AFDC has been abandoned, and food stamp support has been slashed.

While the congressional war on welfare continued unabated through 1996, so did the politics of greed. Sociologists have long known that big business and corporations have exercised a profound influence on national politics and formation of policy preferences. Much of this is accomplished through private policy-planning groups funded and sponsored mostly by the nation's largest corporations. The boards of directors of these advisory groups, hidden from the limelight of public scrutiny, read like a *Who's Who* of CEOs from among the very largest industrial and banking firms in the country. Although the great majority of Americans have never heard of these groups, they have the ear of the president and members of Congress. The corporate elite and upper class come together in these associations, review and publish their in-house research, talk about policy alternatives, and work toward consensus on important issues.[47] Much of the consensus-seeking on policy recommendations, of course, centers around issues salient to business, such as minimum wage rates, NAFTA, health-care systems, trade policies with other nations, and the like. The twelve leading policy planning groups include the American Enterprise Institute, Brookings Institution, Business Council, Business Roundtable, Chamber of Commerce, Committee for Economic Development, Conference Board, Council on Foreign Relations, Heritage Foundation, Hoover Institution, National Association of Manufacturers, and Trilateral Commission. There is an extensive overlap of members, and the same powerful executives sit on the boards of many of these organizations, although some groups are more important than others even among this elite dozen highly influential organizations. The Business Council and the Business Roundtable have been at the apogee of all these groups, at least for the past two decades. Research has firmly established the pivotal role of big business leaders in these groups and the dominance they exercise in defining (and limiting) options for Congress and the president. Moreover, the evidence also indicates that there was a sharp increase of interlocking and cohesion among the twelve groups, especially among the ultraconservative groups, just prior to the 1980 election of Ronald Reagan to the presidency.[48] This was a period marked by an increasing conservative consensus among business leaders, and their efforts help to explain the corresponding shift to the right in national policy during the past decade.

The influence of big business stems even more directly from its disproportionate money contributions to politicians who are agreeable to advancing the agenda of America's corporations. Although there is a pattern of giving to candidates of both parties (above everything else, all bases must be covered), big business is much more generous to Republican contenders than to Democrats.[49] We are also now well aware that the larger the company and the more tied-in it is with government via regulation or contracts, the higher its level of political giving.[50] Heavy political contributions are not out-and-out bribery, although the CEOs who form Political Action Committees (PACs) and contribute heavily to them are well aware that their generosity buys access to politicans who are writing laws that affect their firms. Their largesse builds a sense of indebtedness among lawmakers and is a chip that can be called in at a later time to make "changes in a bill, which exempt a particular company or industry."[51] These types of activities nearly always take place outside the light of public scrutiny, while loopholes in bills (such as the Clean Air Act) become so routine as to gut the effectiveness of the very legislation that is passed to remedy such problems as industrial pollution. Even when PAC contributions from corporations equal the amount given by other sources (such as unions), big business PACs have more clout because they are seen as having greater legitimacy and authority. Politicians are more inclined to defer to them than to any other special interest.

The large campaign contributions from big business are coveted by most politicians as well. At the start of the 1990s, an average representative in the House had to raise $3,700 per week to win reelection, while incumbent senators needed over $12,000 per week. An interesting new wrinkle to the money chase, which also helps explain why Republicans were so successful at wresting control of Congress in 1994 and keeping it in 1996, is the fact that Republican challengers are now more well-heeled then ever before. While campaign spending by the average Republican challenger for a House seat went up 40 percent between 1992 and 1994, spending by the average Democrat stayed flat. We should not be surprised to learn, therefore, that while thirty-four Republican challengers were successful, none of the Democrats were.[52]

Once in power, the newly dominant 1994 Republican Party in Congress became increasingly adept at leveraging ever more gargantuan contributions from corporate coffers and wealthy donors. By 1996 Wall Street contributors were figuring prominently in the Republican primary race for the presidential nomination, and fundraisers were seen pulling in half a million to a million dollars per evening.[53] According to a report by Citizen Action, during the first nine months after the Republican takeover of Congress in 1994, corporate PACS were giving three times as much money to the GOP than they gave to the Democratic Party. With the exception of House Minority Leader Richard Gephardt (fifth highest), the top fifteen PAC recipients were all Republicans. After the takeover, half of all PAC contributions were coming from corporations:

> Corporate PAC contributions during the period of Democratic control were designed more to stalemate the legislative process than to achieve major legislative and ideological change. . . . Now, with Republicans in charge, corporate America has seized the opportunity to move beyond gridlock in an effort to demolish programs that protect the health, safety and environment of ordinary Americans. The corporate and special interests have succeeded in buying the kind of Congress they have always wanted. . . . Under the Republicans, Congress has become nothing more than a bazaar where favors and prizes are peddled to the highest bidder.[54]

To keep the money gushing in, corporate and business PACS that contributed to Democrats prior to the 1994 takeover have been specifically targeted by the GOP to "put up, or shut up." In political terms, this is known as "catching the late train." Business lobbyists and contributors who were reluctant to contribute before the takeover were expected to help pay off the campaign debts of successfully elected Republican freshmen, or else—according to House Speaker Newt Gingrich—they would suffer from "the two coldest years in Washington."[55] In fact, later publication of *The PAC List of Shame* by the GOP which named firms, lobbyists, and trade groups whose PACS gave too much to the Democrats in 1992 sharpened the threat of nonsupport by the newly elected lawmakers.

The intimidation, bullying, and threats were successful—despite the fact that the givers were on the same side of the ideological fence as the takers. In the first half of 1995, House Republicans received nearly 60 percent of the money from the most benevolent four hundred PACS; the average Republican freshman raised $123,000 in this period, nearly double the amount captured by Democrats:

> Money is at the center of Gingrich's transformation of the House. With the new alignment of ideological allies in the business and political worlds, there are unparalleled opportunities for both the people who give the money and the people who receive it. It is such an obvious quid pro quo that it goes almost unnoticed. From House Republicans come measures that gratify industry: weakening environmental standards, loosening workplace safety rules, limiting the legal liability of corporations, defunding nonprofit groups that present an opposing view. From the beneficiaries of that legislation come millions of dollars in campaign contributions.[56]

Strings are attached to both ends of this increased corporate funding. One of the first things the new Republican majority in 1994 did in the House was pass a moratorium that effectively halted government directives to ensure a safe and clean society—measures which big business rates as oppressive and counterproductive. By 1995, the line between corporate lobbyists and politicians had all but disappeared in Congress. A group of lobbyists under the name of *Project Relief* had actually started drafting legislation, which was then introduced by Republican lawmakers as bills:

> Project Relief sounded more like a Third World humanitarian effort than a corporate alliance with a half-million-dollar communications budget. On key amendments, the coalition provided the draftsman. And once the bill and the debate moved to the House floor, lobbyists hovered nearby, tapping out talking points on a laptop computer for delivery to Republican floor leaders. Many of Project Relief's 350 industry members . . . were in the position of being courted and consulted by newly empowered Republicans dedicated to cutting government regulation and eager to share the job.[57]

The rollbacks and deregulation of safety rules and inspections may have a devastating impact upon life as we know it. One rider attached to an agricultural appropriations bill, for example, has been pushed by lobbyists from the National Meat Association and Kraft General Foods (the owner of such meat and poultry processors as Oscar Mayer and Louis Rich). The rider eases proposed new meat inspection regulations that, according to consumer groups, will block protection from *Salmonella* and *E. coli* bacteria in our food.[58] In the meantime, what has been termed "corporate welfare" escalated with the 1994 Republican congressional takeover. While austere budgets were advanced in 1996 by both the president and Congress, including deep cuts in education, health, welfare, and social spending items, the financial pork for America's largest corporations continued unabated and in some cases even grew. The Center on Budget and Policy Priorities estimated that if business subsidies were cut to the same degree as social spending was cut in the 1994–96 congressional session, $100 billion could have been saved in the federal budget over the following seven years.[59] Critics disagree somewhat on what exactly constitutes corporate welfare, but there is a consensus that it includes cash payments, special rules and exemptions, and tax subsidies that favor specific businesses and industries:

> Most major thinktanks . . . spotlighted projects such as the Small Business Administration, which gives loans to a tiny percentage of companies; export aid to giants like General Electric and Westinghouse; farm subsidies that largely go to agribusiness enterprises; government funds to prop up the floundering merchant marine industry; antiquated mining laws that allow corporations to buy federal lands for a pittance; tax subsidies to sustain pharmaceutical companies in Puerto Rico and, everyone's favorite, the Market Promotion Program, which pays Gallo to advertise its wine in France and McDonald's to air french fry jingles in Singapore. The lists totaled vast amounts and dozens of programs, which Stephen Moore of the Cato Institute summed up as "incestuous . . . anti-consumer, anti-capitalist and unconstitutional."[60]

Nonetheless, the fervor to cut corporate welfare has been rare to almost nonexistent. This is precisely due to the cozy relationship between wealthy corporate PACs that donate substantial sums to the reelection campaigns of congressional representatives who favor big business. The fat is definitely there. A report by Essential Information in Washington, D.C., identified 153 sources of federal business welfare in fiscal year 1995, totalling $167.2 billion ($1,388 per individual taxpayer).[61] Put another way, the $50 billion per year ($415 per individual taxpayer) spent on what is commonly regarded as traditonal welfare—AFDC, food stamps, housing assistance, and child nutrition—is peanuts in comparison to the beneficence awarded to rich corporations by Congress. Every year we are spending $830 million to irrigate land where surplus crops are grown, $500 million to promote arms exports, $1.1 billion in forfeited federal taxes from foreign profits of U.S. multinational corporations—the list goes on and on. Cargill Inc., a company that dominates the world's grain markets and which has a net worth of $3.6 billion, has received $1.29 billion in Export Enhancement bonuses over the past decade. Recently, firms have been allowed to use an accelerated depreciation subsidy that costs the federal government $32 billion per year while $200 million a year is given in agribusiness subsidies to corporate farms that each earn over $5 million a year.[62]

Another study by the Progressive Policy Institute in Washington, D.C., goes even further, specifically identifying 120 federal programs that provide tax havens and subsidies to powerful industry groups, and which will cost our government $265 billion over five years.[63] In an analysis of who gains or loses from the subsidies, it was shown that 80 percent of Americans actually come out on the short end (–$6.64 billion per year). While the top quintile shows a net gain in family income from the subsidies, mostly through higher rates of return from the firms they own stock in, the figures become very large for the ninety-fifth through ninety-ninth percentiles (+$11.8 billion) and gigantic for the richest 1 percent of American families (+$10.5 billion), who also happen to own 35 percent of all corporate stock in the country.

Even at the state and local level, corporate subsidies and give-backs have mushroomed to attract new industries or simply to bribe old ones to stay put. In 1977 only eight states allowed cities and counties to lend corporations money for construction (compared to forty-five in 1995), only twenty states gave low-interest, tax-exempt revenue bond loans (compared to forty-four in 1995), and only twenty-one states gave corporate income tax exemptions (compared to thirty-six in 1995).[64] "Wealthfare" such as this can reach astronomical proportions at the local level, as shown by Minnesota's $828 million bailout of Northwest Airlines. While the company did promise to create 1,500 new jobs at two repair facilities ($558,000 subsidy per job), the expansion was delayed for many years after the deal was made.

Corporate welfare was not invented with the Republican takeover of Congress in 1994, although some might argue it achieved a new art form under their mentorship. The point that really needs to be recognized is that bias toward corporations and big business favoritism is a systemic feature of our capitalist society that has been with us for a long time. It goes unrecognized, despite the high cost of dubious projects of questionable value, because we have been socialized to not even see it at the same time we have been taught through endless vilification to hate people who receive AFDC or food-stamp assistance. Not being able to see the unfairness and injustice of corporate welfare is a learned distortion that can be remedied through education and objective information.

MASS MEDIA AND MANIPULATION

We can ask how these negative stereotypes, false ideas, and distortions emerged in the first place. To begin with, the public does not generally use the media it has access to. In many respects, this can be viewed as a chicken-or-egg dilemma: which came first—the public's disdain for in-depth news analyses or the media's concentrated offerings of fluff because they believe the public does not want serious content? At any rate, much has been written about the move away from reading newspapers that tend to have more thoughtful substance and objective examination of important political and economic issues. Broadcast media, especially television, tend toward sensationalistic and superficial treatment, avoiding big issues and societal problems in lieu of personal themes (murders, fires, crime victims, and the like). As the public makes more use of television as opposed to print journalism for its source of news, it can be argued that a "dumbing down" of America is taking place. According to General Social Survey data, adults watch an average of three hours of television per day. While this did not change substantially between 1975 and 1993, newspaper readership went down considerably. In 1975 nearly two of every three adults read a newspaper every day; by 1993 fewer than half did (46.0 percent). While the nation's top newspapers such as the *New York Times* or *Wall Street Journal* may have circulations of one to two million, the three network newscasts on CBS, ABC, and NBC reach more than 50 million people each night.[65] In a half hour of television news broadcasting, the twenty-two minutes that remain after commercials must be divided into fifteen to twenty stories. Thus, the origin of the "sound bite." It is literally impossible to offer more than a snippet in the ninety or so seconds given to each story—so a viewer is never made aware of complexities, nuances, shadings, ambiguities, or alternatives. Moreover, in keeping with visual impact (the bottom line is to attract and retain an audience), dramatic footage which frequently emphasizes violence has become the norm. The information content of broadcast journalism, in short, is almost nonexistent. When this is coupled with the well-known tendency of the broadcast media to cover political campaigns as horse races, where hoopla rather than differences on substantive issues is stressed, the danger of an ill-informed public becomes obvious. Indeed, even as early as the start of the 1970s nearly two-thirds of the American public were using television for much of their political information. Many of these viewers are not well-educated, and do not use any competing sources of information, such as magazines or newspapers.[66]

It should come as no surprise, then, that the average political IQ of the electorate is fairly low. A recent poll by the *Washington Post*, Kaiser Family Foundation, and Harvard University found that most Americans lack a basic knowledge of how our political system functions and who is in control:

> Two-thirds of those interviewed could not name the person who serves in the U.S. House of Representatives from their congressional district. Half

did not know whether their representative was a Republican or a Democrat. . . . Who's the vice president of the United States? Four in 10 Americans surveyed did not know, or got it wrong. Two out of three could not name the majority leader of the U.S. Senate (Robert I. Dole of Kansas, a Republican candidate for president). Nearly half—46 percent—did not know the name of the U.S. House of Representatives (Newt Gingrich, whom *Time* magazine crowned "Man of the Year" for 1995). . . . Nearly half did not know that the U.S. Supreme Court has the final responsibility for deciding whether a law is constitutional. Three out of four were unaware that U.S. senators are elected to serve six-year terms. Many don't know basic facts about the political parties. Four in 10 Americans were unaware that Republicans control both chambers of Congress.[67]

To dwell on the political and economic ignorance of the American public, however, is a little like blaming the victim. The very alienation spoken of earlier would drive many away from any interest in political developments. If the system is deemed irrelevant or incapable of meeting your needs, why pay any attention to it? But that is only part of the story. While people can be partially faulted for their lack of knowledge, can we also ask whether the ideological bias of the American mass media keeps them in ignorance? Ben Bagdikian, in his various editions of *The Media Monopoly*,[68] has argued persuasively that the growing concentration of 25,000 U.S. media outlets in the hands of just fifty corporations cannot help but introduce a probusiness, conservative bias into what we read, listen to, or watch. Even though only two companies (Tele-Communications Inc. and Time-Warner) control more than 40 percent of U.S. cable receivers, recent congressional bills threaten to remove most restraints on media monopolies.[69] When the news is filtered and distorted to reflect favorably on corporate interests, we are deprived of the essential information needed to maintain a vital society and a viable democracy. We are less likely, for example, to view programs on NBC that sharply question appropriations for defense or nuclear energy. General Electric, the tenth largest U.S. corporation and a major defense contractor, bought RCA (which owns NBC) in 1986.[70]

Martin A. Lee and Norman Solomon have offered a detailed description of how our mass media reflect a probusiness, high-income bias that denies the reality of poverty, gaps between rich and poor, and the very existence of social classes in American life.[71] When issues that have a macroeconomic base and are generally caused by preexisting inequality are examined, such as debates over the capital gains tax or raising the minimum wage, the media frequently offer explanations and simplifications that obscure wide economic divisions in our society. Not raising the minimum wage, for example, is framed as a way of fighting inflation or helping to build more jobs, rather than as perpetuating misery for the working class laboring for wages below a subsistence level. The tax giveaways to the very rich sponsored by Reaganomic policies during the 1980s are labelled "tax reform." The savings and loan scandal is written as if it were caused by a few uncontrolled, super-greedy bankers rather than by a systematic deregulation of the industry by a consortium of political and economic elites bent upon maximizing further profits. Unions are nearly always portrayed as corrupt, graft-ridden organizations interested mainly in featherbedding jobs and grabbing huge wages that end up costing consumers through higher prices. In fact, a *Los Angeles Times* survey found that 54 percent of newspaper editors said they generally took the side of business in a dispute, while only 7 percent regularly sided with labor.[72]

A great number of other studies corroborate the conservative, enhanced corporate image, big business favoritism of our mass media.[73] There are still occasional arguments about whether the media is liberal or conservative. Yet the leading organization which seeks to expose an alleged liberal bias (The Media Research Center) has been reduced to using ridicule and anecdotal evidence. Fairness and Accuracy in Media (FAIR) portrays the media as a bastion of conservatism where genuine criticism of the corporate status quo is essentially absent. Overall, mass media offer the public either mainstream pundits or right wing conservatives who love to scapegoat (such as Rush Limbaugh, Oliver North, G. Gordon Liddy, Robert Novak, etc). Almost always left out are progressive populists who challenge the probusiness, corporate agenda. Even the editor of the Heritage Foundation's *Policy Review* admits that today's op-ed pages are dominated by conservatives.[74] One of the best examples of corporate dominance in media punditry and news analysis can be seen in the twenty-year-old MacNeil-Lehrer (now Lehrer) *News Hour.* The show attracts 5 million viewers each night on three hundred Public Broadcasting System (PBS) TV stations. Although PBS was high on the hit-list for zero funding by the 1994 Republican majority, especially because of its so-called liberal bias, a FAIR content analysis study done of the *News Hour* shows over half a year found:

- The program's guest list was dominated by think-tank experts from such conservative, corporate-funded groups as the American Enterprise Institute.
- Virtually no representation was given to experts from progressive centers.
- Nine-tenths of guests were white and 87 percent were male.
- Almost half of the U.S. guests were current or former government officials.
- Only 6 percent of all guests were from public interest groups (labor, consumer-rights, civil-rights) critical of government policies.

- Two major corporations (Archer Daniels Midland and the New York Life Insurance Company) bankroll half of the program's total budget.[75]

Given the rise of talk radio in the past few years, with its consistent harsh rhetoric and vituperation of liberals, it is difficult to go on believing that our mass media has a left-wing bias. To begin with, the impact of talk radio should not be glibly dismissed. Its audience is surprisingly massive. Half of the American public listens to talk radio on a relatively frequent basis, with one in six listening regularly. According to an in-depth study by the Times Mirror Center for the People and the Press, both the audiences for these shows and a survey of 112 talk show hosts in the major markets revealed deep-seated conservatism.[76] Republicans and self-rated conservatives were twice as likely as Democrats and self-rated liberals to be regular listeners. Not surprisingly, listeners were far more critical of Clinton's job performance as president, his economic plan, Congress (dominated by Democrats at the time of the research), and a federal government they believed was overly intrusive in private life.

The talk jocks are typically white, male, middle-aged, and ultra-conservative. They are nearly always harshly condemnatory of Democrats, liberals, government, and any policies that might assist minorities, women, and/or the economically disadvantaged. Their usual tactics of disagreement are intense negativism, wholesale sarcasm, ridicule, name-calling, scapegoating, and the most vulgar stereotyping conceivable, whereby a feminist becomes a "femi-Nazi" or an environmentalist is an "eco-terrorist." They are loud and angry, speaking with bitter voices that frequently spew raw hatred. Their effect has been so pervasive that in the aftermath of the Oklahoma City bombing, President Clinton was moved to warn that their "sole goal seems to be to try to keep some people as paranoid as possible and the rest of us all torn up and upset with each other."[77] The talkmeisters themselves are eager to admit they have the ability to get people angry while happily stirring the pot of disgust with government. And they are proud to claim a large share in the Republican victories of 1994, accepting at least partial credit for shaping public opinion that led to the GOP victories.[78] (By the 1996 elections, their influence became less powerful as a populist, negative reaction set in with the public's view of the GOP agenda.)

The king of talk radio, of course, is Rush Limbaugh. His opinion reach is massive, broadcast to 20 million listeners over 650 radio and 250 TV stations around the country:

> Limbaugh instructs and guides his listeners into the Republican establishment camp, educating them to the new realities of congressional power.

The combination of Limbaugh's loyal audience and his praise of the GOP leadership has made him irresistible to candidates peddling an idea or running for office. As Republican political strategist Eddie Mahe puts it: "Limbaugh on any given day will have significantly more influence than any of the networks. The audience is huge, and they are paying attention." As Dole's press secretary Nelson Warfield puts it, Limbaugh serves as "the telegraph line of the conservative movement."[79]

The seventy-three Republican freshmen elected to Congress in 1994 attributed their sweeping victories to Rush Limbaugh, citing polls that showed heavy listeners of talk shows (ten or more hours per week) voted Republican by a margin of three to one.[80] These same freshmen anointed Limbaugh the "majority-maker" and inducted him as an honorary member of the 104th Congress. Although right-wing, demagogic radio broadcasting can be traced as far back as the 1930s to extremism fostered by the Great Depression, the "nonguested confrontation" that Limbaugh and his clones use is entirely new. In this format, the host can pontificate, lecture, joke, bluster, and intimidate to his heart's content without fear of contradiction or counterargument. Although there are no guests, a few heavily screened callers may be allowed to question the opinion. But the cutoff switch is always in the hands of the talk jock if things get out of hand.

The one-way flow of information without retribution has produced two major consequences for Limbaugh's shows: cruel, offensive humor and factual distortion that some critics unhesitatingly label outright lies. Limbaugh is famed for his jokes about the homeless, gays, and minorities. Texas columnist Molly Ivins objects:

> Because he consistently targets dead people, little girls, and the homeless—none of whom are in a particularly good position to answer back. Satire is a weapon. . . . When you use satire against powerless people, as Limbaugh does, it is not only cruel, it's profoundly vulgar. It is like kicking a cripple. On his TV show, early in the Clinton administration, Limbaugh put up a picture of Socks, the White House cat, and asked, "Did you know there's a White House dog?" Then he put up a picture of Chelsea Clinton, who was 13 years old at the time and as far as I know had never done any harm to anyone. . . . On another occasion, Limbaugh put up a picture of Labor Secretary Robert Reich that showed him from the forehead up, as though that were all the camera could get. Reich is indeed a very short man as a result of a bone disease he had as a child. Somehow the effect of bone disease in children has never struck me as an appropriate topic for humor.[81]

Fairness and Accuracy in Reporting (FAIR) has also compiled a list of over one hundred "false and foolish statements" by Limbaugh that they claim could easily have been one thousand. The researchers at FAIR are especially disturbed that Limbaugh "does lie and does make things up to advance his cause" because millions of people believe him.[82] Aside from his clout with voters:

> Limbaugh's chronic inaccuracy, and his lack of accountability, wouldn't be such a problem if he were just an obnoxious entertainer, like Howard Stern. But Limbaugh is taken seriously by "serious" media: In addition to repeat "Nightline" appearances, he's been an "expert" on such chat shows as "Today," "MacNeil/Lehrer NewsHour," "Charlie Rose," and "This Week with David Brinkley." The *New York Times, Los Angeles Times,* and *Newsweek* have published his columns. . . . Limbaugh is also taken seriously as a political figure. . . . A cover story in the conservative *National Review* (9/6/93) declared Limbaugh "the leader of the opposition." Ronald Reagan called him "the number one voice for conservatism in our country." Columnist George Will referred to him as "the fourth branch of government." Former Secretary of Education William Bennett went beyond politics, describing Limbaugh as "possibly our greatest living American."[83]

What types of "facts" advanced by Limbaugh are untrue? Without going into great detail, a number concern major themes in this book. For example, "The poorest people in America are better off than the mainstream families of Europe," is manifestly untrue. Limbaugh claims that liberals have deceived the public into believing that there was a bigger gap between the haves and the have-nots at the end of the 1980s than at the start. He even claims that Congressional Budget Office data dispel this myth, which is false. Data from the CBO consistently support the growth of income inequality. In speaking of the homeless, he claims that the Reagan administration did not cut the budget for public housing, and that the construction of public housing actually increased during the Reagan years. In point of fact, public housing units under construction actually declined by more than half between 1980 and 1988, while the HUD budget was slashed from $3.7 billion to $573 million.[84]

WHAT IS TO BE DONE?

By now it is obvious that a great number of things are wrong with the American system of distributive justice. Despite the growth of inequality and unfairness, mainstream politicans and media monopolies have steadfastly refused to register concern for the core problems we all face. In the light of what one might see as insurmountable forces, where does an average person begin? The situation is not hopeless. First of all, it is important to remember that no trend will go on forever. This thought can help us to at least open the door to positive change. There are also a number of dif-

ferent actions we can take to make a difference. We can prevent the further erosion of real income and the growth of income inequality by supporting certain strategies.

A variety of journals and magazines can help detach us from the
falsehoods and hate of our talk-radio culture while offering a refreshing
and healthy substitute to big business media. For an exhaustive list, there
is no substitute for Michael Parenti's guide to alternative media in his
book *Inventing Reality*.[85] This list is an excellent first step in getting
acquainted with the candid, honest, and independent media that exist in
the United States. These news sources are not beholden to large business
interests nor are they greatly dependent upon mass advertising revenues—both of which can influence what stories are covered and how
they are written. Such periodicals as *Mother Jones, The Multinational
Monitor, The American Prospect, The Nation, The Progressive,* and *Z
Magazine* are famous for in-depth analyses and investigative journalism.
Another welcome alternative to official government pronouncements on
economic trends, one that tends to question the political agenda of who-
ever is in power, is *Dollars & Sense*.[86] This journal focuses on economic
issues and attempts to describe them in a jargon-free manner that is simple to understand. It also emphasizes economic justice, the fight against
inequality, and problems with fairness in the distribution of income.

A large segment of the solution is by now quite obvious. Our problems must be addressed in a political context. But meaningful change cannot occur simply by substituting Democrats for Republicans. Both parties
have consistently refused to deal with concerns that strike to the very
heart of the American middle class—economic security. While some
issues that receive attention by politicians and the media perhaps nip at
the heels of our problems, such as tax policies or Social Security/Medicare
issues, most of our difficulties are never confronted head-on. The mainstream media are reluctant to talk about the harshness of unfettered
capitalism, although more stories about CEO greed and corporate downsizings are now beginning to emerge. Aside from a few lone politicians,
such as Ross Perot and his stand against including Mexico under NAFTA
or Patrick Buchanan and his criticism of American corporations for letting greed outweigh what is good for the country as a whole, there is not
much to choose from. It seems, however, as if the country is ripe for a
third party movement. Should it occur, voters may get to exercise more
choice than they currently have.

In the meantime, we can inform ourselves as best we can, be as
politically active as possible, and closely scrutinize candidates' views on
important issues. There should be a number of items on our checklist
(recognizing full well that we will never get everything we wish for). Any
progressive candidate should be in favor of intensifying taxes on corpora-

tions, on those with high incomes, on the very rich and the super-wealthy, who have benefitted enormously from the so-called tax reforms of the 1980s. We should oppose any flat tax, which would undeniably worsen the burden of the middle class; any reduction in the capital-gains tax, which again benefits the wealthy; and any effort to cut the Earned-Income Tax Credit that is currently helping the working poor.

There are numerous reasons to oppose any cut in the capital gains tax. An analysis by the Congressional Joint Committee on Taxation found that 80 percent of the benefit of this cut would go to persons who earn over $100,000 per year.[87] The idea behind a cut in the capital gains tax is to encourage domestic investment. The concept, the twin brother of supply-side economics, is that cutting taxes for higher income people would create more jobs and industries through greater investment. Without any mechanism for accountability, however, this would be another in a long line of handouts to the very wealthy in our society. It would be equivalent to giving extra recess time to F students while forcing the A students to stay after school. Why continue to support an idea that failed in the 1980s?

Most Americans are tragically unaware of what would really take place with more tax reform. For example, the tax cuts passed in the Reagan administration figure heavily in today's federal deficit. One estimate puts the drop in federal revenues at $1.8 trillion over the 1982–1990 period.[88] For politicians, opposing tax cuts is like refusing to kiss babies. On the surface it is tantamount to being against mom and apple pie, though in reality the roll back in taxes has damaged the economy. The promised buildup of our infrastructure never followed, but a tide of red ink did. The devastation of tax cuts was even worse for the poor. In spite of the 1986 Tax Reform Act (or perhaps because of it), the poor have had to pay more out of their own pockets. A Congressional Budget Office study showed that the poorest 10 percent of our population was paying 20 percent more in taxes at the end of the 1980s than a decade earlier.[89] The CBO study included the rise in gasoline, alcohol, and tobacco taxes along with the real rise in state income taxes caused by cuts in federal social expenditures. The reforms have all but destroyed progressive income tax rates, paring them to only 15 percent and 28 percent. What this all added up to for the richest tenth of America's population is a 16 percent decline in their effective tax burdens over a twenty year period.[90] Despite the hoopla of so-called tax reform surrounding the 1980s' tax legislation, the new laws were revenue-neutral. This means that no new tax money came out of all the changes.

In the end, most Reagan tax-cut policies remain in effect today. Such is the backdrop against which we evaluate the call for a further roll-back in the capital gains tax, additional exemptions for corporate taxes, a flat tax whereby all pay the same percentage, or a cancelling of the

Earned-Income Tax Credit. This is being sold as a means to get our country back on the productivity track again, so the revenue saved will be reinvested in new plants and updated technology. The litany is familiar, and the results will be as well. Tax savings failed to be used to buy new plants and equipment in the 1980s. Why should we expect anything different in the present? Since the rich have won such disproportionate benefits from prior tax cuts, it would be refreshing for a change to advocate well-deserved *tax increases* for this group.[91] Many well-planned schemes never see the light of day in today's Congress. Ross Perot advocates taxing Social Security benefits of the well-off, raising tobacco taxes, reducing the home mortgage deduction to no more than $250,000, and raising the gasoline tax. The Concord Coalition's plan advanced by Paul Tsongas and Warren Rudman essentially echoes Perot in calling for such increases.[92]

A predictable outcome of the tax-cut giveaway to the wealthy during the 1980s was a crippled budget-balancing mechanism for the 1990s. The revenue was no longer there, but expenditures were. In short, we have inherited a manufactured crisis of government insolvency that has almost nothing to do with how the nation spends its money. Nonetheless, since higher taxes are taboo as an item of discussion, this means budgets must be cut. Again, how a political candidate approaches this question should serve as a litmus test for our support. Nearly everyone regardless of political persuasion believes that changes have to be made. Conservative thought, however, continues to blame the victim. Americans are accused of being fat and lazy, of having grandiose wage expectations, of being unproductive, of spending too much (especially on imports), and of not saving enough. What we need, according to some critics, is to get "lean and mean" again so we can compete effectively in the global economy. We need *austerity*. Does this sound familiar?

Advocates of a new austerity for the American public have now set their sights on middle-class entitlements such as Social Security and Medicare since they have nearly exhausted any further cuts in AFDC and welfare transfer programs. The fashionable targets of conservative politicians are now the allegedly "bloated budgets" connected with civil service retirement, farm subsidies, student loans, Social Security, and Medicare. The huge run-up in defense spending over the past decade has been conveniently forgotten. Jeff Faux, in a devastating critique of the call for austerity at home, points out that the total government outlay for entitlement programs actually dropped over eight years in the the 1980s while military spending rose by one-fourth.[93] Suggestions for Social Security reform have ranged from setting cost-of-living adjustments (COLAs) lower than the inflation rate to raising the age of retirement. Faux points out, however, that almost 60 percent of the elderly now survive on less than $10,000 a year—most of which comes from Social Security. There is the added benefit of a rapid buildup in the Social Security fund because of the recent overhaul of the system. It was supposed to amass money quickly to pay for the retirement of rapidly aging baby boomers. The plan has worked well, yet the money set aside for retirement for those who are also currently paying the added tax is being circled by financial vultures. While our national savings rate is anemic, this represents one area that can be used for future investment. The pity is that even now the "surplus" funds are being used to buy U.S. Treasury notes to subsidize the federal deficit.

Of course, the priorities we set for the federal budget are of major importance. And savings must be instituted because we do have a very real shortfall. But how this should be done is the question. Conservatives would cut Social Security and Medicare to the bone while giving tax cuts to the rich, all the while increasing corporate welfare and military spending. This script does not have to be followed. The politicians we support should offer fresh alternatives that deal with the actual illness rather than with its symptoms. NAFTA and GATT should be revisited, and the wisdom of free trade needs to be questioned and openly discussed. While all of these forces may benefit multinational corporations, they actually work against most Americans and Third World countries caught up in the web of the global economy.

A few lawmakers who try to curb corporate welfare are starting to emerge. We should encourage them and persuade their colleagues to go along with such efforts. A great deal can be done at the state and local levels that is more difficult to implement at the national level. The Minnesota House and Senate passed a bill in 1996 requiring any corporation receiving subsidies from state and local governments (grants, loans, tax increment financing) to pay at least a poverty level wage ($7.28), which is more than the minimum wage.[94] While Republican Governor Arne Carlson vetoed this bill once it reached his desk, the writing is on the wall. Enough Republicans in the state legislature joined this effort to ensure its passage through both houses, and the governor himself admitted he might accept a more watered-down version (although none was forthcoming). In the meantime, national attention from the *New York Times* and two major television networks added an impact to this legislation, which advanced further than any other statewide initiative.

The National Lawyers Guild of Los Angeles has also developed model laws for state and local governments that advocates can use to muzzle corporate welfare abuse. In their publication, *Getting Business Off the Public Dole*, they present examples of acts which seek: to prevent luring business through taxes and other incentives; to assure that business pays its fair share of public infrastructure improvements; to require business to pay for its destruction of natural resources; and to require business to assume financial responsibility for its own meals, private club dues, and lobbying expenses.[95]

Another initiative brought before the U.S. House by Representative Martin Olav Sabo (D-Minnesota) was HR 3278, known as the Income Equity Act.[96] We all know that the minimum wage needed to be raised. While Sabo's bill proposed to do exactly this, bringing the level up to $6.50 from $4.25 per hour, it went a step further. It linked the tax deductibility of CEO salaries to the pay of other workers in the same business. This law, once in effect, would deny tax deductions for executive salaries that are more than twenty-five times the pay of the company's lowest paid worker. As an example, if a night cleaner earned $10,000 a year any amount above $250,000 of an executive's salary could not be deducted as a business expense. In essence, this law would create an incentive for private sector employers to take a close look at how inequitable their salary structure really is. But it also would take government out of the business of subsidizing such inequities through tax breaks. Although the bill never passed, it serves to indicate which direction political winds are now blowing.

Secretary of Labor Robert Reich advocated tax breaks for corporations that do the right thing by helping their workers learn new skills in an effort to prevent downsizing, or for firms that undertake retraining workers they have laid off. While stopping short of embracing this idea, President Clinton called on corporate America to exercise a higher standard of citizenship than just immediate profit and bottom-line consideration.[97] He pushed Congress to pass legislation that raised the minimum wage, provide federal vouchers for worker retraining, and make health insurance more affordable and accessible to the large minority of Americans who do not presently have it. The minimum wage was raised at the end of the 1996 congressional session. Congress also improved the adequacy of health insurance somewhat by increasing its portability. Workers who are laid off or who leave their jobs can elect to maintain their health insurance package.

Republican Patrick Buchanan, who aspired to be his party's presidential nominee, opened the proverbial can of worms. His call for less open immigration and steep tariffs on imported goods to protect U.S. jobs resonated with many working Americans. Nor did he hesitate to bash corporations and the CEOs who seek a quick jump in a company's stock price by laying off workers—regardless of healthy profits.[98] Buchanan's early success in the primaries and the public opinion surveys that showed support for some of his suggested remedies indicate that the issues of economic justice and job security will not simply vanish. It has finally become a political issue that could galvanize the American electorate. Get government off our backs? Not when it comes to the possibility of your job disappearing, in many cases because of unneeded downsizing. Nearly two out of three Americans believe Congress can do something about layoffs and the loss of jobs, while 57 percent of *Republican* voters believe that restrictions on free trade are needed to protect domestic industries.[99]

No analysis relating the woes of the American economy should fail to mention how well the country's best interests could be served by finally bringing the federal budget back into balance. This should be a basic requirement for the political support of voters. The loss caused by the debt hole our country has dug itself into is staggering. The interest alone can stop us from doing what we must to protect our standard of living. The longer it goes on, the more damaging it is to our real income. The funds so badly needed to make us productive once again are flowing to foreign banks and lenders, as well as to the ultra-rich within our own country. We need to stop the loss of our personal income, which is partially caused by this deficit spending.

Many good ideas for solving this problem never receive a fair hearing by those in power. A federal budget that is carefully crafted to reflect the priorities of workers and citizens over the interests of corporations and the rich would go a long way toward alleviating economic distress. The World Policy Institute has come up with a detailed plan to do just this, while addressing nearly all of the issues discussed in this book.[100] Their suggested program would allocate $215 billion per year in new public investment for the next ten years, while military spending would be reduced to $125 billion annually. Among the domestic initiatives funded on a yearly basis would be $44 billion for new investment in civilian R&D, $30 billion for better education, and $22 billion for the modernizing of our infrastructure. Other domestic areas addressed would be funds needed to clean up nuclear weapons plant sites and toxic waste dumps, a doubling of our committment to build public housing, renewed Head Start and other social programs aimed to help poor children, and law enforcement/prevention programs aimed at drug abuse. In essence, their suggested budget concerns itself with actual people, with real problems, and with genuine needs that it attempts to meet. The Congressional Black Caucus also has zeroed in on a bloated defense budget in need of a crash diet, suggesting that cuts for obsolete weapons contracts and questionable hardware would save $60 billion per year.[101] The Caucus believes that considerably more savings could be found if Congress changed the foreign tax credit for multinational corporations to a deduction ($71 billion), reformed taxation of multinationals ($86 billion), eliminated favorable tax treatment of capital gains ($67.4 billion), and ended special interest decuctions ($142.5 billion).

Voters should oppose any politician or political party that supports either maintaining or increasing Pentagon funding. What should be at the top of any agenda to get America back on its economic feet are hefty cuts in the defense budget. Much of this spending has been sheer waste with absolutely no payoff in protecting our nation's vital economic productivity. Rather, huge military budgets year after year have been largely responsible for American decline. It is becoming very apparent as the century draws to a close that a nation's strength can no longer be measured in military hardware and stockpiled arms. On the contrary, countries with flourishing export markets—healthy industries manufacturing high quality products at a competitive price—are now receiving respect and deference.

The sheer waste of resources is stupefying. Although $1 billion spent for guided missiles creates 21,000 jobs, over 71,000 jobs would result if that same amount were spent on education.[102] The vague communist threat that was used in the past to justify these gargantuan purchases no longer exists. The former Soviet Union has collapsed and is no longer a menace. An unrealistically high military budget can actually be more dangerous to a country's well-being in the long run. Burdensome defense expenditures have wreaked havoc upon the economies of both Russia and America. In Russia until recently, one-seventh of the nation's resources went to the military—leaving economic growth and technological advancement virtually stagnant for decades.[103] While Russia and the United States were locked in mortal combat slamming shut "windows of vulnerability" and closing alleged missile gaps, Japan quietly walked away with all the financial marbles for two decades. Since Japan spends less than 1 percent of its GNP on defense (while the United States and Russia spent 7 percent and 14 percent, respectively, during the past decade), it could afford to invest in state-of-the art factories. In the United States, on the other hand, over $8.4 trillion was drained off by the military in the quarter century after World War II—more than the total money value of all human-made wealth in our country.[104] Between the Korean War and the collapse of the Berlin Wall, America spent over $10 trillion on defense (about 8 percent of our GDP) to maintain military readiness.[105] Yet today, two-thirds of our tangible civilian wealth (which includes factories and machinery, bridges and dams, railroads and airports, etc.) is so worn out that it needs replacing. It could easily have been renewed were it not for the loss of so much money to the military.

By the start of the 1990s, the Soviet Union had finally learned a lesson about how destructive the insatiable appetite of its generals could be. It simply chose to opt out of the escalating arms race spiral by actively demilitarizing much of its economy. Our own economy would be far healthier if we were to do so as well, despite the resistance of the defense industry and Pentagon brass who see their own suns setting. Incredibly, our nation's response to this unilateral arms cutback has produced no equivalent reduction in the U.S. budget for defense spending. It is true that there has been some reduction in military spending since Reagan's presidency. During the 1980s, military appropriations peaked at $355 billion per year and were $294 billion by 1993, after adjusting for inflation.[106] This is a 17 percent cut, which is minimal considering the current budget reductions for social, health, welfare, and educational expenditures. The slight decline should also be put in a different comparative context. There is no doubt that the former Reagan administration went overboard on increasing military funding, expanding the Pentagon's allocation by 50 percent between 1980 and 1985. A better comparison would consider the defense budget as it has stood for most of the Cold War. Even with the so-called 17 percent cut, the United States today is spending more on defense than it did in 1955 or 1975 or 1980 or for most years of the Cold War with the exception of the Vietnam War and the Reagan peak.[107]

While Russia and our allies have cut their military budgets to the bone since the Cold War ended, the United States has perversely maintained its exorbitant spending level. Today, administration and congressional budget plans call for an outlay of $1.6 trillion over the next six years. The Republican majority in Congress plans to increase the Pentagon's annual budget to $265 billion, $7.5 billion more than the generals requested. The *combined* military budgets of Iran, Iraq, North Korea, Libya, Syria, and Cuba—the most frequently used case studies in likely war scenarios simulated by the Pentagon—is less than 6 percent of the U.S. defense budget.[108] The United States essentially spends more on the military in three weeks than the combined total of what these six "enemy" nations spend in a year. Currently, America's military budget is nearly as large as the military budgets of all the other countries in the world combined.[109] The rate of our military spending is almost beyond comprehension. Take your pick: $5 billion is spent every week, $700 million every day, $500,000 every minute, or $8,000 every second. The bill for this insatiability comes to $4,000 per year for the typical American family.

How much is an adequate defense budget? Opinions vary greatly, of course, but there has emerged an interesting consensus along the political spectrum from far left to far right—however you measure it, today's current level is too high. The smallest cutback is recommended by the General Accounting Office (GAO), which believes $25 billion could be saved over five years simply by eliminating unecessary weapons systems.[110] Indeed, this is a perennial sore point with the Department of Defense, since pork barrel politics continues to foist projects and products upon the military that it does not want or need (V22 assault planes, more B2 bombers, Seawolf attack submarines, etc.). The *Washington Post* in 1995 rightly asked:

How does it happen that the world's only military superpower—which already spends about five times as much on defense as its closest competitor and almost as much as the rest of the world combined—takes defense off the table while planning draconian cuts in health and welfare to deal with its deficit problem?

- First, given his avoidance of military service and lack of foreign policy experience, Clinton may feel he is in no position to stand up to the military.
- The Clinton administration and the Democratic Party remain fearful about the Republicans playing the readiness card again.
- Politicians of both parties see the defense budget as a jobs program. No longer does the defense debate take place between hawks and doves, but between those who have defense facilities in their districts and those who don't.
- The military has succeeded in inflating the threat from our foes and downplaying the contributions of our allies.
- Unlike most businesses and other federal agencies, the Pentagon has not yet fully re-engineered itself. It still keeps several hundred thousand troops regularly deployed around the world as it did when it was containing the Soviet threat.
- The debate on defense spending has been characterized by a number of misleading indicators. Supporters of increased defense spending compare today's military to that of Reagan or Bush.[111]

How much could actually be cut while still maintaining a strong but realistic defense? From the conservative right, the Cato Institute believes military spending could be reduced to $205 billion and then gradually to $140 billion per year by the turn of the cenutry. Former CIA chief William Colby believes the military budget can easily be cut in half from Cold War levels with no significant weakening of our national security.[112] The Brookings Institution, the Program in Science and Technology for International Security at MIT, and Reagan Defense Planner Lawrence Korb all assert that $60 billion could be cut in just a few years.[113] One of the most respected and knowledgeable sources on military budget issues is the Center for Defense Information (CDI). This organization is made up primarily of high-ranking ex-officers who have retired from the armed forces after long service to our nation. Their patriotism and concern for the welfare of our country is beyond question. According to the CDI's analysis, the United States can reduce its annual military expenditure to $175 billion without endangering national security.[114] This would save half a trillion dollars ($500 billion) over the highly inflated figures the president and Congress say we need, put a major dent in our federal deficit, and prevent further painful cuts in health, education, retirement, and social assistance.

If there is need for more persuasion, research now indicates that higher military spending also leads to greater relative income inequality.[115] This remains true even when other variables that could cause rising Gini ratios and an increase in the gap between the highest and lowest income quintiles are held constant—such as taxes, interest rates, economic growth rates, inflation, and nonmilitary spending. The very nature of the military economy gives rise to highly complex, technical, and scientific occupations relatively closed to women and minorities. This effect is amplified by civilian jobs forgone in favor of the military focus. For example, every $1 billion increase in the Pentagon budget means that 1,300 African-American jobs disappear. In the end, increases in military spending are associated with a widening gap between the rich and poor.

Because of the enormous bills for our military might, American leadership is crumbling. As a nation we seem blinded to reality, preferring to see enemies who no longer exist. Our allies are shaken by our stubborn obsession and dismayed by our arrogant, go-it-alone attitude in world affairs. Most important, they are mystified by our apparent willingness to destroy our own economy through fear of a nonexistent ghost in the closet:

> In particular, militarism is a central aspect of contemporary American government policy. It provides profits and some jobs, as well as channeling frustrations and insecurities against outside enemies. The *Rambo* phenomenon symptomizes the high level of xenophobia and the continuing love affair with empire that characterizes contemporary America. . . . The justifications for increased American military spending are unclear or fictive except in the contexts of "power projection" and economic stimulation. The sorry saga of Soviet defeat and humiliation in every aspect of its foreign policy since 1948—the loss of China as an ally, the open resistance to Soviet domination in Poland, the defection of Egypt and much of the Arab world, the economic bankruptcy of most of its client states such as Vietnam and Cuba—let alone its own internal woes, offers little justification for the portrayal of the Soviet Union as a successful superpower. Indeed, one can make little sense of the arguments of either Soviet or American leaders except as cynical attempts to hold or recruit allies and provide a defensive screen behind which other imperatives—economic and psychological—are operative.[116]

Ironically, it was one of the greatest American military heroes of this century who warned our nation against the damaging excesses of the military/industrial complex. Dwight D. Eisenhower, two-term Republican president of the United States and Supreme Allied Commander during World War II, delivered an important message to Americans in his Presidential Farewell Address to the nation in 1960. He told of his fear of the danger coming from within, of his concerns that the military needs of our country were overstated as a result of a cozy relationship between the Pentagon and the powerful defense-industry establishment. Each fed the other's power in an escalating dependency. President Eisenhower was fond of pointing out that every gun made or warship launched was a theft from those without food and shelter. For the most part, the indefensible waste of defense funds has left the average American a great deal poorer.

Also beware of politicians who are constantly harping that our tax structure is too high or that we need more tax reforms. One of the most odious proposals, introduced by Republicans in both houses of Congress, is the *flat tax*. The claim is made that if we all pay the same rate, most often cited as 17 percent of our income, life would be more simple, complexity would be removed from the tax codes, and the system would be fairer. In reality, however, the various plans have hidden costs. Income from interest, capital gains, and stock dividends would not be taxed at all—which would completely remove most of the wealthy from much if not all of their current tax obligations. Representative Dick Armey's 1995 plan would have resulted in an annual revenue shortage of $186 billion per year, and taxes would have gone up for just about every family earning less than $100,000 per year.[117] For families making $500,000 or more, however, taxes owed would have been cut by anywhere between $78,000 and $93,000. Millionaire Steven Forbes, contender for the presidential nomination of the Republican Party in 1996, is one of the most outspoken advocates of a flat tax. This is understandable, since he would see his current tax bill cut from $300,000 to $100,000 while the rest of us would end up paying more.[118] A rate of 21 percent would be required for a flat tax to equal current tax revenues, and this is with no deductions whatsoever. Even if only the mortgage deduction were kept, this would push the rate as high as 25 percent.

What is worse, virtually all progressivity in our current tax structure would vanish overnight. Those with a greater ability to pay simply *should* pay more—which has been the American tradition since well before World War II. If anything, the wealthy ought to pay more in taxes because they have been the ones to benefit from the tax giveaway which began in 1980. Some change could be made quite painlessly, for instance, by substituting inheritance taxes for estate taxes upon death for those with an average net worth of $600,000 or more (this 1 percent of our population controls one-third of all U.S. wealth).[119] Additional tax brackets for high income taxpayers could be restored to the tax structure. Financial transfers could be taxed at 5 percent to slow the destructive merger mania. Most important, tax loopholes which encourage U.S. corporations to relocate overseas can and should be removed.

The proposed halving of the capital gains tax is a remnant of the Reagan/Bush era that simply refuses to die. If instituted, this would cost the nation's treasury $45 to $57 billion over five years.[120] According to estimates, more than two-thirds of the benefit would go to the richest 1 percent of our population, while only 6 percent of the benefit would descend to the bottom 60 percent of Americans. Such blatant favoritism exhibited toward the rich should be resisted.

Much can still be done to make our tax laws more fair, especially to middle-income and poor people. As the Social Security (FICA) tax is currently set up, the rate of 7.65 percent applies only to earnings up to $60,600. This means the rich have been given an immense loophole. While the majority of us have to pay this tax on all or most of our income, the wealthy do not pay this tax at all on the great bulk of their income. Simply removing the ceiling on the Social Security tax would add literally billions of dollars per year to the national budget.

Thus, we can and should raise taxes, but only for that segment of our population which has been so lavishly enriched by the biased Reagonomic, prorich policies of the last decade that have been resurrected in Congress. This seems only just, given the rapid rise of relative and absolute income inequality that has taken place in the past two decades. Since the vested interests of the wealthy are at stake, expect resistance. There will no doubt be a great protest from conservative forces who might have to forgo a portion of the questionable gains derived from the status quo. Yet a budget with less defense spending and a tax plan that goes after the very rich at least has the courage to address the burning issues facing our nation. We cannot go on with business as usual, which is virtually destroying our country. These policies have the merit of pinpointing where the real fat in the budget is. The plan is also honest in admitting that America must brace itself for more taxes to pay for the excesses of the past. In the last analysis, such political priorities would set a course toward a sustainable economy for our country.

Electing the right people to office can go a long way in confronting our difficulties. Our leaders should not be afraid to try new programs and ideas. Big government is not the problem. To begin with, a huge portion of government in modern times has stemmed from pumped-up defense budgets. Other costly programs are quite popular with voters, such as Social Security and Medicare. It is also true that an activist and humane government can protect the weak and vulnerable, the poor and homeless, while limiting the destructive power of the very rich. Ask yourself who benefits the next time you hear a conservative politican or radio talk-show host complain about the excesses of big government—and above all, check to see if the complaints are true. Such horror stories are often distorted, half-baked caricatures of the truth put out by big business interests that want deregulation in order to enhance their profit margin. Molly Ivins points out that among the worst abusers are the pharmaceutical companies, which trash the U.S. Food and Drug Administration, and the National Rifle Association (NRA), which vilifies the Bureau of Alcohol, Tobacco, and Firearms. Timber, mining, and ranching interest have all coalesced to criticize environmental regulations and programs. Businesses now want to limit product liability lawsuits, so we are told about the McDonald's customer who won a $3 million settlement for coffee-spill burns. We are not told that she had severe second- and third-degree burns requiring skin grafts, that she tried to settle for $20,000 (the cost of her medical bill), and that seven hundred other burn claims were also pending. These anecdotes sometimes reach the height of absurdity; for example, Representative David McIntosh solemnly told the House that the Occupational Safety and Health Administration (OSHA) has mandated that all buckets be made with a hole in the bottom so children will not drown in them.[121]

One can sympathize with the view that it is difficult to get anything done about income inequality through the political system, especially given the countervailing thrust of the 1994 Republican takeover of Congress and the conservative bias of our media. But it is not impossible, despite pitfalls and hardships. S. M. Miller and Jaqueline Ortiz offer several effective techniques to use in pursuing such a path, based upon the social scientific knowledge we now have.[122] They believe four questions need to be stressed in presenting the case for a more equitable distribution: (1) Is there really equality of *opportunity* in the U.S. today? (2) Who gets hurt by increasing impoverishment? (3) Do those at the top deserve their increasing wealth? (4) Is the overall economy weakend by the growth of inequality? A good case can be made by highlighting the dangers present in all four themes, from providing a fairer distribution of the economic pie to preventing violence, crime, racism, and social deterioration within our country. Among specific issues the authors believe should be addressed are the decline of democracy itself (as the poor and alienated drop out of a system that does not meet their needs) and the corrupting power of PACS and big money. Campaign finance reform is a virtual necessity. Certain themes resonate with the public and may be used to build support for redistribution—changing Social Security taxation, increasing the number of beneficiaries (more middle-class relief), emphasizing help for the working poor, targeting sympathetic voter blocs (such as women, Latinos, and African Americans), promoting high wage/high employment programs, and underscoring the growth of the entire economy as a result of greater income equality. Above all, Miller and Ortiz believe the moral issue can be used effectively as well, since Americans do have a deep-seated belief that the distribution of income should be *fair* to all. By continually calling attention to the growth of inequities, educating the public and politically potent organizations with factual information, and focusing on a few important areas in which change is feasible, it is possible to significantly challenge the trend toward more income inequality. There is a limited amount of support for egalitarian attitudes among leaders in several different spheres of influence (business, liberal and conservative political parties, media, intellectuals, minority groups, etc). Such leaders favor policies that include reducing the absolute income of top earners and raising the absolute income of those at the bottom of the scale.[123] In the end, change really is possible.

In order to be politically effective, however, we need to confront our own dark side and those of others on a personal level. All too frequently, negative stereotypes and group jealousies hinder coalition building. This weakens the political pressures necessary to bring about fundamental change. Such bias is particularly malevolent as it involves issues of sexism and racism.

One such damaging prejudice is a rigid feminist view that refuses to see male job loss and poverty as a problem. There is certainly a great deal of truth in the concept of the feminization of poverty and of the exploitation of women in the labor force. It is undeniable that deindustrialization was not defined as a problem until white males started to lose their highly paid manufacturing jobs in record numbers. Sexism is an ugly fact of life that we should all work to eradicate. Yet this should not blind us to the fact that men and women are together in the same rapidly sinking boat. Families suffer equally, whether from American plant shutdowns that throw men out of work or from the traditional low wages for women and African Americans.

At bottom, the destruction of the social wage and the slide in real income is all part of the same process brought on by a system that takes advantage of any chink in the armor of worker unity. It is a system that has effectively pitted African Americans against whites, men against women, the native-born against recent immigrants, residents of the sunbelt against workers in the frostbelt, and so on. The constant, ever-present goal has been to pay the lowest wage possible. To attain this objective, our prejudices, fears, and biases have been set against our own best interests by unscrupulous firms. When we come to define each other as the enemy, as the cause of loss of income, we fail to grasp who really benefits from our declining standard of living.

One of the most distressing tendencies that accompanies a lowered standard of living is racial bigotry. America is now witnessing an epidemic of right-wing paramilitary Fascist groups basing their existence on hatred of African Americans and Jews. In particular, skinheads (so called for their closely shaven heads), young adult and teenage whites, are becoming a major force in our large cities.[124] These angry young men, incapable of earning a decent living because of poor education and lack of jobs that pay a living wage, are taking out their frustrations on innocent African Americans and immigrants. Low-scale warfare has broken out in our inner cities. Gangs of skinheads have been known to roam the streets with chains and baseball bats, falling on their defenseless prey with relish. Such victimization invites retaliation from African-American ghetto youth, who have attacked their white targets with the same ferocity.

In the meantime, in places like debt-ridden Louisiana where a deep and prolonged recession has slashed personal income to the bone, the Ku Klux Klan has made a comeback. David Duke, one of the state's elected legislators, emerged as a folk hero to many white residents in the area.[125] He was elected on the basis of fear and racial scapegoating. Before his election, voters had full knowledge of his ties to the KKK and his active involvement in the National Association for the Advancement of White People. Duke ran for president in 1988 under a far-right Populist Party umbrella, which stressed that the major issue in America is "preserving our very bloodline." He has been active in promoting the Holocaust-as-myth theory which argues that the Nazi slaughter of Jews was an historical hoax. Political contributions from ultraconservative interests have poured into Duke's campaign chest from all over the country. With a lot of money backing him plus racial hatred fueled by declining income in the state, he seemed destined for a national position of power until Republican leaders publicly distanced themselves from his policies. Although no longer a political threat, his initial success underlines how far bigotry can advance a candidate.

One of the most recent, semisuccessful political attempts to wed racial antagonism and income inequality was Patrick Buchanan's bid to garner the Republican nomination for president in 1996. Buchanan's early victories can be linked to economic populism, since he believes American workers have taken it on the chin because of free trade and the corporate greed of U.S. multinationals. This message resonated with a large proportion of voters. Yet Buchanan's campaign was haunted by charges of racism. Several of his top-level advisors were forced to resign because of allegations they were aligned with Ku Klux Klan groups. A major problem that Buchanan's candidacy suffered from was

> all the divisive garbage he brings with him. It's exactly what has been used to destroy populist surges in the past—setting whites against Blacks, natives against immigrants, men against women, straights against gays, Christians against Jews—divide, divide, divide—and lose. Look, Hispanic farm workers are not responsible for the savings and loan mess, Blacks on welfare are not moving factories to Taiwan, lowering the tax on capital gains is not part of the "gay agenda," and Jews, having been historically discriminated against, by and large support raising the minimum wage.[126]

These are the consequences which befall a nation that suffers from a decline of income and rising inequality. The similarities to Germany, before Hitler was swept to power through a Nazi party that feasted on racial hatred, is truly frightening. The fallout always seems to promote violence, bigotry, and fighting among the people of a country who are forced to undergo an economic meltdown. One of the greatest social achievements to occur in the United States after the end of World War II was a growing tolerance of African Americans by white Americans. Although this does not mean all bigotry and racism disappeared, over nearly a half century there was a massive shift away from segregationist views. On a wide range of issues, white Americans grew significantly more tolerant of sharing public spaces, employment, schools, neighborhoods, housing, and marriage with African Americans—rejecting the automatically inferior status assigned to African Americans in the past in favor of government efforts to enforce legal equality.[127]

Now, in the wake of declining fortunes for the majority of Americans, racial inolerance is once again on the rise. By the mid-1990s, a series of public opinion polls had documented the growing racial gap between African Americans and whites. An Associated Press poll revealed that six out of ten whites think African Americans have the same opportunities as whites, while seven out of ten African Americans disagree.[128] A majority of whites now oppose affirmative action and other programs aimed at helping African Americans, whom they now believe are getting more breaks in employment opportunities and educational benefits. A 1994 Times Mirror Center for the People and the Press study discovered that among whites, 51 percent believed that equal rights had been pushed too far. While this bare majority may not seem earthshaking, the rapidity of change is: only 42 percent thought so in 1992 and a bare 16 percent felt that way in 1987. In a recent Washington Post survey, an overwhelming majority of whites said African Americans have an equal chance to succeed, that whites bear no responsibility for problems African Americans encounter, that it is not the federal government's role to ensure that all races have equal jobs, pay, or housing, and that African Americans hold jobs that are as good or better than whites'.[129] White ignorance of African-American lives is shocking. Again, a majority of whites believe African Americans are faring as well as they are on education and health care, but the reality is that whites earn 60 percent more than African Americans, are twice as likely to graduate from college, and are far more likely to have health insurance. The gulf between the races is widest regarding the role of the federal government. In the wake of burgeoning inequality, whites believe we need to repeal affirmative action and government programs for minorities while African Americans believe we need an even stronger effort in this regard to counteract centuries of racial discrimination. The threat is real. Some efforts to reverse affirmative action have been partially effective, such as its elimination in the Texas and California university systems—but such laws are being appealed. Not surprisingly, nearly half (45 percent) of whites believe racial tension has increased in the last decade, compared to 59 percent of African Americans. Among the quarter of the sample experiencing the most economic distress, 57 percent see more racial hostility.

One of the saddest manifestations of racial antagonism acquired a surface veneer of respectability because it was based on "science." In 1994, Richard J. Herrnstein and Charles Murray published *The Bell Curve: Intelligence and Class Structure in American Life*.[130] Their book release was a carefully orchestrated media event to ensure maximum publicity and an initially favorable response by sympathetic reviewers. At the time their findings were covered in the mainstream press with widespread exposure, the academic community had not yet had an opportunity to review their book. Thus, no replies or rebuttals to their message were immediately available. In essence, the public was told that their data offered concrete proof of racial differences in IQ (African Americans were 15 points lower, on average, than whites). Moreover, class differences in intelligence were becoming even more polarized with the passage of time. A major message was that poor people as well as people of color simply lacked the innate cognitive ability to achieve high income via demanding, intellectual jobs—now that equality of opportunity for advanced education had been realized in our culture.

The Bell Curve has been criticized to the point where almost no biological or social scientist would today accept its findings. In addition to flawed methodology and data, incorrect conceptualization of what constitutes intelligence, failure to seriously address the impact of educational, social, or environmental effects upon achievement, and overstatement of the genetic heritability of intelligence (Herrnstein and Murray say 70%, one critic argues 25%), the book has a hidden political agenda.[131] The authors state that low IQs contribute to many of today's worst social problems, such as out-of-wedlock births, crime, and welfare dependency. But if African Americans, the poor, and low-income people are really genetically inferior to whites and those with high incomes, none of our social programs will actually work. Herrnstein and Murray argue that welfare and affirmative action should be eliminated because these innate, genetically based IQ differences doom the participants to failure. Such beliefs are totally unsupported by their findings. But the very topic is reminiscent of the pseudoscientific eugenics movement at the start of the twentieth century—which resulted in a racist quota system in U.S. immigration policy during the 1920s and the rise of Nazism in Germany in the 1930s.

Every time arguments about genes or intelligence have arisen in American politics, they did so to blunt the drive for "some sort of redistribution." That is why their argument is not at all new. . . . Whenever we are exhausted with reform, we shrug our shoulders and say, "there's nothing we can do for that poor guy down the street." Thus was pseudo-science about racial differences used to justify the end of Reconstruction and the reimposition of a segregated caste system on the American South. . . . The Herrnstein-Murray book is not a "scientific" book at all but a political argument offered by skilled polemicists aimed at defeating egalitarians. It is gaining attention because . . . it's a lot easier to blame somebody else's genes or brain cells than to improve society.[132]

The appearance and ready acceptance of *The Bell Curve* is indicative of how deep America's present pessemism runs.[133] In a sense, the belief that genetics makes all the difference can provide a ready-made excuse to embrace racism, fatalism, and passivity. If we really believe nothing can be done, we will make no attempt to help, encourage, or protect the vulnerable. Perhaps worse, our current climate of economic meltdown may give rise to scapegoating and witch-hunts. One panel of scientists is fearful that the current economic and political uncertainty, coupled with basic ignorance of genetics, may give rise to a situation where minorities are blamed for society's problems.[134] Thus, we come full circle to politicians adept at the blame game who refuse to focus upon the dark side of unrestrained capitalism, but who are eager to single out welfare mothers, immigrants, and minorities as the culprits who have caused our economic decline. It is those people who prey upon our darkest fears and prejudices, to whom we must turn a deaf ear.

At the start of this book, research evidence was cited which points to increasing violence and political instability for societies that experience high degrees of inequality. The signs show that this is starting to happen in the United States. America is fraying at the edges. With the powder keg of an increasingly impoverished African-American underclass in our metropolitan areas together with neo-Nazi white supremacist skinheads, our nation could be in for a round of incredible violence. We have already suffered from the devastating Los Angeles riots of 1992, which were caused by huge income inequities overlaid by a wide racial gap. The acquittal of those who beat Rodney King provided the spark which set off the explosion. Hate groups and well-armed vigilante militias now dot the country. The climate of fear and distrust culminated in the Oklahoma City bombing of a federal building, a massacre resulting in the deaths of innocent adults and children. Trains are derailed, the Olympic games are disrupted by bombs, and violent paramilitary groups gain new adherents.

Such violence may be a partial response to the decline of income. Families and children are suffering now more than ever. Health care, food and nutrition, shelter, and other basic human needs go increasingly unmet. Good jobs with decent wages are harder to find. The executives of our major corporations, who have never been accountable to their workers or the American electorate, have virtually abandoned the country in a nonstop search for higher profits. If such trends continue, they will cause the downfall of all that we value in the American way of life.

 Whatever plan eventually takes shape to preserve the American economy and our personal standard of living, there is a grave need to deal from the top of the deck. It is by now more obvious than ever that economic decisions affect us all. At this point in our democracy, however, citizens, voters, and workers have been shut out of the major policy-making that affects our economy. The plans of presidents and politicans, CEOs and bankers, all seem to ignore the people who are most affected by their decisions. Yet the financial status quo is not chiseled in stone. There is a more fair and equitable way to generate and share income in our society. Again, many contemporary analyses of the malaise of the U.S. economy are emphatic in their call for more *industrial democracy*.

 This concept goes far beyond simply involving workers in decisions about production or permitting unionization. It involves every one of us as workers and citizens in all walks of life. How our economy is structured and how American firms conduct business, both in the world and here at home, has major repercussions for every one of us. We are in urgent need of economic democracy. Democratic economic planning at its base raises an annoying question which challenges monolithic economic powers. Whose interests are served by the current decisions being made? It questions the belief that a totally unfettered and unplanned market economy is the best means of meeting the social and economic needs of the nation. It asks which corporations and industries are rewarded by development and expansion policies (generally via the Pentagon) and who pays the bills once they come due (e.g., the taxpayer bailout of the savings and loan crisis). What it offers, in the view of Falk and Lyson, is the chance for ordinary people to gain more control of their lives:

> In its boldest form, economic democracy calls for a rising standard of living for working people; an adequate supply of socially useful goods and services, unmindful of their profitability; a more hospitable, less authoritarian, and safer work environment; and increased participation by workers in the day-to-day running of the economy.[135]

 This thinking is indeed different from the profit-driven model so dominant in the market system. It raises the point that all should benefit from the workings of the economy, if not equally, at least to the point where minimal needs are met. It focuses on matters of concern to all of us that the private sector has been unable or unwilling to consider. The list is lengthy, including recycling, developing alternative energy sources, promoting mass transportation, guaranteeing basic health care for every person regardless of ability to pay, providing a widespread system of day care, promoting a sustainable and ecologically responsible system of agriculture geared to family farms, promoting sunrise industries by guaranteeing start-up/development capital for promising technology, cushioning the fall of dying industries through policies regulating plant closures, retraining, aid for employee buyouts of folding businesses, regional and local revitalization, and the like.

Above all, economic democracy is skeptical of the notion that "What is good for General Motors is good for America." This shopworn idea, prevalent in the 1950s, no longer plays well in Peoria. Middle Americans are becoming more astute and aware that the military/industrial complex is failing them. As an engine of economic growth, it is inefficient and can no longer earn its way in a global economy. People question whether the real interests of Americans can be served by policies aimed at maximizing the profits of huge conglomerates. They ask the embarassing question: Is the record overseas expansion of American multinational firms of any benefit to the average U.S. citizen? Or does it instead harm us, reduce our income, and increase inequality?

We can no longer assume that U.S. corporations doing a land-office business abroad will have the best interests of the American people and their government at heart. In fact, there is evidence to the contrary. The heads of multinational firms have been remarkably candid in admitting that their corporations come first. These global enterprises are now shedding any vestige of American identity as they develop world markets and reduce their dependency on home sales. One *New York Times* article found that the executives of American multinationals

> increasingly speak as if the United States were no longer home port. "The United States does not have an automatic call on our resources," said Cyrill Siewart, chief financial officer at the Colgate-Palmolive Co., which sells more toothpaste, soaps, and other toiletries outside the United States than inside. "There is no mindset that puts this country first." . . . More and more high-paying jobs, including those for engineers and other professionals, are going abroad, instead of being kept at home. . . . Many executives say the global strategy supersedes preferential treatment for U.S. employees.[136]

Such policies are a direct threat to the economic well-being of the great majority of Americans. In effect, our country has helped pay for the overseas expansion of multinationals through generous tax policies. As a result, investment capital needed at home is now pouring out of the country at an alarming rate. Along with the capital flight have gone the countless jobs that would have been available if more modern, up-to-date, competitive factories had been built inside our borders. Deindustrialization has set in. One direct consequence has been the decline in our standard of living. Personal income in America is in a tailspin, especially in comparison to other industrial countries. A tide of income inequality is sweeping over us as a result. As voting citizens in a democracy, we need to resist the continuation of such de facto policies which directly threaten our financial security. We need to explicitly support political candidates and policies which aim to rebuild the manufacturing base here at home. We must call for a more active involvement of government—at all levels—to ensure that the living standards of our citizens are no longer sacrificed to the profits of huge corporations with little allegiance to our country.

Sociology teaches us that we are not on a level playing field. Not all economic actors participate as equals. In the competitive struggle for income, this is a fact we should never forget. Huge and powerful vested interests help the wealthy survive and flourish, sometimes at the expense of the middle class and poor. Yet, history tells us that needed changes can be won. Various social movements—such as unionism, civil rights, and feminism—have met with overall success. We are still a democracy, and as citizens we still have reasonably open access to political influence. Although individually we lack the influence of Political Action Committees (PACS) sponsored by corporations, we wield a collective force through our votes. Therein lies our major means of curbing the damaging extremes of American capitalism. In the long run, both as a nation and as individual citizens, we are capable of turning these negative trends around. We do not need to accept rampant income inequality as a permanent feature of our nation. The power we have as workers, consumers, investors, and citizens remains with our votes and our pocketbooks. By being aware, joining forces, confronting these issues, and taking personal action, it is still possible to turn back the rising tide of inequality which threatens to engulf our nation. There is, in essence, still hope. If we relinquish this basic tenet for change—the belief that we can make a difference—then the forces of greed will have won. To paraphrase Franklin Delano Roosevelt: "The only thing we have to fear is believing we are helpless!"

In the end, if we are to remain a viable democracy, people have to matter. We must believe in ourselves. We must reach within to find the strength to be compassionate and caring. The malice generated by envy, causing us to fall upon one another with knives unsheathed, has to be stopped. We can do this by working hard to assure that every person in our society can lead a productive life free from want. We need to guarantee that every one of us will have basic health care, food, education, and shelter. This *can* be accomplished—if we are willing to lay claim to the power that is within us.

APPENDIX

TABLE A1   Stepwise Regressions of Median Household Income with Standardized Betas of Human Capital Variables, 1990

Predictive Variables	All U.S. Counties	Southern Counties	Nonsouthern Counties
Percent high school graduates	.489	.502	.399
Living in metro area	.344	.291	.395
Percent rural/farm	−.233	−.057	−.246
Percent households headed by a female	−.188	−.213	−.139
Percent total payroll in manufacturing	.111	.109	.117
African-American dominance	.068	.082	.057
Percent unemployed	−.053	−.091	Not Significant
Adjusted R-square	.620	.659	.569
Sample size	3,129	1,419	1,709

Source: Data from U.S. Census Bureau, *USA Counties, 1994,* CD-ROM.

TABLE A2   Stepwise Regressions of Income Disparity Index with Standardized Betas of Human Capital Variables, 1990

Predictive Variables	All U.S. Counties	Southern Counties	Nonsouthern counties
Percent high school graduates	.112	Not Significant	Not Significant
Living in metro area	−.038	−.048	Not Significant
Percent rural/farm	Not Significant	Not Significant	−.137
Percent households headed by a female	−.478	−.411	−.497
Percent total payroll in manufacturing	.194	.232	.144
African-American dominance	.040	Not Significant	Not Significant
Percent unemployed	Not Significant	Not Significant	−.047
Median family income	2.818	3.034	2.709
Median family income squared	−3.102	−3.012	−3.209
Adjusted R-square	.774	.725	.802
Sample size	3,129	1,419	1,709

Source: Data from U.S. Census Bureau, *USA Counties, 1994,* CD-ROM.

NOTES

CHAPTER 1

1. Bergen Evans, *Dictionary of Quotations* (New York: Delacorte, 1968), p. 85. The quotation was first recorded by Rousseau and attributed to an unknown princess, who said it well before Marie Antoinette was born. Even though Marie may not have originated these words, another calloused member of the aristocracy did. These words serve to illustrate the indifference of a privileged economic elite to the suffering of common people.

2. Barry Bluestone and Bennett Harrison, *The Deindustrialization of America* (New York: Basic Books, 1982).

3. Monroe W. Karmin, "Is Middle Class Really Doomed to Shrivel Away?" *U.S. News and World Report,* August 20, 1984, p. 65.

4. Robert J. Samuelson, "The Myth of the Missing Middle," *Newsweek,* July 1, 1985, p. 50.

5. William Baldwin, "Chicken Little's Income Statistics," *Forbes,* March 24, 1986, pp. 68–69.

6. Jerry Flint, "Too Much Ain't Enough," *Forbes,* July 13, 1987, pp. 92 ff.

7. Charlotte Salkowski, "Growth in Living Standard Slows for the American Middle Class," *Christian Science Monitor,* January 8, 1986, p. 1 ff.

8. Robert Kuttner, "A Shrinking Middle Class Is a Call to Action," *Business Week,* September 6, 1985, p. 16.

9. "Is the Middle Class Shrinking?" *Time,* November 3, 1986, pp. 54–56.

10. Paulette Thomas, "Widening Rich-Poor Gap Is a Threat to the 'Social Fabric,' White House Says," *Wall Street Journal,* February 15, 1994, p. A2.

11. Paul Krugman, "Disparity and Despair," *U.S. News and World Report,* March 23, 1992, pp. 54–55.

12. Aaron Bernstein, "Inequality: How the Gap Between Rich and Poor Hurts the Economy," *Business Week,* August 15, 1994, pp. 78–81.

13. Richard Harwood, "The Rich and Poor Problem," *The Washington Post National Weekly Edition,* June 19–25, 1995, p. 29.

14. Robert Kuttner, "Kids, Parents and the Economy," *The Washington Post National Weekly Edition,* July 3–9, 1995, p. 5.

15. Robert Kuttner, "The Real Class War," *The Washington Post National Weekly Edition,* July 31–August 6, 1995, p. 5.

16. Barbara Ehrenreich, "Is the Middle Class Doomed?" *New York Times Magazine,* September 7, 1986, pp. 44 ff.

17. For a detailed discussion of consumerism in America and its associated insatiability, together with the spiritual malaise and accompanying huge environmental costs connected with its development, see Alan Thein Durning, *How Much Is Enough?* (New York: Norton, 1992).

18. Ehrenreich, "Is the Middle Class Doomed?" p. 63.

19. James C. Davies, "Toward a Theory of Revolution," *American Sociological Review* 6 (1962): 5–19; James C. Davies, "The J-Curve of Rising and Declining Satisfactions as a Cause of Some Great Revolutions and a Contained Rebellion," in H.D. Graham and T.R. Gurr (eds.), *Violence in America* (New York: Praeger, 1969).

20. T. R. Gurr, *Why Men Rebel* (Princeton, NJ: Princeton University Press, 1970).

21. Joan Neff Gurney and Kathleen J. Tierney, "Relative Deprivation and Social Movements: A Critical Look at Twenty Years of Theory and Research," *Sociological Quarterly* 23 (1982): 33–47.

22. Theda Skocpol, "Explaining Revolutions: In Quest of a Social Structural Approach," in L.A. Coser and O.N. Larsen (eds.), *The Uses of Controversy in Sociology* (New York: Free Press, 1976), p. 158.

23. Adam Szirmai, *Inequality Observed: A Study of Attitudes Towards Income Inequality* (Brookfield, VT: Avebury, 1988), p. 305.

24. Nathaniel Sheppard Jr., "In Panama, Gap Between Rich and Poor Grows at Alarming Rate," *Chicago Tribune,* May 7, 1992, pp. 1, 32.

25. News Services, "Bush Announces Riot Crimes Probe," *Minneapolis Star Tribune,* May 6, 1992, 1A, p. 12A.

26. David Rieff, *Los Angeles: Capital of the Third World* (New York: Simon and Schuster, 1991).

27. Mike Davis, *City of Quartz: Examining the Future in Los Angeles* (New York: Verso, 1990).

28. Denny Braun, *The Rich Get Richer: The Rise of Income Inequality in the United States and the World* (Chicago: Nelson-Hall, 1990), p. 238.

29. Paul Ong, *The Widening Divide: Income Inequality and Poverty in Los Angeles* (Los Angeles: UCLA Graduate School of Architecture and Urban Planning, 1989), p. 119.

30. Ibid., pp. 15–16.

31. David Jacobs and David Britt, "Inequality and Police Use of Deadly Force: An Empirical Assessment of a Conflict Hypothesis," *Social Problems* 26,4 (1979): 403–412.

32. Edward N. Muller, "Income Inequality, Regime Repressiveness, and Political Violence," *American Sociological Review* 50 (1985): 47–61; Edward N. Muller and Mitchell A. Seligson, "Inequality and Insurgency," *American Political*

Science Review 81,2 (June 1987): 427–51. See also Michael Timberlake and Kirk R. Williams, "Structural Position in the World-System, Inequality and Political Violence," *Journal of Political and Military Sociology* 15,1 (1987): 1–15; Sunil Kukreja and James D. Miley, "Government Repression: A Test of the Conflict, World-System Position, and Modernization Hypotheses," *International Journal of Contemporary Sociology* 26,3/4 (1989): 147–57.

33. Edward N. Muller, "Economic Determinants of Democracy," *American Sociological Review* 60,6 (1995): 980–81.

34. Edward N. Muller, "Democracy, Economic Development, and Income Inequality," *American Sociological Review* 53 (1988): 66.

35. Heinrich Zwicky, "Income Inequality and Violent Conflicts in Developing Countries," *Research in Inequality and Social Conflict* 1 (1989): 78.

36. Erich Weede, "Some New Evidence on Correlates of Political Violence: Income Inequality, Regime Repressiveness, and Economic Development," *European Sociological Review* 3,2 (1987): 106.

37. Erich Weede, "Democracy and Income Inequality Reconsidered," *American Sociological Review* 54,5 (1989): 865–68.

38. Kenneth A. Bollen and Robert W. Jackman, "Democracy, Stability, and Dichotomies," *American Sociological Review* 54,4 (1989): 612–13; John Hartman and Wey Hsiao, "Inequality and Violence Issues: Issues of Theory and Measurement," *American Sociological Review* 53,5 (1988): 795.

39. Edward N. Muller, "Inequality, Repression, and Violence: Issues of Theory and Research Design," *American Sociological Review* 53,5 (1988): 799–806; Edward A. Muller, "Democracy and Inequality (Reply to Weede)," *American Sociological Review* 54,5 (1989): 868–71.

40. Mark Irving Lichbach, "An Evaluation of 'Does Economic Inequality Breed Political Conflict?' Studies," *World Politics* 41,4 (July, 1989): 431–70.

41. Ibid., p. 451.

42. Terry Boswell and William J. Dixon, "Dependency and Rebellion: A Cross-National Analysis," *American Sociological Review* 55 (August 1990): 549–50.

43. Terry Boswell and William J. Dixon, "Marx's Theory of Rebellion: A Cross-National Analysis of Class Exploitation, Economic Development, and Violent Revolt," *American Sociological Review* 58 (October 1993): 681–702.

44. Judith R. Blau and Peter M. Blau, "The Cost of Inequality: Metropolitan Structure and Violent Crime," *American Sociological Review* 47 (1982): 114–29.

45. Emile Durkheim, *Suicide*, trans. George Simpson (New York: Free Press, 1964). Originally published in 1867.

46. Steven F. Messner, "Social Development, Social Equality, and Homicide: A Cross-National Test of a Durkheimian Model," *Social Forces* 61 (1982): 225–40.

47. International Economics Department, *World Development Indicators 1994* (Washington, DC: World Bank, 1995), table 1.

48. Ibid., table 30.

49. Roger Cohen, "Rio's Murder Wave Takes on the Aura of a Class Struggle," *Wall Street Journal*, May 9, 1989, pp. A1, A15.

50. Harvey Krahn, Timothy F. Hartnagel, and John W. Gartrell, "Income Inequality and Homicide Rates: Cross-National Data and Criminological Theories," *Criminology* 24 (1986): 269–95.

51. Steven F. Messner, "Economic Discrimination and Societal Homicide Rates: Further Evidence on the Cost of Inequality," *American Sociological Review* 54 (1989): 606.

52. Miles D. Harer, Relative Deprivation and Crime: The Effects of Income Inequality on Black and White Arrest Rates, Ph.D. diss., Pennsylvania State University, 1987.

53. Joanne Belknap, The Effects of Poverty, Income Inequality, and Unemployment on Crime Rates, Ph.D. diss., Michigan State University, 1986.

54. Denny Braun, "Negative Consequences to the Rise of Income Inequality," *Research in Politics and Society* 5 (1995): 23.

55. U.S. Bureau of the Census, *Statistical Abstract of the United States 1994*, CD-ROM (Washington, DC: U.S. Government Printing Office, 1995), table 301.

56. Jim Dawson, "Youth Homicide Reaching Grim Highs," *Minneapolis Star Tribune*, February 18, 1995, p. 6A.

57. "Violent Juvenile Crime Up Sharply," *Minneapolis Star Tribune*, September 8, 1995, p. 7A.

58. Pierre Thomas, "Getting to the Bottom Line on Crime," *The Washington Post National Weekly Edition*, July 18–24, 1995, p. 31.

59. "Inmate Census Highest of All Time," *Minneapolis Star Tribune*, September 13, 1994, p. 7A.

60. "Federal and State Prisons Getting Even More Crowded," *Minneapolis Star Tribune*, August 10, 1995, p. 19A.

61. Sharon Schmickle, "Crime Experts See Flaws in Bill's Focus," *Minneapolis Star Tribune*, August 14, 1994, p. 16A.

62. Martha A. Myers, "Economic Inequality and Discrimination in Sentencing," *Social Forces* 65,3 (1987): 754–55.

63. Thomas, "Bottom Line," p. 31.

64. Katherine McFate, "The Grim Economics of Violence," *Focus*, October 1994, p. 7.

65. Bennett Harrison and Barry Bluestone, *The Great U-Turn: Corporate Restructuring and the Polarizing of America* (New York: Basic Books, 1988), pp. 112–13. See also David Kotz, "Feeling Overworked? Here's Why," *Utne Reader*, July/August 1988, pp. 56–60.

66. Lawrence Mishel and Jared Bernstein, *The State of Working America 1994–95* (New York: M. E. Sharpe, 1994), p. 113.

67. Sheldon Danziger and Peter Gottschalk, "Families with Children Have Fared Worse," *Challenge* 29 (March–April 1986): 40–47.

68. Sandra K. Danziger and Sheldon Danziger, "Child Poverty and Public Policy: Toward a Comprehensive Antipoverty Agenda," *Daedalus* 122 (Winter 1993): 63.

69. Lenore Weitzman, *The Divorce Revolution* (New York: Free Press, 1987), p. 323.

70. William P. O'Hare, *America's Welfare Population: Who Gets What?* (Washington, DC: Population Reference Bureau, 1987); William P. O'Hare, "Poverty in America: Trends and New Patterns," *Population Bulletin* 40,3 (1985).

71. Morton Owen Schapiro, "Socio-Economic Effects of Relative Income and Relative Cohort Size," *Social Science Research* 17 (1988): 377.

72. Graham B. Spanier and Paul C. Glick, "Marital Infidelity in the United States: Some Correlates and Recent Changes," *Family Relations* 31 (July 1981): 329–38.

73. Arloc Sherman, *Wasting America's Future: The Children's Defense Fund Report on the Costs of Child Poverty* (Boston: Beacon Press, 1994), p. 37.

74. U.S. Bureau of the Census, *Statistical Abstract of the United States 1994*, CD-ROM (Washington, DC: U.S. Government Printing Office, 1994), table 604.

75. Sylvia Ann Hewlett, *When the Bough Breaks: The Cost of Neglecting Our Children* (New York: Basic Books, 1991), pp. 88–94, 107–16.

76. Carnegie Task Force, *Starting Points: Meeting the Needs of Young Children* (New York: Carnegie Corporation, 1994), pp. 3–22.

77. Sherman, *Wasting America's Future,* pp. 30–36.

78. Ibid., p. 85.

79. Joan I. Vondra, "Childhood Poverty and Child Maltreatment," in Judith A. Chafel (ed.), *Child Poverty and Public Policy* (Washington, DC: Urban Institute Press, 1993), p. 128.

80. Jane D. McLeod and Michael J. Shanahan, "Poverty, Parenting, and Children's Mental Health," *American Sociological Review* 58 (June 1993): 357.

81. Clifford M. Johnson et al., *Child Poverty in America* (Washington, DC: Children's Defense Fund, 1991), p. 16.

82. *The State of America's Children Yearbook 1995* (Washington, DC: Children's Defense Fund, 1995), p. 82.

83. Carnegie Task Force, *Starting Points,* p. 21.

84. M. D. Nelson, Jr., "Socioeconomic Status and Childhood Mortality in North Carolina," *American Journal of Public Health* 82 (August 1992): 1131–33.

85. Johnson, *Child Poverty in America,* p. 16.

86. Katherin O'Regan and Michael Wiseman, "Birth Weights and the Geography of Poverty," *Focus* 12,2 (Winter 1989): 17.

87. Hewlett, *When the Bough Breaks,* p. 34.

88. Children's Defense Fund, *America's Children 1995,* p. 28.

89. Ibid.; U.S. Bureau of the Census, *Statistical Abstract,* table 1353.

90. Children's Defense Fund, *America's Children 1995,* p. 29.

91. Ibid., p. 27.

92. Gregg M. Olsen, "Locating the Canadian Welfare State: Family Policy and Health Care in Canada, Sweden, and the United States," *Canadian Journal of Sociology* 19,1 (1994): 3, 15–17.

93. Paul L. Menchik, "Permanent and Transitory Economic Status as Determinants of Mortality among Nonwhite and White Older Males: Does Poverty Kill?" Institute for Research on Poverty, Discussion Paper 936 (Madison: University of Wisconsin, 1991), p. 19. There is some evidence that the income/mortality

rate relationship may actually be curvilinear, i.e., a U-shaped curve (Adamchak and Robinson, 1986: 214). This study is based on data drawn in the 1967–72 period, however, at the height of the War on Poverty begun in President Johnson's administration. Since huge inroads were made at this time in reducing poverty, especially for the elderly, the rates calculated for this period for the poor may be more atypical than the current data. This is not to deny that a curvilinear relationship may exist. Wilkinson (1990: 395) found more of a "J-shaped" curve in Great Britain reflecting the effects of income upon health. In short, ill health was especially manifest among the poor, but dropped sharply for the middle-income groups. Ill health climbed somewhat among upper-income groups, but came nowhere close to equalling the high levels of bad health among low-income persons. Researchers need to explore these nuances closely. Ill health among the wealthy may be due to entirely different causes (rich diets, more alcohol consumption, etc.) than ill health among the poor (lack of nutrition, less health insurance coverage, etc.).

94. Nancy Krieger and Elizabeth Fee, "What's Class Got to Do with It? The State of Health Data in the United States Today," *Sociological Review* 23,1 (March, 1993): 61–62.

95. Tabulations from the 1993 General Social Survey. James Davis and Tom W. Smith, *General Social Surveys, 1972–93* [machine-readable data file], National Opinion Research Center, producer (Storrs: The Roper Center for Public Opinion Research, University of Connecticut).

96. Mary Merva and Richard Fowles, *Effects of Diminished Economic Opportunities on Social Stress: Heart Attacks, Strokes, and Crime* (Washington, DC: Economic Policy Institute, 1992).

97. Richard G. Wilkinson, "Income Distribution and Mortality: A 'Natural' Experiment," *Sociology of Health and Illness* 12,4 (December 1990): 406–407.

98. Richard G. Wilkinson, "National Mortality Rates: The Impact of Inequality," *American Journal of Public Health* 82,8 (August 1992): 1083.

99. Frank Levy, *Dollars and Dreams: The Changing American Income Distribution* (New York: Russell Sage, 1987), pp. 79–80.

100. Dowell Myers and Jennifer R. Wolch, "The Polarization of Housing Status," in Reynolds Farley (ed.), *State of the Union: America in the 1990s,* Vol. 1, *Economic Trends* (New York: Russell Sage, 1995), pp. 323–25.

101. Myers and Wolch, "Polarization," p. 294.

102. Children's Defense Fund, *The State of America's Children Yearbook 1994* (Washington, DC: Children's Defense Fund, 1994), p. 39.

103. Johnson, *Child Poverty in America,* p. 14.

104. Children's Defense Fund, *The State of America's Children Yearbook 1994,* p. 42.

105. Sherman, *Wasting America's Future,* pp. 18–23.

106. Danziger, "Child Poverty and Public Policy," p. 71.

107. Sherman, *Wasting America's Future,* pp. 23–28, 78–82.

108. Children's Defense Fund, *The State of America's Children Yearbook 1995,* p. 92.

109. Charles F. Manski, "Income and Higher Education," *Focus* 14,3 (Winter 1993): 15.

110. Ravi Batra, *The Great Depression of 1990* (New York: Dell, 1988).

111. Daniel B. Radner and Denton R. Vaughan, "Wealth, Income, and the Economic Status of Aged Households," in Edward N. Wolff (ed.), *International Comparisons of the Distribution of Household Wealth* (New York: Oxford University Press, 1987), pp. 93–120.

112. Daphne T. Greenwood, "Age, Income, and Household Size: Their Relation to Wealth Distribution in the United States," in Wolff, *Distribution of Household Wealth*, pp. 121–40.

113. Sidney L. Carroll, "American Family Fortunes as Economic Deadweight," *Challenge*, May/June, 1991, p. 13.

114. Edward N. Wolff, "The Rich Get Increasingly Richer: Latest Data on Household Wealth During the 1980s," *Research in Politics and Society* 5 (1995): 38.

115. Harold R. Kerbo, *Social Stratification and Inequality: Class Conflict in the United States* (New York: McGraw-Hill, 1983), pp. 36–37.

116. Batra, *Great Depression of 1990*, pp. 133–34.

117. Ibid., pp. 136–37.

118. Edward N. Wolff, *International Comparisons of the Distribution of Household Wealth* (New York: Oxford University Press, 1987), pp. 5–7.

119. Carroll, "Economic Deadweight," pp. 13–14.

120. Ibid., p. 14.

121. Edward N. Wolff, "How the Pie Is Sliced: America's Growing Concentration of Wealth," *The American Prospect* 22 (Summer 1995): 60.

122. Sheldon Danziger, Peter Gottschalk, and Eugene Smolensky, "How the Rich Have Fared, 1973–87," *American Economic Review* 79,2 (May 1989): 312.

123. Paul Farhi, "Multiplying Millionaires," *Minneapolis Star Tribune*, July 14, 1992, pp. lD ff.

124. Wolff, "Latest Data on Household Wealth," pp. 37, 39.

125. Julie Gozan, "Wealth for the Few," *Multinational Monitor* 13,12 (1992): 6.

126. Melvin L. Oliver, Thomas M. Shapiro, and Julie E. Press, "'Them That's Got Shall Get': Inheritance and Achievement in Wealth Accumulation," *Researh in Politics and Society* 5 (1995): 79.

127. James P. Smith, *Unequal Wealth and Incentives to Save* (Santa Monica, CA: RAND, 1995), p. 5.

128. Richard E. Ratcliff and Suzanne B. Maurer, "Saving and Investment among the Wealthy: The Uses of Assets by High Income Families in 1950 and 1983," *Research in Politics and Society* 5 (1995): 99–125.

129. Aaron Bernstein, "Inequality: How the Gap Between Rich and Poor Hurts the Economy," *Business Week*, August 15, 1994, p. 79.

130. Ibid., p. 82.

131. Andrew Glyn and David Miliband, *Paying for Inequality: The Economic Cost of Social Injustice* (Concord, MA: Paul and Co., 1994).

132. Thorsten Persson and Guido Tabellini, "Is Inequality Harmful for Growth?" *American Economic Review* 84,3 (June 1994): 600–621.

133. Randy Albelda and Chris Tilly, "Unnecessary Evil: Why Inequality Is Bad for Business," *Dollars and Sense* (March/April 1995): 21.

134. Guillermina Jasso and Peter H. Rossi, "Distributive Justice and Earned Income," *American Sociological Review* 42 (August 1977): 639–51; Guillermina Jasso, "On the Justice of Earnings: A New Specification of the Justice Evaluation Function," *American Journal of Sociology* 83,6 (1978): 1398–1419; Guillermina Jasso, "A New Theory of Distributive Justice," *American Sociological Review* 45 (February 1980): 3–32.

135. One of the best available is Harold R. Kerbo, *Social Stratification and Inequality: Class Conflict in Historical and Comparative Perspective,* 3d ed., (New York: McGraw-Hill, 1996).

136. Stock ownership is an indication of capitalism par excellence. A number of studies performed on a much wider scale (Wright 1978; Wright and Peronne 1977) show that income is actually predicted more accurately by Marxian class position, i.e., whether one owns the means of production—capital and stocks—or has to earn one's living by selling one's labor (working for an employer). Status attainments that many of us rely on to increase our income, such as acquiring education or entering a lucrative profession, are simply less relevant in explaining income differences between people. Stock owners have higher incomes than non-stock owners with the same education, occupation, age, job tenure, etc. In essence, being a capitalist—regardless of educational and occupational factors—brings more income (Aldrich and Weiss 1981). Furthermore, when persons are examined within class categories there are not many differences by race and sex with respect to income. The large differences between men and women or blacks and whites are really due to class position: women and blacks are simply a larger proportion of the working class. Empirical research using national samples within the United States and England (Robinson and Kelley 1979) also corroborates Wright's research. With this evidence in mind, the lack of correspondence between performance and education of corporate executives on the one hand, and their job performance on the other, should not be too suprising.

137. Robert McCartney, "Stock Options and the Uneven Paying Field," *The Washington Post National Weekly Edition,* February 10-16, 1992, pp. 23–24.

138. Ibid., p. 23.

139. Robert McCartney, "Pay Dirt: Shining a Light on the Salary Bloat of CEOs," *The Washington Post National Weekly Edition,* February 3–9, 1992, p. 22.

140. Marilyn Geewax, "Let Them Eat Pink Slips," *Minneapolis Star Tribune,* October 14, 1991, p. 11A.

141. "Corporate America's Most Powerful People," *Forbes,* May 30, 1988, pp. 154 ff. Apparently the old axiom that education will pay off in future earnings is a fallacy when it comes to top corporate executives. Forbes reports that the 87 chief executives who hold no college degree earned a median $735,000 in 1987 versus a median of $746,000 for the 378 with an undergraduate college degree. The 323 executives with graduate degrees earned a median of $773,000—leading to the inescapable conclusion that at this level it simply does not pay to pursue higher education. The "increase" in salary between a noncollege and college graduate CEO comes to a paltry 1.5 percent per year.

142. George F. Will, "CEO Megasalaries May Provoke New Bout of Antibusiness Fever," *Minneapolis Star Tribune,* September 1, 1991, p. 18A.

143. Ralph Nader, "General Motors Careful to Protect Bloat at the Top," *Minneapolis Star Tribune,* December 31, 1991, p. 13A.

144. Claudia Deutsch, "Going Away for Big Pay," *Minneapolis Star Tribune,* July 4, 1995, p. 1D, 4D.

145. Joann S. Lublin, "Raking It In," *Wall Street Journal,* April 12, 1995, pp. R1, R13.

146. "CEO Paychecks Keep Getting Bigger," *Minneapolis Star Tribune,* May 9, 1995, p. 1D.

147. Eric S. Hardy, "America's Highest-Paid Bosses," *Forbes,* May 22, 1995, p. 182.

148. "Executive Compensation Up 23%," *Minneapolis Star Tribune,* March 6, 1996, p. 1D.

149. Graef S. Crystal, *In Search of Excess: The Overcompensation of American Executives* (New York: Norton, 1991).

150. Graef S. Crystal, "The Wacky, Wacky World of CEO Pay," *Fortune,* June 6, 1988, pp. 68 ff.

151. For those who are statistical tyros, the very term "multiple regression" conjures up fear and anxiety. This reaction need not be the typical response. One does not need to know complex formulas and mathematical nuances to understand regression analysis. In a word, we try to predict change in a dependent variable (CEO compensation in this example) by use of several independent variables (tenure with the corporation, company performance, CEO's ownership of company stock, etc.). The procedure is useful in isolating what variables are most important in causing change in the subject we are interested in. For the most part, income inequality will nearly always be the dependent variable being predicted in this book. I have tried to interpret regression results throughout this book in generalized lay terms.

152. John Burgess, "The Latest American Export: Higher Executive Salaries are Showing Up Abroad," *The Washington Post National Weekly Edition,* October 28–November 3, 1991, p. 25.

153. Robert B. Reich, *The Work of Nations: Preparing Ourselves for 21st Century Capitalism* (New York: Vintage, 1992), p. 205.

154. Marjorie Kelly, "Mushrooming Executive Pay Prompts Resentment, Problems," *Minneapolis Star Tribune,* October 9, 1995, p. 3D.

155. David Kirkpatrick, "Abroad, It's Another World," *Fortune,* June 1988, p. 78.

156. Crystal, "Wacky World," p. 74.

157. No one, it seems, ever states that "the rich are always with us." The implicit value assumption is that it is in the nature of things to have persons living in poverty, that their presence is perhaps a curse that we must accept with fatalism, that there is nothing we could possibly do to alleviate their suffering, that the natural order precludes a society from doing anything to eradicate poverty. In essence, there is a heavy status-quo orientation to the assumption that "the poor are always with us." Conversely, nothing is ever mentioned about eradicating extreme affluence. Somehow this condition is seen as right and just, and that to have large segments of the population living what could only be described as opulent lifestyles has no bearing upon the presence of the poor. In a word, if someone

gets a bigger piece of the pie, someone else will inevitably get a smaller piece—all things being unequal. One could argue that this is not necessarily the case when the pie itself gets bigger over time. As we will see, however, the pie has been contracting since 1973, making the increasing inequities doubly painful.

CHAPTER 2

1. James R. Kluegel and Eliot R. Smith, *Beliefs about Inequality: Americans' Views of What Is and What Ought to Be* (Hawthorne, NY: Aldine De Gruyter, 1986).

2. Unfortunately, documenting the ideological biases present in our culture would take enough space to form another book. Thankfully, a number of excellent books discuss the ideological mind-set of American life as it is defined for us by our major institutions—particularly in support of free enterprise, the market system, and capitalism. Michael Parenti is keenly insightful about major U.S. institutions, especially education and spectator sports (1978). The media and public opinion also figure prominently in his analysis (1985). Ben H. Bagdikian (1990) has the most thoroughly detailed current account of how major corporations systematically distort the news and includes analyses of major media outlets in support of conservative, big business values. It is also true, however, that ideological bias is present not only in our own society. Shlapentokh (1987) recounts case after case of state pressure, coercion, and intimidation of academicians, the media, and average citizens in the former Soviet Union in favor of Marxist thought. Any deviation from orthodoxy or possible recognition that personal economic self-interest and market forces could motivate people to work harder were rigidly repressed in Russian society before the end of the Cold War.

3. Gerhard Lenski, *Power and Privilege* (New York: McGraw-Hill, 1966).

4. Robert Gilpin, *The Political Economy of International Relations* (Princeton, NJ: Princeton University Press, 1987), p. 41.

5. Michael Parenti, *Inventing Reality: The Politics of News Media,* 2nd ed. (New York: St. Martin's Press, 1993), pp. 112–36.

6. Karl Marx, *Capital: A Critique of Political Economy,* 3 vols. [1867] (New York: Vintage Books, 1981); Karl Marx, *The Grundrisse* (New York: Vintage Books, 1973).

7. Many scholars energetically dispute the notion of a complacent, benign, and nonmilitant labor union movement. The bloody and savage strike-breaking activity in early twentieth century America seems to contradict this image, (Gutman 1988), as does the militancy of some early union organizers. But concessions by big business to labor in the 1930s quickly resulted in a status quo of high wages in heavy industry for low political consciousness and/or a conservative ethos on the part of labor. The ease with which recent conservative presidential administrations were able to encourage an atmosphere that led to a decertification of unions also suggests an absence of working class consciousness and identification within American labor. There is certainly lack of unity, as unions themselves have increasingly permitted two-tier contracts among their own workers, avoided strikes by agreeing to wage rollbacks, crushed wildcat strikes and other opposition initiatives, etc. Moreover, there has always been a division within American labor

between much wealthier, larger "core" firms of heavy industry (e.g., GM or Ford)—whose workers have been granted comparatively high wages—and non-core firms in small, specialized, and transitory industries with less hope of survival (textiles). Even within the core, relatively cohesive unions have not been able to prevent the erosion of earnings of male workers over the past two decades (Tigges 1987).

8. Michael Harrington, *The Twilight of Capitalism* (New York: Simon and Schuster, 1976); Erik O. Wright, *Class, Crisis and the State* (New York: Schocken Books, 1978).

9. Gilpin, *Political Economy of International Relations,* p. 53.

10. Max Weber, *The Protestant Ethic and the Spirit of Capitalism,* trans. Talcott Parsons (New York: Scribner and Sons, 1958).

11. Max Weber, "Class, Status, Party" in Seymour Martin Lipset and Reinhard Bendix (eds.), *Class, Status and Power* (New York: Free Press, 1966).

12. Kingsley Davis and Wilbert E. Moore, "Some Principles of Stratification," *American Sociological Review* 10 (1945): 242–49.

13. Melvin Tumin, "Some Principles of Stratification: A Critical Analysis," *American Sociological Review* 18 (1953): 387–94.

14. Harold R. Kerbo, *Social Stratification and Inequality: Class Conflict in the United States* (New York: McGraw-Hill, 1983). p. 132.

15. Randal Collins, *Conflict Sociology* (New York: Academic Press, 1975).

16. Christopher Jencks et al., *Who Gets Ahead? The Determinants of Economic Success in America* (New York: Basic Books, 1979).

17. J. A. Brittain, *The Inheritance of Economic Status* (Washington, DC: The Brookings Institution, 1977).

18. Milton Friedman, *Capitalism and Freedom* (Chicago: University of Chicago Press, 1962); Paul Samuelson, *Economics,* 11th ed. (New York: McGraw-Hill, 1980).

19. Milton Friedman, "Choice, Chance, and the Personal Distribution of Income," *Journal of Political Economy* 61,4 (August 1953): 277–90.

20. G. Becker, *Human Capital* (New York: Columbia University Press, 1964).

21. Paul Blumberg, *Inequality in an Age of Decline* (New York: Oxford University Press, 1980).

22. Lars Osberg, *Economic Inequality in the United States* (New York: M. E. Sharpe, 1984), p. 136.

23. Ibid., p. 158.

24. Charles Lindblom, *Politics and Markets: The World's Political-Economic Systems* (New York: Basic Books, 1977).

25. For an insightful discussion contrasting the ways sociologists and economists view distribution and redistribution, comparing both assumptions and theoretical stances, see Suzanne Elise Shanahan and Nancy Brandon Tuma, "The Sociology of Distribution and Redistribution," in Neil J. Smelser and Richard Swedberg (eds.), *The Handbook of Economic Sociology* (New York: Russell Sage Foundation, 1994), pp. 433–65.

26. P. B. Doeringer and M. J. Piore, *Internal Labor Markets and Manpower Analysis* (Lexington, MA; Heath, 1971).

27. Lester Thurow, *Generating Inequality: Mechanisms of Distribution in the U.S. Economy* (New York: Basic Books, 1975).

28. Education can be an important means of maintaining class boundaries. The gradual educational upgrading of occupations over time has not been due to any increase in technical skills needed for these jobs. Rather, as more persons attained a high school education in our society, a college degree was redefined as necessary for middle-class occupations in order to keep entrants from flooding the market (Collins 1971). What developed, of course, was a different flooding of the labor market (now composed of college graduates), which resulted in under-employment and a critical questioning of higher education's role in both economic production and occupational attainment (Smith 1986). Social class bias is replete within elementary and secondary education. The home environment of children from higher classes prepares them, through role modeling, encouragement, and resources such as books and educational toys, to do well in school (Jencks 1972). Teachers expect more from children of higher-class families (Stein 1971; Good and Brophy 1973), leading to their better performance in a self-fulfilling prophecy. Tracking students between vocational (blue collar) and college preparatory curriculums (middle and upper class)—practiced in 85% of public schools—is frequently determined by class background (Jencks et al 1972; Alexander, Cook, and McDill 1978; McPortland 1968). Among high intelligence high school students, 91% attend college compared to 40% of highly intelligent lower-class students. Conversely, 58% of students with low intelligence but higher-class backgrounds attend college compared to 9% of low intelligence, lower-class high school students. No matter what their intelligence, 84% of students from higher-class backgrounds go to college but only 21% from lower-class families do so (Featherman and Hauser 1978). Recent evidence comparing twins continues to corroborate an independent effect due to family background—apart from schooling—on subsequent earnings (Hauser and Sewell 1986). Intergenerational data from the Panel Study of Income Dynamics also reveal significant effects of parental family income on the completed schooling and wage rates of adult children (Hill and Duncan 1987). Moreover, the educational payoff is diminished very greatly for Hispanics and nonwhites in labor markets with a large share of minority workers; this drop in earnings is especially sharp for black men and greater among workers with college educations (Tienda and Lii 1987). Even for women, the payoff for enhanced family income via a college education is found more through marriage than through using their education in the labor market (Glenn 1984). College attendance continues to be important, of course, even if a person does not acquire skills needed for the job market. Those who finish college have a nearly 50% occupational advantage over those who do not (Jencks et al. 1979). Differences in personal human capital attributes do help explain initial career position, although not subsequent upward occupational mobility among male workers in low income areas (Rosenberg 1980). Although a college degree has considerable impact upon a worker's subsequent earnings in core industries, this is not true in more marginal, nonunionized, periphery industries (Beck et al. 1978). Lastly, Bowles and Gintis (1976) argue that the most important lessons taught in school are implicit: being on time and obeying authority are emphasized in working-class schools and curriculums, while self-

direction and expectation for success are stressed in the education of upper-class students. Thus, while one segment is socialized for factory labor, the other inculcates the habits of command needed for corporate management.

29. Osberg, *Economic Inequality in the United States,* p. 169.

30. Edward N. Wolff and D. Bushe, *Age, Education and Occupational Earnings Inequality* (New York: National Bureau of Economic Research, 1976); J. C. Riley, "Testing the Educational Screening Hypothesis," *Journal of Political Economy* 87,5 (October, 1979):S227–52; Sam Rosenberg, "Male Occupational Standing and the Dual Labor Market," *Industrial Relations* 19,1 (Winter 1980): 34–48.

31. Arne L. Kalleberg, Michael Wallace, and Robert P. Althauser, "Economic Segmentation, Worker Power, and Income Inequality," *American Journal of Sociology* 87,3 (November 1981): 651–83.

32. E. M. Beck, Patrick M. Horan, and Charles M. Tolbert II, "Stratification in a Dual Economy: A Sectoral Model of Earnings Determination," *American Sociological Review* 43 (October 1978): 704–20; Barry Bluestone, William M. Murphy, and Mary Stevenson, *Low Wages and the Working Poor* (Ann Arbor: Institute of Labor and Industrial Relations, University of Michigan, 1973); R. C. Edwards, M. Reich, and D. Gordon (eds.), *Labor Market Segmentation* (Lexington, MA: Heath, 1975).

33. Lynne G. Zucker and Carolyn Rosenstein, "Taxonomies of Institutional Structure: Dual Economy Reconsidered," *American Sociological Review* 46 (December 1981): 869–84.

34. Bluestone, *Low Wages and the Working Poor,* pp. 28–29.

35. A number of radical writers (Edwards 1975; Bowles and Gintis 1976) agree with the idea that the U.S. labor market is segmented and that many are relegated to low pay on the basis of such ascriptive characteristics as sex and race, but they see more Machiavellian intent with regard to how capitalists develop their production technology. Not only profit motivates business leadership, but also a desire to keep control of production out of the hands of workers. Because capitalism has built giant corporations with centralized control, worker/management relationships must necessarily be authoritarian and remotely impersonal. Work is broken up into very small tasks which are seemingly meaningless to those on the assembly line. The laborer loses touch with the final product, so pride of craftsmanship rapidly disappears. Moreover, it may be that the gaps between differentiated labor in the factory, or even between market segments, is more artificial and arbitrary than real. Such a system helps to perpetuate false barriers between workers, who seem pitted against one another in competition and are thus unable to develop a collective identity or build resistance through class consciousness. While this orientation may seem extreme, it should be remembered that corporations such as Volvo have successfully reconstituted their factory workplaces by forming worker production teams to build cars. Productivity may rise when workers share in decision-making and when production tasks are not so minutely divided, i.e., traditional approaches are not necessary for high efficiency. Recent evidence (Tigges 1986) corroborates the fact that the type of industrial production is an important dimension of workers' earnings over and above a segmented labor force divided between a typology of core and primary industries.

Lastly, U.S. corporations heavily affect the direction of research and development toward high-tech defense projects through their influence on the federal government, i.e., they do "choose" the direction of production technology in determining what and how things will be manufactured.

36. Beck et al., "Stratification in a Dual Economy," p. 78. The challenge did not go unnoticed. One critic who is identified with the human capital school (Hauser 1980) dismissed the study by Beck and his associates as essentially false because they included 99 cases as having zero income when they should have been deleted from the analysis. A reanalysis when this is done destroys many of their findings, although a core vs. periphery distinction in earnings does remain, and at quite a high level at that. Beck et al. (1980) replied that their use of the suspect 99 was theoretically justified, an argument I tend to agree with. At any rate, critics of the dual economy approach have been hard put to explain why core/periphery distinctions remain—especially with reference to income earnings—no matter how the data are manipulated and massaged. When examined in detail, such disagreements tend to hinge mostly upon conceptual nuances and categorization decisions. Raffalovich (1994) discovered that his findings supporting segmentation theory attained statistical significance when he reconstituted the boundaries of industries assigned either to the monopoly (core) or competitive (periphery) sectors. In an earlier study, Raffalovich (1993: 135–37) concluded that changes in the amount of time worked by employees had a major impact upon earnings inequality, quite apart from segmentation theory variables. Yet Hodson (1986: 497) sees segmentation as still operative with this variable, in that larger size of firms (core) leads to higher unionization rates and greater provision of full-time, year-round work—which then leads to higher pay. The importance of size and unionization in producing higher pay is corroborated by Kalleberg and Van Buren as well (1994). Hodson (1984) also points out that segmentation works both at the company level (plant employment size) and at the industry level (capital intensity)—and that the two should remain conceptually distinct along with occupational level. Evidence is accumulating, however, which seems to indicate that the explanatory power of dual economy theory has been eroding over the past two decades and has less impact in untangling earnings inequality today than previously (Tigges 1986, 1987; Raffalovich 1990, 1993; Sakamoto 1988). Yet, in the end the loss of explanatory power for the dual economy has been supplanted by other macro-economic and social-structural variables: deindustrialization, capital flight from the United States, trade and export battles, restructured taxation schemes that benefit the very rich, skewed budgets depleting human services, etc. Very few sociologists, it seems, continue to be swayed by the human capital viewpoint.

37. Nevertheless, there is *some* awareness of class self-interest vis-à-vis political cohesiveness and activism, especially on the part of the upper class (Edsall 1985). Research indicates that greater income consistently results in conservative economic policy preferences (Knocke et al. 1987), whereas lesser income produces a tendency to believe that poverty results from the structure of society and social institutions (Oropesa 1986).

38. Gallup Poll, "U.S. Citizens, British Hold Differing Views of Haves, Have-Nots," *Minneapolis Star Tribune,* August 14, 1988, p. 22A.

39. Thomas R. Dye, *Who's Running America?* (Englewood Cliffs, NJ: Prentice-Hall, 1979); C. Wright Mills, *The Power Elite* (New York: Oxford University Press, 1956).

40. G. William Domhoff, *The Powers That Be* (New York: Vintage Press, 1979); Thomas B. Edsall, *The New Politics of Inequality* (New York: Norton, 1985).

41. G. William Domhoff, *Who Rules America Now?* (Englewood Cliffs, NJ: Prentice-Hall, 1983).

42. G. William Domhoff, *The Higher Circles* (New York: Random House, 1970); David R. Simon and D. Stanley Eitzen, *Elite Deviance* (Boston: Allyn and Bacon, 1982).

43. V. I. Lenin, *Imperialism: The Highest Stage of Capitalism* (New York: International Publishers, 1939).

44. Gilpin, *Political Economy of International Relations,* p. 40.

45. For the reader who is seriously interested in a more detailed and sophisticated theoretical framework regarding the multiplicity of contending economic forces in the modern world, an excellent beginning point would be Chilcote (1984). No full theoretical development is offered here because of space and time limitations. For more information about competing and/or complementary theories to the World System approach, see especially Barran (1960, 1969), Barran and Sweezy (1966), and André Gunder Frank (1967, 1969, 1979, 1981a, 1981b).

46. Immanuel Wallerstein, *The Modern World System: Capitalist Agriculture and the Origins of the European World-Economy in the Sixteenth Century* (New York: Academic Press, 1974).

47. Christopher Chase-Dunn and Richard Rubinson, "Toward a Structural Perspective on the World-System," *Politics and Society* 7,4 (1977): 453–76.

48. Ibid., pp. 472–73.

49. Ibid., pp. 475–76.

50. David Snyder and Edward L. Kick, "Structural Position in the World System and Economic Growth, 1955–1970: A Multiple-Network Analysis of Transnational Interactions," *American Journal of Sociology* 84,5 (1979): 1096–1126.

51. Snyder and Kick, *American Journal of Sociology,* pp. 1106–7.

52. World Bank, calculated from *World Development Indicators 1994: Data on Diskette* (Washington, DC: World Bank, 1994).

53. Interestingly, Snyder and Kick report an unexpected serendipitous finding from their analysis. Separately examining the impact of the four networks upon subsequent economic growth for core countries revealed that only the impact of military interventions was negative. In essence, while trade dominance, diplomatic initiatives, and treaty making all work to enrich core nations, military interventionism within poorer countries actually has the opposite effect—it costs the core country a great deal without yielding any financial benefits. Moreover, the economic impact upon the country receiving the military intervention is negligible. It neither hurts nor helps, at least as measured solely in economic terms. The lesson for core countries should be obvious: military interventionism is a nonproductive form of power maintenance. It should not be used other than in a purely defensive capacity—at least, not if it is seen as a viable means for enriching

the core. Military intervention, of course, may be dictated by less direct economic motivation and in less measurable dollars-and-cents payoffs. It can be used to maintain a nation's political hegemonic dominance, and/or to prevent the "threat of a good example" by labeling a nation's development as communist and thus an evil force that must be resisted at all costs (Chomsky 1988).

54. Kenneth Bollen, "World System Position, Dependency, and Democracy: The Cross-National Evidence," *American Sociological Review* 48 (August 1983): 468–79.

55. Edward L. Kick, "World-System Structure, National Development, and the Prospects for a Socialist World Order," in Terry Boswell and Albert Bergesen (eds.), *America's Changing Role in the World System* (New York: Praeger, 1987), pp. 130–43.

56. Terry Boswell and William J. Dixon, "Dependency and Rebellion: A Cross-National Analysis," *American Sociological Review* 55 (August 1990): 546–47, 556.

57. Giovanni Arrighi and Jessica Drangel, "The Stratification of the World Economy: An Exploration of the Semiperipheral Zone," *Review* 10 (1986): 44.

58. Cornelis Peter Terlouw, *The Regional Geography of the World System: Extended Arena, Periphery, Semiperiphery, Core* (Utrecht, Netherlands: Faculteit Ruimteljke Wetenscheppen, 1992), p. 162.

59. Christopher Chase-Dunn, *Global Formation: Structures of the World Economy* (New York: Basil Blackwell, 1989), pp. 214–15.

60. Terlouw, *Regional Geography*, p. 164.

61. Willy Brandt, *Arms and Hunger* (Cambridge, MA: MIT Press, 1986).

62. Theotonio Dos Santos, "The Structure of Dependence," *The American Economic Review* 60 (May 1970), reprinted in Mitchell A. Seligson, *The Gap Between Rich and Poor* (Boulder, CO: Westview, 1984), pp. 95–104.

63. Dos Santos, *The Gap Between Rich and Poor*, p. 100.

64. Ibid., p. 101.

65. Alejandro Portes, "On the Sociology of National Development: Theories and Issues," *American Journal of Sociology* 82 (July 1976): 55–85.

66. S. Lall, "Is 'Dependence' a Useful Concept in Analyzing Underdevelopment?" *World Development* 3 (1979): 799–810; Tony Smith, "The Underdevelopment of the Development Literature: The Case of Dependency Theory," *World Politics* 31 (1979): 247–88; Richard Rubinson and Deborah Holtzman. "Comparative Dependence and Economic Development," *International Journal of Comparative Sociology* 22 (1981): 86–101.

67. Tony Smith, *The Pattern of Imperialism: The United States, Great Britain, and the Late Industrializing World Since 1815* (Cambridge: Cambridge University Press, 1981).

68. See Evans (1979) for a revision of dependency meant to explain these exceptions and the refutations of Brewer (1980) and Kuo et al. (1981).

69. Henrik Marcussen and Jens Torp, *Internationalization of Capital—Prospects for the Third World: A Re-Examination of Dependency Theory* (London: Zed Press, 1982).

70. Bill Warren, *Imperialism: Pioneer of Capitalism* (London: NLB, 1980); Anthony Brewer, *Marxist Theories of Imperialism* (London: Routledge and Kegan Paul, 1980).

71. Bruce Russett, "International Interactions and Processes: The Internal vs. External Debate Revisited," in Ada W. Finifter (ed.), *Political Science—The State of the Discipline*, Washington, DC: The Political Science Association, 1983), ch. 17.

72. Gary Gereffi, "Rethinking Development Theory: Insights from East Asia and Latin America," in A. Douglas Kincaid and Alejandro Portes (eds.), *Comparative National Development: Society and Economy in the New Global Order* (Chapel Hill: University of North Carolina Press, 1994), pp. 29–30.

73. Gary Gereffi, "The International Economy and Economic Development," in Neil J. Smelser and Richard Swedberg (eds.), *The Handbook of Economic Sociology* (New York: Russell Sage, 1994), pp. 219–25.

74. Giorgio Gagliani, "Income Inequality and Economic Development," *Annual Review of Sociology* 13 (1987): 313–34.

75. Robert W. Jackman, "Dependence on Foreign Investment and Economic Growth in the Third World," *World Politics* 34 (January 1982): 175–97; Paul Streeten et al., *First Things First: Meeting Basic Needs in Developing Countries* (New York: Oxford University Press, 1981).

76. Lloyd G. Reynolds, "The Spread of Economic Growth to the Third World, 1850–1950," *Journal of Economic Literature* 21 (1981): 941–80.

77. David Morawetz, *Twenty-Five Years of Economic Development, 1950–1975* (Baltimore, MD: Johns Hopkins University Press, 1977).

78. Mitchell A. Seligson, *The Gap Between Rich and Poor: Contending Perspectives on the Political Economy of Development* (Boulder, CO: Westview, 1984), p. 3.

79. W. A. Lewis, "Development and Distribution," in A. Cairncross and M. Puri (eds.), *Employment, Income Distribution and Development Strategy* (London: Macmillan, 1976), p. 26; H. F. Lydall, *Income Distribution During the Process of Development* (Geneva: International Labour Office, 1977), pp. 13–14.

80. Simon Kuznets, "Economic Growth and Income Inequality," *American Economic Review* 45 (March 1955): 1–28; Simon Kuznets, "Quantitative Aspects of the Economic Growth of Nations: Distribution of Income by Size," *Economic Development and Cultural Change* 11,2 (January 1963): 1–80.

81. Roger D. Hansen, "The Emerging Challenge: Global Distribution of Income and Economic Opportunity," in James W. Howe (ed.), *The U.S. and World Development Agenda for Action 1975* (New York: Praeger, 1975), p. 61.

82. Montak S. Ahluwalia, "Inequality, Poverty, and Development," *Journal of Development Economics* 3 (1976): 307–42.

83. Ronald H. Chilcote, *Theories of Development and Under-Development* (Boulder, CO: Westview, 1984), p. 11.

84. Shirley Cerresoto, "Socialism, Capitalism, and Inequality," *Insurgent Sociologist* 11,2 (Spring, 1982): 8.

85. Lewis, "Development and Distribution," pp. 28–29.

86. Francis Moore Lappé and Joseph Collins, *World Hunger: Twelve Myths* (New York: Grove Press, 1986), pp. 85–94.

87. Francis Moore Lappé, Rachel Shurman, and Kevin Danaher, *Betraying the National Interest: How U.S. Foreign Aid Threatens Global Security by Undermining the Political and Economic Stability of the Third World* (New York: Grove Press, 1987), pp. 15–55.

88. Gilpin, *Political Economy*, p. 269.

89. Walden Bello and Shea Cunningham, "Reign of Error: The World Bank's Wrongs," *Dollars and Sense* 195 (September/October 1994): 12.

90. Clay Chandler, "A Shrinking Line of Credibility: Critics Claim the World Bank Perpetuates the Poverty It Was Formed to Eliminate," *The Washington Post National Weekly Edition*, June 27–July 3, 1994, p. 21.

91. Susan George, *A Fate Worse Than Debt* (New York: Grove Press, 1988), pp. 58–73.

92. Ibid., p. 10.

93. In fairness, this position is and will continue to be intensely controversial. To begin with, many LDCs are currently attempting to divorce themselves from dependence upon IMF and World Bank loans, multinational corporate investment, etc., in recognition of their capital flight problems (Mason 1988: 117). World Bank (1984) figures suggest the point may become moot, noting that among nonindustrialized countries in the 1970–81 period 79% of gross domestic investment came from gross national savings. Interestingly, criticism of development aid has come from both ends of the political spectrum (Toye 1987). Conservatives argue that aid which focuses upon questions of poverty, basic human needs, and income inequality needs to be challenged; nations can grow economically only through freely operating markets with a hands-off policy by their governments The other problems will then be eventually eradicated through increasing affluence. On the whole, however, some review compendiums of development research find that on balance most recipient countries have benefited from such assistance (Riddell 1987). Although not all countries do benefit and not all types of development aid are helpful, most aid succeeds in meeting its objectives and in yielding respectable rates of economic return (Cassen et al. 1986). This is particularly true with reference to family planning and population control programs that have reduced runaway birth rates, enabling poorer countries to invest more in the future rather than being forced to feed and clothe a burgeoning dependent population. Especially for LDCs, reduced fertility and smaller family size have produced higher savings rates which can be used in national economic development (Mason 1988). The effect of multinational corporate investment, however, on development within periphery countries is much more doubtful and will be addressed in the next chapter. London (1988) found with respect to fertility decline that multinational corporate penetration did significantly retard development in periphery countries. For a radically different view of abuses development aid can produce, i.e., to block fundamental needed change while mounting a pacification effort, U. S. involvement with Central America has provided many recent examples (Barraclough and Scott 1988; Barry and Preusch 1988).

94. Bello and Cunningham, "Reign of Error," p. 11.

95. George, *Fate Worse Than Debt*, pp. 63–73.

96. Ibid., p. 15.

97. Bradley P. Bullock, "Cross-National Research and the Basic Needs Approach to Development: A New Direction," paper presented at the American Sociological Association conference, Chicago, 1987; Paul Streeten et al., *First Things First: Meeting Basic Needs in Developing Countries* (New York: Oxford University Press, 1981).

98. Lee Soltow, *Patterns of Wealthholding in Wisconsin since 1850* (Madison: Wisconsin University Press, 1971).

99. Shirley Kuo, Gustav Ranis, and John C. Fei, *The Taiwan Success Story: Rapid Growth with Improved Distribution in the Republic of China, 1952–1979* (Boulder, CO: Westview, 1981).

100. A Douglas Kincaid and Alejandro Portes (eds.), *Comparative National Development: Society and Economy in the New Global Order* (Chapel Hill: University of North Carolina Press, 1994), p. 11.

101. Terence Moll, "Mickey Mouse Numbers and Inequality Research in Developing Countries," *The Journal of Development Studies* 28,4 (July 1992): 697. Moll argues that Taiwan was egalitarian from the very beginning, but this was camouflaged by notably poor numbers on income distribution gathered in the 1950s. He suggests that the trend toward equality was really due to the end of colonialism in Taiwan, the expropriation of Japanese assets by the state, land reforms (1949–1958), and an emphasis on the "redistribute and educate now, grow later" approach to development.

102. York W. Bradshaw, "Transnational Economic Linkages, the State, and Dependent Development in South Korea, 1966–1988: A Time Series Analysis," *Social Forces* 72,2 (December 1993): 315–46.

103. Dani Rodrik, *Getting Interventions Right: How South Korea and Taiwan Grew Rich* (Cambridge, MA: National Bureau of Economic Research, Inc., 1994), pp. 39–40. All may not be well in these two countries, however, as some evidence indicates that income inequality did start to go up in both areas from the 1960s to the 1980s. There is no argument that both countries started from a base of low-income inequality, but as exports developed inequality started to rise. See especially Walden Bello and Stephanie Rosenfeld, *Dragons in Distress: Asia's Miracle Economies in Crisis* (San Francisco: Institute for Food and Development Policy, 1992), pp. 37–38, 225–27.

CHAPTER 3

1. Richard J. Barnet, "Lords of the Global Economy: Stateless Corporations," *The Nation* 259,21 (December 19, 1994): 754 ff.

2. John Whalley, "The Worldwide Income Distribution: Some Speculative Calculations," *The Review of Income and Wealth* 25 (1979): 261–76.

3. Cuomo Commission on Trade and Competitiveness, *The Cuomo Commission Report: A New American Formula for a Strong Economy* (New York: Simon and Schuster, 1988), pp. 26–27.

4. Ibid., p. 32.

5. Barry Bluestone and Bennett Harrison, *The Deindustrialization of America: Plant Closings, Community Abandonment, and the Dismantling of Basic Industry* (New York: Basic Books, 1982), pp. 113–14.

6. Robert W. Jackman, "Dependence on Foreign Investment and Economic Growth in the Third World," *World Politics* 34 (January 1982): 175–97.

7. U.S. Census Bureau, *Statistical Abstract of the United States 1994*, CD-ROM (Washington, DC: Government Printing Office, 1995), table 1329B

8. Council of Economic Advisors, *Economic Report of the President 1995* (Washington, DC: U.S. Government Printing Office), pp. 235–36.

9. Until recently, the United States used Gross National Product (GNP) rather than Gross Domestic Product (GDP) to measure our nation's economic activity. In 1992, we joined the rest of the world in using Gross Domestic Product as our primary measure. GPD is the total market value of all finished goods and service produced in a country within one year. While GDP measures economic activity within a country, GNP measures the economic production of all the people of a country—whether they reside within its boundaries or not. In short, the economic activity of U. S. citizens working abroad is counted in GNP but it is not added into GDP. Conversely, the income of a foreign person or corporation doing business in the United States is not counted in our GNP, but it is counted in our GDP. In the end, Gross Domestic Product describes the economic activity occurring within the physical boundaries of a nation, while Gross National Product covers the economic activity of all citizens of a country—wherever they may be living. At least for the United States, the distinction between the two measures is not crucial. The value of foreign business production occurring within our boundaries is about the same as the value of U.S. business production that is done abroad (Colander 1993: 158–59).

10. Stephen D. Krasner, *Structural Conflict: The Third World Against Global Liberalism* (Berkeley: University of California Press, 1985), pp. 97, 101.

11. P. T. Bauer, "The Vicious Circle of Poverty," in Mitchell A. Seligson (ed.), *The Gap Between Rich and Poor: Contending Perspectives on the Political Economy of Development* (Boulder, CO: Westview, 1984), p. 324.

12. This is true whether China is seen as a poor, struggling nation seeking economic development or as a political and military monolith completely in charge of its own destiny. Actually, there are elements of truth in both perspectives. The authoritarian government of China has rigidly controlled the pace of development to this point, retaining control of all important financial decisions. Its success has been stupendous. China's annual rate of GNP growth has not sunk under 10 percent since 1980. The impact of state-owned enterprises has declined from 78 percent of all economic activity in 1978 to 43 percent in 1994. More than 100 million Chinese were lifted out of poverty during the 1980s because of development.

Yet not all forces are completely under government control: inflation is relatively high at 20 percent, crime in urban areas is on the rise, tax evasion has become widespread, mass depopulation of the rural hinterland is underway, 30 million state workers are superfluous but draw pay nonetheless (putting a severe drag on the economy), 40 percent of state enterprises continue to lose money, and a massive amount of debt and bad loans have piled up because tottering state enterprises are too unprofitable to pay them back (Mufson 1995: 20).

13. Susan George, *A Fate Worse Than Debt: The World Financial Crisis and the Poor* (New York: Grove Press, 1988), p. 73.

14. Mitchell A. Seligson, "The Dual Gaps: An Overview of Theory and Research," in Mitchell A. Seligson (ed.), *The Gap Between Rich and Poor: Contending Perspectives on the Political Economy of Development* (Boulder, CO: Westview, 1984), p. 3.

15. World Bank, *World Development Indicators 1994* Data Diskette (Washington, DC: World Bank), table 30.

16. Albert Berry, Francois Bourguignon, and Christian Morrisson, "Changes in the World Distribution of Income Between 1950 and 1977," *The Economic Journal* 93 (June 1983): 331–50.

17. Ibid., p. 340.

18. Ibid.

19. N. Kakwani and K. Subbarao, "Global Development: Is the Gap Widening or Closing?" *Policy Studies in Developing Countries* 1 (1994): 65–118.

20. Ibid., pp. 15, 18.

21. World Bank, *The World Development Report 1988* (New York: Oxford University Press, 1988), pp. 46–54, 111–20.

22. Harry Makler, *The New International Economy* (Beverly Hills, CA: Sage, 1982), p. 16.

23. Andrew Reding, "Mexico at a Crossroads," *World Policy Journal* 5,4 (Fall 1988): 633.

24. William R. Cline, *International Debt and the Stability of the World Economy* (Washington, DC: Institute for International Economics, 1983), pp. 20–21.

25. George, *A Fate Worse Than Debt,* pp. 60–61.

26. Ibid., p. 63.

27. Cuomo Commission, *Report,* pp. 53–54.

28. A. Douglas Kincaid and Alejandro Portes (eds.), *Comparative National Development: Society and Economy in the New Global Order* (Chapel Hill: University of North Carolina Press, 1994), pp. 8–9.

29. A. Kent MacDougall, "In Third World, All But the Rich Are Poorer," *Lose Angeles Times,* November 4, 1984, p. 1.

30. Walden Bello and Shea Cunningham, "Reign of Error: The World Bank's Wrongs," *Dollars and Sense* 195 (September/October 1994): 10.

31. Manuel Castells and Roberto Laserna, "The New Dependency: Technological Change and Socioeconomic Restructuring in Latin America," in A. Douglas Kincaid and Alejandro Portes (eds.), *Comparative National Development: Society and Economy in the New Global Order* (Chapel Hill: University of North Carolina Press, 1994), p. 59.

32. International Monetary Fund, *Direction of Trade Statistics Yearbook* (Washington, DC: International Monetary Fund, 1987), p. 405. Some recovery for Mexico's dire situation was witnessed by the early 1990s. Between 1989 to 1993, Mexico increased its exports to the United States by one-half and raised its imports from the United States by nearly two-thirds (*U.S. Statistical Abstract 1994,* table 1329). Up until the NAFTA agreement was signed, which essentially allows free trade between our two countries, Mexico bought nearly as much from the United States as it sold to us. It should be remembered that Mexico is a special case, however, where more strenuous development effort and special trade/financial provisions apply. More details are provided later in the sections on the *maquiladora* and free trade.

33. Bennett Harrison and Barry Bluestone, *The Great U-Turn: Corporate Restructuring and the Polarizing of America* (New York: Basic Books, 1988), pp. 154–55.

34. Robert Gilpin, *The Political Economy of International Relations* (Princeton, NJ: Princeton University Press, 1987), pp. 330, 344.

35. Cuomo Commission, *Report,* p. 5.

36. U.S. Census Bureau, *Statistical Abstract of the United States, 1994,* CD-ROM, table 504.

37. Economy in Numbers, "Debt and Distribution," *Dollars and Sense* 177 (June, 1992): 23.

38. David M. Gordon, "Private Debt Dwarfs Uncle Sam's," *Los Angeles Times,* January 20, 1978, p. 3.

39. Harrison and Bluestone, *The Great U-Turn,* pp. 149, 151–52, report that consumer debt doubled during the Reagan years of 1981–1986 in a desperate attempt by average people to keep abreast of stagnation and decline in real income. The use of "plastic money" expanded at even more astronomical rates. Revolving installment credit owed to Visa, Mastercard, Sears, etc., nearly tripled during the same period. By the mid-1980s, the typical American family owed $11,500, not counting its home mortgage. If this amount is calculated to the norm of a three-year loan at 12 percent interest, the average monthly payment on past debt comes to $380. With an after-tax monthly household income of $1,390 during that period, this meant that the normal American family was paying one-fifth of its monthly income just to pay off its old debts—not including its house payment.

40. Council of Economic Advisors, *Economic Report of the President 1995,* pp. 54–55.

41. Albert J. Crenshaw, "Back on a Borrowing Binge," *The Washington Post National Weekly Edition,* October 3–9, 1994, p. 19.

42. Walter Mossberg, "Cost of Paying the Foreign Piper," *Wall Street Journal,* January 18, 1988, p. 3.

43. Bob Rast, "U.S. Banks Lose Top Status in Global Financial Markets," *Minneapolis Star Tribune,* November 27, 1988, p. 1D.

44. Gilpin, *Political Economy,* p. 329.

45. IMF, *Trade Statistics Yearbook,* pp. 243, 404.

46. U.S. Census Bureau, *Statistical Abstract of the United States, 1994,* CD ROM, table 1329B.

47. Harrison and Bluestone, *The Great U-Turn,* p. 147.

48. David K. Henry, and Richard P. Oliver, "The Defense Buildup, 1977–85: Effects on Production and Employment," *Monthly Labor Review,* August 1987, p. 6.

49. Council of Economic Advisers, *Economic Report of the President* (Washington, DC: U.S. Government Printing Office, 1987), Appendix B.

50. Council of Economic Advisers, *Economic Report of the President* (Washington, DC: U.S. Government Printing Office, 1995), table B-107.

51. "Study Says Deficit Cost 5.1 Million Jobs," *Minneapolis Star Tribune,* October 16, 1988, p. 3A.

52. Hobart Rowen, "Capital Economics: Candidates in Blunderland," *The Washington Post National Weekly Edition,* October 10–16, 1988, p. 5.

53. Gilpin, *Political Economy,* pp. 336, 346–49.

54. Howard M. Wachtel, *The Money Mandarins: The Making of a Supranational Economic Order* (New York: Pantheon Books, 1986), p. 128. It is quite true that the rich do save and invest more than the poor. The implication is that

the poor are less thrifty—in addition to being lazy and shiftless—than a less careless and thoughtless upper class. A moment's reflection leads to the realization, however, that the poor simply do not have excess funds left over for investment after meeting the basic necessities of life.

55. George, *A Fate Worse Than Debt,* pp. 171–77.

56. Ibid., p. 177.

57. Chris Tilly, "Raising Cane in Jamaica," *Dollars and Sense* 169 (September 1991): 16–18, 22.

58. Steven Stack, "The Effect of Direct Government Involvement in the Economy on the Degree of Income Inequality: A Cross-National Study," *American Sociological Review* 43 (December 1978): 880–88.

59. P. A. Della and N. Oguchi, "Distribution, The Aggregate Consumption Function, and the Level of Economic Development: Some Cross-Country Results," *Journal of Political Economy* 84,6 (December 1976): 1325–34; Ashfaque Khan, "Aggregate Consumption Function and Income Distribution Effect: Some Evidence from Developing Countries," *World Development* 15,10/11 (1987): 1369–74.

60. York Bradshaw, Rita Noonan, Laura Gash, and Claudia Buchman Sershen, "Borrowing Against the Future: Children and Third World Indebtedness," *Social Forces* 71,3 (March 1993): 629–57.

61. Manual Pastor, Jr., "The Effects of IMF Programs in the Third World: Debate and Evidence from Latin America," *World Development* 15,2 (1987): 259. One study (Helleiner 1987) even documents the failure of Tanzania, with a government outspokenly sympathetic and concerned for the well-being of its people, to protect its citizens from the erosion and financial deterioration caused by an IMF-induced austerity plan. Despite the best intentions, therefore, some doubt remains whether a country can ward off the evil effects of these adjustment packages and protect its inhabitants—even if it is inclined to do so.

62. Alan Riding, "Debt Fears Realized With Venezuela Unrest," *Minneapolis Star Tribune,* March 2, 1989, p. 4A.

63. News Services, "Venezuela President Blames Debt for Riots," *Minneapolis Star Tribune,* March 4, 1989, p. 3A.

64. Economy in Numbers, "The High Cost of Debt," *Dollars and Sense* 144 (March 1989): 23.

65. John Walton and David Seddon, *Free Markets and Food Riots: The Politics of Global Adjustment* (Cambridge, MA: Blackwell, 1994), p. 42.

66. John Walton and Charles Ragin, "Global and National Sources of Political Protest: Third World Responses to the Debt Crisis," *American Sociological Review* 55 (December 1990): 882.

67. Alejandro Portes, "The Informal Economy and Its Paradoxes," in Neil J. Smelser and Richard Swedberg (eds.), *The Handbook of Economic Sociology* (New York: Russell Sage, 1994), p. 444.

68. UNESCO, *The Use of Socio-Economic Indicators in Development Planning* (Paris: UNESCO, 1976).

69. United Nations Children's Fund, *The State of the World's Children, 1991* (New York: Oxford University Press, 1991).

70. United Nations, *Human Development Report 1990* (New York: Oxford University Press, 1990).

71. Richard J. Estes, "Toward a 'Quality of Life' Index: Empirical Approaches to Assessing Human Welfare Internationally," in Jim Norwine and Alfonso Gonzalez (eds.), *The Third World: States of Mind and Being* (Boston: Unwin Hyman, 1988), ch. 3.

72. Leslie Sklair, *Sociology of the Global System* (Baltimore, MD: Johns Hopkins University Press, 1991), p. 19.

73. Clifford Cobb, Ted Halstead, and Jonathan Rowe, "If the GDP Is Up, Why Is America Down?" *Atlantic Monthly,* October 1995, pp. 59–78.

74. Ibid., p. 72.

75. Ibid., p. 68.

76. John Schwartz, "Pinning a Price Tag on Nations," *The Washington Post National Weekly Edition,* September 25–October 1, 1995, p. 38.

77. World Bank, *World Development Report 1994,* tables 1, 30

78. Terence Moll, "Mickey Mouse Numbers and Inequality Research in Developing Countries," *The Journal of Development Studies* 28,4 (1992): 694.

79. Montek S. Ahluwalia, "Income Inequality: Some Dimensions of the Problem," in Mitchell A. Seligson (ed.), *The Gap Between the Rich and Poor: Contending Perspectives on the Political Economy of Development* (Boulder, CO: Westview, 1984), pp. 14–21.

80. William Loehr, "Some Questions on the Validity of Income Distribution," in Seligson (ed.), *Gap Between the Rich and Poor,* pp. 283–91. The difficulties in drawing international comparisons on income inequality are compounded by shifts in sample coverage (e.g., national, urban only, rural only, etc.), the unit of analysis (families, households, income recipients, economically active population), and a wide variation in the years in which the data are compiled. In a detailed analysis of what biases could result from ill-considered mixing of these data groups, Menard (1986) found that household income and personal income distributions are essentially equivalent, as are personal income distributions based on urban-only samples with national level data. He also discovered very little effect in averaging income inequality data for a single period of time, but only within recent years (estimates based upon data gathered prior to 1960 are largely invalid). On the whole, however, he recommends that researchers avoid combining different data sets on income inequality for the same country that are more than five years apart. Moll (1992) suggests a number of ways researchers could improve the treatment and usage of questionable income distribution data: clarify the concepts (What is income? What unit is being looked at? Before or after taxes? Are income transfers counted? etc.), clarify the methods (What summary measures are being used? Why? If the Gini ratio is used, which calculation formula?), calculate confidence intervals for inequality coefficients, review data collection methodologies (including information on the questionnaires used, interviewer training, etc.), and group data into regions for comparison, rather than using country-to-country comparisons. Although Moll's critique is useful and serves as a caution against sloppy scholarship, not doing analyses because of data imperfections may be worse. The questions that feed the ongoing research are too important to neglect. As he suggests, however, plenty of caveats describing drawbacks to the data are in order—as are some fresh analytical approaches that seek to deal with these deficiencies.

81. Gregg A. Hoover, "Intranational Inequality: A Cross-National Dataset," *Social Forces* 67,4 (June 1989): 1016.

82. World Bank, *World Development Indicators 1994* Data Diskette (Washington, DC: World Bank, 1994). The World Bank correctly points out that the collection of income distribution data is not systematically organized within many countries and that some data may be derived from surveys conducted for other purposes—most frequently consumer buying studies. Although the estimates they use are considered the best available for the most current, up-to-date statistics, in some cases the coverage of the surveys may be too unreliable to make nationwide estimates of income distribution. Nonetheless, the figures compiled by the World Bank are for both rural and urban areas, and a crucial error is injected if both sectors are not represented. The surveys are also relatively recent, which is a major strength in comparison to the ILO study or the carefully compiled data set offered by Greg Hoover (1989), whose figures are circa 1969.

83. Ibid., table 30 footnote. There are some crucial drawbacks which can lead to noncomparability. To begin with, the surveys differ in using income or consumption expenditure as the living standard indicator. For 28 of the 45 low- and middle-income country samples, the data reflect consumption expenditure. This introduces an underestimate bias of inequality, since income figures are typically more unevenly distributed than consumption data. In short, the actual inequality experience for LDCs is probably worse than the statistics indicate. Some surveys also continue to use individuals, rather than households, as their unit of observation.

84. Edward N. Muller, "Democracy, Economic Development, and Income Inequality," *American Sociological Review* 53 (February 1988): 50–68. Strangely, however, Muller chooses not to control for unit of analysis but instead mixes individual and household data. This would be acceptable if the individual data were not only for workers or economically active persons, but for all income recipients. It is impossible to tell directly from his data.

85. Lars Osberg, *Economic Inequality in the United States* (Armonk, NY: M. E. Sharpe, 1984), p. 19.

86. It is crucial to keep in mind that although LDCs have greater *internal* inequality than core countries, the great majority of their problem stems from the huge differences between their average incomes and those of industrial nations. Average per-capita income is 7.6 times higher in developed countries than in LDCs, while over half of worldwide income inequality is created by this gap. Only 18 percent is attributable to internal income inequality within LDCs (Berry et al. 1983). Put another way, Whalley (1979) believes that if one had to choose between attacking income inequality between countries and eradicating income inequality within countries—be they LDCs or industrial nations—equalizing incomes between countries would be the most effective way to reduce worldwide inequality.

87. Economy in Numbers, "U.S. among Worst in Inequality," *Dollars and Sense* 178 (July/August 1992): 23.

88. M. O. Lorenz, "Methods of Measuring the Concentration of Wealth," *Quarterly Publications of the American Statistical Association* 9 (1905): 205–19.

89. Of course, few researchers ever actually draw Lorenz curves to estimate degree of inequality when using the Gini ratio. A full discussion of this and other

income-inequality measurements can be found in Schwarz and Winship (1979). The formulas for the Gini ratio vary, depending upon whether grouped data or continuous data are being used. The calculations for grouped data in my analysis follow Morgan (1962). Aiker (1965) has the clearest explanation of this procedure. Also, the methodology now exists to calculate Gini ratios using grouped data by drawing Lorenz curves with personal computers (Brown and Mazzarino, 1984). For a current argument supporting the use of Lorenz curves rather than Gini ratios in making comparisons between U.S. states or among nations, see Bishop, Formby, and Thistle (1992) and Bishop, Formby, and Smith (1991).

90. Age has an effect upon the ratio, so that the simple aging of a population can increase inequality quite apart from any structural change (Morgan 1962). Although Paglin (1975) introduced a modification of the traditional Gini ratio which eliminates about one-third of the inequality, it has not been widely adopted in subsequent research (for a modification, see Formberg and Seaks 1980). A popular method of computing Gini scores utilizes Census income categories. The assumption is made that the mean income of each income category is the midpoint of that interval. Cumulative percentages are then calculated, frequently with an adjustment using a Pareto curve (Knott, 1970) for the open-ended interval at the top. This is not always done, however, so such calculations can lead to inconsistent comparisons between different sets of data constructed at different times. Moreover, the number of income intervals differs from one census to another (14 in 1950, 13 in 1960, 15 in 1970, 9 in 1980, and 25 in 1990). Even in the same census year, the Gini may vary between publications. In 1970, the Census bureau used 15 income classes in its published state reports but 19 classes were used to compute the Gini (Knott, 1970). It has been shown that the Gini coefficient will be reduced to the degree that the number of income intervals declines (Sale, 1974). Hoover (1989: 1011) also found that as a researcher moves from calculating Gini ratios using quintiles to a more refined decile distribution, the rates indicate a greater degree of inequality. At times, the reduction in accuracy because of the number and/or nature of the income groups may be misleadingly low. Murphy (1985) reports that based upon Irish income data, there is a 1 to 2 percent underestimation of the actual Gini ratio—which seems innocuous. It is especially when Gini ratios are decomposed for separate analytical groups, however, that error becomes quite large. For example, in 1980 the underestimation of the Gini coefficient for Ireland was 1.7 percent for direct income, for households with employee heads it was 3.3 percent, and for households with three or more earners it was 25.5 percent. It is also true that the Gini ratio is not sensitive to nonmoney income components and differential price indices between states. Jonish and Kau (1973: 180) conclude that this exaggerates income inequalities in rural areas. Budd (1970) directs his criticism of the Gini coefficient against its relative insensitiveness, which can be quite serious for researchers making comparisons between various dates. Quite simply, the Gini coefficient is more responsive to changes in the distribution of income among the middle class than among the rich or poor (Osberg 1984: 29; Allison 1978: 868). This may be why the Gini coefficient shows such stability over long periods of time in the United States. Soltow's study of Wisconsin income data (1971), for example, shows no great change in the Gini coefficient over one hundred years while Reynolds and

Smolensky (1977) show no U.S. change from 1950 to 1970 despite massive taxation and welfare shifts. One of the most recent charges against the Gini ratio and other summary measurements (Theil, coefficient of variation, variance of logarithms) is that because of their very nature they cannot adequately measure polarization that has taken place at both ends of the income distribution spectrum (Morris et al. 1994). Rather, the great increases among the poor and the rich tend to be hidden by the averaging effect of these measurements. The authors argue that looking at changes in deciles over time would more sharply focus where changes are taking place. This seems plausible. Although the Gini index is quite sensitive to changes in the middle, i.e., has dutifully registered and recorded the increase of inequality as the middle class has eroded, it cannot directly show where the money has gone. A decile distribution viewed over time shows loss to the poor and gain to the rich.

91. Denny Braun, "Multiple Measurements of United States Income Inequality," *The Review of Economics and Statistics* 70,3 (1988): 398–405. For corroboration through partial replication, see also John A. Bishop, John P. Formby, and Paul D. Thistle, "Explaining Interstate Variation in Income Inequality," *The Review of Economics and Statistics* 74,3 (1992): 553–57.

92. Osberg, *Economic Inequality,* p. 29; Paul D. Allison, "Measures of Inequality," *American Sociological Review* 43 (December 1978): 868.

93. The annual rate does not completely remove bias, however, because some nations in the periphery had very early surveys and were closer to the ill effects of the OPEC oil price hike in 1973 than most nations in the core. Unfortunately, subtracting the early surveys in the periphery would have eliminated a great number of countries—making any comparison meaningless. Thus, a rough comparison with acknowledged distortions became the only viable alternative to no comparison at all.

94. Johan Fritzell, "Income Inequality Trends in the 1980s: A Five-Country Comparison," *Acta Sociologica* 36,1 (1993): 47–62.

95. In a study of Canada, the Netherlands, Sweden, the United Kingdom, and the United States, which decomposed the coefficient of variation as an income inequality measurement, Markus Jantti (1993) also concluded that demographic shifts contributed almost nothing to the growth of income inequality in developed countries. Changes in the age structure of the populations or the rise and fall of single-parent households are not the cause of growing inequality within different nations. With few exceptions, inequality increased within all population subgroups that were studied, including families with different numbers of earners.

96. Peter Gottschalk, "Changes in Inequality of Family Income in Seven Industrialized Countries," *American Economic Review* 83,2 (May 1993): 136–42.

97. Timothy M. Smeeding and John Coder, *Income Inequality in Rich Countries During the 1980s,* Working Paper No. 88 (Syracuse, NY: Syracuse University, Maxwell School of Citizenship and Public Affairs, 1993).

98. A. B. Atkinson, Lee Rainwater, and Timothy Smeeding, *Income Distribution in OECD Countries* (Paris: OECD, 1995).

99. Anthony Atkinson, Lee Rainwater, and Timothy Smeeding, *Income Distribution in Advanced Economies: Evidence from the Luxembourg Income Study,*

Working Paper No. 120 (Syracuse, NY: Syracuse University, Maxwell School of Citizenship and Public Affairs, 1995).

100. Ibid., pp. 36–37.

101. MacDougall, "In Third World," pp. 1, 3, 8.

102. Reding, "Mexico at a Crossroads," p. 630.

103. James M. Cypher, "The Party's Over: Debt, Economic Crisis Undermine Mexico's PRI," *Dollars and Sense* 142 (December 1988): 9–11, 21.

104. Bello and Cunningham, "Reign of Error," pp. 10–11.

105. Joel Millman, "Mexico's Billionaire Pyramid," *The Washington Post National Weekly Edition,* December 5–11, 1994, p. 25.

106. Stephen G. Bunker, *Underdeveloping the Amazon: Extraction, Unequal Exchange, and the Failure of the Modern State* (Chicago: University of Illinois Press, 1985).

107. Benedict J. Clements, *Foreign Trade Strategies, Employment, and Income Distribution in Brazil* (New York: Praeger, 1988), pp. 31–33.

108. David Denslow, Jr., and William Tyler, "Perspectives on Poverty and Income Inequality in Brazil," *World Development* 12,10 (1984): 1019–28.

109. José Camargo, "Income Distribution in Brazil: 1960–1980," monograph, Pontificia Universidade Catolica do Rio de Janeiro, 1984, p. 9.

110. Helga Hoffman, "Poverty and Property in Brazil: What Is Changing?" in Edmar Bocha and Herbert S. Klein (eds.), *Incomplete Transition: Brazil Since 1945* (Albuquerque: University of New Mexico Press, 1989), pp. 197–231.

111. Albert Fishlow, "A Tale of Two Presidents: The Political Economy of Crisis Management," in Alfred Stepan (ed.), *Democratizing Brazil: Problems of Transition and Consolidation* (New York: Oxford University Press, 1989), pp. 83–119.

112. Ibid., p. 99.

113. Seth Racusen, "Lula's Rise," *Dollars and Sense* 195 (September/October 1994): 19. For now, the hyperinflation rate that has haunted Brazil for decades has been reduced to near zero by the success of the *Plano Real*. This plan was introduced by recently elected President Fernando Henrique Cardoso. As a sociologist with Marxist leanings, Cardoso was responsible for finally introducing economic reforms that assured a stable, hard currency (Goertzel 1995). For the first time ever, people earning Brazil's minimum wage can afford to buy the market basket of basic goods used as an index by government planners. For more information, contact Professor Ted Goertzel, Department of Sociology, Rutgers University, Camden, NJ.

114. Peter Evans, *Dependent Development: The Alliance of Multinational, State, and Local Capital in Brazil* (Princeton, NJ: Princeton University Press, 1979), pp. 81–83, 159–62, 214–24, 278–88.

115. Ibid., pp. 96–97, 288.

116. Jan Knippers Black, *United States Penetration of Brazil* (Philadelphia: University of Pennsylvania Press, 1977), p. 241.

117. Clements, *Foreign Trade Strategies,* pp. 143–44.

118. Fishlow, "A Tale of Two Presidents," pp. 83–119.

119. Seth Racusen, "Lula's Rise," *Dollars and Sense* 195 (September/October 1994): 18.

120. MacDougall, "In Third World," pp. 3, 8; World Bank, *World Development Indicators 1994,* table 30.

121. Albert Fishlow, "Brazilian Size Distribution of Income," *American Economic Review* 62 (May 1972): 391–401; Denslow and Tyler, "Perspectives on Inequality in Brazil," p. 1024.

122. Gary S. Fields, "Who Benefits from Economic Development? A Reexamination of Brazilian Growth in the 1960s," *American Economic Review* 67,4 (1977): 570–82.

123. Hoffman, "Poverty and Property in Brazil," pp. 197–231.

124. World Bank, *World Development Indicators 1994,* table 30.

125. Camargo, "Income Distribution in Brazil," pp. 19–20.

126. Black, *United States Penetration of Brazil.*

127. Ibid., p. 236.

128. Ibid., pp. 238–39. Originally in Marcio Moreira Alves, *A Grain of Mustard Seed: The Awakening of the Brazilian Revolution* (New York: Doubleday, 1973), pp. 164–65.

129. Evans, *Dependent Development,* pp. 287–95, 314–29.

130. New York Times, "Income Gaps Pose Threats in 4 Countries, Report Warns," *Minneapolis Star Tribune,* June 2, 1994, p. 13A.

131. Albert Fishlow, "Some Reflections on Post-1964 Brazilian Economic Policy," in Stepan (ed.), *Authoritarian Brazil* (New Haven, CT: Yale University Press, 1973), p. 90.

132. Harry Makler, Alberto Martinelli, and Neil Smelser (eds.), *The New International Economy* (Beverly Hills, CA: Sage, 1982), p. 22.

133. Edmar L. Bacha, "External Shocks and Growth Prospects: The Case of Brazil, 1973–89," *World Development* 14,8 (1986): 919–36.

134. Michael Snellenberger, "Brazil's Debt Debacle," *Dollars and Sense* 199 (May/June 1995): 23.

135. Eul-Soo Pang, "Debt, Adjustment, and Democratic Cacophony in Brazil," in Barbara Stallings and Robert Kaufman (ed.), *Debt and Democracy in Latin America* (Boulder, CO: Westview, 1989), pp. 127–42.

136. Michael Snellenberger, "Brazil's Debt Debacle," *Dollars and Sense* 199 (May/June 1995): 23.

137. George, *A Fate Worse Than Debt,* pp. 122–23.

138. James Brooke, "A Hard Look at Brazil's Surfeits: Food, Hunger and Inequality," *New York Times,* June 6, 1993, S4, p. 7.

139. James Brooke, "Slavelike Conditions Are Proliferating in Brazil," *Minneapolis Star Tribune,* May 30, 1993, p. 8A.

140. Charles H. Wood and Jose Alberto Magno de Carvaho, *The Demography of Inequality in Brazil* (New York: Cambridge University Press, 1988), pp. 184–200.

141. Ibid., p. 258.

142. Walter Russel Mead, "The United States and the World Economy," *World Policy Journal* 6,1 (1989): 1–46; Jerry W. Sanders, "America in the Pacific Century," *World Policy Journal* 6,1 (1989): 47–80; Robert Gilpin, *The Political Economy of International Relations* (Princeton, NJ: Princeton University Press, 1987).

143. Roger Cohen, "Rio's Murder Wave Takes on the Aura of a Class Struggle," *Wall Street Journal,* May 9, 1989, pp. A1, A15.

144. Nancy Gibbs, "So You Think Your City's Got Crime?" *Time,* March 5, 1990, pp. 54–55.

145. Nancy Scheper-Hughes, *Death Without Weeping: The Violence of Everyday Life in Brazil* (Berkeley: University of California Press, 1992).

146. Jean Hopfensperger, "Human Rights of Nation's Poor Are Issue," *Minneapolis Star Tribune,* July 26, 1994, p. 7B.

147. Cohen, "Rio's Murder Wave," p. A15. Ironically, some middle-class Brazilians are actually moving into the *favellas* to escape the crime. Every *favella* is essentially ruled by drug traffickers who will not allow crime in their neighborhoods. The cost of disobedience is immediate execution in a swift, strict code of primitive justice (Harris 1994).

148. For example, Ben Bagdikian (*The Media Monopoly* 1987) points out that General Electric—the tenth largest U.S. corporation—owns RCA, which in turn owns NBC. Bagdikian does a commendable job of providing a great deal of evidence showing a right-wing, pro-business, conservative bias to U.S. news, advertising, and general media content. For further details on the conservative bias of American media caused by big business, see Noam Chomsky and Edward S. Herman, *Manufacturing Consent: The Political Economy of the Mass Media* (New York: Pantheon, 1989), pp. 3–18.

CHAPTER 4

1. Simon Kuznets, "Economic Growth and Income Inequality," *American Economic Review* 45,1 (March, 1955): 1–28.

2. David A. Smith, "Overurbanization Reconceptualized: A Political Economy of the World-System Approach," *Urban Affairs Quarterly* 23,2 (1987): 270–94.

3. Felix Paukert, "Income Distribution at Different Levels of Development: A Survey of the Evidence," *International Labour Review,* August-September 1973, pp. 97–125.

4. Montek S. Ahluwalia, "Income Inequality: Some Dimensions of the Problem," in H. Chenery et al. (eds.), *Redistribution with Growth* (London: Oxford University Press, 1974), pp. 3–37.

5. Jaques Lecaillon, et al. *Income Distribution and Economic Development: An Analytical Survey* (Geneva: International Labour Office, 1984), p. 12.

6. Ibid., p. 41.

7. The statistical significance level of .05 was used to compare the means among all six groups on all three measurements of income inequality. Differences between the groups on the percent of income going to the poorest quintile are not statistically significant, but they are significant when looking at household Gini ratios and at income going to the top 10 percent. For the top 10 percent comparison, the wealthiest core countries are significantly different from countries in the first, third, and fourth groups. When looking at household Gini levels, the wealthiest countries are significantly different from nations in level four ($1,500–$2,999), which are most indicative of periphery countries going through development. Of course, it is at this point that levels of significance do become important,

since the countries in each GDP group are samples representing many more countries from which information is simply not available.

8. W. Beckerman, "Some Reflections on Redistribution and Growth," *World Development,* August 1977, pp. 665–76.

9. Lecaillon, et al., *Income Distribution,* pp. 14–15.

10. Jerry Cromwell, "The Size Distribution of Income: An International Comparison," *Review of Income and Wealth,* September 1977, pp. 291–308.

11. François Nielsen, "Income Inequality and Industrial Development: Dualism Revisited," *American Sociological Review* 59 (October 1994): 654–77; François Nielsen and Arthur S. Alderson, "Income Inequality, Development, and Dualism: Results from an Unbalanced Cross-National Panel," *American Sociological Review* 60 (October 1995): 674–701.

12. Lecaillon, *Income Distribution,* p. 82.

13. Harold R. Kerbo, *Social Stratification and Inequality* (New York: McGraw-Hill, 1983), p. 437.

14. Michael Taylor and Nigel Thrift, *The Geography of Multinationals* (New York: St. Martin's Press, 1982), p. 25.

15. Rahul Jacob, "It was a Banner Year for Profits (The Fortune Global 500)," *Fortune,* August 7, 1995, p. 130ff.

16. Richard J. Barnet, "Lords of the Global Economy: Stateless Corporations," *The Nation,* December 19, 1994, p. 754ff.

17. Paul Graham, "Running Dogs, Lackeys and Imperialism (Multinational Corporations)," *Canadian Dimension* 28,5 (1994): 48.

18. Barry Bluestone and Bennett Harrison, *The Deindustrialization of America* (New York: Basic Books, 1982), pp. 143–47.

19. Taylor and Thrift, *Geography of Multinationals,* p. 1.

20. Robert Gilpin, *The Political Economy of International Relations* (Princeton, NJ: Princeton University Press, 1987), pp. 238–39.

21. U.S. Census Bureau, *Statistical Abstract of the United States 1994,* CD-ROM (Washington, DC: U.S. Government Printing Office, 1995), table 1309.

22. Peter Behr, "Multinationals Once More Are Leading Transformation of the World Economy," *Minneapolis Star Tribune,* September 12, 1994, p. 7D.

23. Raymond J. Mataloni, "U.S. Multinational Companies: Operations in 1993," *Survey of Current Business* 75,6 (1995): 31ff.

24. Raymond J. Mataloni, "A Guide to BEA Statistics on U.S. Multinational Companies," *Survey of Current Business* 75,3 (1995): 38ff.

25. Gilpin, *Political Economy,* p. 257.

26. John Pearson, "Strong Dollar or No, There's Money to Be Made Abroad," *Business Week,* March 1985, p. 155.

27. Brian Zajac, "Getting the Welcome Carpet," *Forbes,* July 18, 1994, p. 276.

28. Brian Zajac, "Weak Dollars, Strong Results," *Forbes,* July 17, 1995, p. 274.

29. Howard M. Wachtel, *The Money Mandarins: The Making of a New Supranational Economic Order* (New York: Pantheon, 1986), pp. 153–78

30. Neil Smelser et al., *The New International Economy* (Beverly Hills, CA: Sage, 1982), pp. 18–19, 21–22.

31. Taylor and Thrift, *Geography of Multinationals,* p. 142.

32. Jacob, "Banner Year," p. 130

33. Ronald Kwan, "Footloose and Country Free," *Dollars and Sense* 164 (March 1991): 6.

34. Steven Pearlstein, "Hard on the Trail of the Up-and-Down Dollar," *The Washington Post National Weekly Edition,* July 25–31, 1994, p. 20.

35. Clay Chandler, "Stepping Back from the Brink," *The Washington Post National Weekly Edition,* February 6–12, 1995, p. 18.

36. John M. Berry and Clay Chandler, "Trying to Cool the 'Hot Money' Game," *The Washington Post National Weekly Edition,* April 24–30, 1995, p. 20.

37. Bluestone and Harrison, *Deindustrialization,* pp. 129–30.

38. Ibid., pp. 130–32.

39. Patricia Horn, "Paying to Lose our Jobs," *Dollars and Sense* 184 (March 1993): 10.

40. Cuomo Commission, *The Cuomo Commission Report* (New York: Simon and Schuster, 1988), p. 41.

41. Bluestone and Harrison, *Deindustrialization,* pp. 171–72.

42. Bennett Harrison and Barry Bluestone, *The Great U-Turn: Corporate Restructuring and the Polarizing of America* (New York: Basic Books, 1988), p. 30.

43. Leslie Sklair, *Sociology of the Global System* (Baltimore, MD: Johns Hopkins University Press, 1991), p. 103.

44. A. Douglas Kincaid and Alejandro Portes, *Comparative National Development: Society and Economy in the New Global Order* (Chapel Hill: University of North Carolina Press, 1994), pp. 68–69, 157.

45. Ibid., p. 158.

46. For excellent reviews of the economic and domestic roles women play in less developed countries, particularly with multinational corporations, see Sklair (1991: 95–100), M. Patricia Fernandez Kelly (1994), and Beneria and Feldman (1992).

47. Bob Hebert, "In Maquiladora Sweatshops: Not a Living Wage," *Minneapolis Star Tribune,* October 12, 1995, p. 23A.

48. Andrew Reding, "Mexico at a Crossroads," *World Policy Journal* 5,4 (1988): 643.

49. U.S. Census Bureau, *Statistical Abstract of the United States 1994,* CD-ROM (Washington, DC: U.S. Government Printing Office, 1995), table 1388. Sri Lanka date is for 1990.

50. Steve Lohr, "'Global Office' Changing White-Collar Work World," *Minneapolis Star Tribune,* October 23, 1988, p. J1.

51. Pete Engardio et al., "High-Tech Jobs All Over the Map," *Business Week,* November 18, 1994, via America Online.

52. Warren Brown and Frank Swoboda, "Ford's Vision: A Company Without Borders," *The Washington Post National Weekly Edition,* October 24–30, 1994, p. 18.

53. Keith Bradsher, "U.S. Losing Skilled Jobs," *Minneapolis Star Tribune,* September 12, 1995, p. 5D.

54. Taylor and Thrift, *Geography of Multinationals,* p. 3.

55. Margaret Shapiro, "Empire of the Sun," *The Washington Post National Weekly Edition,* October 31–November 6, 1988, pp. 6–7.

56. World Bank, *World Development Report 1988* (New York: Oxford University Press, 1988), p. 21.

57. Wachtel, *The Money Mandarins*, p. 162.

58. For one of the more sympathetic portrayals of the difficulties and pitfalls typically encountered by multinational expansion, see Theodore H. Moran, *Multinational Corporations: The Political Economy of Foreign Direct Investment* (Lexington, MA: Heath, 1985) and/or Robert Gilpin, *Political Economy*, pp. 231–305.

59. Francis Moore Lappé and Joseph Collins, *World Hunger: Twelve Myths* (New York: Grove Press, 1986), pp. 85–94.

60. Harriet Friedmann, "Distance and Durability: Shaky Foundations of the Third World," in Philip McMichael (ed.), *The Global Restructuring of Agro-Food Systems* (Ithaca, NY: Cornell University Press, 1994), pp. 258–76.

61. Ibid., p. 259.

62. Harvey Krahn and John W. Gartrell, "Effects of Foreign Trade, Government Spending, and World System Status on Income Distribution," *Rural Sociology* 50,2 (1985): 181–92.

63. World Bank, *Report*, p. 25.

64. Kincaid and Portes, *Comparative National Development*, p. 13.

65. Lappé and Collins, *World Hunger*, pp. 91, 93.

66. Martin Teitel, "Selling Out: The Thriving Business of Patenting Life," *Dollars and Sense* 195 (September/October 1994): 25–27.

67. Moran, *Multinational Corporations*, p. 4.

68. Walden Bello, "U.S.-Phillipine Relations in the Aquino Era," *World Policy Journal* 5,4 (1988): 688.

69. Taylor and Thrift, *Geography of Multinationals*, p. 154.

70. Volker Bornschier, Christopher Chase-Dunn, and Richard Rubinson, "Cross-national Evidence of the Effects of Foreign Investment and Aid on Economic Growth and Inequality: A Survey of Findings and a Reanalysis," *American Journal of Sociology* 84,3 (1978): 651–83.

71. Richard Rubinson, "The World-Economy and the Distribution of Income Within States: A Cross-National Study," *American Sociological Review* 41 (August 1976): 638–59.

72. Michael Timberlake and Kirk R. Williams, "Structural Position in the World-System, Inequality, and Political Violence," *Journal of Political and Military Sociology* 15,1 (1987): 9.

73. Robert W. Jackman, "Dependence on Foreign Investment and Economic Growth in the Third World," *World Politics* 34 (January 1982): 175–97.

74. Erich Weede and Horst Tiefenbach, "Some Recent Explanations of Income Inequality," *The International Studies Quarterly* 25 (June 1981): 255–82; Erich Weede, "Beyond Misspecification in Sociological Analyses of Income Inequality," *American Sociological Review* 45 (June 1980): 497–501.

75. Edward N. Muller, "Financial Dependence in the Capitalist World Economy and the Distribution of Income Within Nations," in Michael Seligson (ed.), *The Gap Between Rich and Poor* (Boulder, CO: Westview, 1984), pp. 256–82.

76. Miles Simpson, "Political Rights and Income Inequality: A Cross-National Test," *American Sociological Review* 55,5 (1990): 689.

77. Bruce London, "Dependence, Distorted Development, and Fertility Trends in Noncore Nations: A Structural Analysis of Cross National Data," *American Sociological Review* 53 (August 1988): 608n.2.

78. Volker Bornschier and Christopher Chase-Dunn, *Transnational Corporations and Underdevelopment* (New York: Praeger, 1985).

79. Specifically replying to the devastating critique of their early work by Weede and Tiefenbach (1981), Bornschier and Chase-Dunn point out that their original research was unfairly replicated. It was not surprising that no effect of multinational penetration was found on income inequality because Weede and Tiefenbach did not use measurement of penetration at an early enough time to actually estimate its effect upon subsequent growth or decline in GNP per person, nor was there any attempt to differentiate between penetration as measured by long-term stock ownership and current flows of investment.

80. Terry Boswell and William J. Dixon, "Dependency and Rebellion: A Cross-National Analysis," *American Sociological Review* 55,4 (1990): 540–59.

81. Glenn Firebaugh, "Growth Effects of Foreign and Domestic Investment," *American Journal of Sociology* 98,1 (1992): 105–30.

82. William J. Dixon and Terry Boswell, "Dependency, Disarticulation and Denominator Effects: Another Look at Foreign Capital Penetration," *American Journal of Sociology* 102,2 (1996): 543–62.

83. Both measurements are less complicated than the original PEN measure used by Bornschier and Chase-Dunn (1985). The first is simply a value of foreign capital stock divided by the value of the total stock of invested capital in a country. The second is even more simple, although perhaps less focused: the value of foreign capital stock divided by the value of the country's GDP. Since the two measurements are new and neither is perfect, both are used in their analysis.

84. Sources for this data include: Organization for Economic Cooperation and Development, *International Direct Investment Statistics Yearbook 1995* (Paris: OECD), tables II and III; World Bank, *World Development Indicators 1994 Data on Diskette* (Washington, DC: World Bank), table 22. Foreign Direct Investment (FDI) is associated with a long-term interest with a view to some control over foreign business enterprises. Investment is categorized as FDI when a foreign investor owns 10 percent or more of voting securities or equity of a business (Frumkin 1994: 207). FDI is different from portfolio investment, which is associated with short-term activity in financial markets that emphasize liquidity and the ability to move funds between countries and different stocks. It is important to note that FDI reflects economic investment activity by corporations or wealthy individual investors (including a parent country owning an affiliate of the parent company in a foreign land—i.e., branch plants of multinationals). As such, it is a plausible measure of multinational penetration. The measure reflects more outward investment of American corporations than foreign domination of U.S. companies. It also separates out an indicator reflecting healthy growth of American multinationals from a more complete and dismal evaluation of our country. A broader definition included under Balance of Payments (BOP) yields an entirely different picture. This concept includes not only capital flow of private assets (including portfolio investment), but also official and government monies (trea-

sury notes, for example). Balance of Payment data includes changes in the prices of bonds and stocks, together with the impact of changing foreign exchange rates. Measured in this way, the United States was a net creditor nation in 1980 by $400 billion (as it had been since 1915). As the 1980s unfolded, our nation's fortunes slipped badly. By 1992 the U.S. was an international debtor nation to the tune of $612 billion.

85. York Bradshaw, Rita Noonan, Laura Gash, and Claudia Buchman Sershen. "Borrowing Against the Future: Children and Third World Indebtedness," *Social Forces* 71,3 (1993): 642.

86. The measure is not perfect, as witnessed by the presence of Gabon, Mali, Panama, and Central African Republic among heavy investor nations. This suggests that for some of these nations, there might be more going on than just a search for better investment opportunities or the routine opening of foreign subsidiaries of giant multinational corporations. There may be another factor operating in some of these countries, such as capital flight caused by poor or unstable economic conditions at home.

87. François Nielsen and Arthur S. Alderson, "Income Inequality, Development, and Dualism: Results from an Unbalanced Cross-National Panel," *American Sociological Review* 60,5 (1995): 674–701.

88. Edward M. Crenshaw, "Democracy and Demographic Inheritance: The Influence of Modernity and Proto-Modernity on Political and Civil Rights, 1965 to 1980," *American Sociological Review* 60,5 (1995): 702–718.

89. Cornelis Peter Terlouw, *The Regional Geography of the World System: External Arena, Periphery, Semiperiphery, Core* (Utrecht, Netherlands: Faculteit Ruimteljke Wetenschppen, 1992).

90. Multinational corporate penetration as measured by Boswell and Dixon's revised foreign stock ownership (1996), degree of coreness as measured by Terlouw's (1992) continuous Z-score variable, Gross Domestic Investment (GDI) as a percent of GDP in 1992, the log of total energy consumed per person in kilograms of oil in 1992, the log of GNP per person in 1992, the square of the log of GNP per person in 1992, and exports as a percent of GDP in 1992. The regression equation predicting the average annual percent growth rate in GNP per person, 1980–1992, is: Y = –.4256 + 1.979 CORENESS + .0747 XPORTGDP — .0888 PEN. Standardized Betas are (respectively) .4387, .4152, and –.2758.

91. World Bank, *World Development Indicators 1994* Data Diskette (Washington, DC: World Bank), table 23. By leaving out the wealthier core countries, the implication is given that Third World countries are really the ones with the debt problems—not industrial nations. Given the performance of the U.S. economy within the past decade, including its staggering leap to a mountain of debt, this assumption is very dubious. Future reports from the World Bank would do well to include data for core countries, if only for greater fairness and objectivity.

92. Rati Ram, "Physical Quality of Life and Inter-Country Economic Inequality," *Economic Letters* 5 (1980): 195–99; Norman L. Hicks and Paul P. Streeten, "Indicators of Development: The Search for a Basic Needs Yardstick," *World Development* 7 (1979): 576–77.

93. Shirley Cerresoto, "Socialism, Capitalism, and Inequality," *Insurgent Sociologist* 11, 2 (1982): 5–29.

94. Bruce London and Bruce A. Williams, "Multinational Corporate Penetration, Protest, and Basic Needs Provision in Non-Core Nations: A Cross-National Analysis," *Social Forces* 66,3 (1988): 747–73.

95. York Bradshaw, Rita Noon, Laura Gash, and Claudia Buchman Sershen, "Borrowing Against the Future: Children and Third World Indebtedness," *Social Forces* 71,3 (1993): 629–57.

96. Norman L. Weatherby et al., "Development, Inequality, Health Care, and Mortality at the Older Ages: A Cross-National Study," *Demography* 20,1 (1983): 27–43. For a detailed discussion of the impact that relative income inequality has as an independent or intervening causative variable, see especially Bornschier and Chase-Dunn, *Transnational Corporations and Underdevelopment*, pp. 131–47.

97. Francis Moore Lappé et al., *Betraying the National Interest* (New York: Grove Press, 1987); Noam Chomsky, *The Culture of Terrorism* (Boston: South End Press, 1988).

98. Bruce London and Thomas D. Robinson, "The Effect of International Dependence on Income Inequality and Political Violence," *American Sociological Review* 54,2 (1989): 305–8.

CHAPTER 5

1. Cuomo Commission, *The Cuomo Commission Report* (New York: Simon and Schuster, 1988), p. 142.

2. John Agnew, *The United States in the World Economy: A Regional Geography* (New York: Cambridge University Press, 1987), p. 70.

3. Frank Levy, *Dollars and Dreams: The Changing American Income Distribution* (New York: Russell Sage Foundation, 1987), pp. 3–4.

4. Bennett Harrison and Barry Bluestone, *The Great U-Turn: Corporate Restructuring and the Polarizing of America* (New York: Basic Books, 1988), pp. 4–5.

5. John M. Berry, "The Legacy of Reagonomics: Underlying Flaws Threaten the Successes," *The Washington Post National Weekly Edition*, December 19–25, 1988, p. 7.

6. Paul Blustein, "Peace for All, Prosperity for Some: The Reagan Expansion Is an Uneven Affair," *The Washington Post National Weekly Edition*, October 3–9, 1988, p. 8.

7. U.S. Census Bureau, *Housing in America, 1985–86*, H 121, No. 19 (Washington, DC: Government Printing Office, 1989), p. 52.

8. U.S. Census Bureau, "Home Sweet Home—America's Housing, 1973 to 1993," *Statistical Brief* SB/95-18, July 1995, p. 2.

9. Blustein, "Peace for All," p. 8.

10. Steven Pearlstein, "We Do the Work, But Who Gets the Money?" *The Washington Post National Weekly Edition*, July 31–August 6, 1995, p. 22.

11. Dean Baker and Lawrence Mishel, *Profits Up, Wages Down: Worker Losses Yield Big Gains for Business* (Washington, DC: Economic Policy Institute, 1995), p. 6.

12. Tom Petruno, *Los Angeles Times*, in *Minneapolis Star Tribune*, September 7, 1995, p. D7.

13. Steven Pearlstein, "A Rising Tide Leaves Some Boats Hard Aground," *The Washington Post National Weekly Edition,* July 3–9, 1995, p. 21.

14. Baker and Mishel, *Profits,* p. 1.

15. Agnew, *United States in the World Economy,* p. 150.

16. David C. Colander, *Economics* (Boston: Irwin, 1993), p. 120.

17. Harrison and Bluestone, *Great U-Turn,* pp. 8–9.

18. Cuomo Commission, *Report,* pp. 13–14.

19. Agnew, *United States and the World Economy,* pp. 142–45.

20. Associated Press, "Despite 4th Quarter Rise, Trade Deficit Down in '88," *Minneapolis Star Tribune,* March 1, 1989, p. 3D.

21. Hobart Rowen, "The Dollar's Silent Tumble," *The Washington Post National Weekly Edition,* November 14–20, 1994, p. 5.

22. Hobart Rowen, "In Defense of the Dollar," *The Washington Post National Weekly Edition,* July 4–10, 1994, p. 5.

23. Mike Meyers, "Nations with Credit Ratings of Banana Republics Fear the U.S. Is Shaky Borrower," *Minneapolis Star Tribune,* December 15, 1995, p. 2D.

24. U.S. Census Bureau, *Statistical Abstract of the United States 1994,* CD-ROM (Washington, DC: U.S. Government Printing Office, 1995), table 970.

25. Up Date, "Financial Games Reduce R&D Spending," *Dollars and Sense* 145 (April 1989): 5.

26. Boyce Rensberger, "No Help Wanted: Young U.S. Scientists Go Begging with Serious Consequences for Our Future," *The Washington Post National Weekly Edition,* January 9–15, 1995, p. 7.

27. Cuomo Commission, *Report,* pp. 10–13.

28. Rensberger, "Scientists Go Begging," p. 6.

29. Ibid., p. 7.

30. I. Magaziner and R. B. Reich, *Minding America's Business: The Decline and Rise of America's Economy* (New York: Vintage Press, 1982), p. 52.

31. Robert Kuttner, "Why We Don't Make TVs Anymore," *Washington Post National Weekly Edition,* October 24–30, 1988, p. 29.

32. Evelyn Richards, "The Pentagon Plans to Get into Television," *Washington Post National Weekly Edition,* December 26, 1988–January 1, 1989, p. 32.

33. Council of Economic Advisors, *Economic Report of the President 1995* (Washington, DC: U.S. Government Printing Office), p. 25.

34. Harrison and Bluestone, *Great U-Turn,* pp. 125, 144–50.

35. Magaziner and Reich, *Minding America's Business,* p. 36

36. U.S. Census Bureau, *Statistical Abstract of the United States 1994,* CD-ROM (Washington, DC: U.S. Government Printing Office, 1995), table 1390. Specifically, average annual increases were in this order: United Kingdom (4.5%), Japan (3.6%), Italy (3.1%), Belgium (2.9%), United States (2.8%), France (2.7%), Sweden (2.6%), Germany (2.0%), Netherlands (1.6%), Canada (0.8%), and Denmark (0.7%).

37. Marc Breslow, "Maintaining Military Might," *Dollars and Sense* 191 (January/February 1994): 42.

38. Ibid., p. 43.

39. Harrison and Bluestone, *Great U-Turn,* p. 178.

40. Tom Riddell, "The Political Economy of Military Spending," in *The Imperiled Economy: Through the Safety Net* (New York: Union for Radical Political Economics, 1988), p. 232.

41. Ibid.

42. Bruce M. Russett, *What Price Vigilance?* (New Haven, CT: Yale University Press, 1970), pp. 140–41.

43. Riddell, "Military Spending," p. 232.

44. For a detailed discussion of this relationship plus earlier data of a similar nature, see R. W. DeGrasse, "The Military: Short-Changing the Economy," *Bulletin of the Atomic Scientists,* May 1984, pp. 39–43.

45. Agnew, *United States in the World Economy,* p. 151.

46. Harrison and Bluestone, *Great U-Turn,* p. 154.

47. U.S. Census Bureau, *Statistical Abstract of the United States 1994,* CD-ROM (Washington, DC: U.S. Government Printing Office, 1995), tables 1311, 1312, 1313, 1316.

48. John Burgess, "When the Product Is America Instead of American Products," *The Washington Post National Weekly Edition,* October 24–30, 1988, p. 21.

49. Cuomo Commission, *Report,* p. 8; U.S. Census Bureau, *Statistical Abstract of the United States 1994,* CD-ROM (Washington, DC: U.S. Government Printing Office, 1995), table 778. According to the General Accounting Office, the $325 billion necessary to close or sell failed savings institutions is greater than the annual Pentagon budget. Comptroller General Charles Brewster believes this could rise to $500 billion. See Associated Press, "Congress Warned S and L Bailout Could Reach $500 Billion," *Minneapolis Star Tribune,* April 7, 1990, p. 12D.

50. Richard C. Michel, *Economic Growth and Income Inequality Since the 1982 Recession* (Washington, DC: The Urban Institute, 1990).

51. Patricia Ruggles, *Drawing the Line: Alternative Poverty Measures and Their Implications for Public Policy* (Washington, DC: The Urban Institute, 1990), pp. 139–46.

52. Leonard Beeghley and Jeffrey W. Dwyer, "Income Transfers and Income Inequality," *Population Research and Policy Review* 8 (1989): 119–42.

53. U.S. Census Bureau, *Current Population Reports: Poverty in the United States: 1992,* Series P-60, No. 185 (Washington, DC: U.S. Government Printing Office), pp. xxii-xxiii.

54. Constance F. Citro and Robert T. Michael (eds.), *Measuring Poverty: A New Approach* (Washington, DC: National Academy Press, 1995).

55. David M. Betson, "Consequences of the Panel's Recommendations," *Focus* 17,1 (1995): 12.

56. William P. O'Hare, *Poverty in America: Trends and New Patterns* (Washington, DC: Population Reference Bureau, 1985), p. 7.

57. Andrew J. Winnick, *Toward Two Societies: The Changing Distribution of Income and Wealth in the U.S. Since 1960* (New York: Praeger, 1989), p. 204.

58. Lee Rainwater, *Poverty in American Eyes,* Working Paper Number 80 (Syracuse, NY: Syracuse University, Maxwell School of Citizenship and Public Affairs, 1992), p. 16.

59. Joan R. Rodgers and John L. Rodgers, "Chronic Poverty in the United States," *The Journal of Human Resources* 28,1 (1993): 25–54.

60. U.S. Census Bureau, *Poverty's Revolving Door*, Statistical Brief SB/95-20, August 1995.

61. Peter Gottschalk, Sara McLanahan, and Gary D. Sandefur, "The Dynamics of Intergenerational Transmission of Poverty and Welfare Participation," in Sheldon H. Danziger, Gary D. Sandefur, and Daniel H. Weinberg (eds.), *Confronting Poverty: Prescriptions for Change* (New York: Russell Sage, 1994), pp. 89–91.

62. Lawrence Mishel and Jared Bernstein, *The State of Working America: 1994–1995* (New York: M.E. Sharpe, 1994), p. 263.

63. Denny Braun, "Negative Consequences to the Rise of Income Inequality," *Research in Politics and Society* 5 (1995): 13.

64. Physician Task Force on Hunger in America, *Hunger in America: The Growing Epidemic* (Middleton, CT: Wesleyan University Press, 1985).

65. U.S. Census Bureau, *Statistical Abstract of the United States 1994*, CD-ROM (Washington, DC: U.S. Government Printing Office, 1995). Data for this graph and the subsequent analyses of poverty trends in the text have been calculated using the working tables on poverty from this source.

66. Clifford M. Johnson, Leticia Miranda, Arloc Sherman, and James D. Weill, *Child Poverty in America* (Washington, DC: Children's Defense Fund, 1991), pp. 12–17.

67. Sheldon H. Danziger and Daniel H. Weinberg, "The Historical Record: Trends in Family Income, Inequality, and Poverty," in Sheldon H. Danziger, Gary D. Sandefur, and Daniel H. Weinberg (eds.), *Confronting Poverty: Prescriptions for Change* (New York: Russell Sage, 1994), p. 33.

68. This surface tranquility is very misleading. A number of reports show great leaps forward for educated blacks, but severe decline for those left in an "underclass" in our major metropolitan areas (Wilson 1987). Wilson points out that the real improvements for some blacks, brought on by the civil rights movement, helped an upwardly mobile group escape inner-city life. What remained were the "truly disadvantaged." These ill-educated poor are haunted by violence, crime, drugs, joblessness, and welfare dependency. A RAND report notes that conditions are worsening for undereducated blacks and single black women with children (Smith and Welch 1986). Such findings are also echoed and corroborated in a study documenting a decline in per capita income in female-headed black households (Farley and Bianchi 1985).

69. Michael Harrington, *The New American Poverty* (New York: Holt, Rinehart and Winston, 1984).

70. Richard Ropers, *Persistent Poverty: The American Dream Turned Nightmare* (New York: Plenum Press, 1991), pp. 115–44.

71. Johnson et al., *Child Poverty*, p. 12.

72. David T. Ellwood, *Poor Support: Poverty in the American Family* (New York: Basic Books, 1988), pp. 81–127.

73. U.S. Bureau of the Census, Current Population Reports, Series P-60, Number 185. *Poverty in the United States: 1992* (Washington, DC: U.S. Government Printing Office, 1993), p. xv.

74. Katherine McFate, Timothy Smeeding, and Lee Rainwater, "Markets and States: Poverty Trends and Transfer System Effectiveness in the 1980s," in Katherine McFate, Roger Lawson, and William Julius Wilson (eds.), *Poverty, Inequality, and the Future of Social Policy* (New York: Russell Sage, 1995), p. 32.

75. Greg J. Duncan et al., "Poverty Dynamics in Eight Countries," *Journal of Population Economics* 6 (Spring 1993): 222.

76. Timothy Smeeding, Barbara Boyle Torrey, and Martin Rein, "Patterns of Income and Poverty: The Economic Status of Children and the Elderly in Eight Countries," in *The Vulnerable* (Washington, DC: Urban Institute Press, 1988), pp. 95–98.

77. Katherine McFate, *Poverty, Inequality, and the Crisis of Social Policy* (Washington, DC: Joint Center for Political and Economic Studies, 1991), pp. 15–16.

78. Lee Rainwater and Timothy M. Smeeding, *Doing Poorly: The Real Income of American Children in a Comparative Perspective*, Working Paper Number 127 (Syracuse, NY: Syracuse University, Maxwell School of Citizenship and Public Affairs, 1995), p. 9.

79. Johnson et al., *Child Poverty*, p. 17.

80. Joint Economic Committee, *Poverty, Income Distribution, the Family and Public Policy* (Washington, DC: U.S. Government Printing Office, 1986), pp. 56–63.

81. U.S. House of Representatives, Committee on Ways and Means, *Where Your Money Goes: The 1994–95 Green Book* (Washington, DC: Brassey's, 1994), pp. 324, 370. Partially offsetting this steep decline was a rise in food stamp benefits. Nonetheless, even with the value of food stamps added to AFDC benefits, the average level of support for a three-member family went from $900 in 1972 to $658 in 1993—in constant 1993 dollars. This 27 percent drop in benefits was almost wholly due to the waning of AFDC support, since the value of food stamp support is indexed to inflation and does not decrease over time (U.S. House of Representatives 1994, p. 374).

82. Julie Strawn, "The States and the Poor: Child Poverty Rises as the Safety Net Shrinks," *Social Policy Report* 6,3 (1992): 10.

83. Danziger and Weinberg, "Historical Record," pp. 46–47.

84. Committee on Ways and Means, U.S. House of Representatives, *Background Material and Data on Programs within the Jurisdiction of the Committee on Ways and Means* (Washington, DC: U.S. Government Printing Office, 1989), pp. 878–908.

85. Kathryn J. Edin, "The Myths of Dependence and Self-Sufficiency: Women, Welfare, and Low-Wage Work," *Focus* 17,2 (1995): 2.

86. Committee on Ways and Means, U.S. House of Representatives, *Background Material and Data on Programs within the Jurisdiction of the Committee on Ways and Means* (Washington, DC: U.S. Government Printing Office, 1989), pp. 958–83.

87. Gwen Hill, "It Is Becoming a Permanent Issue," *The Washington Post National Weekly Edition*, March 27–April 2, 1989, p. 8.

88. Arthur Haupt, "Another Winter for the Homeless," *Population Today* 17,2 (1989): 3–4.

89. "Homelessness: Rare or Commonplace?" *Minneapolis Star Tribune,* December 28, 1994, p. 2A.

90. Tom Hamburger, "New Report Increases Homeless Population to 7 Million," *Minneapolis Star Tribune,* May 18, 1994, p. 7A.

91. Chris Spolar, "No Home—and Not Much More," *The Washington Post National Weekly Edition,* March 27–April 2, p. 7.

92. Coleman McCarthy, "Doing the Right Thing for Hunger," *Minneapolis Star Tribune,* November 28, 1991, p. 38A.

93. Felicity Barringer, "Whether It's Hunger or 'Misnourishment,' It's a National Problem," *New York Times,* December 27, 1992, p. E2.

94 .U.S. Census Bureau, *Statistical Abstract of the United States 1994,* CD-ROM (Washington, DC: U.S. Government Printing Office, 1995), tables 66, 80.

95. Ibid., tables 711, 734.

96. Gottschalk, et al., *Confronting Poverty,* pp. 95–97.

97. Irwin Garfinkel and Sara S. McLanahan, *Single Mothers and Their Children: A New American Dilemma* (Washington, DC: Urban Institute, 1986); William P. O'Hare, *America's Welfare Population: Who Gets What?* (Washington, DC: Population Reference Bureau, 1987), p. 10; Kristin Moore, "The Effect of Government Policies on Out-of-Wedlock Sex and Pregnancy," *Family Planning Perspectives* 94,4 (1977): 164–69.

98. Robert Moffitt, "Incentive Effects of the U.S. Welfare System: A Review," *Journal of Economic Literature* 30 (March 1992): 29.

99. U.S. House of Representatives, *Where Your Money Goes,* p. 400.

100. Isabell V. Sawhill, "Poverty in the U.S.: Why Is It So Persistent? *Journal of Economic Literature* 26 (September 1988): 1104.

101. Irwin Garfinkel and Sara McLanahan, "Single-Mother Families, Economic Insecurity, and Government Policy," in Sheldon H. Danziger, Gary D. Sandefur, and Daniel H. Weinberg (eds.), *Confronting Poverty: Prescriptions for Change* (New York: Russell Sage, 1994), p. 211; Garfinkel and McLanahan, *Single Mothers,* pp. 62–63.

102. O'Hare, *Poverty in America,* pp. 17–18.

103. Michael Harrington, *The New American Poverty* (New York: Holt, Rinehart and Winston, 1984).

104. Sheldon Danziger and Peter Gottschalk, "Do Rising Tides Lift All Boats? The Impact of Secular and Cyclical Changes on Poverty," *American Economic Review* 76,2 (1986): 405.

105. Peter Gottschalk and Sheldon Danziger, "A Framework for Evaluating the Effects of Economic Growth and Transfers on Poverty," *American Economic Review* 75,1 (1985): 153–61.

106. Data from Lawrence Mishel and Jared Bernstein, *The State of Working America: 1994–95* (New York: M.E. Sharpe, 1994), table 6.1.

107. Sawhill, "Poverty in the U.S.," p. 1112.

108. Council of Economic Advisers, *Economic Report of the President, 1989* (Washington, DC: U.S. Government Printing Office, 1989), p. 433; 1980s data from U.S. Census Bureau, *Statistical Abstract of the United States 1994,* CD-ROM (Washington, DC: U.S. Government Printing Office, 1995), table 614.

109. Marc Breslow and Matthew Howard, "The Real Un(der)employment Rate," *Dollars and Sense* 199 (May/June 1995): 35.

110. James R. Wetzel, "Labor Force, Unemployment, and Earnings," in Reynolds Farley (ed.), *State of the Union: America in the 1990s,* Vol. 1: *Economic Trends* (New York: Russell Sage, 1995), p. 61.

111. Edward M. Gramlich and Deborah S. Laren, "How Widespread Are Income Losses in a Recession?" in D. Lee Bawdin (ed.), *The Social Contract Revisited* (Washington, DC: Urban Institute, 1984), pp. 157–80.

112. Charles Murray, *Losing Ground: American Social Policy 1950–1980* (New York: Basic Books, 1984).

113. Richard Vedder and Lowell Gallaway, "AFDC and the Laffer Principle," *Wall Street Journal,* March 26, 1986, p. 30, and Lowell Gallaway, Richard Vedder, and Therese Foster, "The New Structural Poverty: A Quantitative Analysis," in Joint Economic Committee, *War on Poverty: Victory or Defeat* (Washington, DC: U.S. Government Printing Office, 1986).

114. Sanford F. Schram and Paul H. Wilken, "It's No 'Laffer' Matter: Claim That Increasing Welfare Aid Breeds Poverty and Dependence Fails Statistical Test," *American Journal of Economics and Sociology* 48,2 (1989): 203–17.

115. Sawhill, "Poverty in the U.S.," p. 1113.

116. Moffitt, "Incentive Effects," pp. 56–57.

117. U.S. Census Bureau, *Statistical Abstract of the United States 1994,* CD-ROM (Washington, DC: U.S. Government Printing Office, 1995), table 726.

118. U.S. Census Bureau, *Income and Poverty 1993,* CD-ROM (Washington, DC: U.S. Government Printing Office, 1995).

119. Lars Osberg, *Economic Inequality in the United States* (Armonk, NY: Sharpe, 1984), pp. 19–22.

120. Herbert Inhaber and Sidney Carroll, *How Rich Is Too Rich? Income and Wealth in America* (New York: Praeger, 1992).

121. Folke Dovring, *Inequality: The Political Economy of Income Distribution* (New York: Praeger, 1991), pp. 47–66.

122. Patricia Horn, "Measure for Measure: Deciphering the Statistics on Income and Wages," *Dollars and Sense* 143 (January/February 1989): 10.

123. Council of Economic Advisors, *Report of the President,* 1995, table B45.

124. U.S. Census Bureau, *Statistical Abstract of the United States 1994,* CD-ROM (Washington, DC: U.S. Government Printing Office, 1995), table 661.

125. Council of Economic Advisors, *Report of the President,* 1995, table B31.

126. Suzanne M. Bianchi, "Changing Economic Roles of Women and Men," in Reynolds Farley (ed.), *State of the Union: America in the 1990s,* Vol. 1: *Economic Trends* (New York: Russell Sage, 1995), p. 127.

127. New York Times, "Race Income Gap Little Changed, But Black Lives Improving Overall," *Minneapolis Star Tribune,* February 6, 1995, p. 8A. Despite the more upbeat interpretation offered by government agencies, however, a large number of scholars have challenged the interpretation that earnings have either stayed the same or improved for African-Americans compared to whites. It really comes down to how one measures full-time workers and what time frame is used.

Nearly all analyses show African Americans prospering during the late 1960s and throughout most of the 1970s. The skid really starts around 1980 and has continued into the 1990s. Since African Americans (especially young males) have historically had unemployment rates triple that of whites or have become so discouraged that they drop completely out of the labor market, comparing full-time workers obscures the worsening economic picture for African-American males vis-à-vis white males. Bennett Harrison and Lucy Gorham (1992), who "annualize" wage and salary data to compensate for this bias, find a vastly different picture. Taking all workers who earned income and calculating their wages to a full-time, year-round equivalent shows a substantial decline in high-wage workers among both male and female African Americans over the past decade, as well as an explosion of the proportion who are working poor. While whites experienced a 31 percent increase in the number of working poor, African Americans underwent a 44 percent increase. For young (ages twenty-five to thirty-four) African-American males, the shift was even more severe: their share of working poor doubled during the 1980s, going from 18.4 percent to 34 percent (see also Cotton 1989). Much of the decline for African-American males has been due to the loss of manufacturing jobs, the drop in union membership, and the erosion of the minimum wage—all hitting African-American workers harder than their white counterparts because of their initially lower occupational distribution (Breslow 1995, 10). Yet even among well-educated African Americans, there has been more deterioration than among whites. By 1989, African Americans in the same educational stratum earned 18 percent less than whites—compared to 6 percent less in 1976. The failure of higher education to pay off for African Americans has become an embarrassing reflection of racial discrimination. African-American males closed a 3.3-year gap in median years of schooling between 1940 and 1990, yet they have almost nothing to show for it. Especially during the 1980s—and most sharply among African-American college graduates—the gap in hourly earnings between whites and African Americans became a chasm (Bernstein 1995, 33). This pattern may be reversing in the 1990s, but we cannot get a stable picture until data become available.

128. Chris Tilly and Randy Abelda, "Family Structure and Family Earnings: The Determinants of Earnings Differences among Family Types," *Industrial Relations* 33,2 (1994): 156.

129. Horn, "Measure for Measure," pp. 10–11. The growth in numbers of female workers has been said to depress wages while promoting inequality. Such claims are open to severe questioning. In studies over the past decade (Maxwell 1990; Treas 1987, 283), the effect of such changes has been found to have inconsequential to small bearing on the growth of income inequality.

130. Suzanne M. Bianchi, "Changing Economic Roles," p. 117, 145. For detailed reviews of women's earnings in comparison to those of men, see also Margaret Mooney Marini (1989) and/or June O'Neill (1991).

131. Staff Study, *Families on a Treadmill: Work and Income in the 1980s,* Joint Economic Committee, U.S. House of Representatives, January 17, 1992, p. 1.

132. Council of Economic Advisors, *Economic Report of the President,* 1995, p. 401.

133. Alfred L. Malabre, Jr., "The Outlook: Is the Bill Arriving for the Free Lunch?" *Wall Street Journal,* January 9, 1989, p. 1.

134. Mishel and Bernstein, *Working America 1995,* p. 335.

135. Harrison and Bluestone, *Great U-Turn,* pp. 117–28.

136. Nan L. Maxwell, "Demographic and Economic Determinants of United States Income Inequality," *Social Science Quarterly* 70,2 (1989), p. 260.

137. Levy, *Dollars and Dreams,* pp. 78–80.

138. Frank Levy and Richard J. Murnane, "U.S. Earnings Levels and Earnings Inequality: A Review of Recent Trends and Proposed Explanations, *Journal of Economic Literature* 30 (September 1992): 1357–60; Lawrence F. Katz and Kevin M. Murphy, "Changes in Relative Wages, 1963–87: Supply and Demand Factors," *Quarterly Journal of Economics,* February 1992, p. 76.

139. Frank Levy, "Incomes and Income Inequality" in Reynolds Farley (ed.), *State of the Union: America in the 1990s,* Vol. 1 *Economic Trends* (New York: Russell Sage, 1995), pp. 9, 23.

140. Harrison and Bluestone, *Great U-Turn,* p. 120.

141. Mishel and Bernstein, *Working America,* p. 109.

142. Levy, "Incomes and Income Inequality," pp. 43–45.

143. Steven Pearlstein, "Recessions Fade, But Downsizings Are Forever," *The Washington Post National Weekly Edition,* October 3–9, 1994, p. 21.

144. Economy in Numbers, "The New Unemployment," *Dollars and Sense* 181 (November 1992): 23.

145. Associated Press, "U.S. Business Failures Hit Record Level in '91," *Minneapolis Star Tribune,* February 21, 1992, p. 3D; William Raspberry, "How Economy Can Be Doing Very Well and Very Badly," *Minneapolis Star Tribune,* January 9, 1996, p. 9A.

146. John Miller, "As the Economy Expands, Opportunity Contracts," *Dollars and Sense* 193 (May/June 1994): 9.

147. New York Times Poll, "The Downsizing of America: Layoffs," *New York Times* World Wide Web site, March 9, 1996.

148. Dean Baker and Lawrence Mishel, *Profits Up, Wages Down: Worker Losses Yield Big Gains for Business* (Washington, DC: Economic Policy Institute, 1995), pp. 1, 6.

149. Linda LeGrande, *The Service Sector: Employment and Earnings in the 1980s,* Congressional Research Service, Library of Congress, Report No. 85-167 E, August 15, 1985, p. 5.

150. U.S. Census Bureau, *Statistical Abstract of the United States 1994,* CD-ROM, table 654.

151. Barry Bluestone and Bennett Harrison, *The Great American Job Machine: The Proliferation of Low-Wage Employment in the U.S. Economy* (Washington, DC: Joint Economic Committee, 1986), p. 3.

152. U.S. Census Bureau, *Statistical Abstract of the United States 1994,* CD-ROM, table 661.

153. Lester C. Thurow, "A Surge in Inequality," *Scientific American,* May 1987, p. 34.

154. Peter Gottschalk, *Policy Changes and Growing Earnings Inequality in Seven Industrialized Countries,* Working Paper No. 86 (Syracuse, NY: Syracuse

University, Maxwell School of Citizenship and Public Affairs, 1994), p. 13; Peter Gottschalk and Mary Joyce, *The Impact of Technological Change, Deindustrialization, and Internationalization of Trade on Earnings Inequality: An International Perspective,* Working Paper No. 85 (Syracuse, NY: Syracuse University, Maxwell School of Citizenship and Public Affairs, 1992), pp. 22, 24–25; Peter Gottschalk and Mary Joyce, "The Impact of Technological Change, Deindustrialization, and Internationalization of Trade on Earnings Inequality: An International Perspective," in Katherine McFate, Roger Lawson, and William Julius Wilson (eds.), *Poverty, Inequality, and the Future of Social Policy* (New York: Russell Sage, 1995), pp. 197–228.

155. Harrison and Bluestone, *Great U-Turn,* p. 120.

156. Martin Dooley and Peter Gottschalk, "The Increasing Proportion of Men with Low Earnings in the United States," *Demography* 22,1 (1985): 25–34; Martin Dooley and Peter Gottschalk, "Earnings Inequality among Males in the United States: Trends and the Effect of Labor Force Growth," *Journal of Political Economy* 92,1 (1984): 59 ff.

157. Bluestone and Harrison, *Great American Job Machine,* p. 6.

158. Harrison and Bluestone, *Great U-Turn,* pp. 121–23. For an analogous treatment, see Patrick J. McMahon and John H. Tschetter, "The Declining Middle Class: A Further Analysis," *Monthly Labor Review* 1 (September 1986): 22–27.

159. Harrison and Bluestone, *Great U-Turn,* pp. 126–28.

160. Barry Bluestone, "The Great U-Turn Revisited: Economic Restructuring, Jobs, and the Redistribution of Earnings," in John D. Kasarda (ed.), *Jobs, Earnings, and Employment Growth Policies in the United States* (Boston: Kluwer Academic Publishers, 1990), p. 14.

161. Joel I. Nelson and Jon Lorence, "Metropolitan Earnings Inequality and Service Sector Employment," *Social Forces* 67,2 (1988): 492.

162. Jon Lorence and Joel Nelson, "Industrial Restructuring and Metropolitan Earnings Inequality, 1970–1980," *Research in Social Stratification and Mobility* 12 (1993): 145.

163. James R. Wetzel, "Labor Force, Unemployment, and Earnings," in Reynolds Farley (ed.), *State of the Union: America in the 1990s,* Vol. 1: *Economic Trends* (New York: Russell Sage, 1995), pp. 72, 92–93.

164. Michael Wallace and Joyce Rothschild, "Plant Closings, Capital Flight, and Worker Dislocation: The Long Shadow of Deindustrialization," *Research in Politics and Society* 3 (1989): 3.

165. Levy, "Incomes and Income Inequality," p. 11.

166. Mishel and Bernstein, *Working America,* p. 153.

167. U.S. Census Bureau, *Income and Job Mobility in the Early 1990s,* Statistical Brief SB/95-1, March 1995.

168. Martina Morris, Annette D. Bernhardt, and Mark S. Handcock, "Economic Inequality: New Methods for New Trends," *American Sociological Review* 59,2 (1994): 211.

169. Leann M. Tigges, "Age, Earnings, and Change Within the Dual Economy," *Social Forces* 66,3 (1988): 676–98.

170. Leann M. Tigges, "Dueling Sectors: The Role of Service Industries in the Earnings Process of the Dual Economy," in George Larkas and Paula England

(eds.), *Industries, Firms, and Jobs: Sociological and Economic Approaches* (New York: Plenum Press, 1988) , pp. 281–301.

171. Lynn A. Karoly, "Changes in the Distribution of Individual Earnings in the United States: 1967–1986" (Santa Monica, CA: RAND Corp., 1989); W. Norton Grubb and Robert H. Wilson, *The Distribution of Wages and Salaries, 1960–1980: The Contributions of Gender, Race, Sectoral Shifts and Regional Shifts,* Lyndon B. Johnson School of Public Affairs, Working Paper No. 39 (Austin: University of Texas). Grubb and Wilson largely corroborate the fact that earnings inequality has been on the rise for over twenty years in all race, industry, and regional groups. Their data stop before it became possible to observe the increase of earnings inequality for women as well as men. Decomposing inequality with the Theil index led them to agree with Bluestone and Harrison that increases in part-time work were partly responsible for the growth of earnings inequality. They also uncovered much support for the idea that shifts away from industrial production toward service employment has led to an increase in earnings inequality. See also W. Norton Grubb and Robert H. Wilson, "Trends in Wage and Salary Inequality, 1967–88," *Monthly Labor Review,* June 1992, pp. 23–29.

172. For quite some time it seemed that the experience of women in the the U.S. labor force was doubly blessed. Although they have consistently made less money than men, we have seen a continued increase in their real earnings. At least until 1980, their earnings were also much more equal on a relative basis in comparison to men. Looking at Gini scores for earnings of men and women separately shows a decline of inequality for women between 1958 and 1977 (Henley and Ryscavage 1980). Yet figures through 1989 strongly indicate that relative earnings inequality is now probably on the rise for women as well as for men (Harrison and Bluestone 1988; Karoly 1989; Karoly 1993). In fact, recent labor force data on women shows that the very growth in earnings inequality among women has been caused by a large scale improvement in their wages relative to men (Smith 1991, 133).

173. Lynn A. Karoly, "The Trend in Inequality among Families, Individuals, and Workers in the United States: A Twenty-five Year Perspective," in Sheldon Danziger and Peter Gottschalk (eds.), *Uneven Tides: Rising Inequality in America* (New York: Russell Sage, 1993), p. 77.

174. Monica Castillo, *A Profile of the Working Poor, 1993,* Report 896 (Washington, DC: Bureau of Labor Statistics, 1995), p. 1.

175. U.S. Census Bureau, *The Earnings Ladder: Who's at the Bottom? Who's at the Top?* Statistical Brief SB/94-3RV, June 1994, p. 2.

176. Inge O'Connor and Timothy M. Smeeding, *Working But Poor—A Cross-National Comparison of Earnings Adequacy,* Working Paper No. 94 (Syracuse, NY: Syracuse University, Maxwell School of Citizenship and Public Affairs, 1993).

177. The position of the United States improves somewhat when severity and degree of poverty are taken into account. For example, our working poor are closer to the poverty threshold and less widely distributed in their income (Gini ratio among the working poor) than those in the United Kingdom (Delhausse 1995).

178. Isaac Shapiro and Sharon Parrott, *An Unraveling Consensus? An Analysis of the Effect of the New Congressional Agenda on the Working Poor* (Washington, DC: Center on Budget and Policy Priorities, 1995), p. 6.

179. Sylvia Ann Hewlett, *When the Bough Breaks: The Cost of Neglecting Our Children* (New York: Basic Books, 1991), pp. 27–28, 62–98.

180. Juliet B. Schor and Laura Leete-Guy, *The Overworked American: The Unexpected Decline of Leisure* (Washington, DC: Economic Policy Institute, 1992).

181. Hewlett, *When the Bough Breaks,* p. 73.

182. Harrison and Bluestone, *Great U-Turn,* p. 118.

183. Ibid., p. 14.

184. Martin Olav Sabo, "A $6.50 U.S. Minimum Wage Would Help Lessen Income Gap," *Minneapolis Star Tribune,* June 16, 1994, p. 25A.

185. David Card and Alan Krueger, *Myth and Measurement: The New Economics of the Minimum Wage* (Princeton, NJ: Princeton University Press, 1995).

186. Robert Kuttner, "Raising the Minimum Wage Issue," *Washington Post National Weekly Edition,* August 14–20, 1995, p. 5.

187. John McDermott, "Bare Minimum: A Too-Low Minimum Wage Keeps All Wages Down," *Dollars and Sense* 200 (July/August 1995): 28.

188. Katherine Newman and Chauncey Lennon, "The Job Ghetto," *The American Prospect* 22 (Summer 1995): 66–67.

189. Guy Gugliotta, "Making Do—A Piece at a Time," *The Washington Post National Weekly Edition,* October 10–16, 1994, p. 6.

190. Guy Gugliotta, "The Minimum Wage Culture," *The Washington Post National Weekly Edition,* October 3–9, 1994, p. 6.

191. Harrison and Bluestone, *Great U-Turn,* pp. 21–52.

192. "It's Better in the Union—If You Can Find One," *Dollars and Sense* 201 (September/October 1995): 51.

193. Alan Caniglia and Sean Flaherty, "Unionism and Income Inequality: A Comment on Rubin," *Industrial and Labor Relations Review* 43,1 (1989): 131–37; Alan Caniglia and Sean Flaherty, "The Relative Effects of Unionism on the Earnings Distribution of Women and Men," *Industrial Relations* 31,2 (1992): 382.

194. Richard S. Belous, *Two-Tier Wage Systems in the U.S. Economy,* No. 85-165 E (1985), Congressional Research Service, Library of Congress.

195. U.S. Census Bureau, *Statistical Abstract 1987,* pp. 408–9; *Statistical Abstract 1994,* CD-ROM, table 683.

196. Richard B. Freeman, "How Much Has De-Unionization Contributed to the Rise in Male Earnings Inequality?" in Sheldon Danziger and Peter Gottschalk (eds.), *Uneven Tides: Rising Inequality in America* (New York: Russell Sage, 1993), p. 154.

197. Update, "The Booming Temp Industry," *Dollars and Sense* 181 (November 1992): 5.

198. Harrison and Bluestone, *Great U-Turn,* pp. 44–47.

199. U.S. Department of Labor, Bureau of Labor Statistics, *Contingent and Alternative Employment Arrangements,* Report 900, August 1995, p. 1.

200. Camille Colatosti, "A Job Without a Future: Temporary and Contract Workers Battle Permanent Insecurity," *Dollars and Sense* 176 (May 1992): 9.

201. Economy in Numbers, "Cutting Off the Unemployed," *Dollars and Sense* 167 (June 1991): 23.

202. Marc Baldwin, "How Low Can You Go? Why So Few of the Unemployed Receive Benefits," *Dollars and Sense* 177 (June 1992): 20.

203. Ibid., p. 21.

204. U.S. Census Bureau, *Statistical Abstract 1994,* CD-ROM, table 1376.

205. Economy in Numbers, "Taxes: An American Dilemma," *Dollars and Sense* 184 (March 1993): 23.

206. Clay Chandler, "If Taxes Are Cut, Will It Really Make a Difference?" *The Washington Post National Weekly Edition,* December 26, 1994–January 1, 1995, p. 13.

207. Mishel and Bernstein, *Working America,* pp. 90–91.

208. Donald L. Barlett and James B. Steele, *America: Who Really Pays the Taxes?* (New York: Simon and Schuster, 1994), p. 87.

209. Robert S. McIntyre, "Tax Inequality Caused Our Ballooning Budget Deficit," *Challenge* (November/December 1991), p. 27.

210. Ibid., p. 218.

211. Barlett and Steele, *Taxes,* p. 104.

212. McIntyre, "Tax Inequality," p. 25.

213. Ibid., pp. 96–97.

214. Mishel and Bernstein, *Working America,* pp. 100–103.

215. Barlett and Steele, *Taxes,* pp. 23–24.

216. Robert S. McIntyre, Douglas P. Kelly, Michael P. Ettlinger, and Elizabeth A. Fray, *A Far Cry from Fair* (Washington, DC: Citizens for Tax Justice, 1991), p. 18.

217. Council of Economic Advisors, *Report of the President,* 1995, table B104.

218. U.S. Census Bureau, *Income and Poverty 1993,* CD-ROM, table F3.

219. Denny Braun, "Negative Consequences to the Rise of Income Inequality," *Research in Politics and Society* 5 (1995): 6–7.

220. Christopher Niggle, "Monetary Policy and Changes in Income Distribution," *Journal of Economic Issues* 23,3 (1989): 819.

221. U.S. Census Bureau, *Housing in America, 1985–86,* H 121, No. 19 (Washington, DC: U.S. Government Printing Office, 1989), p. 52.

222. U.S. House of Representatives, Committee on Ways and Means, *Where Your Money Goes: The 1994–95 Green Book* (Washington, DC: Brassey's, 1994), calculated from data in table H23.

223. McKinley L. Blackburn and David E. Bloom, *Family Income Inequality in the United States: 1967–1984,* Paper no. 1294 (Cambridge, MA: Harvard Institute of Economic Research, 1987), table 4.

224. Judith Treas, "The Effect of Women's Labor Participation on the Distribution of Income in the United States," *Annual Review of Sociology* 13 (1987): 283.

225. Isaac Shapiro, *Unequal Shares: Recent Trends among the Wealthy* (Washington, DC: Center on Budget and Policy Priorities, 1995), p. 1.

226. Calculated from U.S. House of Representatives, *Where Your Money Goes,* table H25.

227. Blackburn and Bloom, "Family Income Inequality," table 2.

228. McKinley L. Blackburn and David E. Bloom, "What Is Happening to the Middle Class?" *American Demographics,* January 1985, p. 21.

229. Katherine L. Bradbury, "The Shrinking Middle Class," *New England Economic Review,* September/October 1986, pp. 41–55.

230. Lynn A. Karoly, "The Trend in Inequality among Families, Individuals, and Workers in the United States: A Twenty-Five Year Perspective," in Sheldon Danziger and Peter Gottschalk (eds.), *Uneven Tides: Rising Inequality in America* (New York: Russell Sage, 1993), p. 40.

231. Kevin Phillips, *Boiling Point: Democrats, Republicans, and the Decline of Middle-Class Prosperity* (New York: HarperCollins, 1993), p. 31.

232. Tim Knapp, "The Declining Middle Class in an Age of Deindustrialization," *Research in Inequality and Social Conflict* 2 (1992): 114–15.

233. Greg J. Duncan, Timothy M. Smeeding, and Willard Rodgers, "W(h)ither the Middle Class? A Dynamic View," in Dimitri B. Papadimitriou and Edward N. Wolff (eds.), *Poverty and Prosperity in the USA in the Late Twentieth Century* (New York: St. Martin's, 1993), pp. 265–67.

234. Levy, *Dollars and Dreams,* pp. 151–91.

235. For an analogous treatment, see Congressional Budget Office, *Trends in Family Income: 1970–1986* (Washington, DC: U.S. Government Printing Office, 1988).

236. Levy, "Incomes and Inequality," p. 1.

237. Maria Cancian, Sheldon Danziger, and Peter Gottschalk, "Working Wives and Family Income Inequality among Married Couples," in Sheldon Danziger and Peter Gottschalk (eds.), *Uneven Tides: Rising Inequality in America* (New York: Russell Sage, 1993), pp. 14, 216.

238. Judith Treas, "The Effect of Women's Labor Force Participation on the Distribution of Income in the United States," *Annual Review of Sociology* 3 (1987): 278, 282.

239. Judith Treas, "Trickle Down or Transfers? Postwar Determinants of Famliy Income Inequality," *American Sociological Review* 48 (August 1983): 546–59; Judith Treas, "U.S. Income Stratification: Bringing Families Back In," *Sociology and Social Research* 66,3 (1982): 231–51.

240. Congressional Budget Office, *Trends in Family Income,* pp. 6–9.

241. Katherine L. Bradbury, "The Shrinking Middle Class," *New England Economic Review,* September/October 1986, pp. 41–55.

242. Lynn A. Karoly, "The Trend in Inequality among Families, Individuals, and Workers in the United States: A Twenty-five Year Perspective," in Sheldon Danziger and Peter Gottschalk (eds.), *Uneven Tides: Rising Inequality in America* (New York: Russell Sage, 1993), pp. 19–97.

CHAPTER 6

1. U.S. Census Bureau, "Press Briefing on 1994 Income and Poverty Estimates," Census Bureau World Wide Web Homepage, October 5, 1995, graphic 3.

2. Using a measure that calculates the proportion of income actually received in terms of what a region *should* receive based upon its population size,

Nissan and Carter (1993) also corroborate the trend of declining inequality between regions comparing eight state clusters designated by the Bureau of Economic Analysis. Over a half century period, from 1929 through 1979, there was a substantial decline in inequality between regions—with a slight rise thereafter. The authors particularly note a dramatic reduction of inequality for the Plains, Southeast, and Southwest regions which represents a continuation of the convergence between North and South.

3. Kevin Phillips, *The Politics of Rich and Poor* (New York: Random House, 1990), p. 191. As the 1990s unfold, however, this surge of prosperity—nicknamed "bicoastalism"—may be on the wane because of the white-collar, middle-management downsizing and recession alluded to earlier. Real estate, financial services, and defense contracting have shrunk while a shakeout is occurring in the computer industry (Levy 1995, p. 19).

4. James C. Cobb, *Industrialization and Southern Society, 1877–1984* (Lexington: University of Kentucky Press, 1984), pp. 57–58.

5. Gary P. Green, *Finance Capital and Uneven Development* (Boulder, CO: Westview, 1987), p. 86.

6. William Julius Wilson, *The Declining Significance of Race: Blacks and Changing American Institutions* (Chicago: University of Chicago Press, 1980). See especially chapters 2 and 3, which contain a cogent summary of race relations and the southern economy after the Civil War.

7. Cobb, *Industrialization and Southern Society,* p. 63.

8. John Agnew, *The United States in the World-Economy: A Regional Geography* (New York: Cambridge University Press, 1987), pp. 110–29.

9. Gary P. Green, *Finance Capital and Uneven Development* (Boulder, CO: Westview, 1987), pp. 80–83.

10. William W. Falk and Thomas A. Lyson, *High Tech, Low Tech, No Tech: Recent Industrial and Occuptional Change in the South* (Albany: State University of New York Press, 1988), p. 85.

11. The attraction of a lower paid, non-unionized work force, however, can be oversold. Romo et al. (1989: 44) point out as an example that the decline of the steel industry was not due to high wages among its unionized work force—nor a matter of choosing Japan, South Africa, or Texas as an alternative to the frostbelt. Although steel mills in the Northeast operated at a higher $300 average cost per ton than those in the Southwest, only 6 percent of this gap could be accounted for by wage differential. Their seminal article offers complementary insights into plant closings in addition to the deindustrialization perspective. In particular, three relational forces can encourage plant departures: the network of resource dependencies, involvement in local growth machine politics, and participation in national and international business networks. See also Romo and Schwartz (1995).

12. Falk and Lyson, *High Tech, Low Tech, No Tech,* p. 2.

13. Candace Howes, "Transplants No Cure," *Dollars and Sense* (July/August 1991): 16–18.

14. Douglas S. Massey and Mitchell L. Eggers, "The Ecology of Inequality: Minorities and the Concentration of Poverty, 1970–1980," *American Journal of Sociology* 95,5 (March 1990): 1185.

15. Barry Bluestone and Bennett Harrison, *The Deindustrialization of America* (New York: Basic Books, 1982), pp. 266–69.

16. Jon Lorence and Joel Nelson, "Industrial Restructuring and Metropolitan Earnings Inequality, 1970–1980," *Research in Social Stratification and Mobility* 12 (1993): 170–72.

17. Linda LeGrande and Mark Jickling, "Earnings as a Measure of Regional Economic Performance," No. 87–377E, Congressional Research Service, Library of Congress, p. 2. Earnings in this report were defined as wage and salary disbursements, other labor income (e.g., employer contributions to pension funds), and proprietors' income (income of the self-employed). In essence, this study tracks income earnings from a much wider variety of sources than we have been considering to this point. The widened spectrum also includes the private (including farm) and public (including military) sectors.

18. Agnew, *United States in the World Economy,* p. 172.

19. Falk and Lyson, *High Tech, Low Tech, No Tech,* pp. 27–37.

20. To be fair, such advanced industrial foreign countries have only been responding to the constant wooing of southern industrial expansionists. By the 1970s, European and Japanese industrialists were as likely to be courted by southern development agencies as the reverse. Georgia has actually established state development branch offices in Brussels, Tokyo, São Paulo, and Toronto. In the South, both North and South Carolina have been the most successful in attracting foreign plants. The latter saw 40 percent of its yearly industrial investment come from abroad. By the end of the 1970s, there was more West German capital in South Carolina than anywhere else in the world except West Germany (Cobb, *Industrialization and Southern Society,* p. 58).

21. Falk and Lyson, *High Tech, Low Tech, No Tech,* p. 22.

22. Charles M. Tolbert and Thomas A. Lyson, "Earnings Inequality in the Nonmetropolitan United States: 1967–1990," *Rural Sociology* 57,4 (1992): 494–511.

23. Falk and Lyson, *High Tech, Low Tech, No Tech,* p. 22.

24. Cobb, *Industrialization and Southern Society,* p. 161.

25. Ibid., p. 136.

26. U.S. Census Bureau, *U.S. Statistical Abstract of 1994* CD-ROM, table 699.

27. Cobb, *Industrialization and Southern Society,* p. 137.

28. U.S. Census Bureau, *Current Population Reports: Income, Poverty, and Valuation of Noncash Benefits: 1993,* Series P60-188, February 1995 (Washington, DC: U.S. Government Printing Office), p. xv.

29. William H. Frey and Alden Speare, Jr., *Regional and Metropolitan Growth and Decline in the United States* (New York: Russell Sage Foundation, 1988), p. 474.

30. Cobb, *Industrialization and Southern Society,* p. 139.

31. Agnew, *The United States in the World Economy,* pp. 199–200.

32. Falk and Lyson, *High Tech, Low Tech, No Tech,* p. 152.

33. Green, *Finance Capital and Uneven Development,* p. 93.

34. Ibid., p. 93.

35. Cobb, *Industrialization and Southern Society,* p. 60.

36. K. Sale, *Power Shift: The Rise of the Southern Rim and Its Challenge to the Eastern Establishment* (New York: Vintage, 1975), pp. 17–53.

37. Falk and Lyson, *High Tech, Low Tech, No Tech*, pp. 152–53.

38. Morris S. Thompson, "Clouds in the Sunbelt: Signs of Economic Deceleration Appear in the Southeast," *The Washington Post National Weekly Edition*, November 21–27, 1988, p. 21.

39. Green, *Finance Capital and Uneven Development*, p. 92.

40. Falk and Lyson, *High Tech, Low Tech, No Tech*, pp. 135–36.

41. Gunnar Myrdal, *Economic Theory and Underdeveloped Regions* (London: Duckworth and Co, 1957).

42. David M. Smith, *Where the Grass Is Greener: Living in an Unequal World* (Baltimore, MD: Johns Hopkins University Press, 1982), p. 119.

43. Lionel J. Beaulieu, (ed.). *The Rural South in Crisis: Challenges for the Future* (Boulder, CO: Westview, 1988), pp. 2–4.

44. Stuart A. Rosenfeld, "The Tale of Two Souths," in Lionel J. Beaulieu (ed.), *The Rural South in Crisis* (Boulder, CO: Westview, 1988), pp. 51–71.

45. Stuart A. Rosenfeld, Edward Bergman, and Sara Rabin, *After the Factories: Changing Employment Patterns in the Rural South* (Research Triangle Park, NC: Southern Growth Policies Board, 1985), p. 53.

46. Daniel T. Lichter, "Race and Unemployment: Black Employment Hardship in the Rural South," in Lionel J. Beaulieu (ed.), *The Rural South in Crisis* (Boulder, CO: Westview, 1988), pp. 181–97.

47. Jerry Wilcox and Wade Clark Roof, "Percent Black and Black-White Status Inequality: Southern Versus Nonsouthern Patterns," *Social Science Quarterly* 59,3 (1978): 421–34.

48. Glenna S. Colclough, "Uneven Development, Racial Composition, and Areal Income Inequality in the Deep South: 1970–1980," *Economic Development Quarterly* 4,1 (1990): 51.

49. Stuart Holland, *The Regional Problem* (New York: St. Martin's Press, 1976), pp. 96–120.

50. Eric A. Hanushek, "Regional Differences in the Structure of Earnings," *The Review of Economics and Statistics* 55,2 (1973): 204–13.

51. William W. Falk and Bruce H. Rankin, "The Cost of Being Black in the Black Belt," *Social Problems* 39,3 (1992): 304.

52. The definition of a Metropolitan Statistical Area for the most part consists of a central city of at least 50,000 people, the county it is located in, and any adjacent counties that display a high degree of social and economic integration with the central city nucleus. Part of this integration is measured by commuting ties. In 1990 there were 830 metropolitan counties, or slightly over one-fourth of all U.S. counties.

53. According to research by Morton Winsberg (1989), this difference has increased over a thirty-year period. Comparing central cities and suburbs of the 37 largest metropolitan statistical areas in the United States, Winsberg found that from 1950 to 1980 the median share of the poorest of all households had grown and was double that of the poorest households in the suburbs. Conversely, the share of the wealthiest households in the suburbs rose to double that of the wealthiest of all households in central cities.

54. Toby L. Parcel, "Race, Regional Labor Markets and Earnings," *American Sociological Review* 44 (April 1979): 262–79. See also Falk and Rankin, "The Cost of Being Black," pp. 304–9.

55. The Census Bureau defines people as urban if they are living in a place with 2,500 or more people. To confuse matters, the concept is different from the metropolitan definition, which throws all residents of a county into the category whether they live in an urban place or not. It is theoretically possible to have a heavily urbanized county that is not metropolitan, or a metropolitan county that is not very urbanized. Fortunately, however, this is seldom the case. It is especially at the other end that the concepts are in agreement. Rural counties are rarely very urban or connected with metropolitan areas. They also stand in sharp contrast to areas dominated by city life. They are qualitatively different from urban and/or metropolitan counties. Rural counties are deeply branded with low income.

56. Jerry R. Skees and Louis E. Swanson, "Farm Structure and Local Society Well-Being in the South," in Lionel J. Beaulieu (ed.), *The Rural South in Crisis* (Boulder, CO: Westview, 1988), pp. 141–57.

57. E. Yvonne Beauford and Mack C. Nelson, "Social and Economic Conditions of Black Farm Households: Status and Prospects," in Lionel J. Beaulieu (ed.), *The Rural South in Crisis* (Boulder, CO: Westview, 1988), pp. 99–119.

58. Louis E. Swanson, "The Human Dimension of the Rural South in Crisis," in Lionel J. Beaulieu (ed.), *The Rural South in Crisis* (Boulder, CO: Westview, 1988), pp. 91–93.

59. Levy, *Dollars and Dreams*, pp. 103–6, 164. The impact has been especially positive for African Americans starting during World War II (pp. 31–33, 136).

60. David M. Smith, *Where the Grass Is Greener: Living in an Unequal World* (Baltimore, MD: Johns Hopkins University Press, 1982).

61. Denny Braun, "Multiple Measurements of United States Income Inequality," *The Review of Economics and Statistics* 70,3 (1988): 398–405. See also Tom S. Sale, "Interstate Analysis of the Size Distribution of Family Income, 1950–1970," *Southern Economic Journal* 40 (January 1974): 434–41, plus James F. Jonish and James B. Kau, "State Differentials in Income Inequality," *Review of Social Economy* 31 (October 1973): 179–90. In particular, 17 income groupings were used to calculate household income Gini ratios for 1980 and 25 income categories were employed with the 1990 data, which made available a more detailed breakdown of the needed figures.

62. Denny Braun, *The Rich Get Richer* (Chicago: Nelson-Hall, 1990), p. 232.

63. The analysis by households and in terms of Gini ratios had to be abandoned at this stage. The electronic data was simply not available to calculate Gini rates for either households or for families at the county level. Furthermore, the Income Disparity Index had to be calculated for families rather than households, again due to lack of appropriate data. In the data set analyzed, which was derived from the *USA Counties 1994* CD-ROM, only family-income figures were available for percent in poverty. As explained before, families are units whose members are related by blood ties. They are in a household if they live with one another. In contrast, households can also be made up of people who are not necessarily

related: roommates, cohabiting couples who are not married, one person living alone, etc. Although important differences exist between the two units on a number of different dimensions, for the purpose of this analysis the differences are not crucial. For example, the correlation between household-income Gini scores and family-income Gini scores among U.S. states was .9433 in 1980.

64. Joel I. Nelson, "Income Inequality: The American States," *Social Science Quarterly* 65,3 (1984): 854–60; David Ruthenberg and Miron Stano, "The Determinants of Interstate Variation in Income Distribution," *Review of Social Economy* 35,1 (1977): 55–66; Tom S. Sale, "Interstate Analysis of the Size Distribution of Family Income 1950–1970," *Southern Economic Journal* 40 (January 1974): 434–41; James F. Jonish and James B. Kau, "State Differentials in Income Inequality," *Review of Social Economy* 31 (October 1973): 179–90; D. J. Aigner and A. J. Heins, "On the Determinants of Income Inequality," *American Economic Review* 57 (March 1967): 175–84; Ahmad Al-Samarrie and Herman P. Miller, "State Differentials in Income Concentration," *American Economic Review* 57 (March 1967): 59–72; John Conlisk, "Some Cross-State Evidence on Income Inequality," *Review of Economics and Statistics* 49 (February 1967): 115–18; David I. Verway, "A Ranking of States by Inequality Using Census and Tax Data," *Review of Economics and Statistics* 48,3 (1966): 314–21.

65. H. E. Frech and L. S. Burns, "Metropolitan Interpersonal Income Inequality: A Comment," *Land Economics* 47 (February 1971): 104–6; Michael D. Betz, "The City as a System Generating Income Inequality," *Social Forces* 51 (December 1972): 192–98; Sheldon Danziger, "Determinants of the Level and Distribution of Family Income in Metropolitan Areas, 1969," *Land Economics* 52 (November 1976): 467–78.

66. Charles T. Haworth, James E. Long, and David W. Rasmussen, "Income Distribution, City Size, and Urban Growth," *Urban Studies* 15 (February 1978): 1–7; James E. Long, David W. Rasmussen, and Charles T. Haworth, "Income Inequality and City Size," *Review of Economics and Statistics* 59 (May 1977): 244–46; Gasper Garofalo and Michael S. Fogarty, "Urban Income Distribution and the Urban Hierarchy-Equality Hypothesis," *Review of Economics and Statistics* 61 (August 1979): 381–88; Stephen Nord, "Income Inequality and City Size: An Examination of Alternative Hypotheses for Large and Small Cities," *Review of Economics and Statistics* 62 (November 1980): 502–8.

67. Garey C. Durden and Ann V. Schwarz-Miller, "The Distribution of Individual Income in the United States and Public Sector Employment," *Social Science Quarterly* 63,1 (1982): 39–47; John W. Foley, "Trends, Determinants and Policy Implications of Income Inequality in United States Counties," *Sociology and Social Research* 61,4 (1977): 441–61.

68. It is important to point out that this graph resembles earlier research (including my own) which corroborates the existence and strength of the Kuznets curve (see especially Braun 1991a, p. 528, and Braun 1991b, p. 247). What is really involved is the change of signs of the dependent variable, mentioned earlier, involving the Income Disparity Index. With Gini ratios, used in prior research, higher scores indicate *more* inequality. With the Income Disparity Index employed in this analysis, high positive scores mean *less* inequality. Thus, when

both are plotted against family income, they yield the same curve outline (an inverted-U) but have exactly the opposite meaning. In my earlier research, no Kuznets curve was found among all U.S. counties using 1980 household income data and measuring inequality with Gini ratios—but a curve was found among southern counties where it was theoretically least expected.

69. Orley M. Amos, Jr., "Unbalanced Regional Growth and Regional Income Inequality in the Later Stages of Development," *Regional Science and Urban Economics* 18,4 (1988): 565.

70. Smith, *Where the Grass Is Greener,* p. 121.

71. Marc Breslow, "How Free Trade Fails," *Dollars and Sense* 180 (October 1992): 7.

72. Ravi Batra, *The Pooring of America: Competition and the Myth of Free Trade* (New York: Macmillan, 1993), p. 47.

73. Arthur MacEwan, "The Gospel of Free Trade: The New Evangelists," *Dollars and Sense* 171 (November 1991): 6–8.

74. Bloomberg Business News, "Mexican Output Drops 10.5% in 2nd Quarter," *Minneapolis Star Tribune,* August 17, 1995, p. 3D.

75. Associated Press, "Thousands in Mexico Join Economic Protests," *Minneapolis Star Tribune,* March 21, 1995, p. 3D.

76. Tod Robberson, "For Mexico, NAFTA Spells a Rising Deficit," *The Washington Post National Weekly Edition,* August 15–21, 1994, p. 21.

77. Tod Robberson, "The Mexican Miracle Unravels," *The Washington Post National Weekly Edition,* January 16–22, 1995, p. 20.

78. Clay Chandler and Daniel Williams, "Who Wins in the Peso Bailout?" *The Washington Post National Weekly Edition,* January 30–February 5, 1995, p. 16.

79. James Cypher, "More Trade Talk," *Dollars and Sense* 192 (March/April 1994): 42.

80. James Cypher, "NAFTA Shock: Mexico's Free Market Meltdown," *Dollars and Sense* (March/April, 1995): 39.

81. Patrick J. Buchanan, "Cancel the North American Free Trade Agreement Now," *Minneapolis Star Tribune,* September 25, 1995, p. 11A.

82. Jeff Faux and Thea Lee, "Implications of NAFTA for the United States: Investment, Jobs, and Productivity," in Ricardo Grinspun and Maxwell A. Cameron (eds.), *The Political Economy of North American Free Trade* (New York: St. Martin's Press, 1993), p. 244.

83. Peter Behr, "In the NAFTAmath, Some Texas-Sized Gains," *The Washington Post National Weekly Edition,* August 29–September 4, 1994, p. 19.

84. "NAFTA Hasn't Fulfilled Promises, Study Says," *Minneapolis Star Tribune,* September 4, 1994, p. 6A.

85. Thea Lee, "Happily Never NAFTA," *Dollars and Sense* 183 (January/February, 1993): 13.

86. "NAFTA Hasn't Fulfilled Promises, Study Says," *Minneapolis Star Tribune,* September 4, 1994, p. 6A.

87. Tod Robberson, "NAFTA's One-Way Street," *The Washington Post National Weekly Edition,* August 28–September 3, 1995, p. 17.

CHAPTER 7

1. Robert W. Cox, "The Core-Periphery Structure of Production and Jobs: The Internationalizing of Production," in Edward Weisband (ed.), *Poverty Amidst Plenty: World Political Economy and Distributive Justice* (Boulder, CO: Westview, 1989), pp. 186–96.

2. Daniel W. Rossides, *Social Stratification: The American Class System in Comparative Perspective* (Englewood Cliffs, NJ: Prentice-Hall, 1990), p. 259.

3. Terry Nichols Clark and Seymour Martin Lipset, "Are Social Classes Dying?" *International Sociology* 6,4 (1991): 397–410; Terry Nichols Clark, Seymour Martin Lipset, and Michael Rempel, "The Declining Political Significance of Social Class," *International Sociology* 8,3 (1993): 293–316; Jan Pakulski, "The Dying of Class or Marxist Class Theory?" *International Sociology* 8,3 (1993): 279–92; Aage B. Sorensen, "On the Usefulness of Class Analysis in Research on Social Mobility and Socioeconomic Inequality," *Acta Sociologica* 34,2 (1991): 71–87.

4. Mike Hout, Clem Brooks, and Jeff Manza, "The Persistence of Classes in Post-Industrial Societies," *International Sociology* 8,3 (1993): 259–77.

5. James R. Kluegel and Eliot R. Smith, "Stratification Beliefs," *Annual Review of Sociology* 7 (1981): 29–56; Mary R. Jackman and Robert W. Jackman, *Class Awareness in the United States* (Berkeley: University of California Press, 1983).

6. James R. Kluegel and Eliot R. Smith, *Beliefs about Inequality: Americans' Views of What Is and What Ought to Be* (New York: Aldine, 1986), p. 5.

7. Ibid., p. 29.

8. Adam Swift, Gordon Marshall, Carol Burgoyne, and David Routh, "Distributive Justice: Does It Matter What People Think?" in James R. Kluegel, David S. Mason, and Bernd Wegener (eds.), *Social Justice and Political Change: Public Opinion in Capitalist and Post-Communist States* (New York: Aldine de Gruyter, 1995), p. 35.

9. James R. Kluegel and Masaru Miyano, "Justice Beliefs and Support for the Welfare State in Advanced Capitalism," in James R. Kluegel, David S. Mason, and Bernd Wegener (eds.), *Social Justice and Political Change: Public Opinion in Capitalist and Post-Communist States* (New York: Aldine de Gruyter, 1995), p. 91.

10. James R. Kluegel, Gyorgy Csepeli, Tamas Kolosi, Antal Orkeny, and Maria Nemenyi, "Accounting for the Rich and the Poor: Existential Justice in Comparative Perspective," in James R. Kluegel, David S. Mason, and Bernd Wegener (eds.), *Social Justice and Political Change: Public Opinion in Capitalist and Post-Communist States* (New York: Aldine de Gruyter, 1995), p. 194.

11. James R. Kluegel and Peter Mateju, "Egalitarian vs. Inegalitarian Principles of Distributive Justice," in James R. Kluegel, David S. Mason, and Bernd Wegener (eds.), *Social Justice and Political Change: Public Opinion in Capitalist and Post-Communist States* (New York: Aldine de Gruyter, 1995), pp. 215, 219–20.

12. James A. Davis and Tom W. Smith, *The NORC General Social Survey: A User's Guide* (Newbury Park, CA: Sage, 1992), p. 1.

13. Kluegel and Smith, *Beliefs about Inequality,* pp. 43–44, 52, 78–80.

14. Benjamin I. Page and Robert Y. Shapiro, *The Rational Public: Fifty Years of Trends in Americans' Policy Preferences* (Chicago: University of Chicago Press, 1992), p. 45.

15. A question inquiring whether the government in Washington ought to reduce income differences between the rich and the poor has been asked in the GSS nearly every year since 1978, using a seven-point Likert scale (1=strong support for government intervention; 7=strong support for no government intervention). Theoretically, if all respondents were spread evenly on this item, the mean should be 4.0. Instead, it has fluctuated around 1.9 for the past fifteen years, but has not varied much. In other words, more people are in favor of government intervention than oppose it, but the average has not really budged despite the huge growth of inequality during the same time period. Moreover, the proportion who favor government intervention (persons choosing a "1" or "2" on the item) has also been quite stable and is not substantially different from the proportion who oppose government stepping in (those indicating a "6" or "7" on the item). By using these cutoffs to illustrate the point, it can be seen that in 1978—before the gargantuan run-up of income inequality—30.5 percent of the public favored government intervention while 20.2 percent were opposed to it (the rest were in the middle, choosing 3, 4, or 5). By 1993, when the reality of income chasms had become glaring in our society, 29.6 percent were in favor of government doing something about it but 20.4 percent were still opposed. The moral of the story is that Americans may recognize and be unhappy about growing inequality, but they tend to balk at getting the government involved in strategies to reduce differences. A further impediment to government intervention may be the strengthening of individualistic explanations for poverty among the American public. According to a CBS/*New York Times* poll (1994), 30 percent of the public saw "lack of effort" as causing people to be poor in 1990 but 44 percent held this belief in 1994. Conversely, while almost half of 1990 survey respondents saw "circumstances beyond one's control" as causing poverty in 1990, this had dropped to one-third by 1994.

16. Kevin Phillips, *The Politics of Rich and Poor* (New York: Random House, 1990); Kevin Phillips, *Boiling Point: Democrats, Republicans, and the Decline of Middle-Class Prosperity* (New York: HarperCollins, 1993).

17. David S. Broder, "The GOP Earthquake," *The Washington Post National Weekly Edition*, November 14–20, 1994, p. 6.

18. Dan Balz and Richard Morin, "They're Still Angry Out There," *The Washington Post National Weekly Edition*, May 30–June 5, 1994, p. 12.

19. George Pettinico, "1994 Vote: Where Was Public Thinking at Vote Time?" in Everett Carll Ladd (ed.), *America at the Polls 1994* (Storrs: University of Connecticut, Roper Center, 1995), p. 31.

20. Balz and Morin, "Still Angry," p. 12.

21. Richard Morin, "The Message from the Voters: Less Is More," *The Washington Post National Weekly Edition*, November 21–27, 1994, p. 8.

22. Erik Black and Steve Berg, "How Did GOP Do It? Let Us Count the Ways," *Minneapolis Star Tribune*, November 13, 1994, p. 9A.

23. Thomas B. Edsall, "Revolt of the Discontented," *The Washington Post National Weekly Edition*, November 21–27, 1994, p. 28.

24. Kevin Phillips, "The Voters Are Already Tapping Their Feet," *The Washington Post National Weekly Edition*, November 21–27, 1994, pp. 23–24.

25. Robin Toner, "Women in Growing Numbers Say They Favor Democrats," *New York Times*/CBS News Poll, April 21, 1996, *New York Times* on the World Wide Web.

26. CBS News/*New York Times* Poll, *The Public Weighs in on the Republican Agenda*, New York, CBS News, December 14, 1994.

27. Richard Morin, "Myths and Messages in the Election Tea Leaves," *The Washington Post National Weekly Edition*, November 21–27, 1994, p. 37

28. E.J. Dionne, Jr., "A Shift, Not a Mandate," *The Washington Post National Weekly Edition*, November 28–December 4, 1994, p. 28.

29. U.S. Census Bureau, *Turnout in '94 Congressional elections—45 Percent; Young Voter Participation Shows No Gain*, CB95-105 (Washington, DC: U.S. Government Printing Office); "Higher-Income Voters Turned Out in '94," *Minneapolis Star Tribune*, June 8, 1995, p. 9A.

30. Economy in Numbers, "The GOP's 17% Mandate," *Dollars and Sense* 198 (March/April, 1995): 43.

31. "Voters Just as Dissatisfied Now as Before Election, Poll Shows," *Minneapolis Star Tribune*, April 13, 1995, p. 14A.

32. "Poll Finds MostUnimpressed by House So Far," *Minneapolis Star Tribune*, August 11, 1995, p. 16A.

33. Robert Reno, "Polls Suggest Electorate Stands Well to the Left of GOP Tidal Wave," *Minneapolis Star Tribune*, November 16, 1995, p. 24A.

34. Ann Devroy, "Bad News for the GOP and Bob Dole," *The Washington Post National Weekly Edition*, October 9–15, 1995, p. 11.

35. Kevin Phillips, "The Champagne's Gone Flat," *The Washington Post National Weekly Edition*, August 14–20, 1995, p. 23.

36. "Voter Turnout Rose Slightly in November 1994," *Minneapolis Star Tribune*, June 26, 1995, p. 2A.

37. "Public Support for Third-Party Candidate Rises," *Minneapolis Star Tribune*, August 24, 1995, p. 7A.

38. Devroy, "Bad News for the GOP," p. 12.

39. Telephone conversation with Curtis Gans, Committee for the Study of the American Electorate, 421 New Jersey Ave. SE, Washington, DC, December 18, 1996.

40. Jodie T. Allen, "Don't Count Those Budget Chickens before They Hatch," *The Washington Post National Weekly Edition*, May 29–June 4, 1995, p. 23.

41. Page and Shapiro, *The Rational Public*, pp. 123–27; Fay Lomax Cook and Edith J. Barrett, *Support for the American Welfare State: The Views of Congress and the Public* (New York: Columbia University Press, 1992).

42. Steven Kull, *Fighting Poverty in America: A Study of American Public Attitudes* (Washington, DC: Center for the Study of Policy Attitudes, 1994).

43. Richard Morin, "What the Public Really Wants," *The Washington Post National Weekly Edition*, January 9–15, 1995, p. 37.

44. Clay Chandler, "The Committee to Reelect Ronald Reagan," *The Washington Post National Weekly Edition*, October 10–16, 1994, p. 8.

45. Steven Pearlstein, "The Rich Get Richer and...Proposed Tax and Spending Cuts Will Widen the Inequity in U.S. Incomes," *The Washington Post National Weekly Edition,* June 12–18, 1995, p. 6.

46. CBS News/*New York Times* Poll, *The Public Weighs In on the Republican Agenda,* New York, CBS News, December 14, 1994.

47. Harold R. Kerbo, *Social Stratification and Inequality: Class Conflict in Historical and Comparative Perspective* (New York: McGraw-Hill, 1991), p. 223.

48. Val Burris, "Elite Policy-Planning Networks in the United States," *Research in Politics and Society* 4 (1992): 113.

49. Richard E. Ratcliff, Mary Elizabeth Gallagher, and Anthony C. Kouzi, "Political Money and Partisan Clusters in the Capitalist Class," *Research in Politics and Society* 4 (1992): 76.

50. David Jacobs, Michael Useem, and Mayer N. Zald, "Firms, Industries, and Politics," *Research in Political Sociology* 5 (1991): 145–49.

51. Dan Clawson, Alan Neustadtl, and Denise Scott, *Money Talks: Corporate PACS and Political Influence* (New York: Basic Books, 1992), p. 18.

52. Michael J. Malbin, "1994 Vote: The Money Story," in Everett Carll Ladd (ed.), *America at the Polls 1994* (Storrs: Roper Center, University of Connecticut, 1995), p. 127.

53. Brett D. Fromson, "Going Where They Make the Money," *The Washington Post National Weekly Edition,* April 3–9, 1995, p. 14.

54. "Corporate PACs Gave Heavily to GOP," *Minneapolis Star Tribune,* December 5, 1995, p. 10A.

55. Michael Weisskopf, "To the Victors Belong the Pac Checks," *The Washington Post National Weekly Edition,* January 2–8, 1995, p. 13.

56. David Maraniss and Michael Weisskopf, "Cashing In," *The Washington Post National Weekly Edition,* December 4–10, 1995, p. 6.

57. Michael Weisskopf and David Maraniss, "In on the Takeoff: In This Congress, Corporate Lobbyists Are Invited to Sit in the Cockpit and Help Fly the Plane," *The Washington Post National Weekly Edition,* March 20–26, 1995, p. 13.

58. Dan Morgan, "Taking Care of Business," *The Washington Post National Weekly Edition,* July 3–9, 1995, p. 31.

59. Brian Kelly, "The Pork That Just Won't Slice," *The Washington Post National Weekly Edition,* December 18–24, 1995, p. 21.

60. Ibid.

61. Janice Shields, *Aid for Dependent Corporations (AFDC) 1995* (Washington, DC: Essential Information, 1995).

62. Chuck Collins, "Aid to Dependent Corporations," *Dollars and Sense* 199 (May/June, 1995): 17.

63. Robert J. Shapiro, *Cut and Invest: A Budget Strategy for the New Economy* (Washington, DC: Progressive Policy Institute, 1995).

64. Greg Leroy, "No More Candy Store," *Dollars and Sense* 199 (May/June, 1995): 11.

65. Harry Holloway and John George, *Public Opinion: Coalitions, Elites, and Masses,* 2d ed. (New York: St. Martin's Press, 1986). p. 247.

66. Ibid., p. 254.

67. Richard Morin, "Tuned Out, Turned Off," *The Washington Post National Weekly Edition,* February 5–11, 1996, p. 6.

68. Ben H. Bagdikian, *The Media Monopoly,* 3d ed. (Boston: Beacon Press, 1990).

69. Paul Farhi, "Hearst-Case Scenario," *The Washington Post National Weekly Edition,* July 31–August 6, 1995, p. 21.

70. Ibid., p. 11.

71. Martin A. Lee and Norman Solomon, *Unreliable Sources: A Guide to Detecting Bias in News Media* (New York: Carol Publishing, 1990), pp. 175–200.

72. Ibid., p. 188.

73. Edward S. Herman and Noam Chomsky, *Manufacturing Consent: The Political Economy of the Mass Media* (New York: Pantheon Books, 1988); Michael Parenti, *Make Believe Media: The Politics of Entertainment* (New York: St. Martin's Press, 1992); Michael Parenti, *Inventing Reality: The Politics of News Media,* 2d ed. (New York: St. Martin's Press, 1993).

74. News with a View, "Slanted Media," *Minneapolis Star Tribune,* January 1, 1996, p. 9A.

75. Jeff Cohen and Borman Solomon, "Hold Your Applause for 'News-Hour,' Where 'Public' TV Speaks for Big Money," *Minneapolis Star Tribune,* September 29, 1995, p. 27A.

76. Andrew Kohut, *The Vocal Minority in American Politics* (Washington, DC: Times Mirror Center for the People and the Press, 1993).

77. William Raspberry, "Bomb-Throwers and Broadcasters," *The Washington Post National Weekly Edition,* May 1–7, 1995, p. 27.

78. "Angry at Government? Tell Me about It," *Minneapolis Star Tribune,* November 18, 1995, p. 13A.

79. Eleanor Randolph, "Firebrand of the Radio Waves Turns Down the Rhetorical Heat," *Minneapolis Star Tribune,* January 1, 1996, p. A6.

80. Stephen Talbot, "Wizard of Ooze," The MOJO Wire, *Mother Jones Magazine* World Wide Web site, 1995.

81. Molly Ivins, "Lyin' Bully," The MOJO Wire, *Mother Jones Magazine* World Wide Web site, 1995.

82. Steven Rendall, Jim Naureckas, and Jeff Cohen, *The Way Things Aren't* (New York: The New Press, 1995), p. 8.

83. Ibid., pp. 9–10.

84. Ibid., p. 25.

85. Parenti, *Inventing Reality,* pp. 229–32.

86. The Economic Affairs Bureau, Inc., One Summer Street, Somerville, MA 02143, (617) 628-8411.

87. Jeffrey H. Birnbaum, "Merging a Lower Capital Gains Rate with a Rise in Gasoline Tax Would Leave the Poor the Losers," *Wall Street Journal,* May 8, 1989, p. A16.

88. Economy in Numbers, "Where the Federal Deficit Comes From," in *Real World Macro: A Macroeconomics Reader from Dollars and Sense,* 5th ed. (Somerville, MA: Economic Affairs Bureau, 1988), p. 24.

89. Randy Albelda, "Let Them Pay Taxes: The Growing Tax Burden on the Poor," in *Real World Macro: A Macroeconomics Reader from Dollars and Sense,* 5th ed. (Somerville, MA: Economic Affairs Bureau, 1988), p. 29.

90. Albelda, "Let Them Pay Taxes," p. 31.

91. Immediately after assuming the presidency, Bill Clinton was able to persuade Congress to do just this (Associated Press, April 14, 1995). Two more tax brackets were restored, making the tax structure more progressive: 36 percent for the $115,000 to $249,999 bracket and 39.6 percent for the $250,000-and-over bracket. While tax returns that report $100,000 or more as income make up only 4 percent of all returns, one fourth of all taxes paid come from this affluent group. As a result, taxes from well-to-do Americans surged 16 percent in 1993, the first year the new tax law took hold.

92. Jim Lenfestey, "Tax Increase Can Be Good for Everyone," *Minneapolis Star Tribune*, June 26, 1995, p. 9A.

93. Jeff Faux, "The Austerity Trap and the Growth Alternative," *World Policy Journal* 5,3 (1988): 367–414.

94. Dane Smith, "Livable-wage," *Minneapolis Star Tribune*, March 28, 1996, p. 1B.

95. Robert W. Benson, *Getting Business Off the Public Dole* (Los Angeles, CA: Loyola Law School, 1995).

96. Martin Olav Sabo, "A $6.50 U.S. Minimum Wage Would Lessen Income Gap," *Minneapolis Star Tribune*, June 16, 1994, p. 25A.

97. Todd S. Purdum, "Clinton Calls on Companies to Train Workers," *New York Times* World Wide Web site, March 9, 1996.

98. Sharon Schmickle, "Buchanan Is Talking about Plight of the Working Class, and Washington Is Listening," *Minneapolis Star Tribune*, March 4, 1996, p. A3.

99. *New York Times* Poll, "The Downsizing of America: Getting to the Root of the Problem," *The New York Times* World Wide Web site, March 9, 1996; *New York Times*/CBS News Poll, "Social Issues Behind Buchanan's Success, Poll Shows," *The New York Times* World Wide Web site, February 27, 1996.

100. Richard J. Barnet, et al. "American Priorities in a New World Era," *World Policy Journal* 6,2 (1989): 203–38.

101. Paul Seltman, "Congressional Black Caucus Cuts Corporate Welfare to Balance U.S. Budget," Congressional Black Caucus, U.S. Congress, May 1995.

102. Tom Riddell, "The Political Economy of Military Spending," in *The Imperiled Economy, Book II: Through the Safety Net* (New York: Union for Radical Political Economics, 1988), p. 232.

103. Lester R. Brown, "Redefining National Security," in Lester R. Brown (ed.), *State of the World 1986* (New York: Norton, 1986), p. 201.

104. Michael Renner, "Enhancing Global Security," in Lester R. Brown (ed.), *State of the World 1989* (New York: Norton, 1989), p. 139.

105. Lawrence J. Korb, "The Indefensible Defense Budget," *The Washington Post National Weekly Edition*, July 17–23, 1995, p. 19.

106. Economy in Numbers, "Maintaining Military Might," *Dollars and Sense* 191 (January/February 1994): 42. See also Center for Defense Information, "Reduce Military Spending: Create More Jobs," *The Defense Monitor* 23,6 (1994): 2.

107. Karen M. Paget, "Can't Touch This: The Pentagon's Budget Fortress," *The American Prospect* 23 (Fall 1995): 38.

108. J. Michael Cline, "Critiques of Defense Spending Folly Came Too Late," *Minneapolis Star Tribune*, September 6, 1995, p. 19A.

109. Center for Defense Information, "Balanced Budget, Unbalanced Priorities?" *The Defense Monitor* 24 (7): 2.

110. Economy in Numbers, "How to Pare Down the Pentagon," *Dollars and Sense* 200 (July/August 1995): 43. See also Walter Pincus and Dan Morgan, "Pork: The Last Line of the Dense Budget," *The Washington Post National Weekly Edition,* April 3–9, 1995, p. 32, and Dana Priest, "Damn the Defense Budget, Full Speed Ahead," *The Washington Post National Weekly Edition,* August 7–13, 1995, p. 32.

111. Korb, "The Indefensible Defense Budget," p. 19.

112. Paget, "Can't Touch This," pp. 39–40.

113. Mark Mellstrom and Bob Lamb, "Military Budget Throws Away Billions on a Cold War That's Been Won," *Minneapolis Star Tribune,* July 2, 1994, p. 23A.

114. Center for Defense Information, "Balanced Budget, Unbalanced Priorities?" pp. 3–5.

115. John D. Abell, "Military Spending and Income Inequality," *Journal of Peace Research* 31,1 (1994): 35.

116. John Agnew, *The United States in the World-Economy: A Regional Geography* (New York: Cambridge University Press, 1987), p. 216.

117. Robert S. McIntyre, "The Flat Taxers' Flat Deceptions," *The American Prospect* 22 (November 1995): 94.

118. Steven Pearlstein, "Floating the Flat Tax Idea," *The Washington Post National Weekly Edition,* January 22–28, 1996, p. 20.

119. Sidney L. Carroll, "American Family Fortunes as Economic Deadweight," *Challenge* 34,3 (1991): 11–18.

120. John Miller, "The Capital Gains Giveaway," *Dollars and Sense* 198 (March/April 1995): 15.

121. Molly Ivins, "Many Horror Stories of Government Untrue," *Minneapolis Star Tribune,* May 26, 1995, p. 25A.

122. S. M. Miller and Jaqueline Ortiz, "The Policies of Income Distribution," *Research in Politics and Society* 5 (1995): 269–92.

123. Sidney Verba, Steven Kelman, Gary R. Orren, Ichoro Miyake, Joji Watanuki, Ikuo Kabashima, and G. Donald Ferree, Jr., *Elites and the Idea of Equality: A Comparison of Japan, Sweden, and the United States* (Cambridge, MA: Harvard University Press, 1987), p. 141.

124. Jeff Coplon, "Skinhead Reich," *Utne Reader* 33 (May/June 1989): 80–83, 85–89; "The Roots of Skinhead Violence: Dim Economic Prospects for Young Men," *Utne Reader* 33 (May/June 1989): 84.

125. Jason Berry, "In Louisiana, The Hazards of Duke," *The Washington Post National Weekly Edition,* May 22–28, 1989, p. 25.

126. Molly Ivins, "Look Long and Hard to Find Silver Lining in Buchanan's New Hampshire Victory," *Minneapolis Star Tribune,* February 23, 1996, p. 17A.

127. Benjamin I. Page and Robert Y. Shapiro, *The Rational Public: Fifty Years of Trends in Americans' Policy Preferences* (Chicago: University of Chicago Press, 1992), p. 68.

128. Sam Fulwood III, "Minority Efforts Draw Resentment," *Minneapolis Star Tribune,* October 15, 1994, p. 4A.

129. Richard Morin, "Across the Racial Divide," *The Washington Post National Weekly Edition,* October 16–22, 1995, p. 6.

130. Richard J. Herrnstein and Charles Murray, *The Bell Curve: Intelligence and Class Structure in American Life* (New York: The Free Press, 1994).

131. For a detailed critique on scientific grounds, see the separate reviews of Robert M. Hauser, Howard F. Taylor, and Troy Duster in *Contemporary Sociology* (March 1995). For a large variety of broader, more comprehensive criticisms and commentary, see Jacoby and Glauberman (1995). For extended analyses of the limits of IQ testing, see Mensh and Mensh (1991) or Schiff and Lewontin (1986).

132. E. J. Dionne, Jr., "Race and IQ: Stale Notions," *The Washington Post National Weekly Edition,* October 24–30, 1994, p. 29.

133. Robert J. Samuelson, "Bell Curve Ballistics," *The Washington Post National Weekly Edition,* October 31–November 6, 1994, p. 28.

134. Rick Weiss, "A Human Face on Gene Theories," *The Washington Post National Weekly Edition,* February 27–March 5, 1995, p. 37.

135. William W. Falk and Thomas A. Lyson, *High Tech, Low Tech, No Tech: Recent Industrial and Occupational Change in the South* (Albany: State University of New York Press, 1988), pp. 145–46.

136. Louis Uchitelle, Louis, "U.S. Companies Are Globalizing Their Resources," *Minneapolis Star Tribune,* May 30th, 1989, p. 5B.

REFERENCES

Abell, John D. 1994. "Military Spending and Income Inequality." *Journal of Peace Research* 31, 1 (February): 35–43.

Adamchak, Donald J. and James L. Robinson. 1986. "Evidence of a Curvilinear Relationship Between Income and Infant Mortality: Individual Data Do Not Bear Out Ecological Correlations." *Sociology and Social Research* 70,3(April): 214–17.

Agnew, John. 1987. *The United States in the World Economy: A Regional Geography.* New York: Cambridge University Press.

Ahluwalia, Montek S. 1976. "Inequality, Poverty, and Development." *Journal of Development Economics* 3: 307–42.

Ahluwalia, Montek S. 1984. "Income Inequality: Some Dimensions of the Problem." In Mitchell A. Seligson (ed.), *The Gap Between the Rich and Poor: Contending Perspectives on the Political Economy of Development.* Boulder, CO: Westview.

Aigner, D. J. and A. J. Heins. 1967. "On the Determinants of Income Inequality." *American Economic Review* 57 (March): 175–84.

Aiker, Hayward R., Jr. 1965. *Mathematics and Politics.* New York: Macmillan.

Al-Samarrie, Ahmad and Herman P. Miller. 1967. "State Differentials in Income Concentration." *American Economic Review* 57(March): 59–72.

Albelda, Randy. 1988. "Let Them Pay Taxes: The Growing Tax Burden on the Poor." In *Real World Macro: A Macroeconomics Reader from Dollars and Sense,* 5th ed. Somerville, MA: Economics Affairs Bureau.

Albelda, Randy and Chris Tilly. 1995. "Unnecessary Evil: Why Inequality Is Bad for Business." *Dollars and Sense* 198(March/April): 18–21.

Alexander, Karl, Martha Cook, and Edward McDill. 1978. "Curriculum Tracking and Educational Stratification: Some Further Evidence." *American Sociological Review* 43: 47–66.

Allen, Jodie T. 1995. "Don't Count Those Budget Chickens before They Hatch." *The Washington Post National Weekly Edition* (May 29–June 4): 23.

Allison, Paul D. 1978. "Measures of Inequality." *American Sociological Review* 43 (December): 865–80.

Amos, Jr., Orley M. 1988. "Unbalanced Regional Growth and Regional Income Inequality in the Later Stages of Development." *Regional Science and Urban Economics* 18,4: 549–66.

Arrighi, Giovanni and Jessica Drangel. 1986. "The Stratification of the World Economy: An Exploration of the Semiperipheral Zone." *Review: A Journal of the Fernand Brandel Center for the Study of Economies, Historical Systems, and Civilizations* 10: 9–74.

Associated Press. 1988. "Study Says Deficit Cost 5.1 Million Jobs." *Minneapolis Star Tribune* (October 16): 3A.

Associated Press. 1989. "Despite 4th Quarter Rise, Trade Deficit Down in '88." *Minneapolis Star Tribune* (March 1): 3D.

Associated Press. 1990. "Congress Warned S and L Bailout Could Reach $500 Billion." *Minneapolis Star Tribune* (April 7): 12D.

Associated Press. 1992. "U.S. Business Failures Hit Record Level in '91." *Minneapolis Star Tribune* (February 21): 3D.

Associated Press. 1994. "NAFTA Hasn't Fulfilled Promises, Study Says." *Minneapolis Star Tribune* (September 4): 6A.

Associated Press. 1994. "Homelessness: Rare or Commonplace?" *Minneapolis Star Tribune* (December 28): 2A.

Associated Press. 1995. "Thousands in Mexico Join Economic Protests." *Minneapolis Star Tribune* (March 21): 3D.

Associated Press. 1995. "Tax on the Rich Up 16 percent in First Year of Law Revisions." *Minneapolis Star Tribune* (April 14): 6A.

Associated Press. 1995. "Higher-Income Voters Turned Out in '94." *Minneapolis Star Tribune* (June 8): 9A.

Associated Press. 1995. "Voter Turnout Rose Slightly in November 1994." *Minneapolis Star Tribune* (June 26): 2A.

Associated Press. 1995. "Federal and State Prisons Getting Even More Crowded." *Minneapolis Star Tribune* (August 10): 19A.

Associated Press. 1995. "Poll Finds Most Unimpressed by House So Far." *Minneapolis Star Tribune* (August 11): 16A.

Associated Press. 1995. "NAFTA Hasn't Fulfilled Promises, Study Says." *Minneapolis Star Tribune* (September 4): 6A.

Associated Press. 1996. "Executive Compensation Up 23%." *Minneapolis Star Tribune* (March 6): 1D.

Atkinson, Anthony, Lee Rainwater, and Timothy Smeeding. 1995a. *Income Distribution in Advanced Economies: Evidence from the Luxembourg Income Study*. Working Paper No. 120. Syracuse, NY: Syracuse University, Maxwell School of Citizenship and Public Afairs.

Atkinson, Anthony, Lee Rainwater, and Timothy Smeeding. 1995b. *Income Distribution in OECD Countries*. Paris: OECD.

Bacha, Edmar L. 1986. "External Shocks and Growth Prospects: The Case of Brazil, 1973–89." *World Development* 14, 8: 919–36.

Bagdikian, Ben H. 1990. *The Media Monopoly.* 3rd ed. Boston: Beacon Press.

Baker, Dean and Lawrence Mishel. 1995. *Profits Up, Wages Down: Worker Losses Yield Big Gains for Business.* Washington, DC: Economic Policy Institute.

Baldwin, Marc. 1992. "How Low Can You Go? Why So Few of the Unemployed Receive Benefits." *Dollars and Sense* 177(June): 19–22.

Baldwin, William. 1986. "Chicken Little's Income Statistics." *Forbes* 137 (March 24): 68–69.

Balz, Dan and Richard Morin. 1994. "They're Still Angry Out There." *The Washington Post National Weekly Edition* (May 30–June 5): 12.

Barlett, Donald L. and James B. Steele. 1994. *America: Who Really Pays the Taxes?* New York: Simon and Schuster.

Barnet, Richard J. 1994. "Lords of the Global Economy: Stateless Corporations." *The Nation* 259,21(December 19): 754ff.

Barnet, Richard J. et al. 1989. "American Priorities in a New World Era." *World Policy Journal* 4, 2 (Spring): 203–38.

Barran, Paul A. 1960. *The Political Economy of Growth.* New York: Prometheus.

Barran, Paul A. 1969. *The Longer View: Essays Toward a Critique of Political Economy.* New York: Monthly Review Press.

Barran, Paul A. and Paul M. Sweezy. 1966. *Monopoly Capital: An Essay on the American Economic and Social Order.* New York: Monthly Review Press.

Barringer, Felicity. 1992. "Whether It's Hunger or 'Misnourishment,' It's a National Problem." *The New York Times* (December 27): E2 ff.

Batra, Ravi. 1988. *The Great Depression of 1990.* New York: Dell.

Batra, Ravi. 1993. *The Pooring of America: Competition and the Myth of Free Trade.* New York: Macmillan.

Bauer, P.T. 1984. "The Vicious Circle of Poverty." In Mitchell A. Seligson (ed.), *The Gap Between Rich and Poor.* Boulder, CO:Westview.

Beauford, E. Yvonne and Mack C. Nelson. 1988. "Social and Economic Conditions of Black Farm Households: Status and Prospects." In Lionel J. Beaulieu (ed.), *The Rural South in Crisis.* Boulder, CO: Westview.

Beaulieu, Lionel J. (ed.). 1988. *The Rural South in Crisis: Challenges for the Future.* Boulder, CO: Westview.

Beck, E.M., Patrick M. Horan, and Charles M. Tolbert II. 1978. "Stratification in a Dual Economy: A Sectoral Model of Earnings Determination." *American Sociological Review* 43 (October): 704–20.

Beck, E.M., Patrick M. Horan, and Charles M. Tolbert II. 1980. "Social Stratification in Industrial Society: Further Evidence from a Structural Perspective (Reply to Hauser)." *American Sociological Review* 45, 4 (August): 712–19.

Becker, G. 1964. *Human Capital.* New York: Columbia University Press.

Beckerman, W. 1977. "Some Reflections on Redistribution with Growth." *World Development.* (August): 665–76.

Beeghley, Leonard and Jeffrey W. Dwyer. 1989. "Income Transfers and Income Inequality." *Population Research and Policy Review* 8: 119–42.

Behr, Peter. 1994. "In the NAFTAmath, Some Texas-Sized Gains." *The Washington Post National Weekly Edition* (August 29–September 4): 19.

Behr, Peter. 1994. "Multinationals Once More Are Leading Transformation of the World Economy." *Minneapolis Star Tribune* (September 12): 7D.

Belknap, Joanne. 1986. *The Effects of Poverty, Income Inequality, and Unemployment on Crime Rates.* Ph.D. diss., Michigan State University.

Bello, Walden. 1988. "U.S.-Phillipine Relations in the Aquino Era." *World Policy Journal* 5 (Fall): 688.

Bello, Walden and Shea Cunningham. 1994. "Reign of Error: The World Bank's Wrongs." *Dollars and Sense* 195 (September/October): 10–13, 39–40.

Bello, Walden and Stephanie Rosenfeld. 1992. *Dragons in Distress: Asia's Miracle Economies in Crisis.* San Francisco: Institute for Food and Development Policy.

Belous, Richard S. 1985. *Two-Tier Wage Systems in the U.S. Economy.* Congressional Research Service, Library of Congress, No. 85-165 E.

Beneria, Lourdes and Shelley Beneria (eds.). 1992. *Unequal Burden: Economic Crisis, Persistent Poverty, and Women's Work.* Boulder, CO: Westview.

Benson, Robert W. 1995. *Getting Business Off the Public Dole.* Los Angeles: Loyola Law School.

Bergen, Evans. 1968. *Dictionary of Quotations.* New York: Delacorte.

Bernstein, Aaron. 1994. "Inequality: How the Gap Between Rich and Poor Hurts the Economy." *Business Week* (August 15): 78-81.

Bernstein, Jared. 1995. *Where's the Payoff? The Gap Between Black Academic Progress and Economic Gains.* Washington, DC: Economic Policy Institute.

Berry, Albert. 1983. "The Level of World Inequality: How Much Can One Say?" *The Review of Income and Wealth* 29, 3 (September): 217–41.

Berry, Albert, François Bourguignon, and Christian Morrisson. 1983. "Changes in the World Distribution of Income Between 1950 and 1977." *The Economic Journal* 93 (June): 331–50.

Berry, Jason. 1989. "In Louisiana, the Hazards of Duke." *The Washington Post National Weekly Edition* (May 22–28): 25.

Berry, John M. 1988. "The Legacy of Reagonomics: Underlying Flaws Threaten the Successes." *The Washington Post National Weekly Edition* (December 19–25): 6–7.

Berry, John M. and Clay Chandler. 1995. "Trying to Cool the 'Hot Money' Game." *The Washington Post National Weekly Edition* (April 24–30): 20–21.

Betson, David M. 1995. "Consequences of the Panel's Recommendations." *Focus* 17,1 (Summer): 11–13.

Betz, Michael D. 1972. "The City as a System Generating Income Inequality." *Social Forces* 51 (December): 192–98.

Bianchi, Suzanne M. 1995. "Changing Economic Roles of Women and Men." In Reynolds Farley (ed.), *State of the Union: America in the 1990s,* Vol. 1: *Economic Trends.* New York: Russell Sage.

Birnbaum, Jeffrey H. 1989. "Merging a Lower Capital Gains Rate with a Rise in Gasoline Tax Would Leave the Poor the Losers." *Wall Street Journal* (May 8): A16.

Bishop, John A., John P. Formby, and W. James Smith. 1991. "International Comparisons of Income Inequality: Tests for Lorenz Dominance across Nine Countries." *Economica* 58 (November): 461–77.

Bishop, John A., John P. Formby, and Paul D. Thistle. 1992. "Explaining Interstate Variation in Income Inequality." *The Review of Economics and Statistics* 74,3 (August): 553–57.

Black, Erik and Steve Berg. 1994. "How Did GOP Do It? Let Us Count the Ways." *Minneapolis Star Tribune* (November 13): 1A.

Black, Jan Knippers. 1977. *United States Penetration of Brazil*. Philadelphia: University of Pennsylvania Press.

Blackburn, McKinley L. and David E. Bloom. 1985. "What Is Happening to the Middle Class?" *American Demographics* (January): 18–25.

Blackburn, McKinley L. and David E. Bloom. 1987. "Family Income Inequality in the United States: 1967–1984." Monograph No. 1294. Cambridge, MA: Harvard Institute of Economic Research.

Blau, Judith R. and Peter M. Blau. 1982. "The Cost of Inequality: Metropolitan Structure and Violent Crime." *American Sociological Review*, 47: 114–29.

Bloomberg Business News. 1995. "Mexican Output Drops 10.5% in 2nd Quarter." *Minneapolis Star Tribune* (August 17): D3.

Bluestone, Barry. 1990. "The Great U-Turn Revisited: Economic Restructuring, Jobs, and the Redistribution of Earnings." In John D. Kasarda (ed.), *Jobs, Earnings, and Employment Growth Policies in the United States*. Boston: Kluwer Academic Publishers.

Bluestone, Barry and Bennett Harrison. 1982. *The Deindustrialization of America*. New York: Basic Books.

Bluestone, Barry and Bennett Harrison. 1986. *The Great American Job Machine: The Proliferation of Low Wage Employment in the U.S. Economy*. Washington, DC: Joint Economic Committee.

Bluestone, Barry, William M. Murphy, and Mary Stevenson. 1973. *Low Wages and the Working Poor*. Ann Arbor: Institute of Labor and Industrial Relations, University of Michigan.

Blumberg, P. 1979. "White Collar Status Panic." *The New Republic* (December 1): 21–23.

Blumberg, Paul. 1980. *Inequality in an Age of Decline*. New York: Oxford University Press.

Blustein, Paul. 1988. "Peace for All, Prosperity—For Some." *The Washington Post National Weekly Edition* (October 3–9): 8.

Bollen, Kenneth. 1983. "World System Position, Dependency, and Democracy: The Cross-National Evidence." *American Sociological Review* 48 (August): 468–79.

Bollen, Kenneth A. and Robert W. Jackman. 1989. "Democracy, Stability, and Dichotomies." *American Sociological Review* 54,4: 612–21.

"Booming Temp Industry, The." 1992. *Dollars and Sense* 181 (November): 5.

Bornschier, Volker and Christopher Chase-Dunn. 1985. *Transnational Corporations and Underdevelopment*. New York: Praeger.

Bornschier, Volker, Christopher Chase-Dunn, and Richard Rubinson. 1978. "Cross-National Evidence of the Effects of Foreign Investment and Aid on Economic Growth and Inequality: A Survey of Findings and a Reanalysis." *American Journal of Sociology* 84, 3: 651–83.

Boswell, Terry and William J. Dixon. 1990. "Dependency and Rebellion: A Cross-National Analysis." *American Sociological Review* 55(August): 540–59.

Boswell, Terry and William J. Dixon. 1993. "Marx's Theory of Rebellion: A Cross-National Analysis of Class Exploitation, Economic Development, and Violent Revolt." *American Sociological Review* 58(October): 681–702.

Bowles, S. and H. Gintis. 1976. *Schooling in Capitalist America: Educational Reform and the Contradictions of Economic Life.* New York: Basic Books.

Bradbury, Katherine L. 1986. "The Shrinking Middle Class." *New England Economic Review* (September/October): 41–55.

Bradshaw, York W. 1993. "Transnational Economic Linkages, the State, and Dependent Development in South Korea, 1966–1988: A Time Series Analysis." *Social Forces* 72,2(December): 315–46.

Bradshaw, York, W., Rita Noonan, Laura Gash, and Claudia Buchman Sershen. 1993. "Borrowing Against the Future: Children and Third World Indebtedness." *Social Forces* 71,3(March): 629–57.

Bradsher, Keith. 1995. "U.S. Losing Skilled Jobs." *Minneapolis Star Tribune* (September 12): 5D.

Brandt, Willy. 1986. *Arms and Hunger.* Cambridge, MA: MIT Press.

Braun, Denny. 1988. "Multiple Measurements of U.S. Income Inequality." *The Review of Economics and Statistics,* 70, 3(August): 398–405.

Braun, Denny. 1991a. "Income Inequality and Economic Development: Geographic Divergence." *Social Science Quarterly* 72,3 (September): 520–36.

Braun, Denny. 1991b. *The Rich Get Richer: The Rise of Income Inequality in the United States and the World.* Chicago: Nelson-Hall.

Braun, Denny. 1995. "Negative Consequences to the Rise of Income Inequality." *Research in Politics and Society* 5: 3–31.

Breslow, Marc. 1992. "How Free Trade Fails." *Dollars and Sense* 180(October): 6–9.

Breslow, Marc. 1994. "Maintaining Military Might." *Dollars and Sense* 191 (January/February): 42–43.

Breslow, Marc. 1995. "The Racial Divide Widens: Why African-American Workers Have Lost Ground." *Dollars and Sense* 197 (January/February): 8–11ff.

Breslow, Marc and Matthew Howard. 1995. "The Real Un(der)employment Rate." *Dollars and Sense* 199 (May/June): 35.

Brewer, Anthony. 1980. *Marxist Theories of Imperialism.* London: Routledge and Kegan Paul.

Brittain, J. A. 1977. *The Inheritance of Economic Status.* Washington, DC: The Brookings Institution.

Broder, David S. 1994. "The GOP Earthquake." *The Washington Post National Weekly Edition* (November 14–20): 6.

Brooke, James. 1993. "A Hard Look at Brazil's Surfeits: Food, Hunger and Inequality." *New York Times* (June 6): A7.

Brooke, James. 1993. "Slavelike Conditions Are Proliferating in Brazil." *Minneapolis Star Tribune* (May 30): 8A.

Brown, J. A. C. and G. Mazzarino. 1984. "Drawing the Lorenz Curve and Calculating the Gini Concentration Index from Grouped Data by Computer." *Oxford Bulletin of Economics and Statistics* 46, 3: 273–78.

Brown, Lester R. 1986. "Redefining National Security." In Lester R. Brown (ed.), *State of the World 1986.* New York: Norton.

Brown, Warren and Frank Swoboda. 1994. "Ford's Vision: A Company Without Borders." *The Washington Post National Weekly Edition* (October 24–30): 18.

Buchanan, Patrick J. 1995. "Cancel the North American Free Trade Agreement Now." *Minneapolis Star Tribune* (September 25): 11A.

Budd, Edward C. 1970. "Postwar Changes in the Size Distribution of Income in the U.S." *American Economic Review* (May): 247–60.

Bullock, Bradley P. 1985. "Cross-National Research and the Basic Needs Approach to Development: A New Direction." Paper presented at the American Sociological Association conference, Chicago.

Bunker, Stephen G. 1985. *Underdeveloping the Amazon: Extraction, Unequal Exchange, and the Failure of the Modern State.* Chicago: University of Illinois Press.

Burgess, John. 1988. "When the Product Is America Instead of American Products." *The Washington Post National Weekly* (October 24–30): 21.

Burgess, John. 1991. "The Latest American Export: Higher Executive Salaries Are Showing Up Abroad." *The Washington Post National Weekly Edition* (October 28–November 3): 25.

Burris, Val. 1992. "Elite Policy-Planning Networks in the United States." *Research in Politics and Society* 4: 111–34.

Camargo, José Marcio. 1984. *Income Distribution in Brazil: 1960–1980.* Rio de Janeiro: Pontificia Universidade Catolica do Rio de Janeiro. (Rua Marques de São Vicente, 225 Rio de Janeiro RJ CEP 22453.)

Cancian, Maria, Sheldon Danziger, and Peter Gottschalk. 1993. "Working Wives and Family Income Inequality Among Married Couples." In Sheldon Danziger and Peter Gottschalk (eds.), *Uneven Tides: Rising Inequality in America.* New York: Russell Sage, 1993.

Caniglia, Alan and Sean Flaherty. 1989. "Unionism and Income Inequality: A Comment on Rubin." *Industrial and Labor Relations* Review 43,1(October): 131–37.

Caniglia, Alan and Sean Flaherty. 1992. "The Relative Effects of Unionism on the Earnings Distribution of Women and Men." *Industrial Relations* 31,2(Spring): 382–93.

Card, David and Alan Krueger. 1995. *Myth and Measurement: The New Economics of the Minimum Wage.* Princeton, NJ: Princeton University Press.

Carnegie Task Force. 1994. *Starting Points: Meeting the Needs of Young Children.* New York: Carnegie Corporation.

Carroll, Sidney L. 1991. "American Family Fortunes as Economic Deadweight." *Challenge* (May/June): 11–18.

Castells, Manuel and Roberto Laserna. "The New Dependency: Technological Change and Socioeconomic Restructuring in Latin America." In A. Douglas Kincaid and Alejandro Portes (eds.), *Comparative National Development: Society and Economy in the New Global Order.* Chapel Hill: University of North Carolina Press.

Castillo, Monica. 1995. *A Profile of the Working Poor, 1993.* Report 896. Washington, DC: Bureau of Labor Statistics.

CBS News/New York Times Poll. 1994. *The Public Weighs In on the Republican Agenda.* New York, CBS News, December 14.

Center for Defense Information. 1994. "Reduce Military Spending: Create More Jobs." *The Defense Monitor* 23, 6 (Washington, DC, Center for Defense Information)

Center for Defense Information. 1995. "Balanced Budget, Unbalanced Priorities?" *The Defense Monitor* 24, 7 (Washington, DC, Center for Defense Information)

Cerresoto, Shirley. 1982. "Socialism, Capitalism, and Inequality." *Insurgent Sociologist* 11, 2 (Spring): 5–38.

Chandler, Clay. 1994. "A Shrinking Line of Credibility: Critics Claim the World Bank Perpetuates the Poverty It Was Formed to Eliminate." *The Washington Post National Weekly Edition* (June 27–July 3): 20-21.

Chandler, Clay. 1994. "The Committee to Reelect Ronald Reagan." *The Washington Post National Weekly Edition* (October 10-16): 8.

Chandler, Clay. 1995. "If Taxes Are Cut, Will It Really Make a Difference?" *The Washington Post National Weekly Edition* (December 26, 1994–January 1, 1995): 13–14.

Chandler, Clay. 1995. "Stepping Back from the Brink." *The Washington Post National Weekly Edition* (February 6–12): 18.

Chandler, Clay and Daniel Williams. 1995. "Who Wins in the Peso Bailout?" *The Washington Post National Weekly Edition* (January 30–February 5): 16.

Chase-Dunn, Christopher. 1989. *Global Formation: Structures of the World Economy.* New York: Basil Blackwell.

Chase-Dunn, Christopher and Richard Rubinson. 1977. "Toward a Structural Perspective on the World-System." *Politics and Society* 7, 4: 453–76.

Chilcote, Ronald H. 1984. *Theories of Development and Underdevelopment.* Boulder, CO: Westview.

Children's Defense Fund. 1994. *The State of America's Children Yearbook 1994.* Washington, DC: Children's Defense Fund.

Children's Defense Fund. 1995. *The State of America's Children Yearbook 1995.* Washington, DC: Children's Defense Fund.

Chomsky, Noam. 1988. *The Culture of Terrorism.* Boston: South End Press.

Citro, Constance F. and Robert T. Michael (eds.). 1995. *Measuring Poverty: A New Approach.* Washington, DC: National Academy Press.

Clark, Terry Nichols and Seymour Martin Lipset. 1991. "Are Social Classes Dying?" *International Sociology* 6,4(December): 397–410.

Clark, Terry Nichols, Seymour Martin Lipset, and Michael Rempel. 1993. "The Declining Political Significance of Social Class." *International Sociology* 8,3(September): 293–316.

Clawson, Dan, Alan Neustadtl, and Denise Scott. 1992. *Money Talks: Corporate PACS and Political Influence.* New York: Basic Books.

Clements, Benedict J. 1988. *Foreign Trade Strategies, Employment, and Income Distribution in Brazil.* New York: Praeger.

Cline, J. Michael. 1995. "Critiques of Defense Spending Folly Came Too Late." *Minneapolis Star Tribune* (September 16): 19A.

Cline, William. 1983. *International Debt and the Stability of the World Economy.* Washington, DC: Institute for International Economics.

Cobb, Clifford, Ted Halstead and Jonathan Rowe. 1995. "If the GDP Is Up, Why Is America Down?" *Atlantic Monthly* (October): 59–78.

Cobb, James C. 1984. *Industrialization and Southern Society: 1877–1984.* Lexington: The University Press of Kentucky.

Cohen, Jeff and Borman Solomon. 1995. "Hold Your Applause for 'NewsHour,' Where 'Public' TV Speaks for Big Money." *Minneapolis Star Tribune* (September 29): 27A.

Cohen, Roger. 1989. "Rio's Murder Wave Takes on the Aura of a Class Struggle." *Wall Street Journal* (May 9): A1, A15.

Colander, David. 1993. *Economics.* Boston: Irwin.

Colatosti, Camille. 1992. "A Job Without a Future: Temporary and Contract Workers Battle Permanent Insecurity." *Dollars and Sense* 176 (May): 9–11, 21.

Colclough, Glenna S. 1990. "Uneven Development, Racial Composition, and Areal Income Inequality in the Deep South: 1970–1980." *Economic Development Quarterly* 4,1(February): 47–54.

Collins, Chuck. 1995. "Aid to Dependent Corporations." *Dollars and Sense* 199(May/June): 15–17, 40.

Collins, Randal. 1971. "Functional and Conflict Theories of Educational Stratification." *American Sociological Review* 36: 1002–19.

Collins, Randal. 1975. *Conflict Sociology.* New York: Academic Press.

Conlisk, John. 1967. "Some Cross-State Evidence on Income Inequality." *Review of Economics and Statistics* 49 (February): 115–18.

Cook, Fay Lomax and Edith J. Barrett. 1992. *Support for the American Welfare State: The Views of Congress and the Public.* New York: Columbia University Press.

Coplon, Jeff. 1989. "Skinhead Reich." *Utne Reader* 33 (May/June): 80–83, 85–89.

"Corporate America's Most Powerful People." 1988. *Forbes* (May 30): 154 ff.

Cotton, Jeremiah. 1989. "Opening the Gap: The Decline in Black Economic Indicators in the 1980s." *Social Science Quarterly* 70,4(December): 803–19.

Council of Economic Advisers. 1987. *Economic Report of the President, 1987.* Washington, DC: U.S. Government Printing Office.

Council of Economic Advisers. 1989. *Economic Report of the President, 1989.* Washington, DC: Government Printing Office.

Council of Economic Advisors. 1995. *Economic Report of the President, 1995.* Washington, DC: U.S. Government Printing Office.

Cox News Service. 1995. "Corporate PACs Gave Heavily to GOP." *Minneapolis Star Tribune* (December 5): 10A.

Cox, Robert W. 1989. "The Core-Periphery Structure of Production and Jobs: The Internationalizing of Production." In Edward Weisband (ed.), *Poverty Amidst Plenty: World Political Economy and Distributive Justice.* Boulder, CO: Westview.

Crenshaw, Albert J. 1994. "Back on a Borrowing Binge." *The Washington Post National Weekly Edition* (October 3–9): 19.

Crenshaw, Edward M. 1995. "Democracy and Demographic Inheritance: The Influence of Modernity and Proto-Modernity on Political and Civil Rights, 1965 to 1980." *American Sociological Review* 60,5(October): 702–18.

Cromwell, Jerry. 1977. "The Size Distribution of Income: An International Comparison." *Review of Income and Wealth* (September): 291–308.

Crystal, Graef S. 1988. "The Wacky, Wacky World of CEO Pay." *Fortune* (June 6): 68 ff.

Crystal, Graef S. 1991. *In Search of Excess: The Overcompensation of American Executives.* New York: Norton.

Cuomo Commission on Trade and Competitiveness. 1988. *The Cuomo Commission Report: A New American Formula for a Strong Economy.* New York: Simon and Schuster

Cypher, James M. 1988. "The Party's Over: Debt, Economic Crisis Undermine Mexico's PRI." *Dollars and Sense* (December) 142: 9–11, 21.

Cypher, James M. 1994. "More Trade Talk." *Dollars and Sense* 192(March/April): 34, 42.

Cypher, James M. 1995. "NAFTA Shock: Mexico's Free Market Meltdown." *Dollars and Sense* (March/April): 22–25, 39.

Danziger, Sandra K. and Sheldon Danziger. 1993. "Child Poverty and Public Policy: Toward a Comprehensive Antipoverty Agenda." *Daedalus* 122(Winter): 57–84.

Danziger, Sheldon. 1976. "Determinants of the Level and Distribution of Family Income in Metropolitan Areas, 1969." *Land Economics,* 52 (November): 467–78.

Danziger, Sheldon and Peter Gottschalk. 1986. "Do Rising Tides Lift All Boats? The Impact of Secular and Cyclical Changes on Poverty." *American Economic Review* 76, 2 (May): 405–10.

Danziger, Sheldon, and Peter Gottschalk. 1986. "Families with Children Have Fared Worse." *Challenge* 29 (March–April): 40–47.

Danziger, Sheldon, Peter Gottschalk, and Eugene Smolensky. 1989. "How the Rich Have Fared, 1973–87." *American Economic Review* 79, 2 (May): 310–14.

Danziger, Sheldon H. and Daniel H. Weinberg. 1994. "The Historical Record: Trends in Family Income, Inequality, and Poverty." In Sheldon H. Danziger, Gary D. Sandefur, and Daniel H. Weinberg (eds.), *Confronting Poverty: Prescriptions for Change.* New York: Russell Sage.

Davis, James A. and Tom W. Smith. 1992. *The NORC General Social Survey: A User's Guide.* Newbury Park, CA: Sage.

Davis, James A. and Tom W. Smith. 1993. *General Social Surveys, 1972–93* (machine-readable data file). Chicago: National Opinion Research Center. (Distributed by the Roper Center for Public Opinion, Storrs, CT.)

Davies, James C. 1962. "Toward a Theory of Revolution." *American Sociological Review* 6: 5–19.

Davies, James C. 1969. "The J-Curve of Rising and Declining Satisfactions as a Cause of Some Great Revolutions and a Contained Rebellion." In H.D. Graham and T.R. Gurr (eds.), *Violence in America.* New York: Praeger.

Davis, Kingsley and Wilbert E. Moore. 1945. "Some Principles of Stratification." *American Sociological Review* 10: 242–49.

Davis, Mike. 1990. *City of Quartz: Excavating the Future in Los Angeles.* New York: Verso.

Dawson, Jim. 1995. "Youth Homicide Reaching Grim Highs." *Minneapolis Star Tribune* (February 18): 6A.

DeGrasse, R. W. 1984. "The Military Shortchanging the Economy." *Bulletin of the Atomic Scientists* (May): 39–43.

Delhausse, B. 1995. *Working But Poor: A Reassessment.* Working Paper No. 125. Syracuse, NY: Syracuse University, Maxwell School of Citizenship and Public Affairs.

Della, P. A. and N. Oguchi. 1976. "Distribution, The Aggregate Consumption Function, and the Level of Economic Development: Some Cross-Country Results." *Journal of Political Economy* 84, 6 (December): 1325–34.

Denslow, David, Jr. and William Tyler. 1984. "Perspectives on Poverty and Income Inequality in Brazil." *World Development* 12, 10: 1019–28.

Deutsch, Claudia H. 1995. "Going Away for Big Pay." *Minneapolis Star Tribune* (July 4): 1D,4D.

Devroy, Ann. 1995. "Bad News for the GOP and Bob Dole." *The Washington Post National Weekly Edition* (October 9–15): 11.

Dionne, E. J., Jr. 1994. "Race and IQ: Stale Notions." *The Washington Post National Weekly Edition* (October 24–30): 29.

Dionne, E. J., Jr. 1994. "A Shift, Not a Mandate." *The Washington Post National Weekly Edition* (November 28–December 4): 28.

Dixon, William J. and Terry Boswell. 1996. "Dependency, Disarticulation and Denominator Effects: Another Look at Foreign Capital Penetration." *American Journal of Sociology* 102 (September): 543–62.

Doeringer, P. B. and M. J. Piore. 1971. *Internal Labor Markets and Manpower Analysis.* Lexington, MA: Heath.

Domhoff, G. William. 1970. *The Higher Circles.* New York: Random House.

Domhoff, G. William. 1979. *The Powers That Be.* New York: Vintage Press.

Domhoff, G. William. 1983. *Who Rules America Now?* Englewood Cliffs, N.J.: Prentice-Hall.

Dooley, Martin and Peter Gottschalk. 1984. "Earnings Inequality among Males in the United States: Trends and the Effect of Labor Force Growth." *Journal of Political Economy* 92, 1: 59 ff.

Dooley, Martin and Peter Gottschalk. 1985. "The Increasing Proportion of Men with Low Earnings in the United States." *Demography* 22, 1 (February): 25–34.

Dos Santos, Theotonio. 1970. "The Structure of Dependence." *American Economic Review* 60: 231–36.

Dovring, Folke. 1991. *Inequality: The Political Economy of Income Distribution.* New York: Praeger.

Duncan, Greg J., Bjorn Gustafsson, Richard Hauser, Gunther Schmauss, Hans Messinger, Ruud Muffels, Brian Nolan, and Jean-Claude Ray. 1993. "Poverty Dynamics in Eight Countries." *Journal of Population Economics* 6(Spring): 215–34.

Duncan, Greg J., Timothy M. Smeeding, and Willard Rodgers. 1993. "W(h)ither the Middle Class? A Dynamic View." In Dimitri B. Papadimitriou and Edward N. Wolff (eds.), *Poverty and Prosperity in the USA in the Late Twentieth Century.* New York: St. Martin's Press.

Durden, Garey C. and Ann V. Schwarz-Miller. 1982. "The Distribution of Individual Income in the U.S. and Public Sector Employment." *Social Science Quarterly* 63, 1 (March): 39–47.

Durkheim, Émile. 1964, 1897. *Suicide.* Trans. by George Simpson. New York: Free Press.

Durning, Alan Thein. 1992. *How Much Is Enough?* New York: Norton.

Duster, Troy. 1995. "The Bell Curve." *Contemporary Sociology* 24,2 (March): 158–61.

Dye, Thomas R. 1979. *Who's Running America?* Englewood Cliffs, NJ: Prentice-Hall.

Economic Affairs Bureau. 1988. "Tax Reform Hoopla: Do Lower Rates and Fewer Loopholes Equal Reform?" In *Real World Macro: A Macroeconomics Reader from Dollars and Sense.* 5th ed. Somerville, MA: Economic Affairs Bureau.

Economy in Numbers. 1988. "Where the Federal Deficit Comes From." In *Real World Macro: A Macroeconomics Reader from Dollars and Sense.* 5th ed. Somerville, MA: Economic Affairs Bureau.

Economy in Numbers. 1989. "The High Cost of Debt." *Dollars and Sense* 144 (March): 23.

Economy in Numbers. 1991. "Cutting Off the Unemployed." *Dollars and Sense* 167 (June): 23.

Economy in Numbers. 1992. "Debt and Distribution." *Dollars and Sense* 177 (June) 23.

Economy in Numbers. 1992. "U.S. Among Worst in Inequality." *Dollars and Sense* 178(July/August): 23.

Economy in Numbers. 1992. "The New Unemployment." *Dollars and Sense* 181(November): 23.

Economy in Numbers. 1993. "Taxes: An American Dilemma." *Dollars and Sense* 184(March): 23.

Economy in Numbers. 1994. "Maintaining Military Might." *Dollars and Sense* 191 (January/February): 42.

Economy in Numbers. 1995. "The GOP's 17% Mandate." *Dollars and Sense* 198(March/April): 43.

Economy in Numbers. 1995. "How to Pare Down the Pentagon." *Dollars and Sense* 200 (July/August): 43.

Economy in Numbers. 1995. "It's Better in the Union—If You Can Find One." *Dollars and Sense* 201 (September/October): 51.

Edin, Kathryn J. 1995. "The Myths of Dependence and Self-Sufficiency: Women, Welfare, and Low-Wage Work." *Focus* 17,2 (Fall/Winter): 1–9.

Edsall, Thomas B. 1985. *The New Politics of Inequality.* New York: Norton.

Edsall, Thomas B. 1994. "Revolt of the Discontented." *The Washington Post National Weekly Edition* (November 21–27): 28.

Edwards, R.C., M. Reich, and D. Gordon (eds). 1975. *Labor Market Segmentation*. Lexington, MA: Heath.

Ehrenreich, Barbara. 1986. "Is the Middle Class Doomed?" *New York Times Magazine* 135 (September 7): 44 ff.

Ellwood, David T. 1988. *Poor Support: Poverty in the American Family*. New York: Basic Books.

Engardio, Pete, Rob Hof, Elisabeth Malkin, Neil Gross, and Karen Lowry Miller. 1994. "High-Tech Jobs All Over the Map." *Business Week* (November 18), America Online.

Estes, Richard J. 1988. "Toward a 'Quality of Life' Index: Empirical Approaches to Assessing Human Welfare Internationally." In Jim Norwine and Alfonso Gonzalez (eds.), *The Third World: States of Mind and Being*. Boston: Unwin Hyman.

Evans, Bergen. 1968. *Dictionary of Quotations*. New York: Delacorte.

Evans, Peter. 1979. *Dependent Development: The Alliance of Multinational, State, and Local Capital in Brazil*. Princeton, NJ: Princeton University Press.

Falk, William W. and Thomas A. Lyson. 1988. *High Tech, Low Tech, No Tech: Recent Industrial and Occupational Change in the South*. Albany: State University of New York Press.

Falk, William W. and Bruce H. Rankin. 1992. "The Cost of Being Black in the Black Belt." *Social Problems* 39,3(August): 299–313.

Farhi, Paul. 1992. "Multiplying Millionaires." *Minneapolis Star Tribune* (July 14): 1D.

Farhi, Paul. 1995. "Hearst-Case Scenario." *The Washington Post National Weekly Edition* (July31–August 6): 21.

Farley, Reynolds and Suzanne M. Bianchi. 1985. "Social Class Polarization: Is It Occurring among Blacks?" *Research in Race and Ethnic Relations* (Population Studies Center, University of Michigan) 4: 1-31.

Faux, Jeff. 1988. "The Austerity Trap and the Growth Alternative." *World Policy Journal* 5, 3 (Summer): 367–414.

Faux, Jeff and Thea Lee. 1993. "Implications of NAFTA for the United States: Investment, Jobs, and Productivity." In Ricardo Grinspun and Maxwell A. Cameron (eds.), *The Political Economy of North American Free Trade*. New York: St. Martin's Press.

Featherman, David and Robert Hauser. 1978. *Opportunity and Change*. New York: Academic Press.

Fields, Gary S. 1977. "Who Benefits from Economic Development? A Reexamination of Brazilian Growth in the 1960's." *American Economic Review* 67, 4 (September): 570–82.

"Financial Games Reduce R & D Spending." 1989. *Dollars and Sense* 145 (April): 5.

Firebaugh, Glenn. 1992. "Growth Effects of Foreign and Domestic Investment." *American Journal of Sociology* 98,1 (July): 105–30.

Fishlow, Albert. 1972. "Brazilian Size Distribution of Income." *American Economic Review* 62 (May): 391–401.

Fishlow, Albert. 1973. "Some Reflections on Post-1964 Brazilian Economic Policy." In Alfred Stepan (ed.), *Authoritarian Brazil.* New Haven, CT: Yale University Press.

Fishlow, Albert. 1989. "A Tale of Two Presidents: The Political Economy of Crisis Management." In Alfred Stepan (ed.), *Democratizing Brazil: Problems of Transition and Consolidation.* New York: Oxford University Press.

Flint, Jerry. 1987. "Too Much Ain't Enough." *Forbes* 140 (July 13): 92 ff.

Foley, John W. 1977. "Trends, Determinants and Policy Implications of Income Inequality in U.S. Counties." *Sociology and Social Research* 61, 4: 441-61.

Formberg, J. P. and T. G. Seaks. 1980. "Paglin Gini Measure of Inequality—A Modification." *American Economic Review* 70,3: 479–82.

Frank, André Gunder. 1967. *Capitalism and Underdevelopment in Latin America: Historical Studies of Chile and Brazil.* New York: Monthly Review Press.

Frank, André Gunder. 1969. *Latin America, Underdevelopment or Revolution: Essays on the Development of Underdevelopment and the Immediate Enemy.* New York: Monthly Review Press.

Frank, André Gunder. 1979. *Dependent Accumulation and Underdevelopment.* New York: Monthly Review Press.

Frank, André Gunder. 1981a. *Crisis in the Third World.* New York: Holmes and Meier.

Frank, André Gunder. 1981b. *Reflections on the World Crisis.* New York: Monthly Review Press.

Frech, H. E. and L. S. Burns. 1971. "Metropolitan Interpersonal Income Inequality: A Comment." *Land Economics* 47 (February): 104–6.

Freeman, Richard B. 1993. "How Much Has De-Unionization Contributed to the Rise in Male Earnings Inequality?" In Sheldon Danziger and Peter Gottschalk (eds.), *Uneven Tides: Rising Inequality in America.* New York: Russell Sage.

Frey, William H. and Alden Speare, Jr. 1988. *Regional and Metropolitan Growth and Decline in the United States.* New York: Russell Sage Foundation.

Friedman, Milton. 1953. "Choice, Chance, and the Personal Distribution of Income." *Journal of Political Economy* 61, 4 (August): 277–90.

Friedman, Milton. 1962. *Capitalism and Freedom.* Chicago: University of Chicago Press.

Friedmann, Harriet. 1994. "Distance and Durability: Shaky Foundations of the Third World." In Philip McMichael (ed.), *The Global Restructuring of Agro-Food Systems.* Ithaca, NY: Cornell University Press.

Fritzell, Johan. 1993. "Income Inequality Trends in the 1980s: A Five-Country Comparison." *Acta Sociologica* 36,1: 47–62.

Fromson, Brett D. 1995. "Going Where They Make the Money." *The Washington Post National Weekly Edition* (April 3–9): 14.

Frumkin, Norman. 1994. *Guide to Economic Indicators.* 2nd ed. New York: M.E. Sharpe.

Fulwood III, Sam. 1994. "Minority Efforts Draw Resentment." *Minneapolis Star Tribune* (October 15): 4A.

Gagliani, Giorgio. 1987. "Income Inequality and Economic Development." *Annual Review of Sociology* 13: 313–34.

Gallaway, Lowell, Richard Vedder, and Therese Foster. 1986. "The New Structural Poverty: A Quantitative Analysis." In Joint Economic Committee, *War on Poverty: Victory or Defeat?* Washington, DC: U.S. Government Printing Office.

Gallup Poll. 1988. "U.S. Citizens, British Hold Differing Views of Haves, Have-Nots." *Minneapolis Star Tribune* (August 14).

Garfinkel, Irwin and Sara McLanahan. 1986. *Single Mothers and Their Children: A New American Dilemma.* Washington, DC: Urban Institute.

Garfinkel, Irwin and Sara McLanahan. 1994. "Single-Mother Families, Economic Insecurity, and Government Policy." In Sheldon H. Danziger, Gary D. Sandefur, and Daniel H. Weinberg (eds.), *Confronting Poverty: Prescriptions for Change.* New York: Russell Sage.

Garofalo, Gasper and Michael S. Fogarty. 1979. "Urban Income Distribution and the Urban Hierarchy-Equality Hypothesis." *Review of Economics and Statistics* 61 (August): 381–88.

Geewax, Marilyn. 1991. "Let Them Eat Pink Slips." *Minneapolis Star Tribune* (October 14): 11A.

George, Susan. 1988. *A Fate Worse Than Debt: The World Financial Crisis and the Poor.* New York: Grove Press.

Gereffi, Gary. 1994. "Rethinking Development Theory: Insights from East Asia and Latin America." In A. Doughlas Kincaid and Alejandro Portes (eds.), *Comparative National Development: Society and Economy in the New Global Order.* Chapel Hill: University of North Carolina Press.

Gereffi, Gary. 1994. "The International Economy and Economic Development." In Neil J. Smelser and Richard Swedberg (eds.), *The Handbook of Economic Sociology.* New York: Russell Sage.

Gibbs, Nancy. 1990. "So You Think Your City's Got Crime?" *Time* (March 5): 54–55.

Gilpin, Robert. 1987. *The Political Economy of International Relations.* Princeton, NJ: Princeton University Press.

Glenn, Norvall D. 1984. "Education and Family Income." *Social Forces* 63, 1: 169–83.

Glyn, Andrew and David Miliband. 1994. *Paying for Inequality: The Economic Cost of Social Injustice.* Concord, MA: Paul and Co.

Goertzel, Ted. 1995. "President Fernando Cardoso Reflects on Brazil and Sociology." *Footnotes* 23,8(November): 1, 8.

Good, T. and J. Brophy. 1973. *Looking in Class-Rooms.* New York: Harper and Row.

Gottschalk, Peter. 1993. "Changes in Inequality of Family Income in Seven Industrialized Countries." *American Economic Review* 83,2(May): 136–42.

Gottschalk, Peter. 1994. *Policy Changes and Growing Earnings Inequality in Seven Industrialized Countries.* Working Paper No. 86. Syracuse, NY: Syracuse University, Maxwell School of Citizenship and Public Affairs.

Gottschalk, Peter and Sheldon Danziger. 1985. "A Framework for Evaluating the Effects of Economic Growth and Transfers on Poverty." *American Economic Review* 75, 1 (March): 153–61.

Gottschalk, Peter and Mary Joyce. 1992. *The Impact of Technological Change, Deindustrialization, and Internationalization of Trade on Earnings Inequality: An International Perspective.* Working Paper No. 85. Syracuse, NY: Syracuse University, Maxwell School of Citizenship and Public Affairs.

Gottschalk, Peter and Mary Joyce. 1995. "The Impact of Technological Change, Deindustrialization, and Internationalization of Trade on Earnings Inewuality: An International Perspective." In Katherine McFate, Roger Lawson, and William Julius Wilson (eds.), *Poverty, Inequality, and the Future of Social Policy.* New York: Russell Sage.

Gottschalk, Peter, Sara McLanahan, and Gary D. Sandefur. 1994. "The Dynamics of Intergenerational Transmission of Poverty and Welfare Participation." In Sheldon H.Danziger, Gary D. Sandefur, and Daniel H. Weinberg (eds.), *Confronting Poverty: Prescriptions for Change.* New York: Russell Sage.

Gordon, David M. 1987. "Private Debt Dwarfs Uncle Sam's." *Los Angeles Times* (January 20).

Gozan, Julie. 1992. "Wealth for the Few." *Multinational Monitor* 13,12(December): 6.

Graham, Paul. 1994. "Running Dogs, Lackeys and Imperialism (Multinational Corporations)." *Canadian Dimension* 28,5(October–November): 48.

Gramlich, Edward M. and Deborah S. Larsen. 1984. "How Widespread Are Income Losses in a Recession?" In D. Lee Bawdin (ed.), *The Social Contract Revisited.* Washington, DC: Urban Insitute.

Green, Gary P. 1987. *Finance Capital and Uneven Development.* Boulder, CO: Westview.

Greenwood, Daphne T. 1987. "Age, Income, and Household Size: Their Relation to Wealth Distribution in the United States." In Edward N. Wolff (ed.), *International Comparisons of the Distribution of Household Wealth.* New York: Oxford University Press.

Grubb, W. Norton and Robert H. Wilson. 1987. *The Distribution of Wages and Salaries, 1960-1980: The Contributions of Gender, Race, Sectoral Shifts and Regional Shifts.* Working Paper No. 39. Austin: Lyndon B. Johnson School of Public Affairs, The University of Texas.

Grubb, W. Norton and Robert H. Wilson. 1992. "Trends in Wage and Salary Inequality, 1967–88." *Monthly Labor Review* (June): 23–39.

Gugliotta, Guy. 1994. "The Minimum Wage Culture." *The Washington Post National Weekly Edition* (October 3–9): 6.

Gugliotta, Guy. 1994. "Making Do—A Piece at a Time." *The Washington Post National Weekly Edition* (October 10–16): 6.

Gurney, Joan Neff and Kathleen J. Tierney. 1982. "Relative Deprivation and Social Movements: A Critical Look at Twenty Years of Theory and Research." *Sociological Quarterly* 23: 33–47.

Gurr, Ted. 1970. *Why Men Rebel.* Princeton, NJ: Princeton University Press.

Gutman, Herbert G. 1988. *Power and Culture: Essays on the American Working Class.* New York: Pantheon Books.

Hamburger, Tom. 1994. "New Report Increases Homeless Population to 7 Million." *Minneapolis Star Tribune* (May 18): 7A.

Hansen, Roger D. 1975. "The Emerging Challenge: Global Distribution of Income and Economic Opportunity." In James W. Howe (ed.), *The U.S. and World Development Agenda for Action 1975*. New York: Praeger.

Hanushek, Eric A. 1973. "Regional Differences in the Structure of Earnings." *The Review of Economics and Statistics* 55, 2 (May): 204–13.

Hardy, Eric S. 1995. "America's Highest-Paid Bosses." *Forbes* (May 22): 182.

Harer, Miles D. 1987. *Relative Deprivation and Crime: The Effects of Income Inequality on Black and White Arrest Rates*. Ph.D. diss., State College, Pennsylvania State University.

Harrington, Michael. 1976. *The Twilight of Capitalism*. New York: Simon and Schuster.

Harrington, Michael. 1984. *The New American Poverty*. New York: Holt, Rinehart and Winston.

Harris, Ron. 1994. "Middle-Class Members Retreat to Relative Safety of Brazilian Slum." *Minneapolis Star Tribune* (July 17): 22A.

Harrison, Bennett and Barry Bluestone. 1988. *The Great U-Turn: Corporate Restructuring and the Polarizing of America*. New York: Basic Books.

Harrison, Bennett and Lucy Gorham. 1992. "Growing Inequality in Black Wages in the 1980s and the Emergence of an African-American Middle Class." *Journal of Policy Analysis and Mangement* 11,2: 235–53.

Hartman, John and Wey Hsiao. 1988. "Inequality and Violence Issues: Issues of Theory and Measurement." *American Sociological Review* 53,5: 794–99.

Harwood, Richard. 1995. "The Rich and Poor Problem." *The Washington Post National Weekly Edition* (June 19–25): 29.

Haupt, Arthur. 1989. "Another Winter for the Homeless." *Population Today* 17, 2 (February): 3–4.

Hauser, Robert M. 1980. "On 'Stratification in a Dual Economy' (Comment on Beck et al., ASR, October, 1978)." *American Sociological Review* 45, 4 (August): 702–12.

Hauser, Robert M. 1995. "The Bell Curve." *Contemporary Sociology* 24,2 (March): 149–53.

Hauser, Robert M. and William H. Sewell. 1986. "Family Effects in Simple Models of Education, Occupational Status, and Earnings: Findings from the Wisconsin and Kalamazoo Studies." *Journal of Labor Economics* 4, 3:S83.

Haworth, Charles T., James E. Long, and David W. Rasmussen. 1978. "Income Distribution, City Size, and Urban Growth." *Urban Studies* 15 (February): 1–7.

Helleiner, Gerald K. 1987. "Stabilization, Adjustment, and the Poor." *World Development* 15, 12: 1499–1513.

Henley, Peter and Paul Ryscavage. 1980. "The Distribution of Earned Income among Men and Women, 1958–1977." *Monthly Labor Review* 103, 4: 3–10.

Henry, David K. and Richard P. Oliver. 1987. "The Defense Buildup, 1977–85: Effects on Production and Employment." *Monthly Labor Review* (August).

Herbert, Bob. 1995. "In Maquiladora Sweatshops: Not a Living Wage." *Minneapolis Star Tribune* (October 12): 23A.

Herman, Edward S. and Noam Chomsky. 1988. *Manufacturing Consent: The Political Economy of the Mass Media.* New York: Pantheon Books.

Herrnstein, Richard J. and Charles Murray. 1994. *The Bell Curve: Intelligence and Class Structure in American Life.* New York: The Free Press.

Hewlett, Sylvia Ann. 1991. *When the Bough Breaks: The Cost of Neglecting Our Children.* New York: Basic Books.

Hicks, Norman L. and Paul P. Streeten. 1979. "Indicators of Development: The Search for a Basic Needs Yardstick." *World Development* 7: 576–77.

Hill, Gwen. 1989. "It Is Becoming a Permanent Issue." *The Washington Post National Weekly Edition* (March 27–April 2): 8.

Hill, Martha S. and Greg J. Duncan. 1987. "Parental Family Income and the Socioeconomic Attainment of Children." *Social Science Research* 16, 1 (March): 39–73.

Hodson, Randy. 1984. "Companies, Industries, and the Measurement of Economic Segmentation." *American Sociological Review* 49(June): 335–48.

Hodson, Randy. 1986. "Modeling the Effects of Industrial Structure on Wages and Benefits." *Work and Occupations* 13,4(November): 488–510.

Hoffman, Helga. 1989. "Poverty and Property in Brazil: What Is Changing?" In Edmar Bacha and Herbert S. Klein (eds.), *Incomplete Transition: Brazil Since 1945.* Albuquerque: University of New Mexico Press.

Holland, Stuart. 1976. *The Regional Problem.* New York: St. Martin's Press.

Holloway, Harry and John George. 1986. *Public Opinion: Coalitions, Elites, and Masses.* 2d ed. New York: St. Martin's Press.

Hoover, Gregg A. 1989. "Intranational Inequality: A Cross-National Dataset," *Social Forces* 67,4(June): 1008–26.

Hopfensperger, Jean. 1994. "Human Rights of Nation's Poor Are Issue." *Minneapolis Star Tribune* (July 26): 7B.

Horn, Patricia. 1989. "Measure for Measure: Deciphering the Statistics on Income and Wages." *Dollars and Sense* 143 (January/ February): 10.

Horn, Patricia. 1993. "Paying to Lose our Jobs." *Dollars and Sense* 184(March): 10–11.

Hout, Mike, Clem Brooks, and Jeff Manza. 1993. "The Persistence of Classes in Post-Industrial Societies." *International Sociology* 8, 3 (September): 259–77.

Howes, Candace. 1991. "Transplants No Cure." *Dollars and Sense* 168(July/ August): 16–19.

Inhaber, Herbert and Sidney Carroll. 1992. *How Rich Is Too Rich? Income and Wealth in America.* New York: Praeger.

International Economics Department. 1994. *World Development Indicators 1994.* Washington, DC: World Bank.

International Monetary Fund. 1987. *Direction of Trade Statistics Yearbook.* Washington, DC: International Monetary Fund.

"Is the Middle Class Shrinking?" 1986. *Time* (November 3): 54–56.

Ivins, Molly. 1995. "Lyin' Bully." The MOJO Wire, *Mother Jones Magazine* World Wide Web site.

Ivins, Molly. 1995. "Many Horror Stories of Government Untrue." *Minneapolis Star Tribune* (May 26): 25A.

Ivins, Molly. 1996. "Look Long and Hard to Find Silver Lining in Buchanan's New Hampshire Victory." *Minneapolis Star Tribune* (February 23): A17.

Jackman, Mary R. and Robert W. Jackman. 1983. *Class Awareness in the United States*. Berkeley, CA: University of California Press.

Jackman, Robert W. 1982. "Dependence on Foreign Investment and Economic Growth in the Third World." *World Politics* 34 (January): 175–97.

Jacob, Rahul. 1995. "It was a Banner Year for Profits (The Fortune Global 500)." *Fortune* 132,3(August 7): 130ff.

Jacobs, David and David Britt. 1979. "Inequality and Police Use of Deadly Force: An Empirical Assessment of a Conflict Hypothesis." *Social Problems* 26,4: 403–12.

Jacobs, David, Michael Useem, and Mayer N. Zald. 1991. "Firms, Industries, and Politics." *Research in Political Sociology* 5: 141–65.

Jacoby, Russell and Naomi Glauberman (eds.). 1995. *The Bell Curve Debate: History, Documents, Opinions*. New York: Random House.

Jantti, Markus. 1993. *Changing Inequality in Five Countries: The Role of Markets, Transfers and Taxes*. Working Paper No. 91. Syracuse, NY: Syracuse University, Maxwell School of Citizenship and Public Affairs.

Jasso, Guillermina. 1978. "On the Justice of Earnings: A New Specification of the Justice Evaluation Function." *American Journal of Sociology* 83,6: 1398–1419.

Jasso, Guillermina. 1980. "A New Theory of Distributive Justice." *American Sociological Review* 45(February): 3–32.

Jasso, Guillermina and Peter H. Rossi. 1977. "Distributive Justice and Earned Income." *American Sociological Review* 42(August): 639–51.

Jencks, Christopher et al. 1972. *Inequality: a Reassessment of the Effect of Family and Schooling in America*. New York: Harper.

Jencks, Christopher et al. 1979. *Who Gets Ahead? The Determinants of Economic Success in America*. New York: Basic Books.

Johnson, Clifford M., Leticia Miranda, Arloc Sherman, and James D. Weill. 1991. *Child Poverty in America* Washington, DC: Children's Defense Fund.

Joint Economic Committee. 1986. *Poverty, Income Distribution, The Family and Public Policy*. S.Prt. 99-199, December 19. Washington, DC: U.S. Government Printing Office.

Joint Economic Committee. U.S. House of Representatives. 1992. "Families on a Treadmill: Work and Income in the 1980s." Staff study (January 17).

Jonish, James F., and James B. Kau. 1973. "State Differentials in Income Inequality." *Review of Social Economy* 31 (October): 179–90.

Kalleberg, Arne L., Michael Wallace, and Robert P. Althauser. 1981. "Economic Segmentation, Worker Power, and Income Inequality." *American Journal of Sociology* 87, 3 (November): 651–83.

Kalleberg, Arne L. and Mark E. Van Buren. 1994. "The Structure of Organizational Earnings Inequality." *American Behavioral Scientist* 37,7(June): 930–47.

Karmin, Monroe W. 1984. "Is Middle Class Really Doomed to Shrivel Away?" *U.S. News and World Report* (August 20): 65.

Karoly, Lynn A. 1989. "Changes in the Distribution of Individual Earnings in the United States: 1967–1986." Santa Monica, CA: RAND Corp.

Karoly, Lynn A. 1993. "The Trend in Inequality among Families, Individuals, and Workers in the United States: A Twenty-five Year Perspective." In Sheldon Danziger and Peter Gottschalk (eds.), *Uneven Tides: Rising Inequality in America*. New York: Russell Sage.

Katz, Lawrence F. and Kevin M. Murphy. 1992. "Changes in Relative Wages, 1963–87: Supply and Demand Factors." *Quarterly Journal of Economics* (February): 76.

Kelly, Brian. 1995. "The Pork That Just Won't Slice." *The Washington Post National Weekly Edition* (December 18–24): 21–22.

Kelly, M. Patricia Fernandez. 1994. "Broadening the Scope: Gender and the Study of International Development." In A. Douglas Kincaid and Alejandro Portes (eds.), *Comparative National Development: Society and Economy in the New Global Order*. Chapel Hill: University of North Carolina Press.

Kelly, Marjorie. 1995. "Mushrooming Executive Pay Prompts Resentment, Problems." *Minneapolis Star Tribune* (October 9) :D3.

Kerbo, Harold R. 1983. *Social Stratification and Inequality: Class Conflict in the United States*. New York: McGraw-Hill.

Kerbo, Harold R. 1991. *Social Stratification and Inequality: Class Conflict in Historical and Comparative Perspective*. New York: McGraw-Hill.

Khan, Ashfaque. 1987. "Aggregate Consumption Function and Income Distribution Effect: Some Evidence From Developing Countries." *World Development* 15, 10/11: 1369–74.

Kick, Edward L. 1987. "World-System Structure, National Development, and the Prospects for a Socialist World Order." In Terry Boswell and Albert Bergesen (eds.), *America's Changing Role in the World System*. New York: Praeger.

Kincaid, A. Douglas and Alejandro Portes (eds.). 1994. *Comparative National Development: Society and Economy in the New Global Order*. Chapel Hill: University of North Carolina Press.

Kirkpatrick, David. 1988. "Abroad, It's Another World." *Fortune* (June): 78.

Kluegel, James R., Gyorgy Csepeli, Tamas Kolosi, Antal Orkeny, and Maria Nemenyi. 1995. "Accounting for the Rich and the Poor: Existential Justice in Comparative Perspective." In James R. Kluegel, David S. Mason, and Bernd Wegener (eds.), *Social Justice and Political Change: Public Opinion in Capitalist and Post-Communist States*. New York: Aldine de Gruyter.

Kluegel, James R. and Peter Mateju. 1995. "Egalitarian vs. Inegalitarian Principles of Distributive Justice." In James R. Kluegel, David S. Mason, and Bernd Wegener (eds.), *Social Justice and Political Change: Public Opinion in Capitalist and Post-Communist States*. New York: Aldine de Gruyter.

Kluegel, James R. and Masaru Miyano. 1995. "Justice Beliefs and Support for the Welfare State in Advanced Capitalism." In James R. Kluegel, David S. Mason, and Bernd Wegener (eds.), *Social Justice and Political Change: Public Opinion in Capitalist and Post-Communist States*. New York: Aldine de Gruyter.

Kluegel, James R. and Eliot R. Smith. 1981. "Stratification Beliefs." *Annual Review of Sociology* 7: 29–56

Kluegel, James R. and Eliot R. Smith. 1986. *Beliefs about Inequality: Americans' Views of What Is and What Ought to Be.* Hawthorne, NY: Aldine De Gruyter.

Knapp, Tim. 1992. "The Declining Middle Class in an Age of Deindustrialization." *Research in Inequality and Social Conflict* 2: 113–36.

Knocke, David, Lawrence Raffalovich, and William Erskine. 1987. "Class, Status, and Economic Policy Preferences." *Research in Social Stratification and Mobility* 6: 141–58.

Knott, Joseph J. 1970. "The Index of Income Concentration in the 1970 Census of Population and Housing." Working paper available from the Population Division, U.S. Bureau of the Census.

Kohut, Andrew. 1993. *The Vocal Minority in American Politics.* Washington, DC: Times Mirror Center for The People and The Press.

Korb, Lawrence J. 1995. "The Indefensible Defense Budget." *The Washington Post National Weekly* (July 17–23): 19.

Krahn, Harvey and John W. Gartrell. 1985. "Effects of Foreign Trade, Government Spending, and World System Status on Income Distribution." *Rural Sociology* 50,2(Summer): 181–92.

Krahn, Harvey, Timothy F. Hartnagel, and John W. Gartrell. 1986. "Income Inequality and Homicide Rates: Cross-National Data and Criminological Theories." *Criminology* 24: 269–95.

Krasner, Stepehn D. 1985. *Structural Conflict: The Third World Against Global Liberalism.* Berkeley: University of California Press.

Krieger, Nancy and Elizabeth Fee. 1993. "What's Class Got to Do with It? The State of Health Data in the United States Today." *Socialist Review* 23,1(March): 59 ff.

Krugman, Paul R. 1992. "Disparity and Despair." *U.S. News and World Report* (March 23): 54–55.

Kukreja, Sunil and James D. Miley. 1989. "Government Repression: A Test of the Conflict, World-System Position, and Modernization Hypotheses." *International Journal of Contemporary Sociology* 26,3/4: 147–57.

Kull, Steven. 1994. *Fighting Poverty in America: A Study of American Public Attitudes.* Washington, DC: Center for the Study of Policy Attitudes.

Kuo, Shirley W. Y., Gustav Ranis, and John C. H. Fei. 1984. "Rapid Growth with Improved Income Distribution: The Taiwan Success Story." In Mitchell A. Seligson (ed.), *The Gap Between Rich and Poor.* Boulder, CO: Westview.

Kuo, Shirley, Gustav Ranis, and John C. Fei. 1981. *The Taiwan Success Story: Rapid Growth with Improved Distribution in the Republic of China, 1952-1979.* Boulder, CO: Westview Press.

Kuttner, Robert. 1985. "A Shrinking Middle Class Is a Call to Action." *Business Week* (September 16): 16.

Kuttner, Robert. 1988. "Why We Don't Make TVs Anymore." *The Washington Post National Weekly Edition* (October 24–30): 29.

Kuttner, Robert. 1995. "Kids, Parents and the Economy." *The Washington Post National Weekly Edition* (July 3–9): 5.

Kuttner, Robert. 1995. "The Real Class War." *The Washington Post National Weekly Edition* (July 31–August 6): 5.

Kuttner, Robert. 1995. "Raising the Minimum Wage Issue." *The Washington Post National Weekly Edition* (August 14–20): 5.

Kuznets, Simon. 1955. "Economic Growth and Income Inequality." *American Economic Review* 45 (March): 1–28.

Kuznets, Simon. 1963. "Quantitative Aspects of the Economic Growth of Nations: Distribution of Income by Size." *Economic Development and Cultural Change* 11, 2 (January): 1–80.

Kwan, Ronald. 1991. "Footloose and Country Free." *Dollars and Sense* 164 (March, 1991): 5–9.

Lall, S. 1979. "Is 'Dependence' a Useful Concept in Analyzing Underdevelopment?" *World Development* 3: 799–810.

Lappé, Francis Moore and Joseph Collins. 1986. *World Hunger: Twelve Myths.* New York: Grove Press.

Lappé, Francis Moore, Rachel Schurman, and Kevin Danaher. 1987. *Betraying the National Interest: How U.S. Foreign Aid Threatens Global Security by Undermining the Political and Economic Stability of the Third World.* New York: Grove Press.

Lecaillon, Jacques et al. 1984. *Income Distribution and Economic Development: An Analytical Survey.* Geneva: International Labour Office.

Lee, Martin A. and Norman Solomon. 1990. *Unreliable Sources: A Guide to Detecting Bias in News Media.* New York: Carol Publishing.

Lee, Thea. 1993. "Happily Never NAFTA." *Dollars and Sense* 183 (January/February): 12–15.

LeGrande, Linda. 1985. "The Service Sector: Employment and Earnings in the 1980s." Congressional Research Service, No. 85-167 E (August 15). Washington, DC: The Library of Congress.

LeGrande, Linda and Mark Jickling. 1987. "Earnings as a Measure of Regional Economic Performance." Congressional Research Service, No. 87-377 E (April 27th). Washington, DC: The Library of Congress.

Lenfestey, Jim. 1995. "Tax Increase Can Be Good for Everyone." *Minneapolis Star Tribune* (June 26): 9A.

Lenin, V. I. 1939. *Imperialism: The Highest Stage of Capitalism.* New York: International Publishers.

Lenski, Gerhard. 1966. *Power and Privilege.* New York: McGraw-Hill.

Leroy, Greg. 1995. "No More Candy Store." *Dollars and Sense* 199 (May/June): 10–14.

Levy, Frank. 1987. *Dollars and Dreams: The Changing American Income Distribution.* New York: Russell Sage Foundation.

Levy, Frank. 1995. "Incomes and Income Inequality." In Reynolds Farley (ed.), *State of the Union: America in the 1990s.* Vol. 1: *Economic Trends.* New York: Russell Sage.

Levy, Frank, and Richard J. Murnane. 1992. "U.S. Earnings Levels and Earnings Inequality: A Review of Recent Trends and Proposed Explanations." *Journal of Economic Literature* 30 (September): 1333–81.

Lewis, W. A. 1976. "Development and Distribution." In A. Cairncross and M. Puri (eds.), *Employment, Income Distribution and Development Strategy.* London: Macmillan.

Lichbach, Mark Irving. 1989. "An Evaluation of 'Does Economic Inequality Breed Political Conflict?' Studies." *World Politics* 41,4(July): 431–70.

Lichter, Daniel T. 1988. "Race and Underemployment: Black Employment Hardship in the Rural South." In Lionel J. Beaulieu (ed.), *The Rural South in Crisis.* Boulder, CO: Westview.

Lindblom, Charles. 1977. *Politics and Markets: The World's Political-Economic Systems.* New York: Basic Books.

Loehr, William. 1984. "Some Questions on the Validity of Income Distribution." In Mitchell A. Seligson (ed.), *The Gap Between Rich and Poor.* Boulder, CO: Westview.

Lohr, Steve. 1988. "'Global Office' Changing White-Collar Work World." *Minneapolis Star Tribune* (October 23): 1J.

London, Bruce. 1988. "Dependence, Distorted Development, and Fertility Trends in Noncore Nations: A Structural Analysis of CrossNational Data." *American Sociological Review* 53 (August): 606–18.

London, Bruce and Bruce A. Williams. 1988. "Multinational Corporate Penetration, Protest, and Basic Needs Provision in Non-Core Nations: A Cross-National Analysis." *Social Forces* 66 (March): 747–73.

London, Bruce and Thomas D. Robinson. 1989. "The Effect of International Dependence on Income Inequality and Political Violence." *American Sociological Review* 54 (April): 305–8.

Long, James E., David W. Rasmussen, and Charles T. Haworth. 1977. "Income Inequality and City Size." *Review of Economics and Statistics* 59 (May): 244–46.

Lorence, Jon and Joel Nelson. 1993. "Industrial Restructuring and Metropolitan Earnings Inequality, 1970-1980." *Research in Social Stratification and Mobility* 12: 145–84.

Lorenz, M. O. 1905. "Methods of Measuring the Concentration of Wealth." *Quarterly Publications of the American Statistical Association* 9: 205–19.

Los Angeles Times. 1995. "Voters Just as Dissatisfied Now as Before Election, Poll Shows." *Minneapolis Star Tribune* (April 13): 14A.

Los Angeles Times. 1995. "Public Support for Third-Party Candidate Rises." *Minneapolis Star Tribune* (August 24): 7A.

Lublin, Joann S. 1995. "Raking It In." *Wall Street Journal* (April 12): R1, R13.

Lydall, H. F. 1977. *Income Distribution During the Process of Development.* Geneva: International Labour Office.

McCarthy, Coleman. 1991. "Doing the Right Thing for Hunger." *Minneapolis Star Tribune* (November 28): 38A.

McCartney, Robert. 1992. "Pay Dirt: Shining a Light on the Salary Bloat of CEOs." *The Washington Post National Weekly Edition* (February 3–9): 22.

McCartney, Robert. 1992. "Stock Options and the Uneven Paying Field." *The Washington Post National Weekly Edition* (February 10–16): 23–24.

McDermott, John. 1995. "Bare Minimum: A Too-Low Minimum Wage Keeps All Wages Down." *Dollars and Sense* 200(July/August): 26–29.

MacDougall, A. Kent. 1984. "In Third World, All But the Rich Are Poorer." *Los Angeles Times* (November 4).

MacEwan, Arthur. 1991. "The Gospel of Free Trade: The New Evangelists." *Dollars and Sense* 171(November): 6–9, 20.

McFate, Katherine. 1991. *Poverty, Inequality, and the Crisis of Social Policy.* Washington, DC: Joint Center for Political and Economic Studies.

McFate, Katherine. 1994. "The Grim Economics of Violence." *Focus* (October): 7.

McFate, Katherine, Timothy Smeeding, and Lee Rainwater. 1995. "Markets and States: Poverty Trends and Transfer System Effectiveness in the 1980s." In Katherine McFate, Roger Lawson, and William Julius Wilson (eds.), *Poverty, Inequality, and the Future of Social Policy.* New York: Russell Sage.

McIntyre, Robert S. 1991. "Tax Inequality Caused Our Ballooning Budget Deficit." *Challenge* (November/December): 24–33.

McIntyre, Robert S. 1995. "The Flat Taxers' Flat Deceptions." *The American Prospect* 22 (November): 93–95.

McIntyre, Robert S., Douglas P. Kelly, Michael P. Ettlinger, and Elizabeth A. Fray. 1991. *A Far Cry from Fair.* Washington, DC: Citizens for Tax Justice.

McLeod, Jane D. and Michael J. Shanahan. 1993. "Poverty, Parenting, and Children's Mental Health." *American Sociological Review* 58(June): 351–66.

McMahon, Patrick J. and John H. Tschetter. 1986. "The Declining Middle Class: A Further Analysis." *Monthly Labor Review* 1 (September): 22–27.

McPortland, J. 1968. *The Segregated Students in Desegregated Schools: Sources of Influence on Negro Secondary Students.* Baltimore, MD: Johns Hopkins University Press.

Magaziner, I. and R. B. Reich. 1982. *Minding America's Business: The Decline and Rise of America's Economy.* New York: Vintage Press.

Makler, Harry, Alberto Martinelli, and Neil Smelser (eds.). 1982. *The New International Economy.* Beverly Hills, CA: Sage.

Malabre, Alfred L., Jr. 1989. "The Outlook: Is the Bill Arriving for the Free Lunch?" *Wall Street Journal* (January 9): 1.

Malbin, Michael J. 1995. "1994 Vote: The Money Story." In Everett Carll Ladd (ed.), *America at the Polls 1994.* Storrs: Roper Center, University of Connecticut.

Manski, Charles F. 1993. "Income and Higher Education." *Focus* 14,3(Winter).

Maraniss, David and Michael Weisskopf. 1995. "Cashing In." *The Washington Post National Weekly Edition* (December 4–10): 6–9.

Marcussen, Henrik and Jens Torp. 1982. *Internationalization of Capital— Prospects for the Third World: A Re-Examination of Dependency Theory.* London: Zed Press.

Marini, Margaret Mooney. 1989. "Sex Differences in Earnings in the United States." *Annual Review of Sociology* 15: 343–80.

Marx, Karl. 1973. *The Grundrisse.* New York: Vintage Books.

Marx, Karl. 1981. *Capital: A Critique of Political Economy.* 3 Vols. New York: Vintage Books.

Mason, Andrew. 1988. "Saving, Economic Growth, and Demographic Change." *Population and Development Review* 14, 1 (March): 113–44.

Massey, Douglas S. and Mitchell L. Eggers. 1990. "The Ecology of Inequality: Minorities and the Concentration of Poverty, 1970–1980." *American Journal of Sociology* 95,5(March): 1153–88.

Mataloni, Raymond J. 1995. "A Guide to BEA Statistics on U.S. Multinational Companies." *Survey of Current Business* 75,3(March): 38 ff.

Mataloni, Raymond J. 1995. "U.S. Multinational Companies: Operations in 1993." *Survey of Current Business* 75,6(June): 31ff.

Maxwell, Nan L. 1989. "Demograpic and Economic Determinants of United States Income Inequality." *Social Science Quarterly* 70,2 (June): 245–64.

Maxwell, Nan L. 1990. "Changing Female Labor Force Participation: Influences on Income Inequality and Distribution." *Social Forces* 68, 4 (June): 1251–66.

Mead, Walter Russell. 1989. "The United States and the World Economy." *World Policy Journal* 6, 1 (Winter): 1–46.

Mellstrom, Mark and Bob Lamb. 1994. "Military Budget Throws Away Billions on a Cold War That's Been Won." *Minneapolis Star Tribune* (July 2): 23A.

Menard, Scott. 1986. "A Research Note on International Comparisons of Inequality of Income." *Social Forces* 64,3 (March): 778–93.

Menchik, Paul L. 1991. "Permanent and Transitory Economic Status as Determinants of Mortality among Nonwhite and White Older Males: Does Poverty Kill?" Discussion Paper No. 936. Madison: University of Wisconsin, Institute for Research on Poverty.

Mensh, Elaine and Harry Mensh. 1991. *The IQ Mythology: Class, Race, Gender, and Inequality*. Carbondale: Southern Illinois University Press.

Merva, Mary and Richard Fowles. 1992. *Effects of Diminished Economic Opportunities on Social Stress: Heart Attacks, Strokes, and Crime*. Washington, DC: Economic Policy Institute.

Messner, Steven F. 1982. "Social Development, Social Equality, and Homicide: A Cross-National Test of a Durkheimian Model." *Social Forces* 61: 225–40.

Messner, Steven F. 1989. "Economic Discrimination and Societal Homicide Rates: Further Evidence of the Cost of Inequality." *American Sociological Review* 54: 597–611.

Meyers, Mike. 1995. "Nations with Credit Ratings of Banana Republics Fear the U.S. Is Shaky Borrower." *Minneapolis Star Tribune* (December 15): 2D.

Michel, Richard C. 1990. *Economic Growth and Income Inequality Since the 1982 Recession*. Washington, DC: Urban Institute.

Miller, John. 1994. "As the Economy Expands, Opportunity Contracts." *Dollars and Sense* 193(May/June): 8-11ff..

Miller, John. 1995. "The Capital Gains Giveaway." *Dollars and Sense* 198 (March/April): 14–17, 35.

Miller, S. M. and Jaqueline Ortiz. 1995. "The Politics of Income Distribution." *Research in Politics and Society* 5: 269–92.

Millman, Joel. 1994. "Mexico's Billionaire Pyramid." *The Washington Post National Weekly Edition* (December 5–11): 25.

Mills, C. Wright. 1956. *The Power Elite*. New York: Oxford University Press.

Mishel, Lawrence and Jared Bernstein. 1995. *The State of Working America: 1994–95*. New York: M.E. Sharpe.

Moffitt, Robert. 1992. "Incentive Effects of the U.S. Welfare System: A Review." *Journal of Economic Literature* 30(March): 1–61.

Moll, Terence. 1992. "Mickey Mouse Numbers and Inequality Research in Developing Countries." *The Journal of Development Studies* 28,4(July): 689–704.

Moore, Kristin. 1977. "The Effect of Government Policies on Out-of-Wedlock Sex and Pregnancy." *Family Planning Perspectives.* 94, 4: 164–69.

Moran, Theodore H. 1985. *Multinational Corporations: The Political Economy of Foreign Direct Investment.* Lexington, MA: Heath.

Morawetz, David. 1977. *Twenty-Five Years of Economic Development, 1950–1975.* Baltimore, MD: Johns Hopkins University Press.

Morgan, Dan. 1995. "Taking Care of Business." *The Washington Post National Weekly Edition* (July 3–9): 31.

Morin, Richard. 1994. "The Message from the Voters: Less Is More." *The Washington Post National Weekly Edition* (November 21–27): 8.

Morin, Richard. 1994. "Myths and Messages in the Election Tea Leaves." *The Washington Post National Weekly Edition* (November 21–27): 37.

Morin, Richard. 1995. "What the Public Really Wants." *The Washington Post National Weekly Edition* (January 9–15): 37.

Morin, Richard. 1995. "Across the Racial Divide." *The Washington Post National Weekly Edition* (October 16–22): 6-7.

Morin, Richard. 1996. "Tuned Out, Turned Off." *The Washington Post National Weekly Edition* (February 5–11): 6.

Morris, Martina, Annette D. Bernhardt, and Mark S. Handcock. 1994. "Economic Inequality: New Methods for New Trends." *American Sociological Review* 59,2(April): 205–19.

Mossberg, Walter. 1988. "Cost of Paying the Foreign Piper." *Wall Street Journal* (January 18).

Mufson, Steven. 1995. "The Princes and Paupers of Reform." *The Washington Post National Weekly Edition* (August 7–13): 20.

Muller, Edward N. 1984. "Financial Dependence in the Capitalist World Economy and the Distribution of Income within Nations. In Mitchel A. Seligson, (ed.), *The Gap Between Rich and Poor.* Boulder, CO: Westview.

Muller, Edward N. 1985. "Income Inequality, Regime Repressiveness, and Political Violence." *American Sociological Review* 50: 47–61.

Muller, Edward N. 1988. "Democracy, Economic Development, and Income Inequality." *American Sociological Review* 53: 50–68.

Muller, Edward N. 1988. "Inequality, Repression, and Violence: Issues of Theory and Research Design." *American Sociological Review* 53,5: 799–806.

Muller, Edward N. 1988. "Democracy, Economic Development, and Income Inequality." *American Sociological Review* 53: 66.

Muller, Edward N. 1989. "Democracy and Inequality (Reply to Weede)." *American Sociological Review* 54,5: 868–71.

Muller, Edward N. 1995. "Economic Determinants of Democracy." *American Sociological Review* 60, 6 (December): 966–82.

Muller, Edward N. and Mitchell A. Seligson. 1987. "Inequality and Insurgency." *American Political Science Review* 81,2(June): 427–51.

Murphy, D. C. 1985. "Calculation of Gini and Theil Inequality Coefficients for Irish Household Incomes in 1973 and 1980." *Economic and Social Review* 16 (April): 225–49.

Murray, Charles. 1986. *Losing Ground: American Social Policy 1950–1980.* New York: Basic Books.

Myers, Dowell and Jennifer R. Wolch. 1995. "The Polarization of Housing Status." In Reynolds Farley (ed.), *State of the Union: America in the 1990s.* Vol. 1: *Economic Trends.* New York: Russell Sage.

Myers, Martha A. 1987. "Economic Inequality and Discrimination in Sentencing." *Social Forces* 65,3(March): 746–66.

Myrdal, Gunnar. 1957. *Economic Theory and Underdeveloped Regions.* London: Duckworth.

Nader, Ralph. 1991. "General Motors Careful to Protect Bloat at the Top." *Minneapolis Star Tribune* (December 31): 13A.

Nelson, Joel I. 1984. "Income Inequality: The American States." *Social Science Quarterly* 65, 3: 854–60.

Nelson, Joel I. and Jon Lorence. 1988. "Metropolitan Earnings Inequality and Service Sector Employment." *Social Forces* 67,2 (December): 492–511.

Nelson, M.D., Jr. 1992. "Socioeconomic Status and Childhood Mortality in North Carolina." *American Journal of Public Health* 82(August): 1131–33.

New York Times. 1994. "Income Gaps Pose Threats in 4 Countries, Report Warns." *Minneapolis Star Tribune* (June 2): 13A.

New York Times. 1995. "Race Income Gap Little Changed, But Black Lives Improving Overall." *Minneapolis Star Tribune* (February 6): 8A.

New York Times/CBS News Poll. 1996. "Social Issues Behind Buchanan's Success, Poll Shows." *The New York Times* World Wide Web site (February 27).

New York Times Poll. 1996. "The Downsizing of America: Getting to the Root of the Problem." *The New York Times* World Wide Web site (March 9).

New York Times Poll. 1996. "The Downsizing of America: Layoffs." *The New York Times* World Wide Web site (March 9).

Newman, Katherine and Chauncey Lennon. 1995. "The Job Ghetto." *The American Prospect* 22 (Summer): 66–67.

News Services. 1989. "Venezuela President Blames Debt for Riots." *Minneapolis Star Tribune* (March 4): 3A.

News Services. 1992. "Bush Announces Riot Crimes Probe." *Minneapolis Star Tribune* (May 6): 1A, 12A.

News Services. 1994. "Inmate Census Highest of All Time." *Minneapolis Star Tribune* (September 13): 7A.

News with a View. 1996. "Slanted Media." *Minneapolis Star Tribune* (January 1): 9A.

Nielsen, François. 1994. "Income Inequality and Industrial Development: Dualsim Revisited." *American Sociological Review* 59(October): 654–77.

Nielsen, François and Arthur S. Alderson. 1995. "Income Inequality, Development, and Dualism: Results from an Unbalanced Cross-National Panel." *American Sociological Review* 60(October): 674–701.

Niggle, Christopher. 1989. "Monetary Policy and Changes in Income Distribution." *Journal of Economic Issues* 23,3: 809–22.

Nissan, Edward and George Carter. 1993. "Income Inequality across Regions over Time." *Growth and Change* 24(Summer): 303–19.

Nord, Stephen. 1980. "Income Inequality and City Size: An Examination of Alternative Hypotheses for Large and Small Cities." *Review of Economics and Statistics* 62 (November): 502–8.

O'Connor, Inge and Timothy M. Smeeding. 1993. *Working But Poor—A Cross-National Comparison of Earnings Adequacy.* Working Paper No. 94. Syracuse, NY: Syracuse University, Maxwell School of Citizenship and Public Affairs.

O'Hare, William P. 1985. "Poverty in America: Trends and New Patterns." *Population Bulletin* 40, 3. Washington, DC: Population Reference Bureau.

O'Hare, William P. 1987. *America's Welfare Population: Who Gets What.* Washington, DC: Population Reference Bureau.

Oliver, Melvin L., Thomas M. Shapiro, and Julie E. Press. 1995. "'Them That's Got Shall Get': Inheritance and Achievement in Wealth Accumulation." *Research in Politics and Society* 5: 69–95.

Olsen, Gregg M. 1994. "Locating the Canadian Welfare State: Family Policy and Health Care in Canada, Sweden, and the United States." *Canadian Journal of Sociology* 19,1: 1–20.

Ong, Paul. 1989. *The Widening Divide: Income Inequality and Poverty in Los Angeles.* Los Angeles: UCLA Graduate School of Architecture and Urban Planning.

O'Neill, June. 1991. "The Wage Gap Between Men and Women in the United States." *Review of Comparative Public Policy* 3: 353–69.

O'Regan, Katherine and Michael Wiseman. 1989. "Birth Weights and the Geography of Poverty." *Focus* 12,2(Winter): 16–22.

Organization for Economic Cooperation and Development. 1995. *International Direct Investment Statistics Yearbook 1995.* Paris: OECD.

Oropesa, R. S. 1986. "Social Class, Economic Marginality, and the Image of Stratification." *Sociological Focus* 19, 3 (August): 229–43.

Osberg, Lars. 1984. *Economic Inequality in the United States.* New York: M. E. Sharpe.

Page, Benjamin I. and Robert Y. Shapiro. 1992. *The Rational Public: Fifty Years of Trends in Americans' Policy Preferences.* Chicago: University of Chicago Press.

Paget, Karen M. 1995. "Can't Touch This: The Pentagon's Budget Fortress." *The American Prospect* 23 (Fall).

Paglin, M. 1975. "The Measurement and Trend of Inequality: A Basic Revision," *American Economic Review* 65,4 (September): 598–609.

Pang, Eul-Soo. 1989. "Debt, Adjustment, and Democratic Cacophony in Brazil." In Barbara Stallings and Robert Kaufman (eds.), *Debt and Democracy in Latin America.* Boulder, CO: Westview.

Parcel, Toby L. 1979. "Race, Regional Labor Markets and Earnings." *American Sociological Review* 44 (April): 262–79.

Pakulski, Jan. 1993. "The Dying of Class or Marxist Class Theory?" *International Sociology* 8,3(September): 279–92.

Parenti, Michael. 1978. *Power and the Powerless.* New York: St. Martin's Press.

Parenti, Michael. 1985. *Inventing Reality: Politics and the Mass Media.* New York: St. Martin's Press.

Parenti, Michael. 1992. *Make Believe Media: The Politics of Entertainment.* New York: St. Martin's Press.

Parenti, Michael. 1993. *Inventing Reality: The Politics of News Media.* 2d ed. New York: St. Martin's Press.

Pastor, Manuel, Jr. 1987. "The Effects of IMF Programs in the Third World: Debate and Evidence from Latin America." *World Development* 15, 2.

Paukert, Felix. 1973. "Income Distribution at Different Levels of Development: A Survey of the Evidence." *International Labour Review* (August/September): 97–125.

Pearlstein, Steven. 1994. "Hard on the Trail of the Up-and-Down Dollar." *The Washington Post National Weekly Edition* (July 25–31): 20–21.

Pearlstein, Steven. 1994. "Recessions Fade, But Downsizings are Forever." *The Washington Post National Weekly Edition* (October 3–9): 21.

Pearlstein, Steven. 1995. "The Rich Get Richer and...Proposed Tax and Spending Cuts Will Widen the Inequity in U.S. Incomes." *The Washington Post National Weekly Edition* (June 12–18): 6.

Pearlstein, Steven. 1995. "A Rising Tide Leaves Some Boats Hard Aground." *The Washington Post National Weekly Edition* (July 3–9): 21.

Pearlstein, Steven. 1995. "We Do the Work, But Who Gets the Money?" *The Washington Post National Weekly Edition* (July 31–August 6): 22.

Pearlstein, Steven. 1996. "Floating the Flat Tax Idea." *The Washington Post National Weekly Edition* (January 22–28): 20.

Pearson, John. 1985. "Strong Dollar or No, There's Money to Be Made Abroad." *Business Week* 22 (March): 155.

Pen, Jan. 1971. *Income Distribution.* Harmondsworth: Penguin Books.

Persson, Thorsten and Guido Tabellini. 1994. "Is Inequality Harmful for Growth?" *The American Economic Review* 84,3: 600–21.

Petruno, Tom. 1995. "Profits Soar, Workers Lose." *Minneapolis Star Tribune* (September 7): 1D,7D.

Pettinico, George. 1995. "1994 Vote: Where Was Public Thinking at Vote Time?" In Everett Carll Ladd (ed.), *America at the Polls 1994* Storrs: Roper Center, University of Connecticut.

Phillips, Kevin. 1990. *The Politics of Rich and Poor.* New York: Random House.

Phillips, Kevin. 1993. *Boiling Point: Democrats, Republicans, and the Decline of Middle-Class Prosperity.* New York: HarperCollins.

Phillips, Kevin. 1994. "The Voters Are Already Tapping Their Feet." *The Washington Post National Weekly Edition* (November 21–27): 23–24.

Phillips, Kevin. 1995. "The Champagne's Gone Flat." *The Washington Post National Weekly Edition* (August 14–20): 23.

Physician Task Force on Hunger in America. 1985. *Hunger in America: The Growing Epidemic.* Middleton, CT: Wesleyan University Press.

Pincus, Walter and Dan Morgan. 1995. "Pork: The Last Line of the Dense Budget." *The Washington Post National Weekly Edition* (April 3–9): 32

Portes, Alejandro. 1976. "On the Sociology of National Development: Theories and Issues." *American Journal of Sociology* 82 (July): 55–85.

Portes, Alejandro. 1994. "The Informal Economy and Its Paradoxes." In Neil J. Smelser and Richard Swedberg (eds.), *The Handbook of Economic Sociology*. New York: Russell Sage.

Priest, Dana. 1995. "Damn the Defense Budget, Full Speed Ahead." *The Washington Post National Weekly Edition* (August 7–13): 32.

Purdum, Todd S. 1996. "Clinton Calls on Companies to Train Workers." *New York Times* World Wide Web site (March 9).

Racusen, Seth. 1994. "Lula's Rise." *Dollars and Sense* 195 (September/October): 18.

Radner, Daniel B. and Denton R. Vaughan. 1987. "Wealth, Income, and the Economic Status of Aged Households." In Edward N. Wolff (ed.), *International Comparisons of the Distribution of Household Wealth*. New York: Oxford University Press.

Raffalovich, Lawrence E. 1990. "Segmentation Theory, Economic Performance, and Earnings Inequality." *Research in Social Stratification and Mobility* 9: 251–82.

Raffalovich, Lawrence E. 1993. "Structural Sources of Change in Earnings Inequality: Evidence from the Current Population Survey, 1967–1981." *Research in Social Stratification and Mobility* 12: 113–44.

Raffalovich, Lawrence E. 1994. "Stability and Variability of Between-Occupation Earnings Inequality: A Segmented Labor-Market Approach." *Research in Social Stratification and Mobility* 13: 265–85.

Rainwater, Lee. 1992. *Poverty in American Eyes*. Working Paper No. 80. Syracuse, NY: Syracuse University, Maxwell School of Citizenship and Public Affairs.

Rainwater, Lee and Timothy M. Smeeding. 1995. *Doing Poorly: The Real Income of American Children in a Comparative Perspective*. Working Paper No. 127. Syracuse, NY: Syracuse University, Maxwell School of Citizenship and Public Affairs.

Ram, Rati. 1980. "Physical Quality of Life and Inter-Country Economic Inequality." *Economic Letters* 5: 195–99.

Randolph, Eleanor. 1996. "Firebrand of the Radio Waves Turns Down the Rhetorical Heat." *Minneapolis Star Tribune* (January 1): A6.

Raspberry, William. 1995. "Bomb-Throwers and Broadcasters." *The Washington Post National Weekly Edition* (May 1–7): 27.

Raspberry, William. 1996. "How Economy Can Be Doing Very Well and Very Badly." *Minneapolis Star Tribune* (January 9): A9.

Rast, Bob. 1988. "U.S. Banks Lose Top Status in Global Financial Markets." *Minneapolis Star Tribune* (November 27): 1D.

Ratcliff, Richard E., Mary Elizabeth Gallagher, and Anthony C. Kouzi. 1992. "Political Money and Partisan Clusters in the Capitalist Class." *Research in Politics and Society* 4: 63–85.

Ratcliff, Richard E. and Suzanne B. Maurer. 1995. "Saving and Investment among the Wealthy: The Uses of Assets by High Income Families in 1950 and 1983." *Research in Politics and Society* 5: 99–125.

Reding, Andrew. 1988. "Mexico at a Crossroads." *World Policy Journal* 5, 4 (Fall): 615–50.

Reich, Robert B. 1992. *The Work of Nations: Preparing Ourselves for 21st Century Capitalism.* New York: Vintage.

Rendall, Steven, Jim Naureckas, and Jeff Cohen. 1995. *The Way Things Aren't.* New York: The New Press.

Renner, Michael. 1989. "Enhancing Global Security." In Lester R. Brown (Ed.), *State of the World 1989.* New York: Norton.

Reno, Robert. 1995. "Polls Suggest Electorate Stands Well to the Left of GOP 'Tidal Wave'" *Minneapolis Star Tribune* (November 16): A24.

Rensberger, Boyce. 1995. "No Help Wanted: Young U.S. Scientists Go Begging with Serious Consequences for our Future." *The Washington Post National Weekly Edition* (January 9–15): 7.

Reuter. 1995. "CEO Paychecks Keep Getting Bigger." *Minneapolis Star Tribune* (May 9): 1D.

Reynolds, Lloyd G. 1983. "The Spread of Economic Growth to the Third World, 1850–1950." *Journal of Economic Literature* 21: 941–80.

Reynolds, Morgan and Eugene Smolensky. 1977. *Public Expenditures, Taxes, and the Distribution of Income.* New York: Academic Press.

Richards, Evelyn. 1989. "The Pentagon Plans to Get into Television." *The Washington Post National Weekly Edition* (December 26, 1988–January 1, 1989).

Riddell, Tom. "The Political Economy of Military Spending." *The Imperiled Economy: Through the Safety Net.* New York: Union for Radical Political Economics.

Riding, Alan. 1989. "Debt Fears Realized with Venezuela Unrest." *Minneapolis Star Tribune.* (March 2): 4A.

Rieff, David. 1991. *Los Angeles: Capital of the Third World.* New York: Simon and Schuster.

Riley, J.C. 1979. "Testing the Educational Screening Hypothesis." *Journal of Political Economy* 87,5 (October):S227–52.

Robberson, Tod. 1994. "For Mexico, NAFTA Spells a Rising Deficit." *The Washington Post National Weekly Edition* (August 15–21): 21.

Robberson, Tod. 1995. "The Mexican Miracle Unravels." *The Washington Post National Weekly Edition* (January 16–22): 20.

Robberson, Tod. 1995. "NAFTA's One-Way Street," *The Washington Post National Weekly Edition* (August 28–September 3): 17.

Rodgers, Joan R. and John L. Rodgers. 1993. "Chronic Poverty in the United States." *Journal of Human Resources* 28,1(Winter).

Rodrik, Dani. 1994. *Getting Interventions Right: How South Korea and Taiwan Grew Rich.* Cambridge, MA: National Bureau of Economic Research.

Romo, Frank P., Hyman Korman, Peter Brantley, and Michael Schwartz. 1989. "The Rise and Fall of Regional Political Economies: A Theory of the Core." *Research in Politics and Society* 3: 37–64.

Romo, Frank P. and Michael Schwartz. 1995. "The Structural Embeddedness of Business Decisions: The Migration of Manufacturing Plants in New York State, 1960–1985." *American Sociological Review* 60,6(December): 874–907.

"Roots of Skinhead Violence: Dim Economic Prospects for Young Men." 1989. *Utne Reader* 33 (May/June): 84.

Ropers, Richard. 1991. *Persistent Poverty: The American Dream Turned Nightmare.* New York: Plenum Press.

Rosenberg, Sam. 1980. "Male Occupational Standing and the Dual Labor Market." *Industrial Relations* 19, 1 (Winter): 34–48.

Rosenfeld, Stuart A. 1988. "The Tale of Two Souths." In Lionel J. Beaulieu (ed.), *The Rural South in Crisis.* Boulder, CO: Westview.

Rosenfeld, Stuart A., Edward Bergman, and Sara Rabin. 1985. *After the Factories: Changing Employment Patterns in the Rural South.* Research Triangle Park, NC: Southern Growth Policies Board.

Rossides, Daniel W. 1990. *Social Stratification: The American Class System in Comparative Perspective.* Englewood Cliffs, NJ: Prentice-Hall.

Rowen, Hobart. 1988. "Capital Economics: Candidates in Blunderland." *The Washington Post National Weekly Edition* (October 10–16).

Rowen, Hobart. 1994. "The Dollar's Silent Tumble." *The Washington Post National Weekly Edition* (November 14–20): 5.

Rowen, Hobart. 1994. "In Defense of the Dollar." *The Washington Post National Weekly Edition* (July 4–10): 5.

Rubinson, Richard. 1976. "The World-Economy and the Distribution of Income within States: A Cross-National Study." *American Sociological Review* 41 (August): 638–59.

Rubinson, Richard and Deborah Holtzman. 1981. "Comparative Dependence and Economic Development." *International Journal of Comparative Sociology* 22: 86–101.

Ruggles, Patricia. 1990. *Drawing the Line: Alternative Poverty Measures and Their Implications for Public Policy.* Washington, DC: Urban Institute.

Russett, Bruce. 1983. "International Interactions and Processes: The Internal vs. External Debate Revisited." In Ada W. Finifter (ed.), *Political Science—The State of the Discipline.* Washington, DC: Political Science Association.

Ruthenberg, David and Miron Stano. 1977. "The Determinants of Interstate Variation in Income Distribution." *Review of Social Economy* 35, 1 (April): 55–66.

Sabo, Martin Olav. 1994. "A $6.50 U.S. Minimum Wage Would Help Lessen Income Gap." *Minneapolis Star Tribune* (June 16): 25A.

Sale, K. 1975. *Power Shift: The Rise of the Southern Rim and Its Challenge to the Eastern Establishment.* New York: Vintage.

Sale, Tom S. 1974. "Interstate Analysis of the Size Distribution of Family Income 1950–1970," *Southern Economic Journal* 40 (January): 434–41.

Sakamoto, Arthur. 1988. "Labor Market Structure, Human Capital, and Earnings Inequality in Metropolitan Areas." *Social Forces* 67,1(September): 86–107.

Salkowski, Charlotte. 1986. "Growth in Living Standard Slows for the American Middle Class." *Christian Science Monitor* (January 8): 1ff.

Samuelson, Paul. 1980. Economics. 11th ed. New York: McGraw-Hill.

Samuelson, Robert J. 1985. "The Myth of the Missing Middle." *Newsweek* (July 1): 50.

Samuelson, Robert J. 1994. "Bell Curve Ballistics." *The Washington Post National Weekly Edition* (October 31–November 6): 28.

Sanders, Jerry W. 1989. "America in the Pacific Century." *World Policy Journal* 6, 1 (Winter): 47–80.

Sawhill, Isabel V. 1988. "Poverty in the U.S.: Why Is It So Persistent?" *Journal of Economic Literature* 26 (September): 1073–1119.

Schapiro, Morton Owen. 1988. "Socio-Economic Effects of Relative Income and Relative Cohort Size." *Social Science Research* 17: 362–83.

Scheper-Hughes, Nancy. 1992. *Death without Weeping: The Violence of Everyday Life in Brazil*. Berkeley: University of California Press.

Schiff, Michael and Richard Lewontin. 1986. *Education and Class: The Irrelevance of IQ Genetic Studies*. New York: Oxford University Press.

Schmickle, Sharon. 1994. "Crime Experts See Flaws in Bill's Focus." *Minneapolis Star Tribune* (August 14): 16A.

Schmickle, Sharon. 1996. "Buchanan Is Talking about Plight of the Working Class, and Washington Is Listening." *Minneapolis Star Tribune* (March 4): A3.

Schor, Juliet B. and Laura Leete-Guy. 1992. *The Overworked American: The Unexpected Decline of Leisure*. Washington, DC: Economic Policy Institute.

Schram, Sanford F. and Paul H. Wilken. 1989. "It's No 'Laffer' Matter: Claim That Increasing Welfare Aid Breeds Poverty and Dependence Fails Statistical Test." *American Journal of Economics and Sociology* 48, 2 (April): 203–17.

Schwartz, John. 1995. "Pinning a Price Tag on Nations." *The Washington Post National Weekly Edition* (September 25–October 1): 38.

Schwarz, Joseph, and Christopher Winship. 1979. "The Welfare Approach to Measuring Inequality." In Karl F. Schuessler (ed.), *Sociological Methodology 1980*. San Francisco: Jossey-Bass.

Seligson, Mitchell A. 1984. *The Gap Between Rich and Poor: Contending Perspectives on the Political Economy of Development*. Boulder, CO: Westview.

Seltman, Paul. 1995. "Congressional Black Caucus Cuts Corporate Welfare to Balance U.S. Budget." Congressional Black Caucus, U.S. Congress, Washington, DC, May.

Shanahan, Suzanne Elise and Nancy Brandon Tuma. 1994. "The Sociology of Distribution and Redistribution." In Neil J. Smelser and Richard Swedberg (eds.), *The Handbook of Economic Sociology*. New York: Russell Sage Foundation.

Shapiro, Isaac. 1995. *Unequal Shares: Recent Trends among the Wealthy*. Washington, DC: Center on Budget and Policy Priorities.

Shapiro, Margaret. 1988. "Empire of the Sun." *The Washington Post National Weekly Edition* (October 31–November 6): 6–7.

Sherman, Arloc. 1994. *Wasting America's Future: The Children's Defense Fund Report on the Costs of Child Poverty*. Boston: Beacon Press.

Shapiro, Isaac and Sharon Parrott. 1995. *An Unraveling Consensus? An Analysis of the Effect of the New Congressional Agenda on the Working Poor*. Washington, DC: Center on Budget and Policy Priorities.

Shapiro, Robert J. 1995. *Cut and Invest: A Budget Strategy for the New Economy*. Washington, DC: Progressive Policy Institute.

Shellenberger, Michael. 1995. "Brazil's Debt Debacle." *Dollars and Sense* 199 (May/June): 23.

Sheppard, Nathaniel, Jr. 1992. "In Panama, Gap Between Rich and Poor Grows at Alarming Rate." *Chicago Tribune* (May 7): 1A.

Shields, Janice. 1995. *Aid for Dependent Corporations (AFDC) 1995*. Washington, DC: Essential Information.

Shlapentokh, Vladimir. 1987. *The Politics of Sociology in the Soviet Union*. Boulder, CO: Westview.

Simon, David R. and D. Stanley Eitzen. 1982. *Elite Deviance*. Boston: Allyn and Bacon.

Simpson, Miles. 1990. "Political Rights and Income Inequality: A Cross-National Test." *American Sociological Review* 55,5(October): 682–93.

Skees, Jerry R. and Louis E. Swanson. 1988. "Farm Structure and Local Society Well-Being in the South." In Lionel J. Beaulieu (ed.), *The Rural South in Crisis*. Boulder, CO: Westview.

Sklair, Leslie. 1991. *Sociology of the Global System*. Baltimore, MD: Johns Hopkins University Press.

Skocpol, Theda. 1976. "Explaining Revolutions: In Quest of a Social Structural Approach." In L.A. Coser and O.N. Larsen (eds.), *The Uses of Controversy in Sociology*. New York: Free Press..

Smeeding, Timothy, Barabara Boyle Torrey, and Martin Rein. 1988. "Patterns of Income and Poverty: The Economic Status of Children and the Elderly in Eight Countries." In *The Vulnerable*. Washington, DC: Urban Institute Press.

Smeeding, Timothy M. and John Coder. 1993. *Income Inequality in Rich Countries during the 1980s.* Working Paper No. 88. Syracuse, NY: Syracuse University, Maxwell School of Citizenship and Public Affairs.

Smelser, Neil et al. 1982. *The New International Economy*. Beverly Hills, CA: Sage.

Smith, Dane. 1996. "Livable-wage." *Minneapolis Star Tribune* (March 28): B1.

Smith, David A. 1987. "Overurbanization Reconceptualized: A Political Economy of the World-System Approach." *Urban Affairs Quarterly* 23, 2 (December): 270–94.

Smith, David M. 1982. *Where the Grass Is Greener: Living in an Unequal World*. Baltimore, MD: Johns Hopkins University Press

Smith, James and Finis R. Welch. 1986. *Closing the Gap: Forty Years of Economic Progress for Blacks*. Santa Monica, CA: RAND Corp.

Smith, James P. 1995. *Unequal Wealth and Incentives to Save*. Santa Monica, CA: RAND Corp.

Smith, Shelley A. 1991. "Sources of Earnings Inequality in the Black and White Female Labor Forces." *Sociological Quarterly* 32,1: 117–38.

Smith, Tony. 1979. "The Underdevelopment of the Development Literature: The Case of Dependency Theory." *World Politics* 31: 247–88.

Smith, Tony. 1981. *The Pattern of Imperialism: The United States, Great Britain, and the Late Industrializing World Since 1815*. Cambridge: Cambridge University Press.

Snyder, David and Edward L. Kick. 1979. "Structural Position in the World System and Economic Growth, 1955–1970: A Multiple-Network Analysis of Transnational Interactions." *American Journal of Sociology* 84, 5: 1096–1126.

Soltow, Lee. 1971. *Patterns of Wealthholding in Wisconsin Since 1850*. Madison: University of Wisconsin Press.

Sorensen, Aage B. 1991. "On the Usefulness of Class Analysis in Research on Social Mobility and Socioeconomic Inequality." *Acta Sociologica* 34,2 (Summer): 71–87.

Spanier, Graham B. and Paul C. Glick. 1981. "Marital Instability in the United States: Some Correlates and Recent Changes." *Family Relations* 31(July): 329–38.

Spolar, Chris. "No Home—and Not Much More." *The Washington Post National Weekly Edition* (March 27–April 2): 6–7.

Stack, Steven. 1978. "The Effect of Direct Government Involvement in the Economy on the Degree of Income Inequality: A Cross-National Study." *American Sociological Review* 43 (December): 880–88.

Stein, A. 1971. "Strategies for Failure." *Harvard Educational Review* 41: 158–204.

Strawn, Julie. 1992. "The States and the Poor: Child Poverty Rises as the Safety Net Shrinks." *Social Policy Report* 6,3(Fall).

Streeten, Paul et al. 1981. *First Things First: Meeting Basic Needs in Developing Countries*. New York: Oxford University Press.

Swanson, Louis E. 1988. "The Human Dimension of the Rural South in Crisis." In Lionel J. Beaulieu (ed.), *The Rural South in Crisis*. Boulder, CO: Westview.

Swift, Adam, Gordon Marshall, Carol Burgoyne, and David Routh. 1995. "Distributive Justice: Does It Matter What People Think?" In James R. Kluegel, David S. Mason, and Bernd Wegener (eds.), *Social Justice and Political Change: Public Opinion in Capitalist and Post-Communist States*. New York: Aldine de Gruyter.

Szirmai, Adam. 1988. *Inequality Observed: A Study of Attitudes Towards Income Inequality*. Brookfield, VT: Avebury.

Talbot, Stephen. 1995. "Wizard of Ooze." The MOJO Wire, *Mother Jones Magazine* World Wide Web site.

Taylor, Howard F. 1995. "The Bell Curve." *Contemporary Sociology* 24,2 (March): 153–58.

Taylor, Michael and Nigel Thrift. 1982. *The Geography of Multinationals*. New York: St. Martin's Press.

Teitel, Martin. 1994. "Selling Out: The Thriving Business of Patenting Life," *Dollars and Sense* 195(September/October): 25–27.

Terlouw, Cornelis Peter. 1992. *The Regional Geography of the World System: External Arena, Periphery, Semiperiphery, Core*. Utrecht, Netherlands: Faculteit Ruimteljke Wetenschppen.

Thomas, Paulette. 1994. "Widening Rich-Poor Gap Is a Threat to the 'Social Fabric,' White House Says." *Wall Street Journal* (February 15):A2.

Thomas, Pierre. 1995. "Getting to the Bottom Line on Crime." *The Washington Post National Weekly Edition* (July 18–24): 31.

Thompson, Morris S. 1988. "Clouds in the Sunbelt: Signs of Economic Deceleration Appear in the Southeast." *The Washington Post National Weekly Edition* (November 21–27): 21.

Thurow, Lester. 1975. *Generating Inequality: Mechanisms of Distribution in the U.S. Economy*. New York: Basic Books.

Thurow, Lester. 1987. "A Surge in Inequality." *Scientific American* 256 (May): 30–37.

Tienda, Marta and Ding-Tzann Lii. 1987. "Minority Concentration and Earnings Inequality: Blacks, Hispanics, and Asians Compared." *American Journal of Sociology* 93, 1 (July): 141–65.

Tigges, Leann M. 1986. "Dueling Sectors: The Role of Service Industries in the Earnings Process of the Dual Economy." Paper presented at the American Sociological Association Conference.

Tigges, Leann M. 1987. "Age, Earnings, and Change Within the Dual Economy." Paper presented at the American Sociological Association Conference.

Tigges, Leann M. 1988. "Age, Earnings, and Change within the Dual Economy." *Social Forces* 66, 3 (March): 676–98.

Tigges, Leann M. 1988. "Dueling Sectors: The Role of Service Industries in the Earnings Process of the Dual Economy. In George Iarkas and Paula England (eds.), *Industries, Firms, and Jobs: Sociological and Economic Approaches.* New York: Plenum Press.

Tilly, Chris. 1991. "Raising Cane in Jamaica." *Dollars and Sense* 169(September): 16–18, 22.

Tilly, Chris and Randy Abelda. 1994. "Family Structure and Family Earnings: The Determinants of Earnings Differences among Family Types." *Industrial Relations* 33,2(April): 151–67.

Timberlake, Michael and Kirk R. Williams. 1987. "Structural Position in the World-System, Inequality, and Political Violence." *Journal of Political and Military Sociology* 15,1: 1–15

Tolbert, Charles M. and Thomas A. Lyson. 1992. "Earnings Inequality in the Nonmetropolitan United States: 1967–1990." *Rural Sociology* 57,4: 494–511.

Treas, Judith. 1982. "U.S. Income Stratification: Bringing Families Back In." *Sociology and Social Research* 66, 3 (March): 231–51.

Treas, Judith. 1983. "Trickle Down or Transfers? Postwar Determinants of Family Income Inequality." *American Sociological Review* 48 (August): 546–59.

Treas, Judith. 1987. "The Effect of Women's Labor Force Participation on the Distribution of Income in the United States." *Annual Review of Sociology* 1987 13: 259–88.

Tumin, Melvin. 1953. "Some Principles of Stratification: A Critical Analysis." *American Sociological Review* 18: 387–94.

Uchitelle, Louis. 1989. "U.S. Companies Are Globalizing Their Resources." *Minneapolis Star Tribune* (May 30): 5B.

UNESCO. 1976. *The Use of Socio-Economic Indicators in Development Planning.* Paris: UNESCO.

United Nations Children's Fund. 1991. *The State of the World's Children, 1991.* New York: Oxford University Press.

U.S. Census Bureau. 1988. *Statistical Abstract of the United States, 1987.* Washington, DC: U.S. Government Printing Office.

U.S. Census Bureau. 1989. *Housing in America, 1985–86.* Washington, DC: U.S. Government Printing Office.

U.S. Census Bureau. 1993. *Current Population Reports: Poverty in the United States: 1992.* Series P-60, No. 185 (September). Washington, DC: U.S. Government Printing Office.

U.S. Census Bureau. 1994. *The Earnings Ladder: Who's at the Bottom? Who's at the Top?* Statistical Brief SB/94-3RV (June).

U.S. Census Bureau. 1995. *Current Population Reports: Income, Poverty, and Valuation of Noncash Benefits, 1993.* Series P60-188. Washington, DC: U.S. Government Printing Office.

U.S. Census Bureau. 1995. *Home Sweet Home—America's Housing, 1973 to 1993.* Statistical Brief SB/95-18 (July).

U.S. Census Bureau. 1995. *Income and Job Mobility in the Early 1990s.* Statistical Brief SB/95-1 (March).

U.S. Census Bureau. 1995. *Income and Poverty 1993.* CD-ROM. Washington, DC: U.S. Government Printing Office.

U.S. Census Bureau. 1995. *Poverty's Revolving Door.* Statistical Brief SB/95-20 (August).

U.S. Census Bureau. 1995. Press Briefing on 1994 Income and Poverty Estimates. Census Bureau World Wide Web Homepage (October 5).

U.S. Census Bureau. 1995. *Statistical Abstract of the United States 1994.* CD-ROM. Washington, DC: U.S. Government Printing Office.

U.S. Census Bureau. 1995. *Turnout in '94 Congressional elections—45 Percent; Young Voter Participation Shows No Gain.* CB95-105 (June 8). Washington, DC: U.S. Government Printing Office.

U.S. Department of Labor, Bureau of Labor Statistics. 1995. *Contingent and Alternative Employment Arrangements.* Report 900 (August). Washington, DC: U.S. Government Printing Office.

U.S. House of Representatives, Committee on Ways and Means. 1994. *Where Your Money Goes: The 1994–95 Green Book.* Washington, DC: Brassey's.

Vedder, Richard and Lowell Gallaway. 1986. "AFDC and the Laffer Principle." *Wall Street Journal* (March 26).

Verba, Sidney, Steven Kelman, Gary R. Orren, Ichoro Miyake, Joji Watanuki, Ikuo Kabashima, and G. Donald Ferree, Jr. 1987. *Elites and the Idea of Equality: A Comparison of Japan, Sweden, and the United States.* Cambridge, MA: Harvard University Press.

Verway, David I. 1966. "A Ranking of States by Inequality Using Census and Tax Data." *Review of Economics and Statistics 48,* 3 (August): 314–21.

"Violent Juvenile Crime Up Sharply." 1995. *Minneapolis Star Tribune* (September 8): 7A.

Vondra, Joan I. 1993. *"Childhood Poverty and Child Maltreatment."* In Judith A. Chafel (ed.), Child Poverty and Public Policy. Washington, DC: Urban Institute .

Wachtel, Howard M. 1986. *The Money Mandarins: The Making of a Supranational Economic Order.* New York: Pantheon Books.

Wallace, Michael and Joyce Rothschild. 1989. "Plant Closings, Capital Flight, and Worker Dislocation: The Long Shadow of Deindustrialization." *Research in Politics and Society* 3: 1–35.

Wallerstein, Immanuel. 1974. *The Modern World System: Capitalist Agriculture and the Origins of the European World-Economy in the Sixteenth Century.* New York: Academic Press.

Walton, John and Charles Ragin. 1990. "Global and National Sources of Political Protest: Third World Responses to the Debt Crisis." *American Sociological Review* 55(December): 876–90.

Walton, John and David Seddon. 1994. *Free Martkets and Food Riots: The Politics of Global Adjustment.* Cambridge, MA: Blackwell.

Warren, Bill. 1980. *Imperialism: Pioneer of Capitalism.* London: NLB.

Weatherby, Norman L. et al. 1983. "Development, Inequality, Health Care, and Mortality at the Older Ages: A Cross-National Study." *Demography* 20, 1 (February): 27–43.

Weber, Max. 1958. *The Protestant Ethic and the Spirit of Capitalism.* Trans. Talcott Parsons. New York: Scribner and Sons.

Weber, Max. 1958. "Class, Status, Party." In Seymour Martin Lipset and Reinhard Bendix (eds.), *Class, Status and Power.* New York: Free Press.

Weede, Erich. 1980. "Beyond Misspecification in Sociological Analyses of Income Inequality." *American Sociological Review* 45 (June): 497–501.

Weede, Erich. 1987. "Some New Evidence on Correlates of Political Violence: Income Inequality, Regime Repressiveness, and Economic Development." *European Sociological Review* 3,2(September): 97–108.

Weede, Erich. 1989. "Democracy and Income Inequality Reconsidered." *American Sociological Review* 54,5: 865–68.

Weede, Erich and Horst Tiefenbach. 1981. "Some Recent Explanations of Income Inequality." *International Studies Quarterly* 25 (June): 255–82.

Weiss, Rick. 1995. "A Human Face on Gene Theories." *The Washington Post National Weekly Edition* (February 27–March 5): 37.

Weisskopf, Michael. 1995. "To the Victors Belong the Pac Checks." *The Washington Post National Weekly Edition* (January 2–8): 13.

Weisskopf, Michael and David Maraniss. 1995. "In On the Takeoff: In This Congress, Corporate Lobbyists Are Invited to Sit in the Cockpit and Help Fly the Plane." *The Washington Post National Weekly Edition* (March 20–26): 13–14.

Weitzman, Lenore. 1987. *The Divorce Revolution.* New York: Free Press.

Wetzel, James R. 1995. "Labor Force, Unemployment, and Earnings." In Reynolds Farley (ed.), *State of the Union: America in the 1990s.* Vol. 1: *Economic Trends.* New York: Russell Sage.

Whalley, John. 1979. "The Worldwide Income Distribution: Some Speculative Calculations." *Review of Income and Wealth* 25: 261–76.

Wilcox, Jerry and Wade Clark Roof. 1978. "Percent Black and Black-White Status Inequality: Southern Versus Nonsouthern Patterns." *Social Science Quarterly* 59, 3 (December): 421–34.

Wilkinson, Richard G. 1990. "Income Distribution and Mortality: A 'Natural' Experiment." *Sociology of Health and Illness* 12,4(December): 391–412.

Wilkinson, Richard G. 1992. "National Mortality Rates: The Impact of Inequality." *American Journal of Public Health* 82,8(August): 1082–84.

Will, George F. 1991. "CEO Megasalaries May Provoke New Bout of Antibusiness Fever." *Minneapolis Star Tribune* (September 1): 18A.

Wilson, William Julius. 1980. *The Declining Significance of Race: Blacks and Changing American Institutions.* Chicago: University of Chicago Press.

Wilson, William Julius. 1987. *The Truly Disadvantaged: The Inner City, the Underclass, and Public Policy.* Chicago: University of Chicago Press.

Winnick, Andrew J. 1989. *Toward Two Societies: The Changing Distribution of Income and Wealth in the U.S. Since 1960.* New York: Praeger.

Winsberg, Morton D. 1989. "Income Polarization Between the Central Cities and Suburbs of U.S. Metropolises, 1950–1980." *American Journal of Economics and Sociology* 48,1(January): 3–10.

Wolff, Edward N. 1987. *International Comparisons of the Distribution of Household Wealth.* New York: Oxford University Press.

Wolff, Edward N. 1995. "How the Pie Is Sliced: America's Growing Concentration of Wealth." *The American Prospect* 22(Summer): 58–64.

Wolff, Edward N. 1995. "The Rich Get Increasingly Richer: Latest Data on Household Wealth during the 1980s." *Research in Politics and Society* 5: 33–68.

Wolff, Edward N. and D. Bushe. 1976. *Age, Education and Occupational Earnings Inequality.* New York: National Bureau of Economic Research.

Wood, Charles H. and José Alberto Magno de Carvaho. 1988. *The Demography of Inequality in Brazil.* New York: Cambridge University Press.

World Bank. 1988. *World Development Report, 1988.* New York: Oxford University Press.

World Bank. 1994. *World Development Indicators, 1994.* Data on Diskette. Washington, DC: World Bank.

Wright, Erick O. 1978. "Race, Class, and Income Inequality." *American Journal of Sociology* 83: 1368–88.

Wright, Erick O. 1978. *Class, Crisis and the State.* New York: Schocken Books.

Wright, Erick O. and Lucca Perrone. 1977. "Marxist Class Categories and Income Inequality." *American Sociological Review* 42: 32–55.

Zajac, Brian. 1994. "Getting the Welcome Carpet." *Forbes* (July 18): 276–77.

Zajac, Brian. 1995. "Weak Dollars, Strong Results." *Forbes* (July 17): 274–76.

Zucker, Lynne G. and Carolyn Rosenstein. 1981. "Taxonomies of Institutional Structure: Dual Economy Reconsidered." *American Sociological Review* 46 (December): 869–84.

Zwicky, Heinrich. 1989. "Income Inequality and Violent Conflicts in Developing Countries." *Research in Inequality and Social Conflict* 1: 67–93.

INDEX